Compound Semiconductor Transistors

Compound Semiconductor Transistors

Physics and Technology

Edited by

Sandip Tiwari

IBM Research Division
Thomas J. Watson Research Center

A Selected Reprint Volume
IEEE Electron Devices Society, *Sponsor*

The Institute of Electrical and Electronics Engineers, Inc., New York

This book may be purchased at a discount from the publisher
when ordered in bulk quantities. For more information contact:

IEEE PRESS Marketing
Attn: Special Sales
PO Box 1331
445 Hoes Lane
Piscataway, NJ 08855-1331
Fax: (908) 981-8062

Printed in the United States of America
10 9 8 7 6 5 4 3 2 1

ISBN 0-7803-0417-9
IEEE Order Number: PC0313-7

Library of Congress Cataloging-in-Publication Data

Tiwari, Sandip,
    Compound semiconductor transistors : physics and technology /
Sandip Tiwari.
        p.   cm.
    "IEEE order number: PC0313-7"—T.p. verso.
    Includes bibliographical references and indexes.
    ISBN 0-7803-0417-9
    1. Transistors. 2. Compound semiconductors. I. Title.
TK7871.9.T57 1993
621.3815′28—dc20   92-46775
                                                        CIP

Contents

Preface

This volume is a compilation of selected papers on compound semiconductor transistors, emphasizing their operational physics and a recognition of their importance in semiconductor applications requiring high-frequency operation, high-speed operation, or optoelectronic compatibility. It should serve as a reference volume for engineers and scientists engaged in this field, both with material and device interests. Graduate students who are considering entering this field will also find it of interest.

Selection of papers is a hazardous task and it has been made particularly difficult by the need to make this compact volume reflect the depth of knowledge, the diversity of the various phenomena, and the achievements in this field. Countless important contributions are not represented here because papers that best served the objective of a clear, concise, and comprehensive treatment of individual subjects were chosen. This volume is made possible by the creative and sustained contributions of the many who have helped shape the development of this field; I hope that my colleagues, whose significant work has been overlooked, will understand these limitations.

The volume is divided into four parts; an introductory part with invited papers provides an overview, and the following three parts compile information on metal–semiconductor field-effect transistors (MESFETs), heterostructure field-effect transistors (HFETs), and heterostructure bipolar transistors (HBTs). These have been and are expected to continue to be the mainstream devices in this field. Each has unique attributes that make it suitable for specific applications.

I am indebted to my colleagues at IBM and elsewhere for discussions and help with final selections, N. Moll and L. H. Camnitz for their comprehensive overviews, the editorial board and staff of IEEE Press for their assistance, and the Electron Device Society Administrative Committee for presenting me with this opportunity and honor.

——SANDIP TIWARI

Introduction

The volume includes two invited papers on heterostructure field-effect and bipolar transistors; these serve as an introduction to the related devices and as an additional reference source for the numerous important papers that could not be included in this volume. The paper by Moll emphasizes the underlying basis of HFETs that contributes to their unique behavior; a subject not widely appreciated. The paper by Camnitz et al. analyzes the physical behavior of HBTs as it relates to high-frequency operation. For a lucid introduction to the operation and attributes of HBTs, the reader is referred to the paper by Kroemer in the section on HBTs. MESFETs, unlike the heterostructure devices, have seen a sustained effort and use for a much longer period of time; they are indeed mainstream devices with many books devoted to various aspects of their operation. The bibliography at the end of this introduction lists many of the longer review papers and books pertinent to the subject of this book.

The papers in this volume emphasize those subjects related to device physics that are important to both the understanding of the operation as well as the use of the devices. Therefore, while papers related to the operational aspects of quasi-static, small-signal, and transient behavior are included, an equal amount of space is also devoted to issues that are specific to the use of compound semiconductors themselves. For field-effect transistors, these include consequences of trap states at the surface and in the bulk in the form of anomalous low-frequency effects, sidegating and backgating effects, and noise; and piezoelectric effects that lead to changes in threshold voltage with gate length. In the case of HFETs utilizing n-type (Ga,Al)As barrier regions, the DX centers lead to anomalous time-dependent effects due to capture and emission of electrons. For HBTs, the consequence of trap states at the surface and in the bulk is recombination current that results in current density-dependent gain. Unusual transport and storage effects also occur in HBTs due to the use of heterostructures. The reader will find here papers addressing the operational physics as well as the physics underlying several of these phenomena that device designers have to contend with in judiciously utilizing these high-speed and high-frequency devices. Finally, toward the end of each section, there are papers that represent the current state of device technology, with emphasis on the fabrication aspects.

BIBLIOGRAPHY

Introductory References

[1] P. M. Solomon, "A comparison of semiconductor devices for high-speed logic," *Proc. IEEE,* vol. 70, no. 5, p. 489, 1982.
[2] H. Morkoç and P. Solomon, "The HEMT: A superfast transistor," *IEEE Spectrum,* vol. 21, p. 28, Feb. 1984.
[3] R. W. Keyes, *The physics of VLSI Systems.* Wokingham, England: Addison-Wesley, 1987.

MESFETs

[1] R. A. Pucel, H. A. Haus, and H. Statz, "Signal and noise properties of Gallium Arsenide microwave field-effect transistors," in *Advances in Electronics and Electron Physics*. New York, NY: Academic Press, 1975.

[2] S. I. Long and S. E. Butner, *Gallium Arsenide Digital Integrated Circuit Design*. New York, NY: McGraw-Hill, 1989.

[3] R. K. Watts, Ed., *Submicron Integrated Circuits*. New York, NY: John Wiley, 1989.

[4] P. H. Ladbrooke, *MMIC Design: GaAs FETs and HEMTs*. Boston, MA: Artech House, 1989.

[5] J. V. DiLorenzo and D. D. Khandelwal, Eds., *GaAs FET Principles and Technology*. Boston, MA: Artech House, 1982.

[6] R. A. Pucel, Ed., *Monolithic Microwave Integrated Circuits*. New York, NY: IEEE Press, 1985.

[7] D. K. Ferry, Ed., *Gallium Arsenide Technology*. Indianapolis, IN: Howard W. Sams, 1985.

[8] M. S. Shur, *GaAs Devices and Circuits*. New York, NY: Plenum, 1987.

[9] R. S. Pengelly, *Microwave Field-Effect Transistors-Theory, Design, and Applications*. Chichester, England: Research Studies, 1982.

[10] S. M. Sze, *Physics of Semiconductor Devices*. New York, NY: John Wiley, 1981.

[11] S. M. Sze, Ed., *High-Speed Semiconductor Devices*. New York, NY: John Wiley, 1990.

[12] C. T. Wang, Ed., *Introduction to Semiconductor Technology: GaAs and Related Compounds*. New York, NY: John Wiley, 1990.

[13] S. Tiwari, *Compound Semiconductor Device Physics*. Boston, MA: Academic Press, 1992.

HFETs

[1] T. Ando, A. B. Fowler, and F. Stern, "Electronic properties of two-dimensional systems," *Rev. Modern Phys.* vol. 54, no. 2, p. 437, 1982.

[2] T. J. Drummond, W. T. Masselink, and H. Morkoç, "Modulation-doped GaAs/(Al,Ga)As heterojunction field-effect transistors: MODFETs," *Proc. IEEE,* vol. 74, no. 6, p. 773, 1986.

[3] P. M. Mooney, "Deep donor levels (DX Centers) in III-V semiconductors," *J. Appl. Phys.,* vol. 67, no. 3, p. R1, 1990.

[4] H. Daembkes, Ed., *Modulation-Doped Field-Effect Transistors: Principles/Design/and Technology,* New York, NY: *IEEE Press,* 1991.

[5] H. Daembkes, Ed., *Modulation-Doped Field-Effect Transistors: Applications and Circuits*. New York, NY: IEEE Press, 1991.

[6] D. K. Ferry, Ed., *Gallium Arsenide Technology*. Indianapolis, IN: Howard W. Sams, 1985.

[7] D. K. Ferry, Ed., *Gallium Arsenide Technology,* Vol. II. Indianapolis, IN: Howard W. Sams, 1985.

[8] K. Heime, *InGaAs Field-Effect Transistors*. Somerset, England: Taunton, 1989.

[9] S. M. Sze, Ed., *High-Speed Semiconductor Devices*. New York, NY: John Wiley, 1990.

[10] C. T. Wang, Ed., *Introduction to Semiconductor Technology: GaAs and Related Compounds*. New York, NY: John Wiley, 1990.

[11] S. Tiwari, *Compound Semiconductor Device Physics*. Boston, MA: Academic Press, 1991.

HBTs

[1] M. S. Shur, *GaAs Devices and Circuits*. New York, NY: Plenum, 1987.

[2] S. M. Sze, Ed., *High-Speed Semiconductor Devices*. New York, NY: John Wiley, 1990.

[3] C. T. Wang, Ed., *Introduction to Semiconductor Technology: GaAs and Related Compounds*. New York, NY: John Wiley, 1990.

[4] S. Tiwari, *Compound Semiconductor Device Physics*. Boston, MA: Academic Press, 1991.

Part 1
Overview

HFETS: A STUDY IN DEVELOPMENTAL DEVICE PHYSICS

Nick Moll
South Fork Scientific
Soda Springs, CA 95728

It is difficult to pinpoint a single event that triggered the avalanche of work on heterostructure field-effect transistors (HFETs). It could be argued that the advent of molecular beam epitaxy and the concomitant well-controlled growth of GaAs started it, or even that early successes with GaAs metal–semiconductor field-effect transistors (MESFETs) played a key role. But from a device physics point of view, the demonstration of enhanced low-temperature mobility in modulation-doped structures cannot be overemphasized. While the direct implications of this mobility enhancement for the performance of short gate FETs are arguably small or nonexistent, the changes in other transport properties that can be surmised from such a result are substantial. As it eventually developed, modulation doping has resulted in material with better room temperature noise properties, and possibly higher peak velocities, than conventional MESFETs, and HFETs have done well in performance areas related to these properties. The higher peak doping available in the HFET has also turned out to have important positive implications for device performance, through its influence on scaling, as has the loosening of the connection of the channel material (e.g., InGaAs) from the materials in the rest of the device.

In this paper, I have set out to examine these effects in enough detail so we can see that they will impact device design, and to develop the necessary insight into the device physics so we can understand exactly why and how this impact occurs. We will begin at what I propose as the historical beginning—demonstration of enhanced mobility in modulation-doped structures—and then move into a detailed discussion of the population and transport properties of the two-dimensional electron gas, which is at the heart of all true HFETs. With this groundwork firmly established, we should be able to gain an understanding of the noise performance and frequency response of these devices. Finally, we will look at how new material systems and technology have provided and will provide a path for the further evolution of these devices.

MODULATION DOPING AND LOW-FIELD MOBILITY

What is modulation doping? As first conceived in 1978 by Dingle et al. [1], a modulation-doped structure is one consisting of alternating layers of low and high bandgap materials, with the doping confined (or modulated) to the high bandgap material. This is shown schematically in Figure 1, taken from [1] along with some information that we will discuss shortly. In order to render this discussion as concrete as possible, we will look at the specific material system employed by Dingle et al., although the concepts may be generalized. Thus, the wide bandgap material is AlGaAs, typically with an aluminum

3

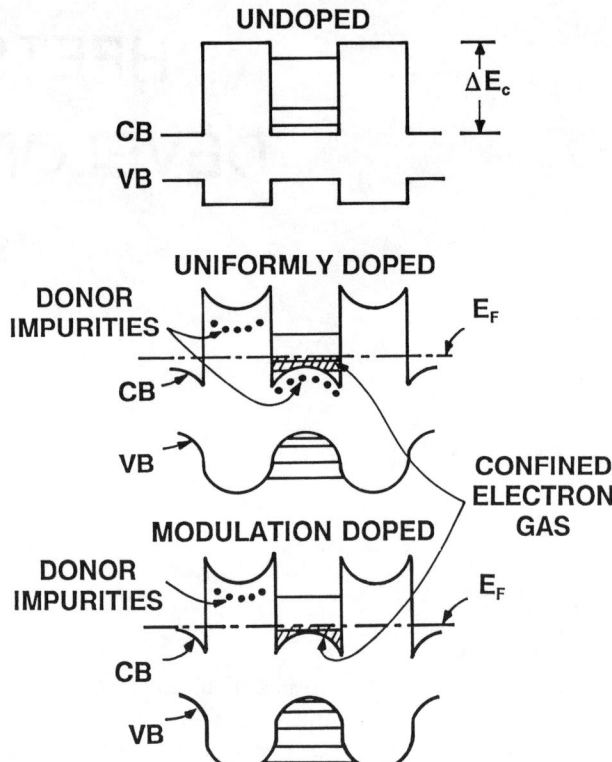

Fig. 1. Energy band diagrams for three superlattice doping schemes: undoped, uniformly doped, and modulation doped. The horizontal lines in the energy wells indicate subband energies. (From [1].)

mole fraction in the range of 30%, the narrow bandgap material is GaAs, and the doping is Si, which in molecular beam epitaxy (MBE) grown AlGaAs produces a relatively shallow donor (though one with some rather odd properties that we must eventually face). This material system in the modulation-doping scheme leads to a conduction band energy diagram as shown in Fig. 1, with alternating layers about 100 Å thick, and a Si concentration in the AlGaAs of about 10^{18} cm^{-3}. One essential property of this system is the virtual absence of interface states between the AlGaAs and the GaAs. Consequently, a fairly straightforward qualitative analysis can be used to get at the energy band diagram of this structure, and to at least hint at the richness of the physics contained therein.

Starting at the top of Fig. 1, an undoped analog of the modulation-doped structure (generally termed a *superlattice*) would form a series of alternating energy wells and barriers, within the GaAs and AlGaAs. For the practical layer thicknesses and energy discontinuities discussed here, quantum mechanical coupling between adjacent wells may be neglected, so that a series of energy levels must form within each GaAs well according to the standard quantum mechanical analysis of a one-dimensional particle in a box. In point of fact, the electrons in a semiconductor begin their conceptual life within our scheme of thought as three-dimensional; since they are confined only in one dimension by the energetic superlattice, each of the energy "levels" shown in Figure 1 in fact represents the bottom of a "two-dimensional subband," in which the electron is unconfined parallel to the superlattice interfaces, but must obey certain quantization rules perpendicular to the interfaces. In the three-dimensional space of energy and two-dimensional space of unquantized momentum then, each subband is a paraboloid of revolution about the energy axis, as shown in Fig. 2.

Returning to Fig. 1, we next ask ourselves what happens when substantial doping is introduced into the AlGaAs layers. We can take the answer to this question in two steps; first, neglecting any change in the shape of the conduction band, we place the Fermi level so as to maintain overall charge neutrality. Clearly, this placement must be somewhere below the donor level in the AlGaAs, or we would have an inordinately large net (integrated over the entire volume of the sample) negative charge. From this requirement, we can deduce that the donors will be nearly empty, and contribute a net positive charge

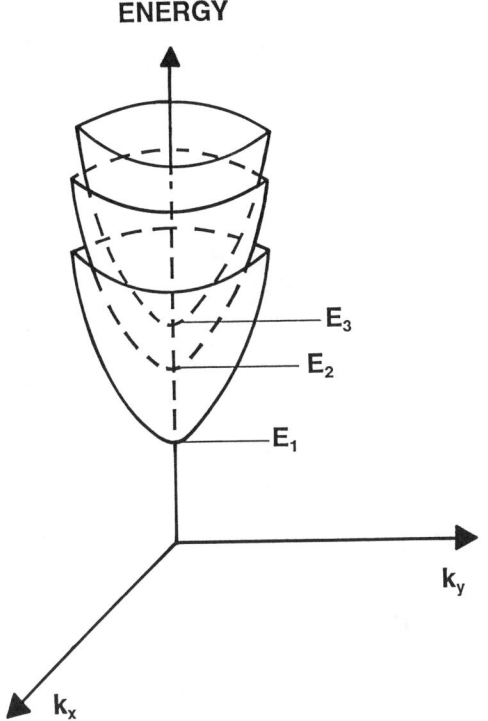

ENERGY

E₃

E₂

E₁

k_y

k_x

Fig. 2. The full *k*-space representation of the first three subbands. Since the electrons are unconstrained in the *x* and *y* directions, they may take on any value of momentum consistent with the effective mass representation. This results in an energy-momentum surface that is a paraboloid of revolution with a vertex at the so-called subband energy.

density equal to the doping density; then in order to maintain overall charge neutrality, and since the densities involved ($> 10^{18}$ cm^{-3}) are somewhat larger than the conduction band density of states in GaAs, the Fermi level must lie somewhere above the bottom of the energy well. Now, we can take the second step in determining the final energy band diagram, and that is to consider the effect of the local space charge associated with the donors in the AlGaAs and the electrons in the GaAs. From Poisson's equation, we see that the energy potential in the AlGaAs must be concave upward due to the positive space charge, while the electrons in the GaAs produce a concave downward potential in that material. The overall effect then is as shown at the bottom of Fig. 1. From a device physicist's point of view, there are a number of interesting features about this diagram, which we will eventually discuss, but one in particular presents itself. The electrons, which exist in a substantial concentration—10^{18} cm^{-3} or more—are spatially separated from the ionized donors. This is exciting, both intellectually and technologically, because at practical FET doping levels in bulk GaAs, ionized impurity scattering has a substantial effect on the transport properties of electrons [2,3]; in the modulation-doped structure this effect should be greatly reduced by the spatial separation of electrons and donors. The consequence, or at least one obvious one, is that the mobility of electrons in a modulation-doped structure should be measurably higher than that of bulk GaAs doped at comparable concentrations; moreover, at temperatures below room temperature, this difference should be far more observable, because of the diminution of phonon scattering, which is the other main determining process in the mobility of electrons in GaAs. These expectations were qualitatively borne out in [1]; as shown in Fig. 3, again taken from that work, the electrons in a modulation-doped structure showed anywhere from a 10% increase in mobility over comparable bulk GaAs to a nearly 200% increase at 77 K.

Of course, there were many details relating to growth technique, capabilities, and structural design that were not understood in this early work, so there have been huge advances in mobility enhancement since then. But that first step was the most important, especially in the history of the HFET; that is, the demonstration that you could use MBE to play with the details of transport physics at technologically interesting electron concentrations. This was a breakthrough whose importance cannot be overplayed.

In due course, most of the experimental effort rooted in Dingle's early work concentrated on a single heterojunction structure consisting of MBE-grown undoped GaAs

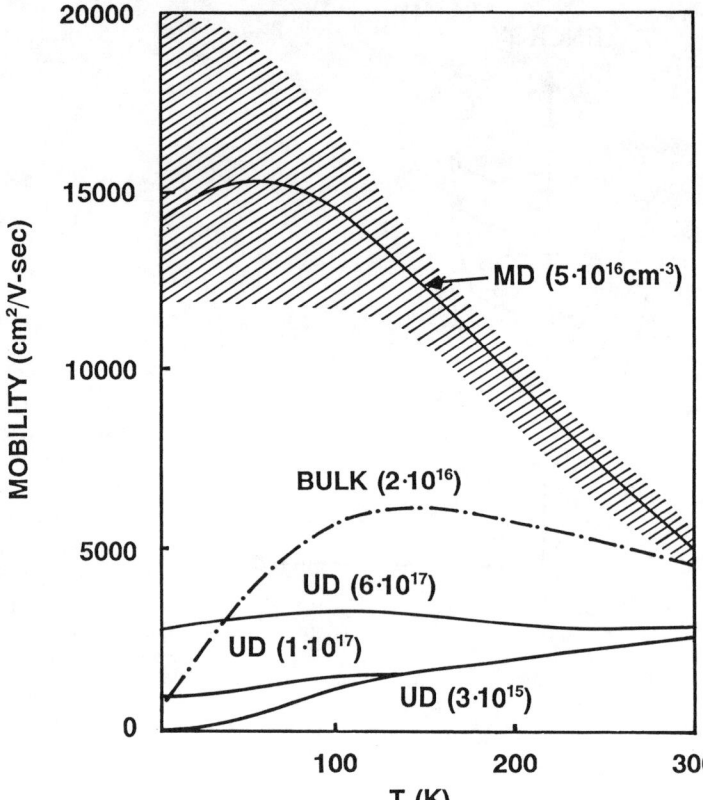

Fig. 3. The first experimental result showing mobility enhancement due to modulation doping. The shaded area represents the locus of different modulation-doped (MD) structures. The samples indicated as UD are uniformly doped superlattices. (After [1].)

capped by AlGaAs with an aluminum mole fraction in the range of 30%. The silicon doping in the AlGaAs was deliberately spaced away from the heterojunction, both to reduce remote ionized impurity scattering (that is, scattering by the Coulomb potential of the randomly placed dopant atoms within the AlGaAs, whose presence could still be felt by the electrons within the GaAs, near the heterojunction) and to eliminate any possibility of Si diffusion back into the GaAs. Spectacular results were obtained, with mobilities eventually reaching low-temperature values well over 10^6 cm²/V-sec [4]; theoretical efforts followed close behind, driven perhaps by the novelty of such large mobilities, but also by the fact that, with ionized impurity scattering largely eliminated, the theoretical ability to describe other scattering mechanisms could now be tested. The suspicion that the confinement of the electrons to two dimensions could sufficiently change some of the scattering integrals to lead to noticeably different (and improved) transport properties through mathematical artifice alone certainly existed as well.

A detailed discussion of these theoretical results is far beyond the scope of this paper, but the interested reader is referred to [5] for an entry into the literature. I will simply summarize the results as I see them from a practical standpoint: electrons in the two-dimensional electron gas of a well-executed heterojunction see scattering that is very close to the scattering they would see in bulk, undoped GaAs. There is one exception to this generalization: the theoretical results that I am aware of uniformly predict a greater scattering rate, as well as a significantly lower mobility, in a room temperature two-dimensional electron gas than in undoped GaAs, due to polar optical phonon scattering. This is not borne out by the experimental situation, which has yielded room temperature mobilities over 8000 cm²/V-sec [6]. This is an interesting discrepancy that could stem from a fatal simplification in the theory (e.g., assuming that scattering rate translates into momentum relaxation in the same way in both bulk and two-dimensional electron gas situations) or from a problem with mobility measurements (which are generally done assuming a Hall factor of 1, which may be incorrect) [7,8,9], or from some combination of the two. There is, in any case, no argument over the essential experimental result that drift mobilities far exceed those that might be obtained in doped material of equivalent electron concentration.

Let us then return for a minute to the historical setting; Dingle et al. [1] have shown that modulation doping can be used to produce structures that exhibit sheet electron concentrations at least on the order of that in GaAs MESFETs, but that show significant increases in mobility, by an order of magnitude, when cooled. The next step was taken by Mimura et al. at Fujitsu [10], who fabricated a transistor with a single modulation-doped heterojunction, which they designated a HEMT, for high electron mobility transistor. Their key result was the demonstration of improved device transconductance, by about three times, at 77 K compared to 300 K operation. Great interest was generated by this work, despite the long gate lengths—400 microns—of their devices. It proved that the two-dimensional electron gas concentration could be efficiently modulated, which set the stage for fabrication of transistors with realistically short gates. This next step was taken at Cornell by L. F. Eastman's group [11], who fabricated a 1.3-micron gate-length device, which they claimed exhibited an "average" electron drift velocity of 1.8×10^7 cm/sec at 77 K. Although some of that group's results were left open to interpretation by the brief description in their paper, and there were no reports of microwave characterization, that publication made the HFET one of the most significant advances in GaAs device technology, and one of the most pursued, during the 1980s. We see why in a later section.

OTHER KEY PROPERTIES OF TWO-DIMENSIONAL ELECTRON GASES

So far, we have examined the historical reasons for the initial interest in HFETs; these might be briefly summarized as

1. Higher electron mobility than in GaAs MESFETs, particularly at 77 K.
2. The demonstration (through FET results) of negligible interface states at the heterojunction.
3. Experimental indications of a larger "average" electron velocity in microwave HFETs.

In those historic times, far and away the greatest weight was placed on the first of these three results, particularly by material scientists and managers. Most if not all device physicists active in the field were, however, well aware that two properties that strongly influence the performance of short-gate FETs were being ignored: the moderate- to high-field transport properties of electrons in an HFET structure, and the areal density of electrons (and hence the sheet conductance) obtainable in such a structure. In the next section we shall see in detail why these properties are important; here we hope to summarize what they are, since they each represent important deviations from standard FETs.

There are two aspects to the high-field (by which we mean fields that cause a deviation from a linear velocity–field relationship) electron transport in modulation-doped structures. The first is a rapid degradation of mobility at low temperature when even a moderate electric field is applied to a sample. As reported by Drummond et al. [12] fields as low as a few tens of volts/centimeter can produce an observable effect at 77 K and below. From a qualitative point of view, the effect is easily understood. The same ultralow scattering rate that leads to electron mobilities as high as 10^6 cm^2/V-sec also means that very low fields will excite the electrons to considerable energies, so that they become subject to various possible emission processes associated with momentum relaxation. Shah et al. [13] showed that, at least in their samples, the effect could be understood as a simple increase in the electron temperature, with temperatures reaching as high as 145 K in a sample held at a 2 K lattice temperature, subjected to 500 V/cm fields. Lei et al. [14] later showed that a careful treatment of energy balance, including both polar optical and acoustic phonon scattering, gives good agreement with experimental results. They make the interesting point that, even at these low fields, the electron distribution function is not a Boltzmann distribution. This observation is at odds with the interpretation of results in [13], which are claimed to show indirectly that the heated electrons have a Fermi–Dirac distribution. The reader is left to ponder this point on his or her own.

The second effect is velocity saturation at higher electric fields. This is, of course, an effect not confined to heterostructures. It is well known that fields of 3–4 kV/cm bring on an onset of intervalley scattering, from the gamma minimum into the higher L valleys, that causes velocity saturation—and negative differential mobility—in GaAs. The final question in the case of a modulation-doped structure is not whether the velocity will saturate, but whether it will saturate at a higher or lower value than in GaAs. This turns out to be a tremendously difficult question, partly because there are more scattering paths that a hot electron at a heterojunction can take, so that theoretical approaches are at best unconvincing, and partly because the experimental determination of electron velocity at high fields is difficult in materials that exhibit negative differential mobility, and moreover may be an ambiguous quantity, depending on experimental detail. Nonetheless, both approaches have been taken.

The theoretical side is well represented by two papers. Brennan and Park compared the results of Monte Carlo calculations of steady-state velocity field curves for undoped bulk material, heterostructures in which the electrons were not confined to two dimensions in the quantum mechanical sense, and heterostructures with two-dimensional confinement [15]. Not surprisingly, the additional scattering path available in the heterostructure without confinement results in a lower peak velocity, at a lower electric field, than in undoped bulk material. For a 32% AlGaAs/GaAs structure, this deficit is found to be about 20%. Interestingly, the addition of quantum mechanical confinement changes this picture very little; the essential effect seems to be a shift of the velocity-field curve toward higher fields, with no change in the peak velocity, as shown in Fig. 4. Also of potential interest is the value predicted for the peak electron velocity, since this may have a good deal to do

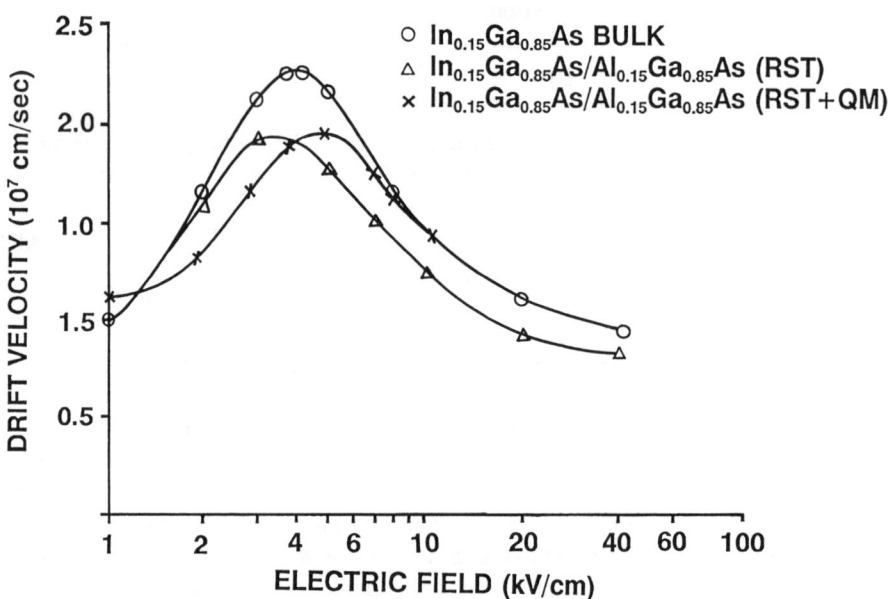

Fig. 4. Results from a Monte Carlo calculation of the steady-state electron drift velocity in an $Al_{0.15}Ga_{0.85}As/In_{0.15}Ga_{0.85}As$ heterostructure, which is energetically quite similar to the standard $Al_{0.3}Ga_{0.7}As/GaAs$ heterostructure. The different curves show the effects of adding real-space transfer (RST) and quantum-mechanical confinement (QM) to the model. Of course in real structures both effects are always present. (These results are from [15].)

with the potential performance limitations of a FET; for the structure studied this is just over 1.6×10^7 cm/sec, slightly larger than the velocity often inferred for conventional GaAs MESFETs of around 1.4×10^7. This is scarcely a compelling improvement, but perhaps precise numerical results from modeling work should not be taken too literally. Indeed, the other theme in modeling work is to model, still using Monte Carlo techniques, the transport in structures more or less meant to resemble physical FETs (although details such as doping levels often vary in order to circumvent numerical difficulties). The general outcome of such work is the prediction of significant velocity overshoot. For example, work by Wang and Hess [16] on 0.5-micron-gate HFETs suggests that electrons reach the drain edge of the gate with velocities around $2.5 \; 10^7$ cm/sec—clearly higher than steady-state velocity predictions, higher than estimated velocities in experimental MESFETs of comparable size, and high enough to be of considerable interest to the device designer. Qualitatively, this is not an unreasonable expectation. The steady-state modeling in [15] shows that both intervalley transfer from the gamma minimum to the L valley, and real-space transfer from the GaAs to the AlGaAs layer, play significant roles in limiting the steady-state electron velocity in an HFET. Both of these processes require participation of some lattice-scattering mechanism to conserve momentum, and lattice scattering is held to a minimum in modulation-doped structures, effectively consisting only of phonon emission (and absorption at room temperature.)

Experimental confirmation of the first sort of theoretical prediction is straightforward in principle, although the experimental techniques required are not trivial. In essence, by measuring the DC conductivity (with small DC fields) as a function of microwave power, it is possible to infer the velocity field characteristic of materials with complex transport properties. This has been done for modulation-doped structures [17], and the results turn out to be in rather fortuitous agreement with the corresponding Monte Carlo calculations for the most commonly encountered situation: a 30% AlGaAs/GaAs structure at room temperature. The peak electron velocity is found to be 1.8×10^7 cm/sec at around 4.5 kV/cm, compared to a Monte Carlo value of 1.7×10^7 at around 3.5 kV/cm [15]. However, the agreement is quite poor for the situation of 50% AlGaAs. There the experimental peak velocity is found to be only 1.3×10^7, while the predicted velocity is 2.0×10^7. Moreover, the experimental trend to lower peak velocities at high aluminum mole fraction is in good agreement with the universally observed poor FET performance at these higher aluminum mole fractions. It seems likely, therefore, that some essential physics, probably involving some detail of real-space transfer important at high aluminum mole fractions but not at lower ones, has been left out of the model. Models are frequently more useful for their explanatory properties than their predictive ones.

Experimental confirmation of the second sort of theory is not straightforward, even in principle. There is no direct way of measuring the electron velocity in the channel of a transistor; some effective value can only be inferred by rather indirect calculations based on various transistor properties. Without exception, these calculations depend on physical measurements such as the gate length, gate-to-channel spacing, and operating electron density, that have a good deal of uncertainty attached to them. For this reason experimentally based claims about electron velocity deserve about equal weight and circumspection as Monte Carlo based predictions. Nonetheless, we would like to observe that there are techniques that tend to statistically iron out some of the experimental uncertainties [18,19,20]. These techniques seem to uniformly indicate peak electron velocities of about 2.2×10^7 cm/sec at room temperature for structures with 30% aluminum mole fractions, and only weak dependence on gate length (for submicron gates) or aluminum mole fraction. This suggests that velocity overshoot does play a role in these devices, since the extracted FET velocity is so much larger than the measured values for uniform electric field (and electron heating), and that work remains to be done in the modeling area before even Monte Carlo programs live up to their potential value as predictive tools. It also suggests that there is some potential performance advantage of modulation-doped structures over conventional ones, based on electron velocity.

Next to the high-field transport properties, the greatest effect on FET performance is exerted by the electron density that can be obtained in a structure, and still be fully depleted by a gate. For a conventional MESFET, this areal electron density is of the order of 2–3 × 10^{12} cm^{-2} [21,22]. This was one of the most serious weaknesses in early modulation-doped field-effect transistors (MODFETs), and remains an area of important progress to this day.

There are various approaches that can be taken to the problem of calculating the electron concentration in the modulation-doped structure, ranging from ignoring quantum effects altogether to generating a self-consistent solution of Schrodinger's and Poisson's equations within the effective mass approximation. As it turns out [23] each of these approaches works well to predict the electron sheet concentration, as long as Fermi statistics are used for the classical calculation. A third technique, where the well is treated as a triangular potential well and the electron occupation is solved for with full quantum effects (still within the effective mass approximation) is fairly common in theoretical papers. This is an attractive approach in that it permits an analytic solution (in terms of Airy functions) of the electron concentration, but it performs poorly as a predictive tool at electron concentrations of any size. This occurs in part because the charge carried by the electrons significantly deforms the shape of the well exactly at the location that exerts the most sensitive influence on the subband eigenenergy, namely the peak in electron concentration. By the same token, higher subbands in which the electrons experience progressively lower electric fields are even more poorly described by this approach. These difficulties can be appreciated by referring to Fig. 5, which shows the results of a full quantum calculation. In this age of relatively cheap computing power, the full quantum approach is fairly easy to implement in a program that functions at a useful speed, and is an invaluable tool for the development of variations on the standard modulation-doped structure, but in terms of developing a basic understanding of the factors that control the electron concentration at the heterointerface, a nearly classic approach combines the advantages of simplicity, familiarity, and acceptable accuracy.

Armed with this knowledge, and the fact that the areal electron concentration is observed and can be shown to be a weak function of temperature, let us examine the question of what sort of electron concentrations we might expect in heterostructures. We will make our considerations at zero temperature, in order to give the Fermi function its simplest form. In that case, the Fermi level clearly will be pinned at the donor level in the AlGaAs, assuming this latter is thick enough not to be quite fully depleted. In the

Fig. 5. The conduction band energy diagram and subband energies calculated by a fully self-consistent model for an $Al_{0.3}Ga_{0.7}As$/GaAs heterostructure with 3 × 10^{18}/cm^3 doping in the AlGaAs. Note that the different subbands each experience different well shapes; at 300 K only 72% of the electrons in the two-dimensional electron gas occupy the lowest subband. (From [23].)

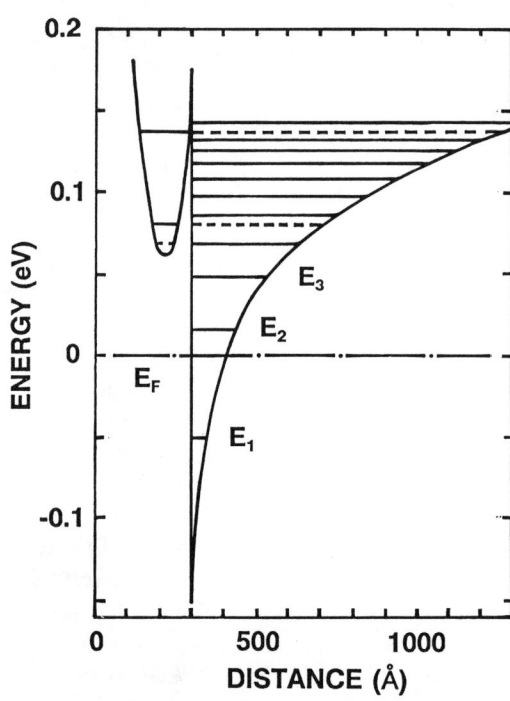

general, quantum mechanical case, there are three relevant equations: filling in the *i*th subband according to

$$(E_f - E_i) \frac{4\pi m_e}{h^2} = N_{si} \tag{1}$$

where E_i is the *i*th subband energy and the other symbols are self-evident, Poisson's equation, which leads to

$$E_0 + \Delta E_c - q \frac{N_s}{\varepsilon} d_{sp} = q \frac{N_d d_i^2}{2\varepsilon} + E_f \tag{2}$$

where the left-hand side contains the effects of the conduction band discontinuity, the field that is terminated by the total electronic charge, and the spacer layer, and the right-hand side represents the effect of the AlGaAs donors, and finally by total charge conservation

$$N_s = \Sigma_i N_{si} = N_a d_i \tag{3}$$

The essential difficulty in either the exact or approximate quantum mechanical approaches lies in the determination of the eigenenergies E_i, which depend on the shape of the potential well and hence the occupation of each and every subband in the exact case. We can develop a qualitative understanding of the factors that control the electron density by assuming that the density of states of the subbands is so large that the Fermi level is close to the bottom of the well; in point of fact that density of states, given implicitly in Eq. (1), is about $2.8 \times 10^{13}/cm^2$ eV. Then the previously given system of equations can be shown to reduce to the following in two extreme cases of interest. If the spacer layer is small,

$$N_s = \sqrt{\frac{2N_d \varepsilon (\Delta E_c - E_d)}{q}} \tag{4}$$

which says essentially that the conduction band discontinuity depletes the calculated amount of charge within the AlGaAs, and the electron population balances that charge. If the spacer layer is large compared to the depletion width of the doped AlGaAs,

$$N_s = \frac{\varepsilon (\Delta E_c - E_d)}{q d_{sp}} \tag{5}$$

To crystallize these considerations, let's take two cases: no spacer layer, a doping of 10^{18} cm^{-3}, and a conduction band discontinuity-donor depth difference of 0.2 eV, which yields an electron density of $1.6 \times 10^{12}/cm^2$, and the case of a 200-Å spacer layer with other parameters the same, which yields an electron density of $6 \times 10^{11}/cm^2$. Note that in the first case, we have probably overestimated the electron density by about 30% because of the Fermi energy required to fill the subband with that many electrons; in the second case, we have also overestimated the electron density by about that much, because of the finite extent of the depletion region into the doped AlGaAs. Applying this correction, we obtain predicted densities of about 1.0 and $0.4 \times 10^{12}/cm^2$, respectively, which agrees well with measured results [24]. There are four points worth making about these densities. First, the electron density is smaller than in a conventional MESFET—this is not advantageous. Second, in order to maximize the electron density, it is necessary to maximize the difference between the conduction band discontinuity and the donor depth in the AlGaAs; this turns out to be a very slow function of aluminum mole fraction beyond about 25% because of the increase in the donor depth, so there is little or no incentive to work at high aluminum mole fractions. Third, the spacer layer should be made as small as possible, consistent with acceptable transport properties. Finally, there are significant advantages, in terms of electron concentration, to maximizing N_d. Nonetheless, most of the early MODFET work was done at AlGaAs doping in the vicinity of 1×10^{18} cm^{-3}, because attempts at higher doping lead to extremely poor gate reverse breakdown behavior. Eventually, workers at Cornell and Hewlett-Packard [25] realized that it was sufficient to place the doping only in the AlGaAs next to the channel. This technique allowed excellent gate breakdown characteristics to be maintained while also permitting the AlGaAs to be doped at the maximum technically feasible concentration. Two-dimensional electron gas concentrations of 1.2–1.5 10^{12} can be obtained in this way.

FET BEHAVIOR

We have spent a good deal of space describing the properties of the two-dimensional electron gas that forms at a selectively doped heterojunction, with the promise that many of these properties can be related to superior FET behavior. The time has come to show how this comes about, with respect to frequency and noise performance.

In order to do this, it is worthwhile to make some observations about the factors that control the speed of FETs, given the amount of misunderstanding that has been fostered on this subject. A transistor's aptness for any given microwave or millimeter wave application cannot be distilled into a single figure of merit; many circuit designers will insist that it cannot be distilled into less than the 10 parameters or so that appear in the equivalent circuit. Nonetheless, two are commonly quoted—f_T and F_{max}—and taken together, they do indicate a good deal about a given device's potential frequency response. Only f_T is under the direct control of the device designer in the sense that it relates directly to the internal physics of the device (though he exerts a great deal of indirect influence over F_{max}), and so we concentrate on this first parameter.

Parameter f_T is defined to be the frequency at which a device, in our case a FET connected in the common source mode, exhibits unity current gain with the output AC shorted. Since improvements in this figure of merit have outstripped the availability of easily accessible network analyzer measurements in the same frequency range, it has become common practice, since the intrinsic equivalent circuit of a FET predicts 20 dB per decade roll-off of the current gain, to quote the current gain–bandwidth product measured at lower frequencies (e.g., 1–25 GHz). We adhere to this practice. With this aside, then, let us determine how the f_T of a FET relates to certain physical quantities in the device.

By definition we must have

$$2\pi f_T \delta Q_G = \delta I_d \qquad (6)$$

at constant drain voltage, so the problem of relating f_T to the physical processes in the transistor comes down to relating the change in gate charge to changes in the drain current. At constant V_{ds} any change in the gate charge must be exactly matched by a change in the channel charge, where we rather loosely define the channel as the semiconductor material between the source and drain. In general, this incremental change in charge might occur either as a change in the electron population, or in the population of occupied donors. In an AlGaAs/GaAs HFET the donors are unable to respond at microwave frequencies, so that the change in channel charge must be purely due to mobile electrons. Let us proceed to estimate the change in this charge associated with changes in the drain current.

There is some point close to the drain end of the gate where the electron concentration is determined by the potential between the gate and channel, and the electrons are traveling at their peak velocity, or nearly so. We are being a little circumspect about the precise location and nature of this point because, as the electrons enter the high-field region associated with the DC gate-drain potential, which may be quite high, some velocity overshoot may occur. But since there must be a substantial potential gradient in the plane of the device associated with this behavior, the electron concentration in the region of maximum velocity overshoot is not necessarily determined by the gate-channel potential. So at our imprecise point close to the drain end of the gate, we can define an "intrinsic" transconductance

$$g_{mi} = \frac{\varepsilon v W}{t} \qquad (7)$$

where t is the gate-channel separation, W is the transistor width, and v is a velocity that must approach the peak velocity, and must certainly be at least the velocity observed in stationary transport experiments.

It is frequently observed that the parallel-plate gate-channel capacitance

$$C_{gs} = \frac{\varepsilon LW}{t} \qquad (8)$$

so that

$$2\pi f_T = \frac{v}{L} \qquad (9)$$

where L is the gate length. From our vantage point near the drain end of the gate, it is clear that while Eq. (7) is the precise relationship between the drain current and the gate-channel potential, Eq. (8) is an underestimate of the channel charge induced by one volt of gate-channel potential and hence Eq. (9) is an overestimate of f_T. We can cite the following phenomena as significant contributors to excess intrinsic gate capacitance:

1. *Channel resistance.* The radio frequency (rf) potential required to drive the drain current through the resistance of the material under the gate means that the gate-channel potential and hence the charge density increase correspondingly as we approach the source. For a device with 6×10^{11} electrons/cm^2, this results in 60% greater charge, for a 1-micron device with a 250-Å gate-channel separation operating at room temperature. It is most serious at long gate lengths, and small gate-channel separations.

2. *Fringing capacitance.* Fringing fields can add 25% to the gate capacitance of a 0.2-micron gate device; the effect is most serious at short gate lengths and large gate-channel separations. It is aggravated by low channel electron concentrations; when the channel resistance is high and the operating frequency is characteristic of 0.25 micron or shorter gate devices, the fringing capacitance shunts the input signal around the active part of the gate.

3. *Charge stored in the drain depletion region.* At high [a few volts for submicron field-effect transistors (FETs)] drain voltages there must be a drain depletion region that is of comparable or larger extent than the gate itself. Electrons stored in this region contribute to the channel charge in an amount related to the time they require to transit this region. The effect is most serious at high drain voltages, in FETs with short gates.

4. *Effects due to a stationary charge domain in the drain depletion region.* [26] This charge domain can produce an apparently negative contribution to the drain-gate capacitance. It does not seem to be very large in terms of C_g; its main effect is on f_{max} through a reduction in the common-source feedback capacitance.

The device designer seeking to wring the maximum performance from an HFET is generally driven toward shorter gate lengths by all the considerations just cited, although there are always technological limitations on that length that rely on the ability to physically realize small dimensions, and on the need to maintain a reasonably low gate resistance and a reasonably high output impedance. For this reason, there is always a second driving force toward increasing the concentration of the two-dimensional electron gas. This increase serves to reduce channel resistance effects, to increase the current drive capabilities of the device, and to reduce the length of the drain depletion region to the extent that the increased electron concentration is associated with a greater fixed charge concentration. Both of these trends can be observed in the historical record; early work at gate lengths of around 1 micron and doping around 1×10^{18} cm^{-3} generally produced transistors with f_T gate-length products of around 20-GHz microns, with the exception of pulse-doped HFETs with high electron concentrations [25], which operated about 30% faster. At the same time, much work was being done to reduce the gate length of HFETs;

13

early work with 0.25-micron gate devices produced devices with an f_T of 30 GHz [27].[1] While an improvement over longer gate devices, this was not a substantial one, due probably to the two-dimensional electron gas concentration of less than $6 \times 10^{11}/cm^2$ and to a large gate-channel spacing. Later work at this same gate length resulted in f_T as high as 70 GHz [28]. Ultimately, devices with f_T of 110 GHz were produced with gate lengths of 0.1 micron [29]. These improved results were achieved with devices made on epitaxial layers of much better design than early ones; in particular by using pulse doping and a thin (20) Å spacer layer, electron concentrations over 10^{12} and a small gate-channel spacing became possible.

The essential elements of the microwave noise performance of HFETs are scarcely different from those of conventional GaAs MESFETs, and this behavior is therefore well described by the empirical theory laid out in Fukui's classic paper [30]. In that work, Fukui showed that the optimal value of the minimum noise figure (that is, the noise figure at the best bias point and terminal impedances) is

$$F_0 = 1 + K_f \frac{f}{f_T} \sqrt{g_m(R_g + R_s)} \qquad (10)$$

where K_f is described as a "fitting factor of approximately 2.5, representing the quality of channel materials" [30]. In this equation, g_m is an "intrinsic" g_m, that is, with the effect of parasitic source resistance removed. [Note that it is not the intrinsic g_m of Eq. (7), however.] Since control of the gate resistance is essentially a lithographic problem, it is the other parameters that the device designer must concentrate on to minimize the noise figure. Other things being equal, we can expect some improvement related to the better frequency response of HFETs compared to MESFETs of comparable gate length; the higher f_T, even at the expense of higher g_m, leads to some decrease in F_0. But what about K_f? A deeper, more physical examination of FET noise behavior [31,32] suggests that the size of K_f is related to excess noise in the drain current, which originates as diffusion noise in the channel. The contribution of any part of the channel to the spectral density of the current noise is [31]

$$S_j \, \delta x = 4q^2 D_l n_{ss}(x) W \qquad (11)$$

where D_l is the longitudinal diffusion coefficient, and n_{ss} the areal electron density. Examining this expression, we see that huge differences in the noise currents of HFETs and MESFETs operating at the same current density are not to be expected. In regions where the transport is still described by the low-field mobility, the higher mobility (and hence higher diffusion coefficient) in the HFET is compensated for by the lower electron density required to support the drain current. Where the velocity begins to saturate, the lower heating rate of HFET electrons more or less compensates for the increased mobility. Then, only to the extent that fewer electrons are required to support the drain current (because of a greater peak velocity) does the HFET have an advantage.

The experimental situation is the following. Over the last decade, 0.25-micron MESFETs have improved from about 1.5 to 1.15 at 10 GHz [33,34]. Comparable values for AlGaAs/GaAs HFETs are 1.14 to 1.086 [35,36]. HFETs have lost some of their performance margin, but they still add less than 60% of the noise than comparable MESFETs. This corresponds to their advantage from f_T. We might fairly summarize the situation, then, by stating that the HFET's noise performance stems primarily from the factors, discussed previously, that lead to improved frequency response, coupled with an absence of other degrading factors.

---

[1]We have revised the figure quoted in [27] as f_T to conform to our convention that it should match the current gain–bandwidth product at lower frequencies. Some early authors were in the habit of extrapolating current gain measurements at slopes considerably less than 20 dB/decade, resulting in inflated values for f_T. This practice has now disappeared.

FET MISBEHAVIOR

No discussion of HFETs would be complete without some reference to the peculiar electrical characteristics of the silicon donor in AlGaAs. This impurity does not behave as a simple hydrogenic donor beyond aluminum mole fractions of about 0.2. Rather, it behaves as a donor-complex (DX) center, with the essential properties as first described extensively by Lang et al. [37]. For many purposes it must be treated as if it has a barrier of several tenths of an eV to electron capture, as well as to emission. The energetics of this behavior can be most easily understood by referring to Fig. 6. In truth, the behavior of DX centers is more complex than this; its exact physical origin remains a point of some controversy [38,39,40], and it seems likely that the physical model originally proposed in [37] is not exact. The phenomenological understanding contained in Fig. 6 is even less exact and does not, for instance, quantitatively describe the optical behavior of DX centers. Still it does serve to qualitatively explain the most important device consequences of DX centers.

There are essentially three effects to be aware of: threshold shift, current–voltage (*I–V*) or drain collapse, and g_m dispersion. All these effects turn on the relatively slow response of the charged and occupied Si donors to changes in the electron imref within the AlGaAs. How slow? The emission barrier and capture cross section are found to be about 0.45 eV and 4.5×10^{-3} cm^2 for 30% AlGaAs [41]; the emission time constant is then

$$\tau = \frac{1}{\sigma N_c v_{th}} \, exp \, \frac{E_a}{kT} \tag{12}$$

or about 3.5 ms at room temperature, and well over one million years at 77 K.

The peak transconductance in almost all, if not all well-designed HFETs occurs at the gate bias that just begins to cause some noticeable occupation of the ionized donors under the gate. At lower bias, the terminal transconductance is always degraded by the resistive drop through the channel, and of course at higher bias it is degraded because the gate charge fills AlGaAs donors rather than the channel with electrons [42]; also there is some saturation of the source resistance at these higher currents. Now, at least for room temperature operation, the state of charge in the AlGaAs will follow changes in the bias point easily enough, but it cannot follow in rf signal because of the long capture time constant. So in an HFET that is biased at its natural bias—the transconductance peak— and is suddenly subjected to an input signal of any size, free electrons can potentially appear in the AlGaAs. Since the population of these electrons must follow

$$n = N_c \, exp \, \frac{q(\phi_n - E_c)}{kT} \tag{13}$$

there is a rectifying effect; the exponential behavior causes the average electron population to rise, in the presence of the rf signal, over its quiescent value. Of course, the

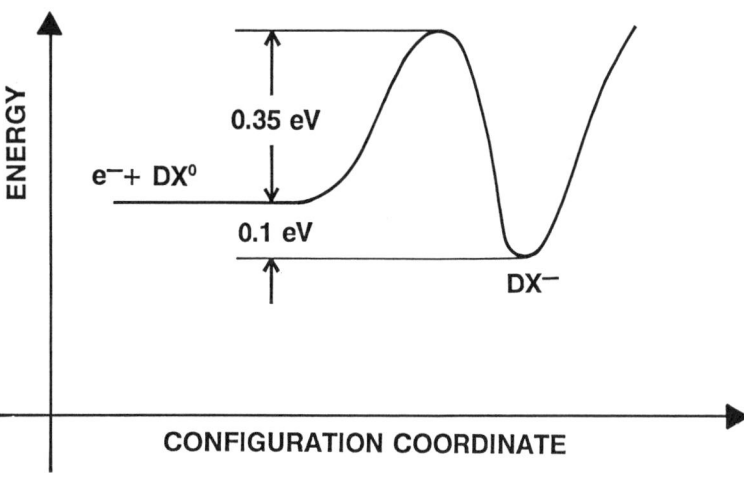

Fig. 6. A schematic representation of the energetic behavior of a conduction band electron and a DX center in Al$_{0.3}$Ga$_{0.7}$As. The electron must overcome a 0.35-eV barrier to be captured by the DX center and a 0.45-eV barrier to be emitted by it. (I have followed the charge assignments (DX$^-$ and DX^0) suggested in [39]; the reader is cautioned that this is a controversial view.)

ionized donors will successfully capture the electrons that, *on the average,* are excess; consequently, the threshold voltage will shift up, that is, the device will tend to turn itself off in the presence of a signal. This effect is devastating for digital applications [43] and must be controlled; it is more of an irritant for most microwave applications, but should be carefully assessed in situations where pulsed signal amplitudes are likely to be encountered. Obviously, the effect can be eliminated by working at sufficiently low Al mole fraction; however, it can also be largely reduced by careful attention to the two-dimensional design of the transistor. It is beyond the scope of this paper to describe these details, but a reconsideration of our heuristic description of the origin of the effect shows that, if the AlGaAs electron concentration can be prevented from exactly following the gate bias by controlling the source of those electrons, the effect can be reduced.

Drain collapse [44] is a more extreme problem than threshold shift, but occurs only during low (e.g., 77 K)-temperature operation. The effect was noticed early in the development of the HFET thanks to the special interest in the low-temperature behavior of these devices. Its signature is the onset of a high resistance state at low drain voltages once the device is operated at drain voltages above a volt or so. Some devices will not pass normal current at higher drain biases. This collapse can only be reversed by warming the device well above 77 K, or by exposure to light. Finally, not every device exhibits drain collapse; depending on the details of design and fabrication, HFETs can be made that are more or less immune to the phenomenon [44]. These details strongly suggest that the collapse is caused by the injection of hot electrons into the AlGaAs, where they are trapped by ionized donors. Note that, since the electrons are hot, they have little difficulty overcoming the barrier to capture by the DX center. Eventually, this process occurs to a sufficient degree that there is not enough ionized charge to create a two-dimensional electron gas in the region of hot electron generation and the device will pass little or no current at low drain voltages. The most likely location for this process to occur is within the drain depletion region at and beyond the drain edge of the gate, although some details of this view can be disputed [45]. The key to avoiding the effect is to provide a more likely alternative path for electrons in this region than injection into the AlGaAs and subsequent capture by the DX centers.

Finally, we come to the question of g_m dispersion. In its most common form, this is a manifestation of the exact same process that produces threshold shift. In essence, g_m measurements made at sufficiently low frequency, near the transconductance peak, tend to be somewhat low because the threshold voltage is shifting during the measurement. The transconductance at sufficiently high frequencies (more than a few megahertz) can be noticeably higher because the DX centers will not charge and discharge at that rate. It is worth noting in this respect that there are typically additional traps in AlGaAs besides the unavoidable DX centers, particularly when the growth of this layer takes place at low temperatures. These traps can lead to g_m dispersion in the opposite direction; that is, the DC transconductance can be higher than the transconductance measured at rf frequencies. This is a fairly common state of affairs, particularly with some early work, and means that estimates of transport phenomena based on the DC transconductance must be viewed with extreme skepticism.

OTHER MATERIAL SYSTEMS

Two areas that tend to limit the performance of HFETs realized in the AlGaAs/GaAs material system are the electron density, through its influence on channel and parasitic resistances and, for some applications, drive current, and the undesirable behavior of donors in AlGaAs. The possibility exists of circumventing these problems by realizing HFETs with different material systems, and in some cases also increasing the electron velocity. In fact, not long after the first microwave HFETs were built using AlGaAs, much attention was focused on heterojunctions in material systems lattice matched to InP, particularly $Al_{0.47}In_{0.53}As/Ga_{0.48}In_{0.52}As$ [46,47]. This early work was driven mainly by the expectation that the electron velocity would be considerably higher in these devices, since bulk InGaAs is known to exhibit a peak electron velocity of 3.5×10^7 cm/sec. The results from those devices were generally disappointing, and apparently were plagued with problems associated with the relative immaturity of the required growth technology. It was eventually realized, and demonstrated, that FETs could be fabricated with a thin layer of InGaAs as the channel material, using GaAs substrates and related growth technology, in spite of the large lattice mismatch between the GaAs and the InGaAs [48,49]. These came to be known as pseudomorphic devices, referring to the fact that the InGaAs layer was compressed in two dimensions to have the same lattice constant as the GaAs it was grown on, rather than its natural, larger, lattice constant. In spite of the rather low In content of these first pseudomorphic devices, and the fact that, since they employed GaAs as the wide-bandgap material in the heterojunction, they had a very small conduction band discontinuity, their performance was somewhere in between that of AlGaAs/GaAs HFETs and InP-based ones. The next step was obvious, at least in hindsight, but not a trivial one in terms of material growth: to make an HFET based on AlGaAs for the wide bandgap material, and InGaAs for the channel material. This was indeed done [50,51], with results that more or less matched those of AlGaAs/GaAs HFETs. Those FETs typically had In and Al contents of around 15%, which produces a structure energetically quite similar to the then (and now) state-of-the-art AlGaAs HFETs. It is therefore no great surprise that they did not surpass the performance of AlGaAs/GaAs HFETs; rather the results suggested that the essential material problems were soluble.

From everything we have discussed so far, it is clear that the key to improved performance lies in increasing the conduction band discontinuity beyond that observed in the conventional AlGaAs/GaAs HFET. This is necessary for two reasons. First, it is the only way to significantly increase the concentration of the two-dimensional electron gas, which plays a role in limiting the conventional device. Second, it is necessary for there to be any likelihood of improved transport in the InGaAs layer because of the limiting effects of real space transfer. Of course, even with a larger band discontinuity, there was no special reason to expect that a pseudomorphic layer, with a lattice the same as that of GaAs in the plane of the substrate, would exhibit transport approaching that of bulk InGaAs.

Indeed, when HFETs were fabricated with higher In and Al mole fractions, and therefore a higher conduction band discontinuity, every indication was that the transport in those devices was identical to that of AlGaAs/GaAs HFETs, although the performance was vastly superior. Those devices were made with 25% In and Al mole fractions, leading to a conduction band discontinuity of 0.46 eV, and an electron density of $2.3 \times 10^{12}/cm^2$, which is about double that obtainable in AlGaAs/GaAs devices. The higher electron density led to a then record f_T of 98 GHz, and significantly improved current drive capabilities [18]. Although that frequency performance has now been far surpassed [19], it serves to drive home the relationship between electron density and f_T.

For devices on GaAs substrates, pseudomorphic HFETs remain the path for future growth. The essential problem that presents itself here is to solve the growth aspects of increasing the In mole fraction significantly beyond 25%.

This may be a rather moot problem, however, as long as InP substrates are acceptable for the application. Considerable progress has been made in the area of InAlAs/InGaAs HFETs since the references cited previously were published, and in many respects this material system has lived up to the promises offered by a 0.5-eV conduction band discontinuity, higher achievable doping than in AlGaAs, and a peak velocity possibly as high as 3.5×10^7 cm/sec. In particular Mishra et al. have obtained attention-getting early results [52,53] with f_T as high as 170 GHz, sheet charge density of 3×10^{12}, and a room temperature electron mobility of 10,000 cm^2/v-sec. The noise figure of this device, scaled to 10 GHz, would be 1.034, a substantial improvement over comparable AlGaAs HFETs. These devices continue to be successfully scaled, with 0.08-micron gate devices yielding f_T as high as 250 GHz [54]. It should be remembered, however, that the material technology behind these devices is not nearly as mature as that for devices based on material systems that can be grown on GaAs substrates. There are clearly difficulties with devices on InP [55] that make DX related problems seem trivial, and these must eventually be solved.

CONCLUSION

It is not possible, in a paper of this length, to adequately cover every major development in a field as rich and broad as that of HFETs. What I hope to have done instead is to share a rather personal perspective on key results concerning these devices. If the reader is left with the impression that advances in device technology are frequently serendipitous, then I will have accomplished half of my task. A significant aspect of the growth of science and technology is the possession of intuitive knowledge of what to try, and how to go about the trying in an accomplished and successful fashion. An equally important aspect, however, is to honestly assess and correctly understand the results obtained in this manner, and that aspect provides the essential balance to the serendipitous search for discovery. If I have conveyed some physical understanding of the processes that lead to the distinctive behavior of HFETs, then I will have accomplished the other half of my task.

ACKNOWLEDGMENTS

Thanks go to my many colleagues in the areas of III–V processing, growth and device physics with whom I have interacted over the years, with special thanks to Mark Hueschen, Hans Rohdin, Lovell Camnitz, and Alice Fischer-Colbrie of Hewlett-Packard. Finally, thanks to Sandip Tiwari for the right combination of encouragement and forbearance during the writing of the paper.

REFERENCES

[1] R. Dingle, H. L. Stormer, A. C. Gossard, and W. Wiegmann, "Electron mobilities in modulation-doped semiconductor heterojunction superlattices," *Appl. Phys. Lett.*, vol. 33, no. 7, pp. 665–667, 1978.

[2] P. A. Folkes, "Measurement of the low-field electron mobility and compensation ratio profiles in GaAs field-effect transistors," *Appl. Phys. Lett.*, vol. 48, no. 6, pp. 431–433, 1986.

[3] D. L. Rode and S. Knight, "Electron transport in GaAs," *Phys. Rev. B*, vol. 3, no. 8, pp. 2534–2540, 1971.

[4] S. Hiyamizu, J. Saito, J. Nanbu, and T. Ishikawa, "Improved electron mobility higher than 10^6 cm^2/Vs in selectively doped GaAs/N-AlGaAs heterostructures grown by MBE," *Japan. J. Appl. Phys.*, vol. 22, pp. L609–L611, 1983.

[5] K. Hirakawa and H. Sakaki, "Mobility of the two-dimensional electron gas at selectively doped n-type $Al_xGa_{1-x}As$/GaAs heterojunctions with controlled electron concentrations," *Phys. Rev. B*, vol. 3, no. 12, pp. 8291–8303, 1986.

[6] H. Morkoç, "Current transport in modulation doped (Al,Ga)As/GaAs heterostructures: applications to high speed FET's," *IEEE Electron Device Lett.*, vol. EDL-2, no. 10, pp. 260–262, 1981.

[7] M. Keever et al., "Hall effect and mobility in heterojunction layers," *J. Appl. Phys.*, vol. 53, no. 2, pp. 1034–1036, 1982.

[8] K. Hess, "Impurity and phonon scattering in layered structures," *Appl. Phys. Lett.*, vol. 35, no. 7, pp. 484–486, 1979.

[9] P. J. Price, "Electron transport in polar heterolayers," *Surface Sci.*, vol. 113, pp. 199–210, 1982.

[10] T. Mimura, S. Hyamizu, T. Fujii, and K. Nanbu "A new field-effect transistor with selectively doped GaAs/n-AlGaAs heterojunctions," *Japan. J. Apl. Phys.*, no. 5, vol. 19, pp. L225–L227, 1980.

[11] S. Judaprawira et al., "Modulation-doped MBE GaAs/n-AlGaAs MESFETs," *IEEE Electron Device Lett.*, vol. EDL-2, no. 1, pp. 14–15, 1981.

[12] T. J. Drummond et al., "Field dependence of mobility in AlGaAs/GaAs heterojunctions at very low fields," *Electron. Lett.*, vol. 17, pp. 545–547, 1981.

[13] J. Shah et al. "Hot electrons in modulation-doped GaAs-AlGaAs heterostructures," *Appl. Phys. Lett.*, vol. 44, no. 3, pp. 322–324, 1984.

[14] X. L. Lei, J. Q. Zhang, J. L. Birman, and C. S. Ting, "Hot-electron transport in GaAs-AlGaAs heterojunctions," *Phys. Rev. B*, vol. 33, no. 6, pp. 4382–4385, 1986.

[15] K. F. Brennan and D. H. Park, "Theoretical comparison of electron real-space transfer in classical and quantum two-dimensional heterostructure systems," *J. Appl. Phys.*, vol. 65, no. 3, pp. 1156–1163, 1989.

[16] T. Wang and K. Hess, "Calculation of the electron velocity distribution in high electron mobility transistors using an ensemble Monte Carlo method," *J. Appl. Phys.*, vol. 57, no. 12, pp. 5336–5339, 1985.

[17] W. T. Masselink, N. Braslau, D. LaTulipe, W. I. Wang, and S. L. Wright, "Electron velocity at high electric fields in AlGaAs/GaAs modulation-doped heterostructures," *Solid-State Electron.*, vol. 31, no. 3/4, pp. 337–340, 1988.

[18] N. Moll, M. Hueschen, and A. Fischer-Colbrie "Pulse-doped AlGaAs/InGaAs pseudomorphic MODFETs," *IEEE Trans. Electron Devices,* vol. ED-35, no. 7, pp. 879–886, 1988.

[19] L. D. Nguyen, P. J. Tasker, D. C. Radelescu, and L. F. Eastman, "Design, fabrication, and characterization of ultra high speed AlGaAs/InGaAs MODFETs," *IEDM Tech. Dig.*, pp. 176–179, 1988.

[20] H. Rohdin "Reverse modelling of E/D logic submicrometer MODFETs and prediction of maximum extrinsic MODFET current gain cutoff frequency," *IEEE Trans. Electron Devices,* vol. ED-37, no. 4, pp. 919–934, 1990.

[21] F. Kharabi and D. R. Decker, "Magnetotransconductance profiling of mobility and doping in GaAs MESFETs," *IEEE Electron Device Lett.*, vol. EDL-11, no. 4, pp. 137–139, 1990.

[22] R. B. Darling, "Subthreshold conduction in uniformly doped epitaxial GaAs MESFETs," *IEEE Trans. Electron Devices*, vol. ED-36, no. 7, pp. 1264–1273, 1989.

[23] J. Yoshida, "Classical versus quantum mechanical calculation of the electron distribution at the n-AlGaAs/GaAs heterointerface," *IEEE Trans. Electron Devices*, vol. ED-33, no. 1, pp. 154–156, 1986.

[24] K. Hirakawa, H. Sakaki, and J. Yoshino, "Concentration of electrons in selectively doped AlGaAs/GaAs heterojunction and its dependence on spacer-layer thickness and gate electric field," *Appl. Phys. Lett.*, vol. 45, vol. 3, pp. 253–255, 1984.

[25] M. Hueschen, N. Moll, E. Gowen, and J. Miller, "Pulse-doped MODFETs," *IEDM Tech. Dig.*, pp. 348–351, 1984.

[26] R. W. H. Engelmann and C. A. Liechti, "Bias dependence of GaAs and InP MESFET parameters," *IEEE Trans. Electron Devices*, vol. ED-24, no. 11, pp. 1288–1296, 1977.

[27] P. C. Chao et al., "Quarter micron gate length microwave high electron mobility transistor," *Electron. Lett.*, vol. 19, pp. 894–896, 1983.

[28] U. K. Mishra et al., "Microwave performance of 0.25 micron gate length high electron mobility transistors," *IEEE Electron Devices Lett.*, vol. EDL-6, no. 3, pp. 142–145, 1985.

[29] A. N. Lepore et al., "0.1 micron gate length MODFETs with unity current gain cutoff frequency above 110 GHz," *Electron. Lett.*, vol. 24, no. 6, pp. 364–366, 1988.

[30] H. Fukui, "Optimal noise figure of microwave GaAs MESFETs," *IEEE Trans. Electron Devices*, vol. ED-26, no. 7, pp. 1032–1037, 1979.

[31] B. Carnez et al., "Noise modelling in submicrometer-gate FETs," *IEEE Trans. Electron Devices*, vol. ED-28, no. 7, pp. 784–795, 1981.

[32] A. Cappy et al., "Noise modelling in submicrometer-gate two-dimensional electron gas field effect transistors," *IEEE Trans. Electron Devices*, vol. ED-32, no. 12, pp. 2787–2796 1985.

[33] S. G. Bandy, D. M. Collins, and C. K. Nishimoto, "Low noise microwave FETs fabricated by molecular-beam epitaxy," *Electron. Lett.*, vol. 15, no. 8, pp. 218–219, 1979.

[34] I. Banerjee, P. W. Chye, and P. Gregory, "Unusual C-V profiles of Si-implanted (211) GaAs substrates and unusually low-noise MESFETs fabricated on them," *IEEE Electron Device Lett.*, vol. 9, no. 1, pp. 10–12, 1988.

[35] K. H. G. Duh et al., "60 GHz low-noise high-electron-mobility transistors," *Electron. Lett.*, vol. 22, no. 12, pp. 647–648, 1986.

[36] K. H. G. Duh et al., "Ultra-low-noise characteristics of millimeter wave high electron mobility transistors," *IEEE Electron Device Lett.*, vol. 9, no. 19, pp. 521–523, 1988.

[37] D. Lang, R. Logan, and M. Jaros, "Trapping characteristics and a donor-complex (DX) model for the persistent-photoconductivity trapping center in Te-doped AlGaAs," *Phys. Rev. B*, vol. 19, no. 2, pp. 1015–1030, 1979.

[38] D. J. Chadi and K. J. Chang, "Energetics of DX center formation in GaAs and $Al_xGa_{1-x}As$ alloys," *Phys. Rev. B*, vol. 39, no. 14, pp. 10063–10074, 1989.

[39] T. N. Morgan, "Theory of the DX center in $Al_xGa_{1-x}As$ and GaAs crystals," *Phys. Rev. B*, vol. 34, no. 4, pp. 2664–2669, 1986.

[40] D. K. Maude et al., "Investigation of the DX center in heavily doped n-type GaAs," *Phys. Rev. Lett.*, vol. 59, no. 7, pp. 815–818, 1987.

[41] A. Valois and G. Robinson, "Characterization of deep levels in modulation-doped AlGaAs FETs," *IEEE Electron Device Lett.*, vol. EDL-4, no. 10, pp. 360–362, 1983.

[42] M. C. Foisy, P. J. Tasker, B. Hughes, and L. F. Eastman, "The role of inefficient charge modulation in limiting the current-gain cutoff frequency of the MODFET," *IEEE Trans. Electron Devices*, vol. ED-35, no. 7, pp. 871–878, 1988.

[43] R. Kaneshiro et al., "Anomalous nanosecond transient component in a GaAs MODFET technology," *IEEE Electron Device Lett.*, vol. EDL-9, no. 5, pp. 250–252, 1988.

[44] R. Fischer et al., "On the collapse of drain I-V characteristics in modulation-doped FETs at cryogenic temperatures," *IEEE Trans. Electron Devices*, vol. ED-31, no. 8, pp. 1028–1032, 1984.

[45] A. Kastalsky and R. Kiehl, "On the low-temperature degradation of (Al,Ga)As/GaAs modulation-doped field-effect transistors," *IEEE Trans. Electron Devices,* vol. ED-33, no. 3, 1986.

[46] C. Y. Chen et al., "Depletion mode modulation doped AlInAs-GaInAs heterojunction field-effect transistors," *IEEE Electron Devices Lett.,* vol. EDL-3, pp. 152–155, 1982.

[47] T. P. Pearsall et al., "Selectively doped AlInAs/GaInAs heterostructure field effect transistor," *IEEE Electron Devices Lett.,* vol. EDL-4, pp. 5–8, 1983.

[48] R. J. Rosenberg et al., "An $In_{0.15}Ga_{0.85}As$/GaAs pseudomorphic single quantum-well HEMT," *IEEE Electron Device Lett.,* vol. EDL-6, pp. 491–493, 1985.

[49] T. E. Zipperian and T. J. Drummond, "Strained quantum-well, modulation doped field effect transistor," *Electron. Lett.,* vol. 21, pp. 823–824, 1985.

[50] A. Ketterson et al., "High transconductance InGaAs/AlGaAs pseudomorphic modulation-doped field effect transistors," *IEEE Electron Device Lett.,* vol. EDL-6, pp. 628–630, 1985.

[51] T. Henderson et al., "DC and microwave characteristics of a high current double interface GaAs/InGaAs/AlGaAs pseudomorphic modulation-doped field effect transistor" *Appl. Phys. Lett.,* vol. 48, pp. 1080–1089, 1980.

[52] U. K. Mishra et al. "High performance sub-micron AlInAs-GaInAs HEMTs," Device Research Conference Paper II a-6, 1987.

[53] U. K. Mishra et al., "Microwave performance of AlInAs-GaInAs HEMTs with 0.2 and 0.1 micron gate length," *IEEE Electron Device Lett.,* vol. 9, no. 12, pp. 647–649, 1988.

[54] L. D. Nguyen, L. M. Jelloian, M. Thompson, and M. Lui, "Fabrication of an 80 nm self-aligned T-gate AlInAs/GaInAs HEMT," *IEDM Tech. Dig.,* pp. 499–502, 1990.

[55] J. B. Kuang et al. "I/V anomaly and device performance of submicrometer-gate $Ga_{0.47}In_{0.53}As/Al_{0.48}In_{0.52}As$ HEMT," *Electron. Lett.,* vol. 24, no. 25, pp. 1571–1572, 1988.

AN ANALYSIS OF THE CUTOFF-FREQUENCY BEHAVIOR OF MICROWAVE HETEROSTRUCTURE BIPOLAR TRANSISTORS

Lovell H. Camnitz
Hewlett-Packard Laboratories
Palo Alto, CA

Nick Moll
South Fork Scientific
Soda Springs, CA

Abstract—The microwave behavior of the heterostructure bipolar transistor (HBT) is reviewed, with an emphasis on the cutoff frequency. With the use of heterostructure emitters and compositionally graded bases, base delays of < 0.5 psec can be realized in III/V HBTs with low base resistance (~1 psec in SiGe HBTs). Closed-form expressions for the collector delay and collector capacitance behavior of GaAs HBTs are derived based on the temperature-dependent static electron velocity-field characteristic. The derivation accounts for electron space-charge and base pushout (Kirk) effects. The model predictions correlate well with measurements of AlGaAs/GaAs HBTs fabricated in the authors' laboratory over a wide range of bias conditions and collector parameters. This agreement is obtained without accounting for velocity overshoot. The dependence of electron velocity on electric field and on junction self-heating explains the decrease in the cutoff frequency with increasing collector voltage. The sensitivity of the velocity profile to the electron space-charge density is responsible for the experimentally observed decrease in collector delay and collector capacitance in high current operation. The collector delay is lowest in high collector current, low collector voltage operation. Under these conditions, the collector delay has been reduced to less than 0.5 psec in special collector structures designed to take advantage of velocity overshoot. If the current is increased further, base pushout (Kirk effect) occurs, increasing the collector capacitance, but not usually lowering the DC current gain. The cutoff frequency of microwave power HBTs, which operate at high collector voltage, is largely determined by the collector delay. The maximum potential breakdown voltage-cutoff frequency products of AlGaAs/GaAs and Si/SiGe/Si power HBTs are roughly equivalent, though device self-heating is of greater concern in the GaAs device.

I. INTRODUCTION

In the past decade, the concept of the heterostructure bipolar transistor (HBT), patented in 1948 by W. Shockley, has been translated into practical microwave and millimeter-wave HBTs [1]. This has extended the frequency range of circuits traditionally designed around silicon bipolar junction transistors (BJTs), and offers a significant alternative to GaAs field-effect transistors (FETs) and heterostructure FETs (HFETs).

An excellent treatment of fundamental HBT theory is given by Kroemer [2,3,4], and will not be repeated here. It is worth noting, however, that several major advantages can result from the use of heterostructures in a microwave bipolar transistor. The most important derive from the wide-bandgap emitter, which results in good emitter injection efficiency regardless of the base and emitter doping. This allows a dramatic increase in base doping (typically $> 10^{19}$ cm^{-3}) and consequently, a reduction in both the base resistance and base width. A relatively low emitter-base capacitance can be maintained, because of the low (typically $\sim 10^{17}$ cm^{-3}) emitter doping and absence of hole storage in the neutral emitter. A graded bandgap base offers high base transit velocity even in moderately wide, heavily doped bases where the minority carrier diffusivity may be low. Finally, a heterostructure collector allows the use of a collector drift region chosen for its superior high-field transport and/or breakdown characteristics.

Though high cutoff frequencies have recently been achieved in Si/SiGe/Si HBTs [5], HBTs were first realized in the III/V semiconductors, and these offer even higher performance. Apart from the advantages of heterostructures mentioned above, the III/V semiconductors such as GaAs and InGaAs offer the additional advantages of high electron mobility, high peak velocity, and the possibility of velocity overshoot [6] over significant distances. When a low-energy electron leaving the base first encounters the high electric field in a GaAs or InGaAs collector, its velocity is temporarily higher than the static value for that electric field until its energy and scattering rate rise to the static values. The velocity overshoot tends to reduce the collector delay time. A cutoff frequency, f_T, of 171 Ghz has been measured in AlGaAs/GaAs HBTs specially designed to take advantage of velocity overshoot [7], corresponding to an effective total delay of only 1 psec. Separately, f_{max} values of 218 Ghz have been achieved with a conservative 1-μm emitter stripe width, demonstrating the benefits of the p$^+$ 10^{20} cm^{-3} base [1]. These frequencies are several times higher than those of microwave Si BJTs.

Here, we focus on the physics of the delay, which is perhaps the most unique and distinguishing feature of the HBT, and gives rise to impressive high-frequency performance. We begin by examining the structure of a representative microwave AlGaAs/GaAs N-p-n HBT fabricated in the authors' laboratory, then move on to the delay analysis, using closed-form expressions and practical examples wherever possible. We include SiGe and InGaAs HBTs in the discussion where it is appropriate. The delay in the collector depletion layer deserves special attention, because the reduction in the resistance–capacitance (RC) time constants, and delays in the emitter and base of the microwave HBT has magnified its relative importance.

The multivalley electron transport in III/V semiconductors is quite unlike that of Si, and is usually analyzed using numerical methods [8]. Here, we derive an analytical solution for the collector transport based on the temperature-dependent static velocity-field curves of GaAs. While perhaps less precise than a full-blown Monte Carlo calculation, it predicts the voltage, current, and temperature dependencies of the cutoff frequency with reasonable agreement with experiment, and illuminates the basic physics underlying HBT delays. Effects such as the dependence of f_T on V_{ce} and collector doping, N_{Dc}, are quantified. It also explains two phenomena unique to III/V HBTs that have been discovered only recently [9]. The first is an oddly sharp peak in the f_T - J_c characteristic, and the second is an unusually low value of active area collector capacitance. Both phenomena result from an interaction between the electron space-charge density profile in the collector and the electric field profile through negative differential mobility.

Next, we explore the implications of the base pushout (the Kirk effect) [10,11] in high-current, low-voltage operation. Surprisingly, in many cases, we find the cutoff frequency is not strongly degraded, and the DC current gain is completely unaffected.

We discuss the implications of our model in the context of velocity overshoot and the continuing need for faster transistors. To assess the potential power-frequency performance, as defined by Johnson [12], of Si/SiGe/Si and AlGaAs/GaAs microwave power HBTs [13], we compare the velocity-breakdown field products of the respective collector semiconductors, Si and GaAs. The advantages of wide-bandgap collectors and the special considerations necessary in their design are outlined.

II. HBT STRUCTURE

Figure 1 shows the schematic cross section of a representative microwave AlGaAs/GaAs HBT fabricated at the authors' laboratory. This device utilizes an 800-Å-wide Be-doped 5×10^{19} cm^{-3} p$^+$ GaAs base in order to minimize the base resistance. At this doping level, the hole mobility of the base is 80 cm^2/V-sec, and the resulting base sheet resistance is a relatively low 200 ohm/□.

The heterojunction N-AlGaAs emitter is employed to maintain useful current gain despite its relatively light 5×10^{17} cm^{-3} doping. The Al mole fraction is graded from 0% to 23% over 500 Å to prevent the formation of a conduction-band potential barrier to electrons under forward bias [14]. This is necessary because about 60% of the band-gap energy difference between $Al_{0.23}Ga_{0.77}As$ and GaAs appears in the conduction band. The GaAs collector drift region is 4000 Å wide, n-doped lightly at 1.5×10^{16} cm^{-3}.

Device fabrication begins with the growth of the entire semiconductor layer structure by molecular beam epitaxy. The device is fabricated by a self-aligned emitter mesa process, for high performance and processing simplicity [15,16]. The base contact is placed directly on the base for low specific contact resistivity. The edges of base contact, the emitter mesa, and the emitter contact are formed during the same masking step in order to eliminate misalignment, and to minimize parasitic resistance and capacitance. The results given here were measured on a transistor with two 2-μm × 14-μm emitter stripes, interdigitated with three base contact stripes. The base collector junction area is 221 μm^2, about four times the emitter-base area.

The cutoff frequency, f_T, extrapolated from high-frequency current gain measurements, is 55 Ghz, corresponding to a total device delay of 2.9 psec. The maximum frequency of oscillation, f_{max}, a measure of the power gain–bandwidth product, is 90 Ghz, reflecting the low value of base resistance and collector capacitance.

III. CUTOFF FREQUENCY

The cutoff frequency behavior of HBTs is both a very important figure of merit and a rich field of academic study. Very high cutoff frequencies have been realized through the ability to tailor conduction-band profiles using heterostructures. Though nonstationary transport is an important factor in the analysis of some of the highest performance III/V HBTs, a purely analytical approach is quite adequate in many cases. A comprehensive treatment of delay in conventional Si bipolar transistors is given by Cooke [17]. Here, we apply the existing closed-form expressions for emitter and base delay, and extend the theory for collector transit delay to account for electron velocity-field behavior in III/V materials, device self-heating, and high injection effects. We correlate these expressions to device measurements in order to validate the model, and to gain physical insight into HBT operation. Finally, we deal with the implications of velocity overshoot.

In the quasi-static approximation, the cutoff frequency of a transistor [17] with a uniform base and a constant electron saturation velocity in the collector is given by

$$\frac{1}{2\pi f_T} = \tau_{ec} = \tau_e + \tau_b + \tau_d + \tau_{cc} \tag{1}$$

$$= r_e(C_e + C_{ci} + C_{cx}) + \frac{W_b^2}{2D_b} + \frac{W_c}{2v_s} + (R_E + R_C)(C_{ci} + C_{cx})$$

The cutoff frequency approximation is adequate for most practical cases, where the operating frequency is lower than the cutoff frequency [18]. The cutoff frequencies of microwave HBTs typically exceed 40 Ghz. We have not accounted for the effect of

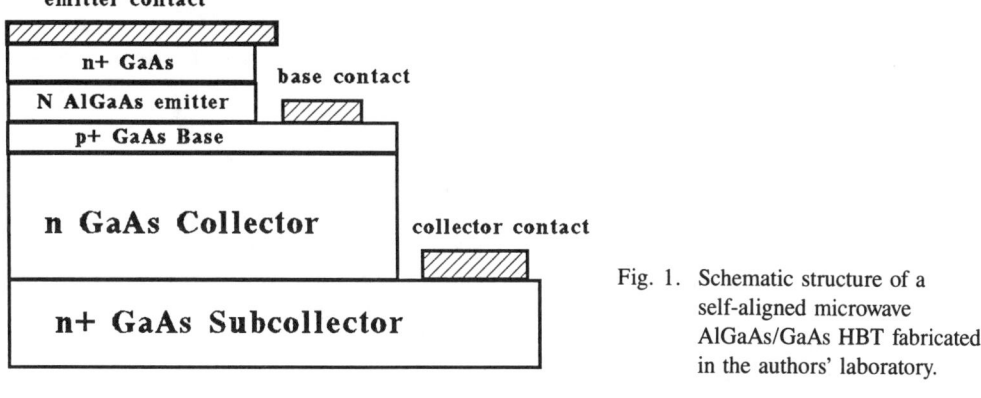

Fig. 1. Schematic structure of a self-aligned microwave AlGaAs/GaAs HBT fabricated in the authors' laboratory.

transit-time-related excess phase shift, because the effect is only significant for operating frequencies several times f_T [19,20].

We proceed by calculating all of the terms in Eq. (1) using one-dimensional analysis, with the exception of the collector charging time, τ_{cc}. In our comparisons to the experimental delay data, we have accounted for τ_{cc} with a hybrid-pi equivalent circuit model.

IV. EMITTER DELAY

According to the depletion approximation, the emitter depletion capacitance, C_e, can be kept small in an HBT with a heavily doped base by doping the emitter lightly ($C_e \alpha N_e^{1/2}$) [3]. The extent of the space-charge region is not well-defined, however, because the conduction band in the emitter-base junction of an operating HBT is nearly flat. As a result, C_e has a fairly weak dependence on N_e. For a given emitter doping, the emitter delay, τ_e, can be reduced by raising the collector current density, J_c. Thus, high-frequency operation demands high collector current densities, and sufficient emitter doping to support this current. Tiwari and Frank [21] have shown that when the emitter doping is insufficient, an increase in base emitter voltage and dynamic emitter resistance may result from a conduction-band energy barrier that forms in the emitter space-charge region. Such an increase in dynamic emitter resistance may degrade $f_{\max}$.

Experimentally, C_e can be determined from the slope of τ_{ec} plotted versus $1/I_c$ [17] (see Fig. 2). For III/V HBTs, the line should be fit to data measured at moderate collector current values. Otherwise, an inaccuracy may be induced by the downward curvature in the data at high current. The curvature reflects a change in the collector delay, τ_d, which we discuss later. For our example device, the slope of the line corresponds $C_c = 0.34$ pF, or 6 fF/cm^2, equivalent to a depletion width of 160 Å. This corresponds to 0.32-psec delay at $J_c = 5 \times 10^4$ A/cm^2, or only about 10% of the total delay in the device.

Equation (1) does not include a delay term for electron and hole storage in the emitter-base space-charge region. Generally, this storage is small in a well-designed heterostructure emitter, but it can become large if an excessively wide "setback" layer is placed between the p$^+$ base and the wideband emitter [22]. The setback layer has been used to compensate for diffusion of the p dopant toward the emitter during the growth [23], but it has become less necessary with improved control of the acceptor doping profile.

V. BASE DELAY

A. Uniform Base

The high electron mobility of the III/V semiconductors gives them a base transit time advantage over Si and SiGe. The base transit time for a uniform base is given by

$$\tau_b = \frac{W_b^2}{2D_b} \qquad (2)$$

Because the minority electron diffusivity in Si and SiGe is fairly low ($D_b \sim 2.6$ cm^2/sec), a uniform p$^+$ Si or SiGe base must be made very thin ($W_b \sim 200$ Å) to realize a desirable value of base delay (1 psec). It is important to remember that, even in HBTs, the requirement for adequate DC gain imposes an upper limit to the base doping. For a base doping of 10^{19} cm^{-3}, the base sheet resistance would be a fairly high 6000 ohm/□, which would result in a high value of base resistance.

While the electron diffusivity in a heavily doped p$^+$ GaAs base is low ($D_b = 26$ cm^2/sec) compared to that of lightly doped GaAs [24], it is about 10 times higher than the value in Si. This allows a GaAs base to be made much wider and more conductive than a Si base. An 800-Å-wide 5×10^{19} cm^{-3} p$^+$-doped GaAs base with a comparable base delay, $\tau_b = 1.24$ psec, has a sheet resistivity of only 200 ohm/□. For a narrower 400-Å GaAs base, the base transit time drops to about 0.31 psec. The results of our comparison of measured delay in 800-Å and 400-Å base HBTs are consistent with this calculation.

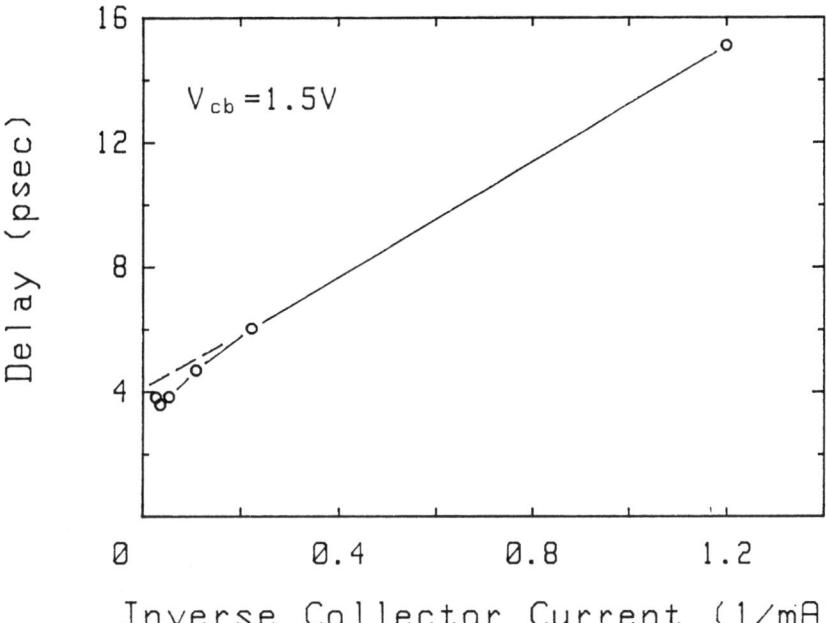

Fig. 2. Plot of measured τ_{ec} vs. J_c^{-1} for an AlGaAs/GaAs HBT with an 800-Å base and a 4000-Å n-type collector doped to 1.5×10^{16} cm^{-3}. The emitter capacitance is computed from the slope of the data, while the deviation from the trend line at high currents is due to electron space-charge effects in the collector. The emitter area is 56 μm^2.

This example points out the relative ease in achieving a small base transit time in high electron mobility III/V materials, while maintaining useful base sheet conductance.

For bases narrower than the mean free path for electrons (about 400 Å at this doping), $\tau_b = W_b/v_{th}$, where v_{th} is the thermal velocity ($\sim 2 \times 10^7$ cm/sec).

B. Graded Base

In order to decrease the base transit time in a base of given width, a drift field for electrons can be introduced into the base through grading of the base doping (drift field base) [25] and/or the band gap (compositional grading) [4]. The built-in quasi-electric field in the conduction band arising from a doping or bandgap grading across the base is

$$\mathscr{E}_{bi} = \frac{d}{dx}\left(\frac{kT}{q}\ln\frac{n_i^2}{N_b}\right) = \frac{d}{dx}(-E_g + \Delta_G^0 + \frac{kT}{q}\ln(N_b)) \qquad (3)$$

While Eq. (3) does not account for Fermi statistics, it illustrates how a quasi-field can be engineered. Fermi statistics (which tend to widen the band gap) and bandgap narrowing [26] become important at high base doping levels. This can be accounted for by adding a single corrective term, Δ_G^0, on the right side of Eq. (3). Bandgap narrowing tends to reduce the effective drift field resulting from an acceptor doping gradient across the base, while Fermi statistics have the opposite effect.

When the drift field is below the critical field for velocity saturation (about 10 kV/cm for electrons in heavily doped GaAs), its magnitude can be defined by the field factor [17]:

$$\eta = \frac{\mu_b W_b}{D_b}\mathscr{E}_{bi} = \frac{qW_b}{kT}\mathscr{E}_{bi} \qquad (4)$$

The rightmost expression in Eq. (4) reflects the Einstein relationship, which applies for small fields where the electrons are not significantly heated by the electric field. In a drift field base with an exponentially graded acceptor profile and uniform composition, in the absence of heavy doping effects, Eq. (4) reduces to

25

$$\eta = \ln \left(\frac{N_{be}}{N_{bc}} \right) \tag{5}$$

The base transit time is given by

$$\tau_b = \frac{W_b^2}{2D_b(0.8 + 0.46\eta)} \tag{6}$$

For Si/SiGe/Si HBTs, compositional grading is of particular interest. Although electrons in this material have a low diffusivity, their saturation velocity should be comparable to that of electrons on GaAs. In practice, much higher quasi-fields can be obtained by compositional grading than by dopant grading, giving these devices a great advantage over their homojunction counterparts. This can be understood by noting that, for a given base width, decreasing base conductivity limits the dopant grading ratio ($N_{be}/N_{bc} < 10$) and the field factor ($\eta < 2$). From Eq. (6), we find the resulting base delay is half that of a uniform base of the same thickness. By contrast, the compositionally graded SiGe base offers a much larger improvement. For example, if the Ge mole fraction were graded from 0% to 30% over a 500-Å base, a 50-kV/cm electric field and a field factor $\eta = 10$ would result, yielding a 5×10^6 cm/sec effective base electron velocity and the desired 1-psec base delay. Such a fast, heavily doped SiGe base seems technologically feasible [27].

On the other hand, both drift field GaAs bases [28] and graded composition AlGaAs bases [29] are practical methods for improving base delay, because uniform GaAs bases are already fast. For an $Al_xGa_{1-x}As$ base graded from 5% to 0% Al composition over 800 Å, $\Delta E_g = 65$ meV, $\mathscr{E}_{bi} = 8$kV/cm, $\eta = 2.5$, and $\tau_b = 0.6$ psec, half the delay of the uniform base. The graded AlGaAs base also reduces the lateral diffusion of electrons and associated surface recombination [30], and maintains the high base velocity even for low temperatures that decrease the electron diffusivity and doping gradient-induced field.

When the drift field exceeds the critical field of the minority electrons, the electron velocity saturates and the diffusivity decreases. The base delay becomes

$$\tau_b = \frac{W_b}{v_s} \tag{7}$$

For heavily doped GaAs bases, the peak velocity is about 10^7 cm/sec [24]. If the quasi-electric field is far above the critical field, the velocity may be somewhat lower [31].

C. Hot Electron Injection into the Base

If an energy barrier is engineered into the conduction band of the emitter-base junction (commonly through the use of an abrupt heterojunction), electrons thermionically emitted over the barrier will enter the base with kinetic energy directed toward the collector [3]. In materials with a very small electron effective mass and a large Γ to L or X valley energy separation, such as $In_{0.53}Ga_{0.47}As$, the electrons will initially travel at nearly the ballistic velocity [32] ($\sim 5 \times 10^7$ cm/sec at 0.3 eV). However, in heavily doped bases, measured ionized impurity and electron-hole plasma scattering rates are quite significant, and the electrons tend to lose their forward "ballistic" momentum and energy [33]. Although the electrons may encounter many scattering events in a fairly wide, heavily doped base designed for low base resistance, the residual excess electron energy increases the electron diffusivity, which is already fairly high. For narrow bases (< 600 Å) an extremely small base transit time results. The reduction of τ_b is one reason for the popularity of the abrupt emitter-base heterojunction in the InP/InGaAs and InAlAs/InGaAs materials systems. The other is simply practical difficulties with the growth of lattice-matched graded quaternary heterojunctions.

One disadvantage of hot electron injection into a narrow InGaAs base is that the electrons carry the extra kinetic energy with them into the collector, degrading the delay and breakdown characteristics of the collector region [34]. A possible disadvantage of abrupt emitter-base heterojunctions in other semiconductors is the smaller remaining energy barrier for confinement of holes in the base [3] compared to the graded emitter case. This is only a minor disadvantage in abrupt InP/InGaAs and InAlAs/InGaAs heterojunctions because the valence-band energy discontinuities are large enough to maintain high emitter injection efficiency.

Whatever the practical details related to particular material systems and structures, HBTs offer low emitter and base delay times along with low base resistance, especially in III/V materials. Now we turn to the collector of the microwave HBT, which in many cases is composed of GaAs or InGaAs.

VI. COLLECTOR TRANSIT TIME

A. Outline of the Model

1. *The Complexity of Transport in GaAs:* The calculation of the collector delay in a Si transistor is straightforward using Eq. (1), once the width of the space-charge region has been determined. Notwithstanding the possible complications introduced by the design of the heterostructure collector, this is also true for Si/SiGe/Si transistors. By contrast, in III/V materials such as GaAs, velocity overshoot and the dependence of the electron velocity on electric field and temperature complicate the analysis of the collector delay. Recently, it has been shown that there is an interaction between the collector space-charge density and velocity overshoot [35]. Self-consistent calculations using Monte Carlo and two valley transport models have qualitatively explained device behavior [9,36,37], but few quantitative correlations have been made to experimental measurements, and it is difficult to gain an intuitive understanding of device behavior from these computer-based models.

Much of the unique physics in GaAs HBTs has its roots in the same phenomena that govern the dependence of the static high-field velocity on electric field. Here, we develop an analytical model for collector behavior based on the static velocity-field characteristics in order to gain insight into the underlying physics and to obtain useful quantitative predictions of delay behavior. After defining a model for the high-field electron velocity and a generalized integral for collector delay, we derive the collector delay and capacitance expressions for three regions of operation: full depletion, partial depletion, and current-induced base pushout (Kirk effect).

2. *Velocity Model for Electrons in GaAs:* A transistor typically operates with a collector electric field >20 kV/cm, which is well above the critical field required to reach the peak electron velocity in GaAs (about 3 kV/cm). The high-field electron velocity in GaAs does not saturate the way it does in Si. As a function of electric field, the electron velocity in GaAs at 300 K falls from its peak of 2×10^7 cm/sec at 4 kV/cm to 6×10^6 cm/sec at 300 kV/cm [38]. Like velocity overshoot, this static negative differential mobility is a consequence of the details of electron scattering and the conduction-band diagram in GaAs. The physics of transport in InGaAs and InP are similar. If the reciprocal of the velocity (or "slowness") of electrons in GaAs is plotted versus electric field in this range, the trend approximates a straight line (see Fig. 3). Similarly, for a given electric field, the slowness is linearly related to the lattice temperature. Thus, we can approximate the inverse of the high-field electron velocity in GaAs by this linear relation:

$$\frac{1}{v(\mathcal{E}, T)} = \kappa_0 + \kappa_1 \mathcal{E} + \kappa_2 (T - T_0) \tag{8}$$

There is some latitude in choosing the fitting parameters. The calculations for GaAs are done here with the following values: $\kappa_1 = 10^{-7}$ sec/cm, $\kappa_2 = 5 \times 10^{-11}$ sec/V, and $\kappa_3 = 2.5 \times 10^{-10}$ sec/(°C cm). One can use this expression with $\kappa_1 = 0$ to account for the temperature dependence of the electron saturation velocity in Si.

3. *Collector Delay Integral:* The collector transit time τ_d must be calculated with a position-dependent weighting [39]. The integral has been generalized here to account for the effect of injected electron space-charge on the velocity profile, $v(x)$:

$$\tau_d = \frac{\partial Q_{bc}}{\partial J_c} = \int_0^{W_{sc}} \frac{\partial}{\partial J_c} \left[\frac{J_c}{v(x)} \right] \left(1 - \frac{x}{W_{sc}} \right) dx \tag{9}$$

where Q_{bc} is the hole charge in the base induced by the collector space-charge and W_{sc} is the width of the space-charge region. Note that when we substitute the linear relationship in Eq. (8) for the reciprocal of velocity (1/v) term in the integrand, the integral has a closed-form solution.

In the active mode of operation, on the basis of the space-charge density configuration, the collector is classified to be either fully depleted or partially depleted. We will first derive the delay and capacitance characteristics in full depletion, which turns out to be the most interesting case.

B. Fully Depleted Collector

1. *Transit Time:* Consider the case of a lightly doped n-type GaAs collector drift region biased so that the space-charge region extends from the base to the n⁺ subcollector (is "fully depleted"). In the case of constant saturation velocity, the conventional expression, $\tau_d = W_c/2v_s$, is applicable, where W_c is the base-subcollector spacing.

In GaAs, we need to know the electric field profile in the collector to calculate the delay. The electric field is

$$\mathscr{E}(x) = \frac{V_{bi} + V_{cb}}{W_c} + \frac{q\,(N_{Dc} - \bar{n})}{\varepsilon}\left(\frac{W_c}{2} - x\right) \tag{10}$$

where we have approximated $n(x)$ by the average electron density, $\bar{n}$, for the purpose of calculating the electric field:

$$\bar{n} = \frac{J_c}{qv(\mathscr{E}_{avg})} \tag{11}$$

and

$$\mathscr{E}_{avg} = (V_{bi} + V_{cb})/W_c \tag{12}$$

Combining Eqs. (8)–(12), the delay in a fully depleted collector is given by

$$\tau_d = [\kappa_0 + \kappa_2\,(T_j - T_0)]\frac{W_c}{2} + \kappa_1\,\frac{V_{bi} + V_{cb}}{2} + \kappa_1\,\frac{q\,(N_{Dc} - 2\bar{n})W_c^2}{12\varepsilon} \tag{13}$$

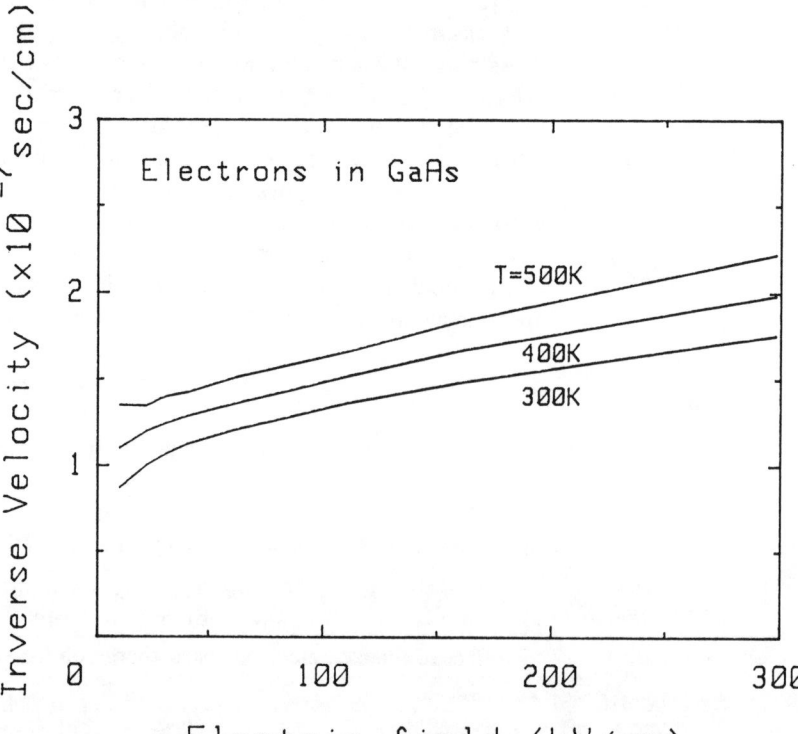

Fig. 3. Plot of v_s^{-1} vs. $\mathscr{E}$ at high fields for electrons in GaAs with the lattice temperature as a parameter. Data are taken from [38].

The first and second terms of Eq. (13) account for the average electron slowness across the depletion region. The third term accounts for the effect of space-charge density on the electron slowness versus position. The second and third terms vanish for Si collectors. We consider the implications of these terms separately below.

The first term of Eq. (13) consists of the conventional delay, plus a correction for the junction temperature rise. It defines the most important aspect of the relationship between τ_d and W_c. It also contains the effect of junction temperature on collector delay, given by

$$\frac{\partial \tau_d}{\partial T_j} = \frac{\kappa_2 W_c}{2} \tag{14}$$

yielding values of 0.005 psec/°C and 0.008 psec/°C for 4000-Å- and 6000-Å-wide GaAs collectors, respectively. We have measured $\partial \tau_{ec}/\partial T_j$ in our AlGaAs HBTs with a temperature-controlled stage and found the temperature coefficients to be 0.009 psec/°C and 0.022 psec/°C, respectively, for these collector widths. It is likely that τ_e and τ_b contribute to the junction temperature dependence of τ_{ec}, and that in the measurement, the actual junction temperature change exceeds the change in stage temperature. This happens because the thermal conductivity of GaAs itself is temperature dependent, resulting in a change in the amount of device self-heating, even with constant power dissipation. A junction temperature rise of 75 °C, which is not unusual for a microwave GaAs HBT, will add about 0.7 psec delay for an HBT with a 4000-Å collector.

It is worth noting that the thermal resistance of a microwave GaAs HBT is larger than a microwave Si BJT of similar active area. The thermal conductivity of GaAs, which is several times smaller than that of Si or InP, is a major issue in the design of microwave power GaAs HBTs, possibly resulting in severe junction overheating and performance degradation unless special precautions are taken to minimize or dissipate the heat [7,8,40].

The second term of Eq. (13), which is proportional to κ_1 accounts for the reduction of the electron velocity induced by the average collector electric field. As a result of the formulation of Eq. (8), this term does not depend on W_c. The combination of the first and second terms of Eq. (13) explains the commonly observed downward sloping f_T versus V_{ce} characteristics of GaAs HBTs. Figure 4 is a plot of τ_{ec} versus V_{cb} with I_c fixed. Unlike Si bipolar transistors, which generally exhibit the opposite trend due to base width modulation, τ_{ec} rises with increasing collector voltage. For this discussion, we must focus on the portion of the curve where V_{cb} is positive (a reversed-biased collector-base junction) and the lightly doped collector is fully depleted ($W_{sc} = W_c$). From Eq. (13), the slope of τ_{ec} versus V_{cb} in full depletion is given by

$$\left. \frac{\partial \tau_{ec}}{\partial V_{cb/e}} \right|_{I_c} \sim \left. \frac{\partial \tau_d}{\partial V_{cb/e}} \right|_{I_c} = \frac{\kappa_1 + \kappa_2 \theta /_c W_c}{2} \tag{15}$$

where θ is the device thermal resistance.

In this case the slope is 0.3 psec/V, of which 0.25 psec/V arises from the increasing electric field and 0.05 psec/V from the increasing junction temperature (the thermal resistance is 1 °C/mW). Thus, the collector delay can be reduced by minimizing the collector bias voltage (while maintaining full depletion in the collector for low capacitance), and minimizing the device thermal resistance. This electric field and temperature dependence of the electron velocity must be accounted for in any attempt to deduce the electron velocity from the collector voltage dependence of f_T.

2. *Velocity Profile Effect:* The third term of Eq. (13) indicates that for a given junction temperature and applied collector voltage, the collector delay depends on the donor and electron densities. We call this the velocity profile effect, because it arises from the spatial variation of electron velocity caused by a spatially varying collector field. We can illustrate the velocity profile effect by comparing the electric field and slowness profiles under conditions of positive and negative total space-charge density. First, note that the electric-field profile depends on the total space-charge density, which is the difference between the donor density, N_{Dc}, and the electron density, $\bar{n}$ [see Eq. (10)]. In this example, we fix the donor density at $N_{Dc} = 2 \times 10^{16}$ cm^{-3} and vary the collector current

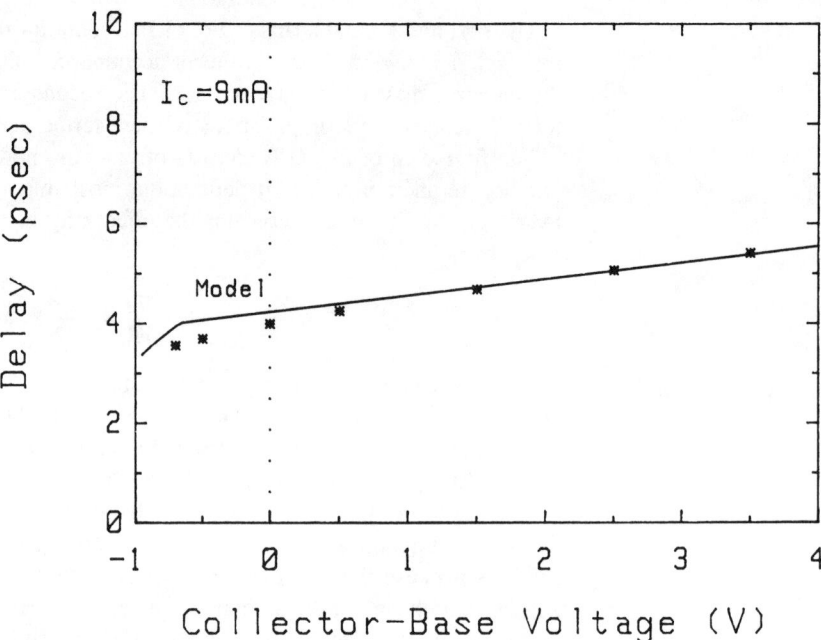

Fig. 4. Plot of measured τ_{ec} vs. V_{cb} at $I_c = 9$ mA for the same device as Fig. 2. The collector is fully depleted for positive V_{cb}. The solid line is the model calculated from Eqs. (1), (13), and (25).

density, which determines the electron density. At a relatively low current density, $J_c = 10$ kA/cm^2, the net collector space-charge is positive, $(N_{Dc} - \bar{n}) \sim 1.4 \times 10^{16}$ cm^{-3}. The electric field is high near the base and low near the subcollector (see Fig. 5). For a high current density, $J_c = 40$ kA/cm^2, the situation is reversed, $(N_{Dc} - \bar{n}) \sim -0.5 \times 10^{16}$ cm^{-3}. Referring to Fig. 6, we see that the slowness profile is qualitatively similar to the electric-field profile because they are linearly related by Eq. (8).

Something interesting happens when we evaluate the integral representing the induced base charge, Q_{bc}, in Eq. (9). In the integrand, the slowness profile, which represents the normalized electron concentration profile, is multiplied by the weighting function, $1 - x/W_c$. The weighting for electrons near the base is higher than those near the subcollector. This weighted profile is plotted for both cases in Fig. 7. The value of the integral in each case is represented by the area under the curves in Fig. 7, respectively. The area is greater in the low current density case because on average, the electrons are closer to the base and are more effective in inducing hole charge, Q_{bc}, in the base. Conversely, the electrons are, on average, further from the base in the high current density case, resulting in a proportionally smaller value of integrated base charge. These weightings are reflected in the collector delay, τ_d. The collector delay is lower in the high current density case, even though the actual total time of flight for electrons from base to subcollector is the same for both cases.

In summary, a negative total space-charge density tends to reduce τ_d, while a positive space-charge density has the opposite effect. A way to make the space-charge more negative, other than to increase the current density, is to reduce the donor density or dope the drift region with acceptors. For a fixed bias voltage and current,

$$\frac{\partial \tau_d}{\partial N_{Dc}} = \frac{q\kappa_1 W_c^2}{12\varepsilon} \tag{16}$$

For a 4000-Å GaAs collector, the delay decreased by 0.067 psec for each 10^{16} cm^{-3} decrease in the donor density or increase in the acceptor density. This explains the popularity of undoped and p-type collector space-charge layers in GaAs HBTs [41]. Unfortunately, p-type collectors suffer from an increased likelihood of base pushout (the Kirk effect).

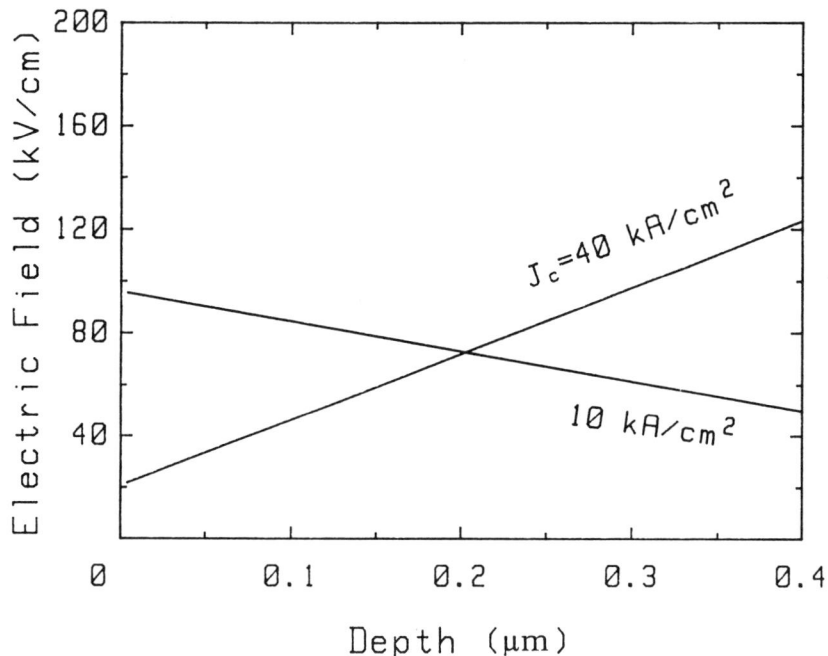

Fig. 5. Plot of $\mathscr{E}$ vs. position calculated for a 4000-Å 1.5 × 10¹⁶ cm⁻³ n-type collector at J_c = 10⁴ and 4 × 10⁴ A/cm². The base and the subcollector are at 0- and 0.4-μm depth, respectively, and V_{cb} = 1.5 V.

normalized electron concentration profile, is multiplied by the weighting function, $1 - x/W_c$. The weighting for electrons near the base is higher than those near the subcollector. This weighted profile is plotted for both cases in Fig. 7. The value of the integral in each case is represented by the area under the curves in Fig. 7, respectively. The area is greater in the low current density case because on average, the electrons are closer to the base and are more effective in inducing hole charge, Q_{bc}, in the base. Conversely, the electrons are, on average, further from the base in the high current density case, resulting in a proportionally smaller value of integrated base charge. These weightings are re-

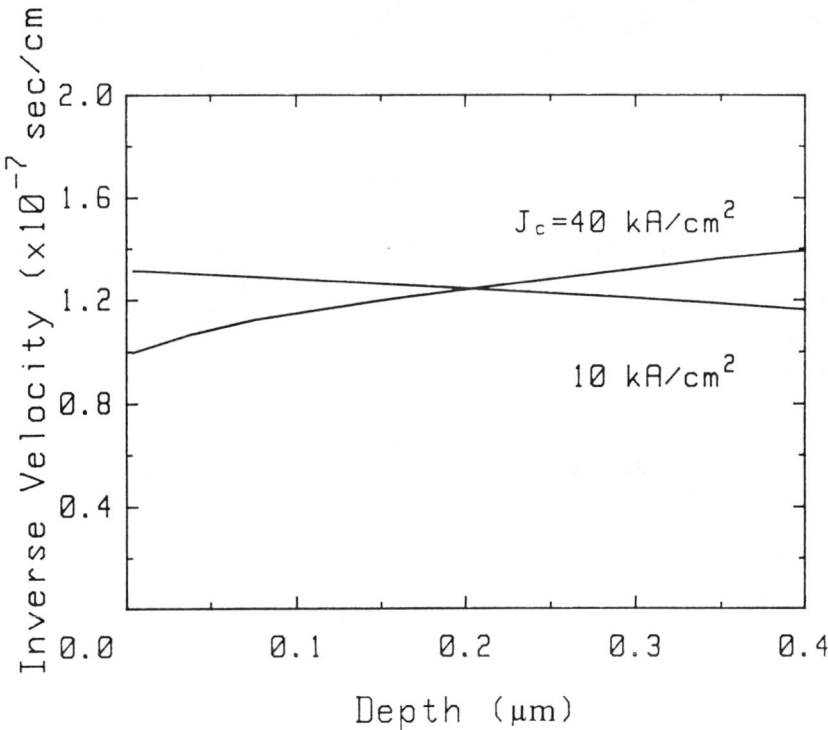

Fig. 6. Plot of v_s^{-1} vs. position in the collector for the conditions given in Fig. 5, calculated using the velocity-field characteristics of Fig. 3.

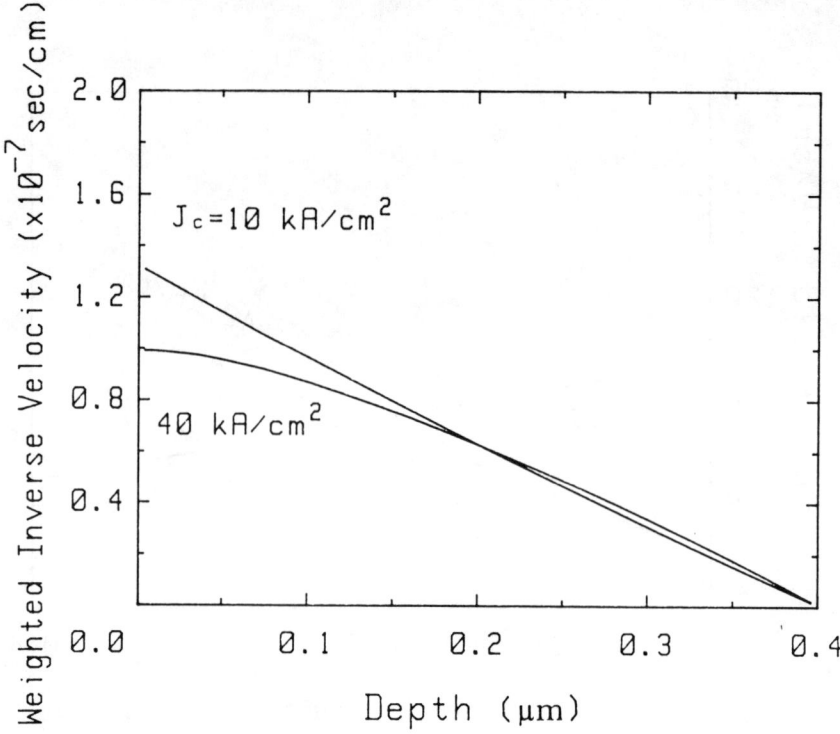

Fig. 7. Plot of v_s^{-1} multiplied by the weighting function, Eq. (9), vs. position for the conditions given in Fig. 5. The area under each curve is the total image charge induced in the base, normalized to the collector current. The effective delay is lower at higher current density.

One might expect, given the simplified explanation given earlier, that additional electrons and donors would affect the collector delay in an equal and opposite way (i.e., τ_d have an $(N_{Dc} - \bar{n})$ dependence). However, this is not precisely correct because in the analysis, while N_{Dc} is a fixed quantity, $\bar{n}$ is variable and dependent on J_c. To see this we return to the mathematical derivation of τ_d, which requires the calculation of $(\partial/\partial J_c)$ $(J_c/v(x))$:

$$\frac{\partial}{\partial J_c}\left(\frac{J_c}{v(x)}\right) = \frac{1}{v(x)} + J_c \frac{\partial}{\partial J_c}\left(\frac{1}{v(x)}\right) \qquad (17)$$

The first term of the expansion in Eq. (17) accounts for the additional electron density needed to support an incremental increase in J_c with the existing velocity profile in the collector drift region. The second term of Eq. (17) reflects the fact that the incremental change in overall space-charge, and therefore field and slowness, brought on by an incremental change in J_c also alters the stored collector charge. When Eq. (17) is recast in terms of $\bar{n}$, N_{Dc}, and v through Eqs. (8) and (10)–(12), the first term results in an expression dependent on $(N_{Dc} - \bar{n})$. This reflects the effects previously described qualitatively. The second term reduces to an expression, with the same coefficient, dependent on $-\bar{n}$. Thus, the sensitivity of the collector field to rf changes in the electron injection turns out to be equally important in reducing τ_d as the DC shape of the electric field. The net result is a single term in $\tau_d(\bar{n},\ N_{Dc},\ v)$ proportional to $N_{Dc} - 2\bar{n}$.

The velocity profile effect is evident in GaAs HBTs upon careful examination of the dependence of the total delay, τ_{ec}, on the inverse of the collector current density, $1/J_c$ (see Fig. 2). The dependence of τ_{ec} on J_c is given by

$$\tau_{ec} = \frac{1}{J_c}\frac{kT_jC_e}{qA_e} - J_c\frac{\kappa_1 W_c^2}{12\varepsilon v(\mathscr{E}_{avg})} + J_c\frac{\kappa_2 A_e \theta W_c V_{ce}}{2} + \cdots \qquad (18)$$

At the highest current densities, τ_{ec} deviates from the linear dependence on $1/J_c$ established at the lower current densities by the first term of Eq. (18), the emitter delay, τ_e. This deviation, a result of the velocity profile effect in the collector, is proportional to $-J_c$, and appears as a negative-going hyperbola on the $1/J_c$ scale. The effect is partially offset by the device self-heating term. The combination of the velocity profile effect and the inverse current dependence of the emitter delay provide a great incentive to operate at the highest possible current densities in order to minimize τ_d. This trend toward smaller collector delays continues until base pushout or excessive nonlinear device self-heating occurs.

The decrease in collector delay at high collector currents has been observed in previous measurements [9]. Katoh [35] has qualitatively explained the effect in terms of a changing thickness of the velocity overshoot region in the collector. Considering that velocity overshoot and negative differential mobility are, respectively, nonequilibrium and equilibrium manifestations of the physics of high-field electron transport in GaAs, it is not surprising that Katoh's analysis resulted in qualitatively similar conclusions. Das and Lundstrom [42] have pointed out that velocity overshoot effects are only significant in GaAs collectors over a fairly narrow range of bias, so velocity overshoot alone is insufficient to explain an effect that occurs over a wide range of bias voltage and doping concentrations. Our derivation demonstrates that this effect can occur in the absence of velocity overshoot; at least in our devices, only static negative differential mobility is necessary to account for the observed decrease of collector transit delay time in high injection.

It is noteworthy that our derivation also shows that the measured small signal collector delay can be substantially lower than it would be if the electron velocity profile were independent of the collector current. One cannot ignore the extra negative contribution from the second term in Eq. (17), which reflects the incremental change in the velocity profile with current. To infer the electron velocity directly from delay data at high currents without taking this effect into account will result in an erroneously high deduced velocity.

3. *Capacitance Cancellation:* Another manifestation of the physics underlying the velocity profile effect involves the collector capacitance. The importance of low collector capacitance to a microwave transistor can hardly be overstated. It is important in determining the output impedance, isolation characteristics, and power gain. Normally, the minimum value of collector capacitance in a Si transistor is determined by the distance between the base and the subcollector, which form the plates of the capacitor. For III/V collectors, the collector capacitance can go below this value.

A negative collector capacitance component arises from the modulation of the electron space-charge by the collector-base voltage at high collector current. It is a result of negative differential mobility. We can calculate the total intrinsic collector capacitance by finding the collector voltage dependence of the electric field at the collector edge of the base:

$$\frac{C_{ci}}{A_e} = \frac{\partial \varepsilon \mathscr{E}(0)}{\partial V_{cb}} = \frac{\varepsilon}{W_c} - \frac{\partial Q_{bc}}{\partial V_{cb}} = \frac{\varepsilon}{W_c} - \frac{\kappa_1 J_c}{2}\left[1 - \frac{\kappa_1 J_c W_c}{6\varepsilon}\right] \qquad (19)$$

The first term on the right side of Eq. (19) is the conventional parallel-plate capacitance, ε/W_c. The second term comes from the response of the collector electron density to small signal variations in the collector electric field. Figure 8 shows the collector capacitance calculated for the parameters given in [9], including and not including capacitance cancellation. In this case, the full depletion capacitance is reduced by as much as 70% at a current density of 8×10^4 A/cm^2, just below onset of the Kirk effect. This result agrees well with the simulation results of [9] that account for velocity overshoot. This effect is related to the reduction in gate-drain capacitance in GaAs FETs attributed to stationary Gunn-domain formation [43].

This capacitance cancellation effect could greatly increase the overall performance of HBTs, but in many present HBT designs, the large value of extrinsic collector-base capacitance would mask the potential improvement. The full benefit of capacitance cancellation will be realized when extrinsic collector capacitance can be substantially reduced or eliminated, such as in a collector-up transistor [3].

C. Partially Depleted Collector

When the collector-base voltage is insufficient to fully extend the space-charge region to the subcollector, and the donor density exceeds the electron density (net positive space-charge), the collector is partially depleted. This second condition is met when the collector current is lower than $J_{fulldep}$:

$$J_{fulldep} \approx v(\mathscr{E}_{avg}) \left[qN_{Dc} - \frac{2\varepsilon(V_{bi} + V_{cb})}{W_c^2} \right] \tag{20}$$

Partial depletion occurs for small collector voltages and currents (see Fig. 9). The space-charge region width, W_{sc} can be calculated by solving

$$V_{bi} + V_{cb} = - \int_{W_{sc}}^{0} \mathscr{E}(x)\, dx \tag{21}$$

where

$$\varepsilon\mathscr{E}(x) = \int_{W_{sc}}^{0} \left[qN_{Dc} - \frac{J_c}{v(\mathscr{E}_0(x))} \right] dx \tag{22}$$

and we have used the following approximation for the electric field in calculating the electron velocity:

$$\mathscr{E}_0(x) = \frac{(V_{bi} + V_{cb})}{W_{sc}^2} x \tag{23}$$

Combining Eqs. (8) and (21)–(23) yields a quadratic equation for the collector space-charge region width that accounts for the injected electron density:

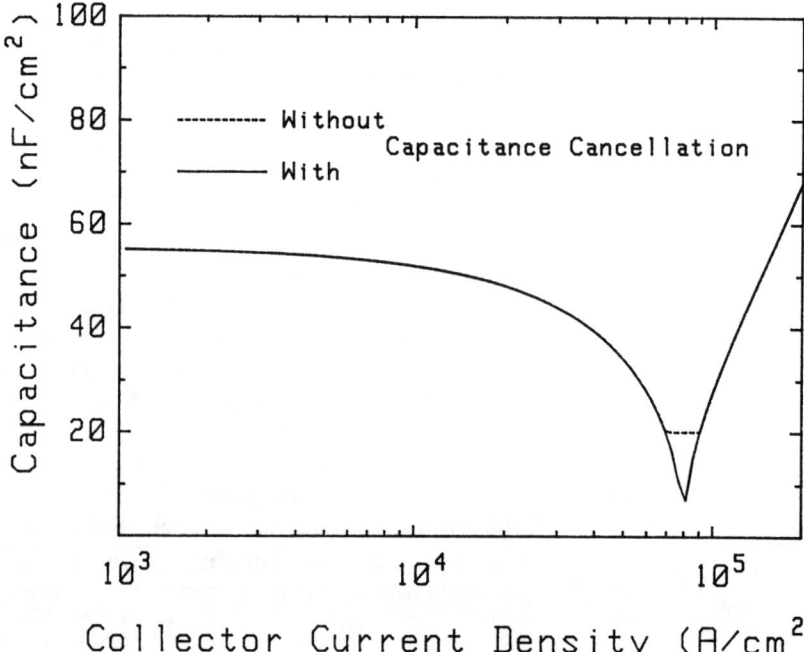

Fig. 8. Plot of C_{cb} calculated for the device presented in [9], showing capacitance cancellation at high current. The solid curve was calculated using Eqs. (19), (24), and (27), while the dotted line is the parallel-plate capacitance. The collector doping and thickness are 5×10^{15} cm^{-3} and 5000 Å, respectively.

$$W_{sc}^2 = \frac{2\varepsilon(V_{bi} + V_{cb})}{qN_{Dc} - J_c \{\kappa_0 + \kappa_1 [2(V_{bi} + V_{cb})/3W_{sc}] + \kappa_2(T_j - T_0)\}} \tag{24}$$

The delay in a partially depleted collector is

$$\tau_d = \frac{\partial Q_{bc}}{\partial J_c} = \frac{\partial \varepsilon \mathscr{E}(0)}{\partial J_c} = [\kappa_0 + \kappa_2(T_j - T_0)] \frac{W_{sc}}{2} + \frac{2\kappa_1}{3} (V_{bi} + V_{cb}) \tag{25}$$

It is instructive to compare Eq. (25) to Eq. (13), taking $W_{sc} = W_c$. The first term on the right-hand side of Eq. (25) is identical to the first term of Eq. (13). However, the second term in Eq. (25) is larger than in Eq. (13), and there is no third term in Eq. (25), while the third term of Eq. (13) can be negative. For a given depletion width and collector voltage, τ_d is greater for a partially depleted collector than for a fully depleted one. Therefore, designing the collector to operate in full depletion is usually more desirable than in partial depletion. Equation (25) works best in low injection because of the approximation made for the electric field.

VII. THE KIRK EFFECT

A. Kirk Effect Threshold Current

Base pushout (the Kirk effect) occurs in high current density operation when the negative space-charge density is high enough (at a given collector-base voltage) for the electric field to vanish in the collector adjacent to the metallurgical base. In response, holes diffuse out from the base to form a quasi-neutral region, called the current-induced base. The threshold current density [10] for this effect, J_{max}, is given by

$$J_{max} = v(\mathscr{E}_{avg})\left[qN_{Dc} + \frac{2\varepsilon(V_{bi} + V_{cb})}{W_c^2}\right] \tag{26}$$

The threshold current for the Kirk effect is depicted by the upward sloping line in Fig. 9. Operation above the threshold current reduces the width of the collector space-charge region.

B. Capacitance in the Kirk Effect

We can calculate the width of the space-charge region by accounting for the negative space-charge density:

$$\frac{\varepsilon}{C_{ci}} = W_{sc} = W_c - W_{cib} = \sqrt{\frac{2\varepsilon(V_{bi} + V_{cb})}{q(\bar{n} - N_{Dc})}} \tag{27}$$

where W_{cib} is the width of the current-induced base region in the collector, and the electron density is given approximately by

$$\bar{n} - \frac{J_c}{qv(\mathscr{E}_{avg})} \tag{28}$$

Figure 8 shows an increase in collector-base capacitance above the Kirk effect threshold current due to the reduced width of the space-charge region. Next, we turn to the analysis of the collector delay in the Kirk effect.

C. Collector Delay in the Kirk Effect

In Si homojunction transistors operating above the Kirk effect threshold current, the low diffusivity of the electrons makes the diffusion velocity across the current-induced base region quite small. The electron and hole densities approach or exceed the level of acceptor doping in the base. The effective basewidth becomes the total width of the two

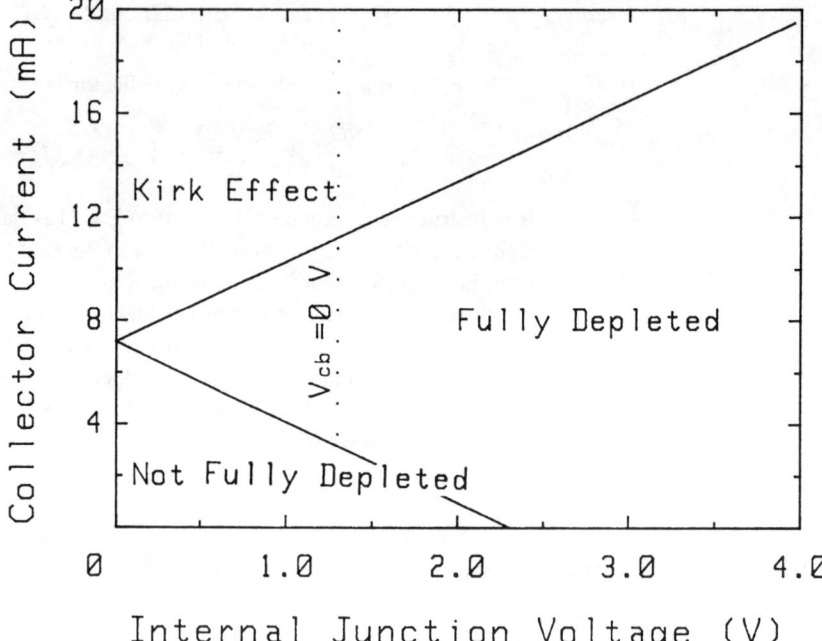

Fig. 9. Graph of I_c vs. $(V_{cb} + V_{bi})$ showing the boundaries between three conditions of the collector space-charge region. The calculation is done for a 56-μm^2 emitter HBT with a 6000-Å 8 × 10^{15} cm^{-3} n-type collector.

regions, resulting in a drastic drop in f_T and h_{fe} [10,11]. This makes operation above the threshold current undesirable.

The situation is quite different in III/V HBTs. Figure 10 shows the simulated electron and hole densities in the base and collector regions of a GaAs HBT for $V_{cb} = 0$ V and $V_{cb} = 7$ V, both at a collector current density of 40 kA/cm^2 and a donor density of 4×10^{16} cm^{-3}. Figure 11 shows the conduction-band energy profile under the same conditions. In this simulation, velocity overshoot effects are accounted for by an electron temperature model [44]. The collector is fully depleted in the $V_{cb} = 7$ V case, while a current-induced base is evident in the $V_{cb} = 0$ V case from a region of nearly equal hole and electron concentrations ($\sim 10^{16}$ cm^{-3}) that extends over about half of the collector. In our analytical model, the hole charge in the current-induced base region in high injection is given by

$$Q_{cib} = J_c \frac{W_{cib}^2}{4D_{nc}} = J_c \frac{(W_c - W_{sc})^2}{4D_{nc}} \tag{29}$$

The effective electron velocity in a 3000-Å-wide current-induced base is nearly 1.7×10^7 cm/sec due to the high diffusivity of electrons (130 cm^2/V-sec), making the electron density here about 1.5×10^{16} cm^{-3}. The velocity here is actually higher, and the electron density lower, than in the $V_{cb} = 7$ V case, because the high field electron velocity is about 6×10^6 cm/sec.

The hole concentration of 5×10^{19} cm^{-3} in the p$^+$ base is about three orders of magnitude higher than the hole density in the current-induced base, forming a 180-meV potential falloff in both the valence and conduction bands between the two regions (see Fig. 11). This potential falloff causes a localized velocity overshoot in GaAs, and maintains a high electron exit velocity from the p$^+$ base; the electron concentration in the p$^+$ base is not influenced by the electron concentration in the current-induced base. However, velocity overshoot is not a necessary condition for this decoupling; the energy barrier will decouple the electron concentrations as long as the base acceptor doping is much higher than the density of holes injected into the collector.

A precise analytical calculation of the collector delay time in the Kirk effect is complicated by the variation of W_{cib} with the collector current. A reasonable approximation can be made by ignoring this change in W_{cib} when calculating $\partial Q/\partial J_c$. In doing this, we are taking advantage of the fact that the velocity in the current-induced base is fairly high,

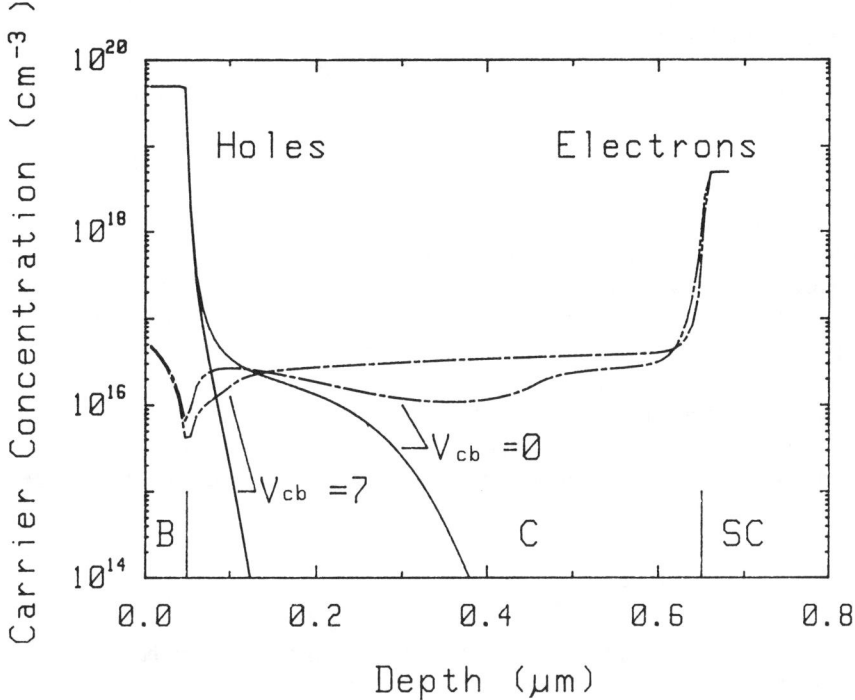

Fig. 10. Electron and hole density profiles in the collector calculated self-consistently using an electron temperature model [44] for a current density of 40 kA/cm². In the $V_{cb} = 0$ V case, a 3000-Å-wide current-induced base has formed.

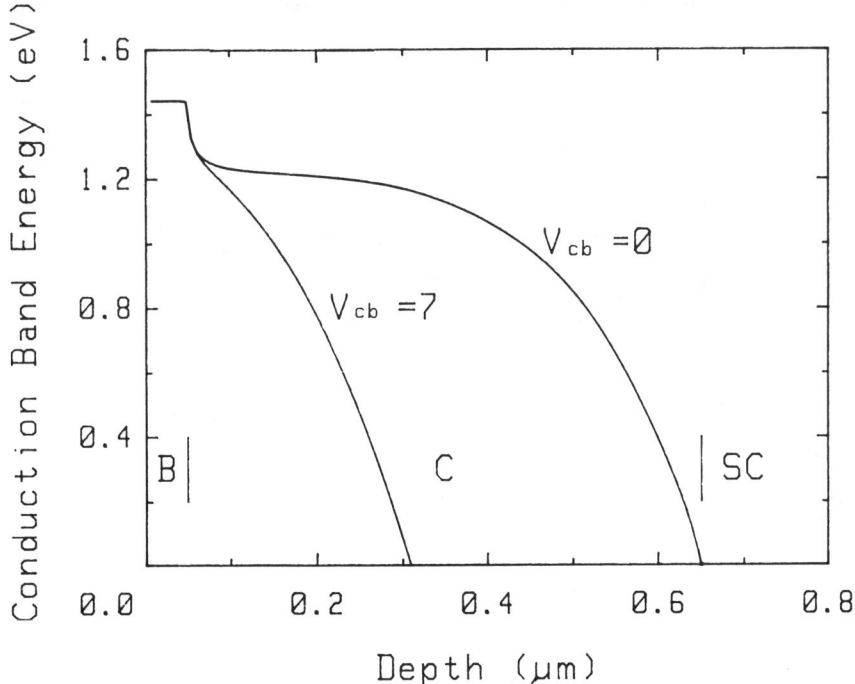

Fig. 11. Conduction-band energy profiles from the same simulation as Fig. 10. In the $V_{cb} = 0$ V case, there is a 180-meV potential falloff from the 5×10^{19} cm⁻³ p⁺ base to the current-induced base.

and W_{cib} has a weak dependence on collector current. The total delay attributable to the collector region is given by

$$\tau_{c,tot} \approx \tau_{cib} + \tau_{sc} = \frac{W_{cib}^2}{4D_{nc}} + (\kappa_0 + \kappa_2(T_j - T_0))\frac{W_{sc}}{2} + \frac{\kappa_1}{3}(V_{bi} + V_{cb}) \qquad (30)$$

The delay added by the current-induced base is at first offset by the reduction in delay of the consequentially smaller collector space-charge region; the collector delay does not increase immediately at the onset of the Kirk effect. This is evident when we compare the trend in the collector capacitance of the 6000-Å collector GaAs HBT of Fig. 12 to its cutoff frequency trend. We can identify the true onset of the Kirk effect to be slightly above 10 mA by the trend of increasing collector capacitance. This onset is also quite evident in f_{max} (Fig. 13), which peaks at 10 mA and decreases at higher currents. In contrast, the cutoff frequency (Fig. 14) peaks at almost 20 mA, partly because τ_e continues to decrease. At higher currents, f_T rolls off fairly slowly. The overall penalty for operating above the critical current density, then, is not large in III/V HBTs. Still, in designing the collector, it is desirable to raise the Kirk effect threshold current in order to take advantage of velocity profile modulation and the declining value of τ_e.

D. DC Current Gain in the Kirk Effect

While we are on the subject of the Kirk effect, let us discuss its effect on the DC current gain. Figure 15 compares the small signal DC current gain characteristics of the 4000-Å and 6000-Å collector HBTs referred to in Figs. 12–14. The devices are identical, with the exception of the collector width and doping. Base pushout begins at approximately $I_c = 10$ mA effect for the 6000-Å unit, and at $I_c > 30$ mA for the 4000 = Å unit. The shapes of the C_c, f_{max}, and f_T curves reflect this difference, but the shapes of the current gain curves are quite similar. This indicates that the current gain is not affected by base pushout.

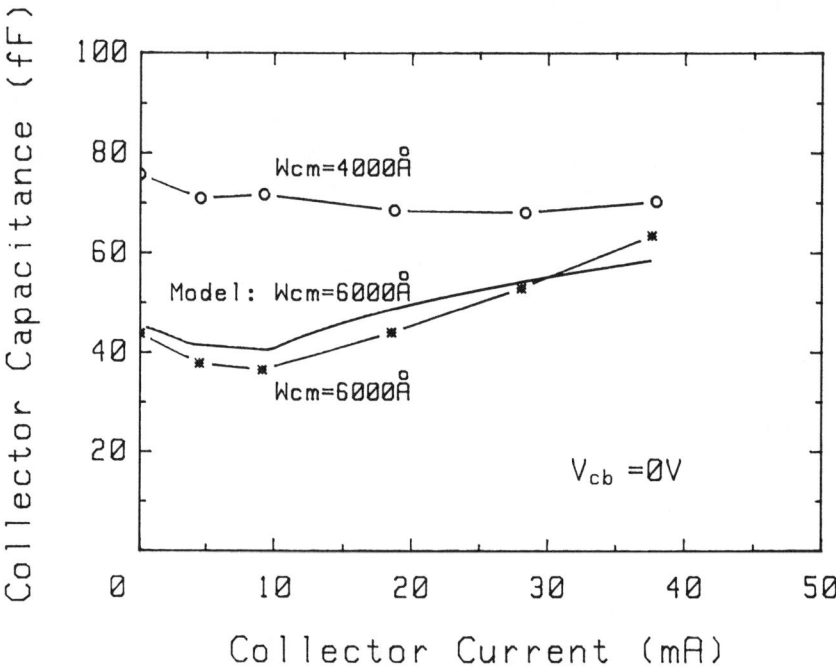

Fig. 12. Plot of measured C_c vs. I_c at $V_{cb} = 0$ V for both a 4000-Å 1.5×10^{16} cm^{-3} collector and a 6000-Å 8×10^{15} cm^{-3} collector, showing the earlier onset of the Kirk effect in the latter device. The collector areas are 221 μm^2 and the emitter areas are 56 μm^2. The model curve was calculated using Eqs. (19), (24), and (27).

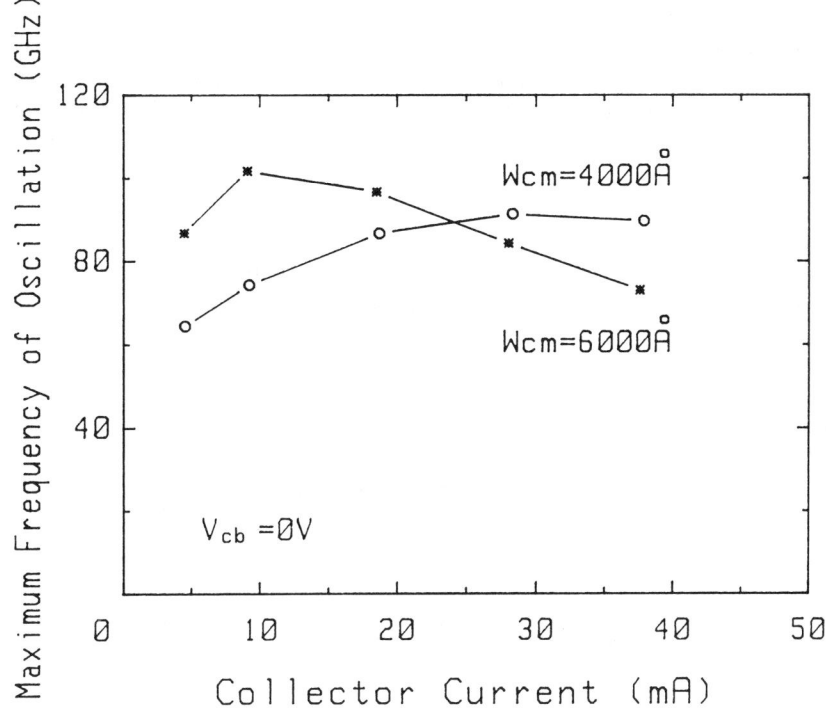

Fig. 13. Plot of f_{max} vs. I_c at $V_{cb} = 0$ V for the devices of Fig. 12.
The peaking of f_{max} for the 6000-Å collector reflects the
influence of the rise in the intrinsic collector capacitance.

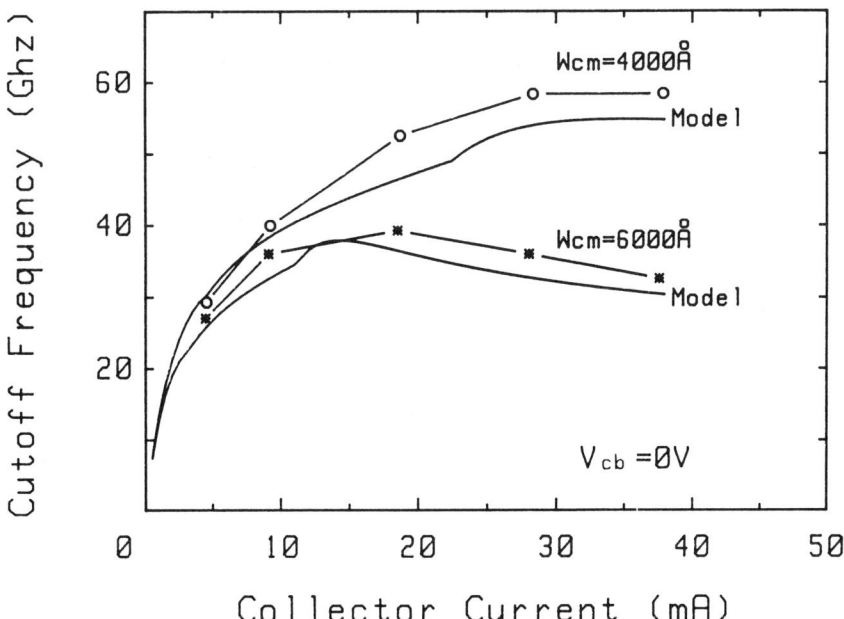

Fig. 14. Plot of measured and calculated f_T vs. I_c at $V_{cb} = 0$ V for
the devices of Figs. 12 and 13, showing how the Kirk effect
limits f_T at high current for the 6000-Å HBT. The
calculations were done using Eqs. (1), (13), (25), and (30).

What causes the flattening of the h_{fe} curves at high current? When a flattening or rolloff in f_T or h_{fe} is observed, a way to determine the cause is to examine the V_{ce} dependence of the parameter at a constant collector current. The parameter will tend to increase with rising V_{ce} if due to base pushout, and decrease if due to self-heating. In the 6000-Å collector device, f_T increases and h_{fe} decreases with rising V_{ce}. The current gain of both devices exhibits a square law dependence on collector current and an inverse dependence on absolute junction temperature. Device self-heating is the source of the flattening in current gain.

Why does base pushout not degrade the DC current gain? We have already shown that the electron concentration in the heavily doped p$^+$ base is decoupled from the quasi-neutral region that forms in the collector; the recombination rate in the p$^+$ base is not changed. The recombination rate in the current-induced base is relatively small, because the hole concentration is much smaller than in the p$^+$ base [45], and the electron lifetime much longer (~ 100 ns versus ~ 0.1 ns for p$^+$ GaAs). Thus, the additional base current that results from recombination in the current-induced base is negligible. A lesson to be learned here is that the DC current gain characteristic is not a reliable indicator of the onset of the Kirk effect in HBTs.

In the case of a wideband heterostructure collector [46] or a homojunction collector with a buried p-layer such as the ballistic collection transistor (BCT) [47], the DC gain can be affected by base pushout, because a local potential barrier may form in the collector, impeding the electron flow.

VIII. VELOCITY OVERSHOOT

When low-energy electrons are injected into the collector, they initially have the high mobility and high velocity associated with the Γ valley. When they are energetic enough (0.3 eV for GaAs, 0.5 eV for In$_{0.53}$Ga$_{0.47}$As), they transfer into the low mobility X or L valleys. In order to observe overshoot effects, the change in potential near the base must be small to prevent the electrons from gaining too much energy too quickly. Early

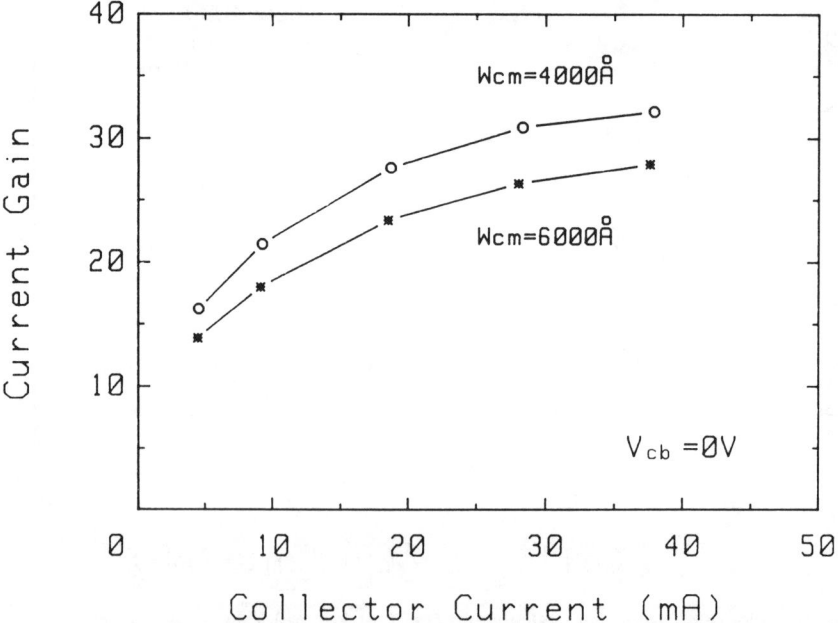

Fig. 15. Plot of measured h_{fe} vs. I_c for the devices of Figs. 12–14, showing a square law dependence on I_c and the effect of device self-heating. The Kirk effect causes no apparent degradation of h_{fe}.

attempts to take advantage of velocity overshoot in GaAs using p-type collector doping were moderately successful [41]. Ishibashi and Yamauchi [47] took the idea one step further in the BCT, placing an acceptor plane near the subcollector to concentrate most of the potential drop there, leaving a small, uniform electric field near the base in order to maximize the spatial extent of velocity overshoot. Bipolar transistor f_T exceeded 100 Ghz for the first time with this device. A disadvantage of both the p-type collector and the early BCT was a fairly low Kirk effect threshold current. Ishibashi et al. [7] later made improvements to the BCT in accordance with the principles we have outlined:

1. they reduced the thickness of the graded AlGaAs base to 400 Å,
2. they reduced the collector width to ~ 1800 Å,
3. they added donors to the collector to raise the Kirk effect threshold current to $1-2 \times 10^5$ A/cm^2.

The resulting collector and base transit times total about 0.5 psec, and f_T reaches 171 Ghz.

A similarly high cutoff frequency, 165 Ghz, has been reported for InP/InGaAs HBTs with an $In_{0.53}Ga_{0.47}As$ collector [32,48], using hot electron injection into a narrow InGaAs base and a fairly narrow collector drift region (3000 Å). While these cutoff frequencies are very impressive, the f_{max} values are modest (<100 Ghz) by comparison. Also, the optimum field configuration can be maintained for a collector voltage swing of only a few volts. Collector voltages outside this range result either in an excessive electric field near the base or a vanishingly small one, which brings on the Kirk effect. Velocity overshoot is more easily realized in $In_{0.53}Ga_{0.47}As$ due to its larger $\Gamma-L$ valley energy separation. This would result in a larger useful collector voltage range for InGaAs than for GaAs, except that InGaAs has a much lower effective avalanche breakdown field. The useful collector voltage swing in either case is not high enough for microwave power applications. We should note that nonequilibrium effects are important factors in determining the breakdown voltage in these structures [34]. Now, we turn to the large signal properties of the HBT.

IX. CUTOFF FREQUENCY-BREAKDOWN VOLTAGE PRODUCT

Johnson [12] showed that transistor gain and power at high frequencies is fundamentally limited by the product of the breakdown field and the carrier saturation velocity in the collector. For example, in theory the highest possible product of the collector breakdown voltage and the cutoff frequency for electrons in GaAs or Si is given by

$$BV_{cb} f_T = \frac{\mathscr{E}_{bd} v_s}{2\pi} \sim 3 \times 10^{11} \frac{V}{s} \tag{31}$$

Microwave power HBTs, having lower base delay and base resistance, come closer to this limit than Si BJTs. In comparing the suitability of Si and GaAs for microwave power amplification, we note that for high electric fields, the electron velocity in Si is greater than that of GaAs (10^7 cm/sec versus 6×10^6 cm/sec, respectively, at 200 kV/cm). For GaAs, velocity overshoot has relatively little overall significance in most microwave power applications, which typically require wide collector drift regions to operate at a large collector voltage. On the other hand, the common-collector breakdown field of GaAs is higher than that of Si, roughly 300 kV/cm and 200 kV/cm, respectively. Thus, the overall potential voltage-frequency products of Si and GaAs are similar. The breakdown field values for both GaAs and Si collectors must be derated for common-emitter operation, due to the amplification of collected holes. For example, the 4000-Å collector GaAs HBT exhibits a common-emitter breakdown voltage of about 10 V, corresponding to 250-kV/cm breakdown field (see [13]).

Although microwave power Si/SiGe/Si HBTs have not yet been reported, the nearly ideal power-frequency performance realized in Si permeable base transistors [49] is

41

strong evidence of the potential. The high thermal conductivity of Si should give Si/SiGe/Si HBTs an advantage over AlGaAs/GaAs HBTs in continuous-wave (CW) operation. Power GaAs HBTs are likely to be mainly useful in pulsed or class C operation unless great pains are taken to dissipate the heat. In class C or switching applications, the GaAs HBT does well [13], because it efficiently alternates between cutoff and favored high-current, low-voltage conditions and dissipates relatively little power. For linear power applications, it has extremely strong competition from GaAs FETs and HFETs, which have lower power densities, and hence a much less severe heat-dissipation problem. Other semiconductors potentially offer greater high-frequency power capability. InP has somewhat higher breakdown fields and better thermal conductivity than GaAs. Because InP has good electron mobility, an HFET based on InP would also be strong competition to an InP HBT. If and when it becomes technologically feasible, an HBT employing a high bandgap collector like α-SiC, with a much higher breakdown field and twice the saturation velocity, would be far superior to any conceivable planar FET. The high electron mobility required by FETs and HBT bases is not needed in the collector, where breakdown field and saturated electron velocity are really the key parameters. Such a device would be a dramatic illustration of the potential advantage offered by a heterostructure collector.

X. HETEROSTRUCTURE COLLECTOR TRANSISTORS

In principle, the bandgap of the collector can be made narrower than that of the base, but most heterostructure collectors have a wider bandgap in order to take advantage of the higher breakdown field or saturation velocity. Wideband collectors are the only practical HBT configuration in some strained-layer systems such as Si/SiGe/Si [27], because the narrow bandgap base is a strained layer of restricted thickness.

A wide-bandgap heterostructure collector must be designed and fabricated with great care, to avoid forming a conduction-band barrier near the edge of the base. The barrier, if present, usually shows up in forward bias and/or high current density operation, causing excessive electron storage, degrading h_{fe} and f_T [46]. The collector current-voltage characteristic may show a strangely rounded saturation. To prevent this, the heterojunction must be graded and correctly positioned outside the base. The required minimum value of donor concentration depends on the alloy gradient and the desired collector current.

XI. CONCLUSION

We have given examples of how bipolar transistor performance can be enhanced through the use of heterostructures. The emitter delay and base transit times are generally reduced to a relatively small percentage of total delay, while base resistance is minimized. High-performance HBTs have been made in both Si/SiGe/Si and III/V semiconductors. The conduction-band electron transport in III/V materials, with its rich physics, has opened up a much higher level of cutoff-frequency performance. The evidence of this physics is seen both in the static velocity-field characteristics and by velocity overshoot. We have derived an analytical model based on the velocity-field characteristic, which largely predicts and explains observed GaAs collector behavior over a range of bias and temperature. Some of the unique characteristics include a reduction of the collector delay at high collector currents, and capacitance cancellation. Base pushout has a fairly weak effect on cutoff frequency and little effect on DC current gain. Over a relatively small range of bias and with specially optimized designs, velocity overshoot has yielded extremely short collector delay times.

Because the emitter delay, base delay, and base resistance have been improved, HBTs approach the fundamental power-frequency limits for transistors more closely than Si BJTs in the microwave range. Further improvement is possible through the use of high-bandgap heterostructure collectors.

ACKNOWLEDGMENTS

We acknowledge the efforts of Arlene Wakita and Ruth Noll for self-aligned HBT processing, Dan Mars and Alice Fischer-Colbrie for MBE growth, and Rita Lee for RF testing. The strong encouragement offered by Sandip Tiwari during the preparation of the manuscript is most appreciated.

REFERENCES

[1] P. M. Asbeck, M. F. Chang, J. A. Higgins, N. H. Sheng, G. J. Sullivan, and K. C. Wang, "GaAlAs/GaAs heterojunction bipolar transistors: Issues and prospects for application," *IEEE Trans. Electron Devices,* vol. 36, pp. 2032–2042, 1989.

[2] H. Kroemer, "Theory of wide-gap emitter for transistors," *Proc. IRE,* vol. 45, pp. 1535–1537, 1957.

[3] H. Kroemer, "Heterostructure bipolar transistors and integrated circuits," *Proc. IEEE,* vol. 70, p. 13, 1982.

[4] H. Kroemer, "Heterostructure bipolar transistors: What should we build?," *J. Vac. Sci. Technol.,* vol. B1, pp. 126–130, 1983.

[5] G. L. Patton et al., "75-GHz f_T SiGe-base heterojunction bipolar transistors," *IEEE Electron Device Lett.,* vol. EDL-11, pp. 171–173, 1990.

[6] C. M. Mazier, M. E. Klausmeier, and M. S. Lundstrom, "A proposed structure for collector transit-time reduction in AlGaAs/GaAs bipolar transistors," *IEEE Electron Device Lett.,* vol. EDL-7, pp. 483–485, 1986.

[7] T. Ishibashi, H. Nakajima, H. Ito, S. Yamahata, and Y. Matsuoka, "Suppressed base widening in AlGaAs/GaAs ballistic collection transistors," In *Proc. 48th Annual Device Research Conference,* Paper VIIB-3, Santa Barbara, CA, June 1990.

[8] R. Katoh and M. Kurata, "Self-consistent particle simulation for (AlGa)As/GaAs HBT's with improved base-collector structures," *IEEE Trans. Electron Devices,* vol. ED-36, pp. 846–853, 1989.

[9] K. Morizuka, R. Katoh, K. Tsuda, M. Asaka, N. Iizuka, and M. Obara, "Electron space-charge effects on high-frequency performance of AlGaAs/GaAs HBT's under high-current-density operation," *IEEE Electron Device Lett.,* vol. EDL-9, pp. 570–572, 1988.

[10] C. T. Kirk, Jr., "A theory of transistor cutoff frequency (F_T) falloff at high current densities," *IRE Trans. Electron Devices,* vol. ED-9, pp. 164–174, 1962.

[11] R. J. Whittier and D. A. Tremere, "Current gain and cutoff frequency falloff at high currents," *IEEE Trans. Electron Devices,* vol. ED-16, pp. 39–57, 1969.

[12] E. O. Johnson, "Physical limitations on frequency and power parameters of transistors," *RCA Rev.,* vol. 26, pp. 163–177, 1965.

[13] N. L. Wang, N. H. Sheng, M. F. Chang, W. J. Ho, G. J. Sullivan, E. A. Sovero, J. A. Higgins, and P. M. Asbeck, "Ultrahigh power efficiency operation of common-emitter and common-base HBT's as 10 Ghz," *IEEE Trans. Microwave Theory Tech.,* vol. MTT-38, pp. 1381–1389, 1990.

[14] J. Yoshida, M. Kurata, K. Morizuka, and A. Hojo, "Emitter-base bandgap grading effects on GaAlAs/GaAs heterojunction bipolar transistor characteristics," *IEEE Trans. Electron Devices,* vol. ED-32, pp. 1714–1716, 1985.

[15] B. Bayraktaroglu, N. Camilleri, H. D. Shih, and H. Q. Tserng, "AlGaAs/GaAs heterojunction bipolar transistors with 4 W/mm power density at X-band," *IEEE MTT-S Dig.* pp. 969–972, 1987.

[16] A Wakita, et al., "Fabrication of a self-aligned heterojunction bipolar transistor utilizing an oxide-assisted base liftoff," in preparation.

[17] H. F. Cooke, "Microwave transistors: Theory and design," *Proc. IEEE,* vol. 59, pp. 1163–1181, 1971.

[18] D. E. Thomas and J. L. Moll, "Junction transistor short-circuit current gain and phase determination," *Proc. IRE,* vol. 46, pp. 1177–1184, 1958.

[19] N. Dagli, "Physical origin of the negative output resistance of heterojunction bipolar transistors," *IEEE Electron Device Lett.*, vol. EDL-9, pp. 113–115, 1988.

[20] S. Tiwari, "Frequency dependence of the unilateral gain in bipolar transistors," *IEEE Trans. Electron Devices,* vol. ED-10, pp. 574–576, 1989.

[21] S. Tiwari and D. J. Frank, "Analysis of the operation of GaAlAs/GaAs HBT's," *IEEE Trans. Electron Devices,* vol. ED-36, pp. 2105–2121, 1989.

[22] N. Moll, "Heterojunction bipolar transistors—a Review," *Proc. IEEE Cornell Conference on Advanced Concepts of High Speed Semiconductor Devices and Circuits,* pp. 35–44, 1985.

[23] R. Malik et al., "High-gain, high-frequency AlGaAs/GaAs graded band-gap base bipolar transistors with a Be diffusion setback layer in the base," *Appl. Phys. Lett.*, vol. 46, pp. 600–602, 1985.

[24] T. Furuta and M. Tomizawa, "Velocity electric field relationship for minority electrons in highly doped p-GaAs," *Appl. Phys. Lett.*, vol. 56, pp. 824–826, 1990.

[25] J. L. Moll and I. M. Ross, "The dependence of transistor parameters on the distribution of base layer resistivity," *Proc. IRE,* vol. 44, pp. 72–78, 1956.

[26] M. E. Klausmeier-Brown, M. S. Lundstrom, and M. R. Melloch, "The effects of heavy impurity doping on AlGaAs/GaAs bipolar transistors," *IEEE Trans. Electron Devices,* vol. ED-36, pp. 2146–2155, 1989.

[27] J. C. Sturm, E. J. Prinz, and C. W. Magee, "Graded-base Si/Si$_{1-x}$Ge$_x$/Si heterojunction bipolar transistors grown by rapid thermal chemical vapor deposition with near-ideal electrical characteristics," *IEEE Electron Device Lett.*, Vol. ED-12, pp. 303–305, 1991.

[28] D. C. Streit et al., "Effect of exponentially graded doping on the performance of GaAs/AlGaAs heterojunction bipolar transistors," *IEEE Electron Device Lett.*, vol. EDL-12, pp. 194–196, 1991.

[29] H. Ito, T. Ishibashi, and T. Sugeta, "Current gain enhancement in graded base AlGaAs/GaAs HBTs associated with electron drift motion," *Japan. J. Appl. Phys.*, vol. 24, pp. L241–L243, 1985.

[30] Y. S. Hiraoka and J. Yoshida, "Two-dimensional analysis of the surface recombination effect on current gain for GaAlAs/GaAs HBT's," *IEEE Trans. Electron Devices,* vol. ED-35, pp. 857–862, 1988.

[31] A. Furukawa, K. Ohta, and T. Baba, "Dependence of base built-in field for InAlAs/InGaAs HBT characteristics," *IEDM Tech. Dig.*, Paper 27.6, pp. 615–618, 1987.

[32] A. F. J. Levi, R. N. Nottenburg, Y. K. Chen, and M. B. Panish, "Ultrahigh-speed bipolar transistors," *Phys. Today,* pp. 58–64, Feb. 1990.

[33] J. R. Hayes, A. F. J. Levi, A. C. Gossard, and J. H. English, "Base transport dynamics in a heterojunction bipolar transistor," *Appl. Phys. Lett.*, vol. 49, pp. 1481–1483, 1986.

[34] B. Jalali, Y. K. Chen, R. N. Nottenburg, D. Sivco, D. A. Humphrey, and A. Y. Cho, "Influence of base thickness on collector breakdown in abrupt AlInAs/InGaAs heterostructure bipolar transistors," *IEEE Electron Device Lett.*, vol. EDL-11, pp. 400–402, 1990.

[35] R. Katoh, "Charge-control analysis of collector transit time for (AlGa)As/GaAs HBT's under a high injection condition," *IEEE Trans. Electron Devices,* vol. ED-37, pp. 2176–2182, 1990.

[36] R. Katoh and M. Kurata, "Self-consistent particle simulation for (AlGa)As/GaAs HBT's under high bias conditions," *IEEE Trans. Electron Devices,* vol. ED-36, pp. 2122–2127, 1989.

[37] R. Katoh and M. Kurata, "Self-consistent particle simulation of heterojunction bipolar transistors under high temperature operating conditions," *1989 IEDM Tech. Dig.*, Paper 18.3, pp. 473–479, 1989.

[38] T. H. Windhorn, T. J. Roth, L. M. Zinkiewicz, O. L. Gaddy, and G. E. Stillman, "High field temperature dependent electron drift velocities in GaAs," *Appl. Phys. Lett.*, vol. 40, pp. 513–515, 1982.

[39] S. E. Laux and W. Lee, "Collector signal delay in the presence of velocity overshoot," *IEEE Electron Device Lett.*, vol. EDL-11, pp. 174–176, 1990.

[40] S. I. Long, "A comparison of the GaAs MESFET and the AlGaAs/GaAs heterojunction bipolar power transistor for power microwave amplification," *IEEE Trans. Electron Devices,* vol. ED-36, pp. 1274–1278, 1989.

[41] K. Morizuka, R. Katoh, M. Asaka, N. Iizuka, K. Tsuda, and M. Obara, "Transit-time reduction in AlGaAs/GaAs HBTs utilizing velocity overshoot in the p-type collector region," *IEEE Electron Device Lett.*, vol. EDL-9, pp. 585–587, 1988.

[42] A. Das and M. Lundstrum, "Does velocity overshoot reduce collector delay time in AlGaAs/GaAs HBT's?," *IEEE Electron Device Lett.*, vol. EDL-12, pp. 335–337, 1991.

[43] R. W. H. Engelmann and C. A. Liechti, "Bias dependence of GaAs and InP MESFET parameters," *IEEE Trans. Electron Devices,* vol. ED-24, pp. 1288–1296, 1977.

[44] R. K. Cook and J. Frey, "Two-dimensional numerical simulation of energy transport effects in Si and GaAs MESFET's." *IEEE Trans. Electron Devices,* vol. ED-29, pp. 970–977, 1982.

[45] D. L. Bowler and F. A. Lindholm, "High current regimes in transistor collector regions," *IEEE Trans. Electron Devices,* vol. ED-20, pp. 257–263, 1973.

[46] S. Tiwari, "A new effect at high currents in heterostructure bipolar transistors," *IEEE Electron Device Lett.*, vol. EDL-9, pp. 142–144, 1988.

[47] T. Ishibashi and Y. Yamauchi, "A possible near-ballistic collection in an AlGaAs/GaAs HBT with a modified collector structure," *IEEE Trans. Electron Devices,* vol. ED-35, pp. 401–414, 1988.

[48] Y. K. Chen, R. N. Nottenburg, M. B. Panish, R. Hamm, and D. A. Humphrey, "Subpicosecond InGaAs/InP heterostructure bipolar transistors," *IEEE Electron Device Lett.,* vol. EDL-10, pp. 267–269, 1989.

[49] D. D. Rathman, "Optimization of the doping profile in Si permeable base transistors for high frequency, high voltage operation," *IEEE Trans. Electron Devices,* vol. ED-37, pp. 2090–2098, 1991.

Part 2
MESFETs

Bias Dependence of GaAs and InP MESFET Parameters

REINHART W. H. ENGELMANN AND CHARLES A. LIECHTI, SENIOR MEMBER, IEEE

Abstract—A comparative analysis of the bias dependence of critical RF parameters in GaAs and InP metal-semiconductor field-effect transistors (MESFET's) led to the following conclusions.

1) The drain-gate feedback capacitance in GaAs MESFET's is lower than in InP MESFET's, because of a stronger tendency in GaAs to form stationary Gunn domains at the typical drain bias levels employed.

2) The drain-source output resistance in InP MESFET's is lower than in GaAs MESFET's mainly for high drain current units, a fact which is linked to a substrate related softer pinch-off behavior in InP.

3) The current-gain cutoff frequency f_T, in the current saturation range of the GaAs MESFET decreases strongly with drain bias as a result of the formation of the stationary Gunn domain. In the InP MESFET, this effect is weaker. At the optimum bias, f_T is only 10–20 percent higher in InP MESFET's than in GaAs ones.

LIST OF SYMBOLS

C_{dg}	Drain-gate capacitance.
C_f	Fringing edge capacitance of gate stripe.
C_{gs}	Gate-source capacitance.
C_h	Capacitance of high-field region.
d_d	Depletion-layer penetration.
d_{eff}	$= (2\epsilon U_{eff}/qn)^{1/2}$, effective epilayer thickness.
E_p	Velocity-peak field.
E_v	Velocity-valley field.
f_{max}	Unilateral-power-gain cutoff frequency.
f_T	Current-gain cutoff frequency.
G_{mo}	Intrinsic dc transconductance.
I_D	Drain current.
L	Metallurgical gate length.
L_s	Parasitic source inductance.
l_f	$= (1 - \alpha)d_d$, length of depletion-layer fringing beyond gate stripe.
l_d	Drain portion of depletion-layer length.
l_h	Length of high-field region.
l_s	Source portion of depletion-layer length (effective length of gradual channel).
n	Epilayer carrier density.
q	Electron charge.
R_d	Parasitic drain resistance.
R_g	Parasitic gate resistance.
R_i	Intrinsic channel resistance.
R_s	Parasitic source resistance.
U_{eff}	$= V_p^{(exp)} + V_B$, effective pinch-off potential.
V_B	Built-in potential.

V_{DG}	$= V_{DS} - V_{GS}$, drain-gate bias.
V_{DS}	drain-source bias.
$V_{D\,sat}^{(exp)}$	Experimental saturation voltage.
V_{GS}	Gate-source bias.
$V_p^{(exp)}$	Experimental pinch-off voltage.
v	Electron velocity.
v_p	Peak velocity.
v_v	Valley velocity.
w	Channel width.
Y_m	$= G_{mo}e^{-j\omega\tau_0}$, intrinsic transconductance.
ϵ	Permittivity.
ω	Radian frequency.

I. INTRODUCTION

INTEREST in metal-semiconductor field-effect transistors (MESFET's) for microwave applications has grown steadily in recent years [1]. After Si devices [2], GaAs realizations were studied extensively because of their higher frequency capabilities [3].

Initial work by Barrera and Archer [4] on the InP MESFET with a 1-μm gate length revealed a 50-percent increase of the current-gain cutoff frequency f_T, beyond best values generally observed in GaAs devices of the same geometry and doping (Table I). This compared to a theoretically predicted 30-percent increase in f_T for the saturated InP MESFET based on the idealized model of Turner and Wilson [6], [7] as a result of a 47 percent higher electron peak velocity in InP. On the other hand, the combined influence of a much larger drain-gate feedback capacitance[1] C_{dg}, a smaller drain-source output resistance R_{ds}, and a somewhat larger input resistance $R_i + R_g + R_s$, in the InP MESFET, gave rise to somewhat lower available power gain and hence a lower unilateral power-gain cutoff frequency f_{max} [4] since [5]

$$f_{max} \approx f_T/[2\sqrt{(R_i + R_g + R_s)/R_{ds}} + 2\pi f_T R_g C_{dg}].$$

It was desirable to determine if the substantially higher C_{dg} and the lower R_{ds} values in InP MESFET's resulted from fundamental material properties or from remediable flaws in the materials and MESFET fabrication technology. In addition, the relatively large discrepancy between experimental and theoretically estimated f_T values (Table I) needed some explanation. The idealized theoretical model is based on the hypothesis that the electron drift velocity saturates in the channel at its peak value v_p [6], [7]. The effect of the actual velocity drop at fields beyond the velocity peak was not considered. In order to arrive at

Manuscript received March 13, 1977; revised July 6, 1977. This work was supported in part by USAECOM, Fort Monmouth, NJ, under Contract DAAB07-75-C-1300.

The authors are with Hewlett Packard Laboratories, Palo Alto, CA 94304.

[1] For the definition of the equivalent circuit elements, see [4].

Reprinted from *IEEE Trans. Electron Dev.*, vol. ED-24, no. 11, pp. 1288–1296, Nov. 1977.

TABLE I
Summary of Initial MESFET Comparison [4]
($L = 1$ μm, $w = 500$ μm, $n = 10^{17}$ cm^{-3}; $V_{DS} = 0$)

	GaAs	InP
f_T (GHz) Experiment	13	20–24
Theory	25	32
v_p (10^7 cm s^{-1})*	1.7	2.5
C_{dg} (pF)	0.014	0.07
R_{ds} (Ω)	300	145
$R_i + R_g + R_s$ (Ω)	7.5	9
f_{max} (GHz)	40	32

*Used in theoretical model for f_T.

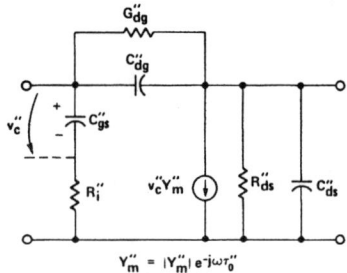

APPROXIMATIONS FOR MESFET IN SATURATION:

$R_i''(11\,\text{GHz}) \approx R_i + R_s + R_g + \dfrac{L_s G_{mo}}{C_{gs}}$

$C_{gs}''(11\,\text{GHz}) \approx C_{gs}$

$C_{dg}''(3\,\text{GHz}) \approx C_{dg}$

$Y_m''(11\,\text{GHz}) \approx Y_m = G_{mo}e^{-j\omega\tau_0}$

$R_{ds}''(3\,\text{GHz}) \approx R_{ds}$

Fig. 1. MESFET π circuit model with approximations for equivalent circuit elements defined in [4].

TABLE II
Experimental MESFET Properties at DC

Device	n (cm^{-3})	μ (cm^{2}V^{-1}s^{-1})	L (μm)	$V_{DS} = 5V$, $V_{GS} = 0$	
				I_D (mA)	I_G (mA)
GaAs:					
SA62-6	1.1×10^{17}	5200	0.9 ±0.1	58.5	0
InP:					
SC126,7-14	1.0×10^{17}	3000	0.9 ±0.1	43.0	0.92
SC126,7-18	1.0×10^{17}	3000	1.0 ±0.1	115.0	1.5

Device	$V_{Dsat}^{(exp)}$ (V)	$I_{Dsat}^{(exp)}$ (mA)	$R_s + R_d$ (Ω)	$V_{Dsat}^{(corr)}$ (V)	$V_p^{(exp)}$ (V)	V_B (V)	U_{eff} (V)	d_{eff} (μm)	d_c (μm)
GaAs:									
SA62-6	1.15	44	4.0	0.97	2.1	0.80	2.9	0.20	0.074
InP:									
SC126,7-14	2.05	38	6.5	1.80	1.9	0.50	2.4	0.18	0.060
SC126,7-18	3.0	104	6.5	2.32	3.6	0.45	4.05	0.24	0.120

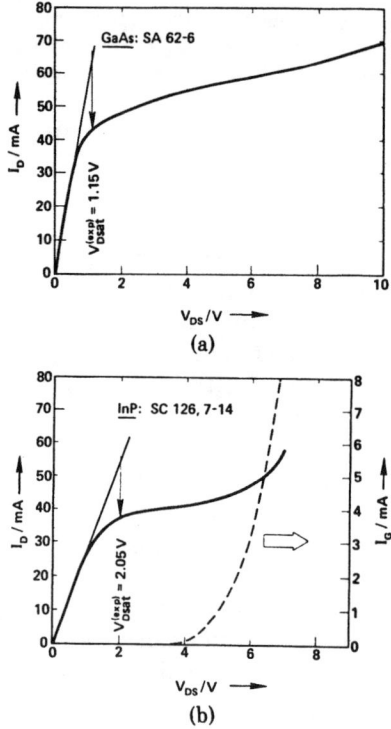

Fig. 2. Drain current characteristics I_D versus V_{DS} at $V_{GS} = 0$ illustrating the definition of $V_{D\,sat}^{(exp)}$. (a) GaAs MESFET. (b) InP MESFET.

a deeper understanding of the striking differences found between GaAs and InP MESFET's and their possible relation to the electron-drift-velocity/field characteristic, we undertook a comprehensive study of important MESFET parameters as functions of drain-source and gate-source bias.

In general, a determination of the element values of the equivalent circuit of a MESFET follows from a model optimization procedure based on measured and calculated s parameters [4], [8]. To avoid this relatively lengthy procedure for each different bias condition, we estimated the crucial equivalent circuit elements from a set of π (circuit) parameters (Fig. 1) that can be derived from the set of s parameters directly. The approximations of Fig. 1 are based on a detailed investigation of the π parameters of a saturated MESFET as a function of frequency [9].

II. PARAMETERS OF MESFET'S STUDIED

The GaAs and InP MESFET's studied were fabricated from the same mask set with a center-fed gate, nominally 1 μm long and 500 μm wide. Equivalent technologies were used including a gate-trough etch to adjust the FET channel thickness [4]. Active n-type epilayers with a

doping density close to 10^{17} cm^{-3} were grown on Cr-doped semi-insulating substrates using liquid-phase epitaxy in both cases. The dc properties of one GaAs and two InP MESFET's, which were investigated in detail, are summarized in Table II. The second InP MESFET is included to show the influence of a large drain saturation current on RF properties. The experimental drain saturation voltage $V_{D\,sat}^{(exp)}$ is defined in a somewhat arbitrary but convenient way, by the drain bias at which the drain current saturates to two-thirds the extrapolated ohmic value (Fig. 2). $V_{D\,sat}^{(corr)}$ is this saturation voltage corrected for the estimated voltage drop across the source and drain parasitic series resistances $R_s + R_d$. A convenient experimental pinch-off voltage $V_p^{(exp)}$ follows from a rough linear extrapolation of the transconductance versus gate bias

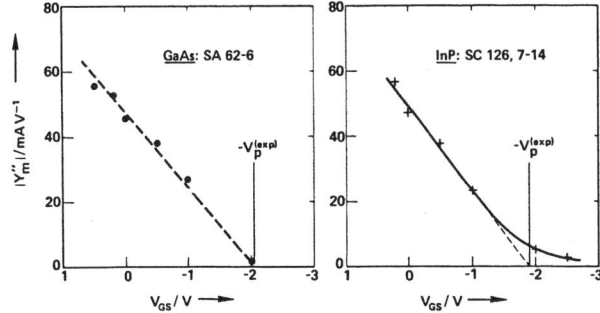

Fig. 3. Transconductance pinch-off behavior at $V_{DS} = 4$ V illustrating the definition of $V_p^{(exp)}$ (intrinsic π circuit transconductance $|Y_m''|$ at 11 GHz). (a) GaAs MESFET. (b) InP MESFET.

TABLE III
Experimental MESFET Parameters at RF, $V_{DS} = 5$ V, $V_{GS} = 0$

Device	MAG @10 GHz (dB)	f_{max} (GHz)	f_T (GHz)	π Parameters (cf. Fig. 1)							
				$\|Y_m''\|$ @11 GHz (mAV^{-1})	τ_0'' @11 GHz (ps)	C_{dg}'' @3 GHz (pF)	R_{dg}'' @3 GHz (kΩ)	R_{ds}'' @3 GHz (Ω)	C_{ds}'' @11 GHz (pF)	C_{gs}'' @11 GHz (pF)	R_i'' @11 GHz (Ω)
GaAs:											
SA62-6	8.8	34	12.5	44.6	3.1	0.0187	-166.0	320	0.047	0.61	12.3
InP:											
SC126,7-14	4.3	18	16.5	50.9	3.9	0.110	1.02	335	0.106	0.49	14.9
SC126,7-18	5.1	21	15.7	54.7	-	0.107	1.56	180	0.053	0.52	17.7

curves[2] (Fig. 3) and the built-in potential V_B from the voltage dependence of the gate capacitance at $V_{DS} = 0$. An effective open channel thickness d_{eff} is then derived from the effective pinch-off potential $U_{eff} = V_p^{(exp)} + V_B$. The channel constriction d_c at the drain end of the gradual channel region is estimated from the Turner–Wilson model [6], [7]. Pertinent RF parameters at the bias condition $V_{DS} = 5$ V, $V_{GS} = 0$ are listed in Table III, including the π parameters of Fig. 1 at the appropriate frequencies.

III. FEEDBACK CAPACITANCE C_{dg}

The dependence of the π circuit feedback capacitance C_{dg}'' (3 GHz) on the bias between the drain and gate V_{DG} is shown in Fig. 4. Fig. 4(a) and (b) compares the GaAs with the equivalent InP MESFET for three bias conditions: 1) the gate shorted to the source ($V_{GS} = 0$) and the drain bias V_{DS} raised into saturation (circles); 2) the channel current kept at zero ($V_{DS} = 0$) and a negative gate bias ($-V_{GS}$) applied toward and beyond channel pinch-off (triangles); 3) the drain bias V_{DS} fixed in saturation and the gate bias V_{GS} varied (dots, dashed lines). Fig. 4(c) summarizes the first case ($V_{GS} = 0$) including the high-current InP MESFET.

Most striking is the strong decrease of C_{dg}'' for the GaAs MESFET at $V_{GS} = 0$, extending beyond saturation and covering almost an order of magnitude in the saturation range (Fig. 4(a)). This is quite contrary to the behavior in the channel current-free state $V_{DS} = 0$. Here C_{dg}'' also decreases strongly as long as the depletion layer moves

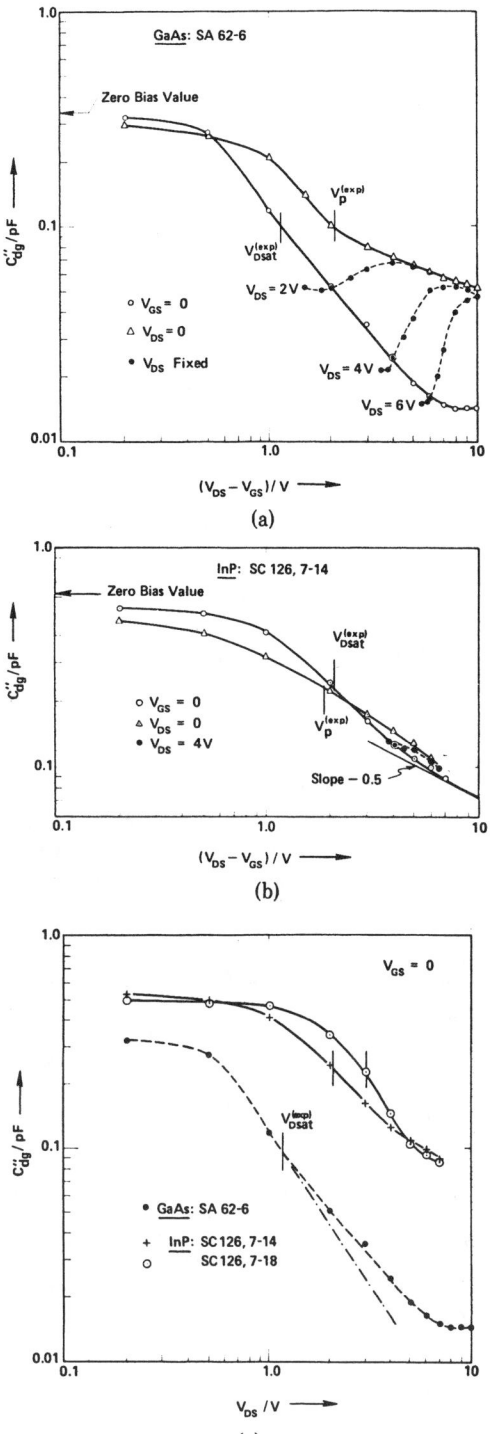

Fig. 4. π circuit drain-gate capacitance C_{dg}'' at 3 GHz versus drain-gate bias, $V_{DG} = V_{DS} - V_{GS}$. (a) GaAs MESFET. (b) InP MESFET. (c) Comparison between GaAs and InP MESFET's at $V_{GS} = 0$.

toward channel pinch-off at $V_p^{(exp)}$, confining the capacitance to the fringing depletion-layer edge. However, a more gradual decrease follows beyond $V_p^{(exp)}$, which is expected from the lateral movement of this depletion-layer edge toward the drain. Thus at a given V_{DG}, the GaAs MESFET has a much smaller C_{dg}'' in saturation than in the pinch-off condition, particularly when V_{DG} is large. This leads to a strong increase of C_{dg}'' when a negative gate bias ($-V_{GS}$)

---

[2] There is a certain dependence of $V_p^{(exp)}$ on V_{DS} [18]; we adopted $V_{DS} = 4$ V as our standard for comparison of devices. The linear extrapolation and the V_{DS} dependence are inconsistent with the Turner–Wilson model [6], [7], but the simplified definition for $V_p^{(exp)}$ is well justified for comparative purposes.

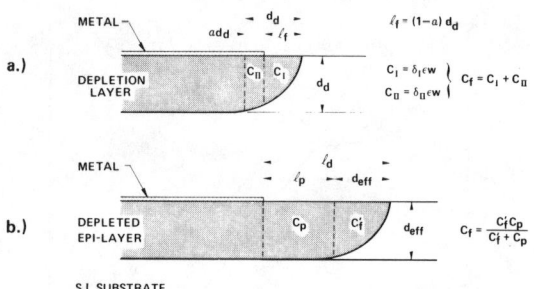

Fig. 5. Depletion-layer fringing. (a) Before pinch-off. (b) After pinch-off.

TABLE IV
Gate Fringing Capacitance at Different Bias Conditions

Bias Condition*	Capacitance Values	GaAs (SA62-6)	InP (SC126,7-14)
①	C''_{dg}/pF	0.10	0.23
	C_f/pF	0.086	0.081
	$C_{f,excess}$/pF	0.01	0.15
②	C''_{gs}/pF	0.30	0.42
	C''_{dg}/pF	0.32	0.63
	$C_{g,exp}$/pF	0.62	1.05
	$C_{g,theor}$/pF	0.58	0.68
	$C_{f,excess}$/pF	0.02	0.19

* Bias condition: ① $V_{DS} = V_{Dsat}^{(exp)}$, $V_{GS} = 0$ or

$V_{DS} = 0$, $V_{GS} = -V_p^{(exp)}$

$C_{f,excess} = C''_{dg} - C_f$

② $V_{DS} = 0$, $V_{GS} = 0$

$C_{f,excess} = \frac{1}{2}(C_{g,exp} - C_{g,theor})$

is applied to the saturated GaAs MESFET toward pinch-off (dashed lines, Fig. 4(a)).

From purely geometrical electrostatic arguments for drain-side fringing of the gate depletion-layer capacitance, no major difference in C_{dg} between the current saturation and the pinch-off state is expected. In fact, the bias dependence of C''_{dg} for the InP MESFET's (Fig. 4(b)) seems to agree at least qualitatively with such a simple geometrical viewpoint. In the case of the GaAs MESFET, however, a fringing capacitance equivalence between the saturated and the pinch-off MESFET is retained only at the limiting bias values, viz., at $V_{DS} = V_{D\,sat}^{(exp)}$, $V_{GS} = 0$, and at $V_{DS} = 0$, $V_{GS} = -V_p^{(exp)}$, respectively. Thus the above results strongly suggest that the flow of charges in the longitudinal channel field between the drain and source is responsible for reducing the feedback capacitance C_{dg} in the saturated GaAs MESFET. Before discussing this question further (Section III-B), a more quantitative analysis of depletion-layer fringing is appropriate.

A. Fringing Capacitance of Gate Stripe

The gate fringing capacitance can be estimated from published calculations of the depletion-layer geometry of a biased metal stripe on a plane semiconductor surface of infinite extension [10]. As indicated in Fig. 5(a), depletion-layer fringing that is relevant for C_{dg} extends underneath the metal stripe by a fraction α of the undisturbed depletion-layer penetration d_d. Deviating from previous estimates [2], [11], the total fringing capacitance C_f is assumed to be composed of this internal region II in addition to the external region I. It follows, independently of d_d,

$$C_f = \delta \epsilon w \qquad (1)$$

where

ϵ permittivity of the semiconductor,
w stripe length (corresponds to "width" of gate),
$\delta = \delta_I + \delta_{II}$. $\qquad (2)$

For region I, including the effect of the air–semiconductor interface [10],

$$\delta_I = 0.71 + 0.86 \, \epsilon_0/\epsilon = 0.78. \qquad (3a)$$

For region II, we estimate

$$\delta_{II} = 2\alpha \approx 0.70 \qquad (3b)$$

by considering the depletion charge increment both

downward and to the left in Fig. 5(a). We conjecture that $C_{dg} \approx C_f$ for the two limiting cases $V_{DS} = V_{D\,sat}^{(exp)}$, $V_{GS} = 0$ and $V_{DS} = 0$, $V_{GS} = -V_p^{(exp)}$ (Table IV, bias condition ①). Experimental (C''_{dg} at 3 GHz) and theoretical (C_f) values are in good agreement for the GaAs MESFET; for the InP MESFET, however, experimental values are more than two times larger. Factors contributing to an increase of the observed C_{dg} values are the presence of a gate trough, possible surface charge effects (both increasing δ_I), and incomplete saturation or channel pinch-off (increasing δ_{II}). In the InP MESFET's, pinch-off behavior is indeed much softer and, hence, less complete at $V_p^{(exp)}$, than in GaAs devices as is evident from Figs. 3 and 4.

To test the importance of gate-trough and/or surface charge effects, we measured the total gate capacitance at zero bias, which can be estimated from $C_{g,exp} \approx C''_{gs} + C''_{dg}$ at 2 GHz (Table IV, bias condition ②). For the GaAs MESFET, again, agreement with the theoretical value, $C_{g,theor}$, is satisfactory[3] when fringing at a planar surface is included [10]. The InP unit on the other hand exhibits experimental values that are about 50 percent larger. Here, the excess in fringing capacitance per side is 0.19 pF, which, within the experimental error, matches with the excess capacitance deduced from C''_{dg} at bias condition ①. SEM studies of the cross-sectioned MESFET chips revealed that most of the excess fringing capacitance results from the gate trough. The GaAs unit had a gate-trough depth of only 0.16 μm, in contrast to a depth of 0.43 μm for the InP unit. In addition, the InP trough exhibited an asymmetry to the metallurgical gate that explains the difference in C''_{gs} and C''_{dg} of Table IV. We thus conclude that gate-trough etching should be kept to a minimum to avoid an

---

[3] Note that the experimental accuracy of L (Table II) is not better than 10 percent.

increase in the drain-gate fringing capacitance. Surface charge effects appear to have little influence if present at all.

B. Influence of a Stationary Gunn Domain Region

After having attained an understanding of the geometrical fringing capacitance, we will now return to the problem of the experimentally observed large C_{dg} difference, for the GaAs MESFET, between the bias cases with and without current flow. The minimum value measured for the GaAs unit is about $C_{dg} = 0.014$ pF in current saturation at $V_{DS} = 8$ V, $V_{GS} = 0$, which compares to $C_{dg} = 0.057$ pF in pinch-off at $V_{DS} = 0$, $V_{GS} = -8$ V (Fig. 4(a)). For a large enough drain-gate bias V_{DG}, the depletion-layer penetration l_d toward the drain can be estimated from C_{dg} assuming a parallel-plate capacitance C_p of separation l_p in series with a fringing capacitance $C_f' \approx (\delta_I + \alpha)\epsilon w$ (Fig. 5(b)). With $C_f' = 0.062$ pF ($\delta_I = 0.71$, $\alpha = (0.35)$ and $d_{eff} = 0.20$ μm (Table II), one obtains $l_d = 0.22$ μm for the pinch-off case, and a four times larger value, viz., $l_d = 0.84$ μm, for the current saturation case. On the other hand, estimating l_d from the drain-gate potential, $U_{DG} = V_{DG} + V_B$ according to electrostatic arguments, viz., $l_d/l_f \approx (U_{DG}/U_{eff})^{1/2}$ [$U_{DG} = 8.8$ V, $U_{eff} = 2.9$ V (Table II), and $l_f = (1 - \alpha)d_{eff} = 0.13$ μm], yields $l_d = 0.23$ μm, which is consistently only with the above value for the pinched-off transistor. This proves quantitatively that the properties of the carrier flow have to be taken into consideration for explaining the low C_{dg} values observed in the saturated GaAs MESFET.

It has been shown by two-dimensional model calculations [12], [13] that velocity saturation in the constricting channel establishes a charge dipole layer across the velocity saturated portion of the channel. The Gunn effect enhances such a dipole-layer formation and rapidly induces fields beyond valley velocity saturation ($E > E_v \approx 12$ kV/cm) [14]. The finite scattering times of the electrons cause velocity overshoot effects [15]–[17] resulting in a positional shift of the charge dipole layer toward the drain [15]. In fact, high fields may even extend as far as directly below the planar metallurgical drain contact and establish valley velocity saturation over a substantial portion of the entire region between gate and metallurgical drain contacts (according to [15] most electrons are lifted to their heavy state in this region). We thus conjecture that the rapid decrease in C_{dg} with drain bias beyond MESFET saturation is a direct result of the growing charge dipole layer.

The situation is qualitatively illustrated in Fig. 6 [18]. The "gradual" portion of the channel terminates at a position where the field E reaches a value $E_s \leq E_p$ corresponding to the sustaining field appropriate for a stationary Gunn domain (E_p is the field at the peak velocity v_p of an effective *dynamic* $v(E)$ characteristic). Further down stream, electron accumulation under the constricting channel rapidly increases the field to $E > E_v$, causing the electron drift velocity to saturate at its valley value, $v \rightarrow v_v$. These high fields can penetrate far toward the metal-

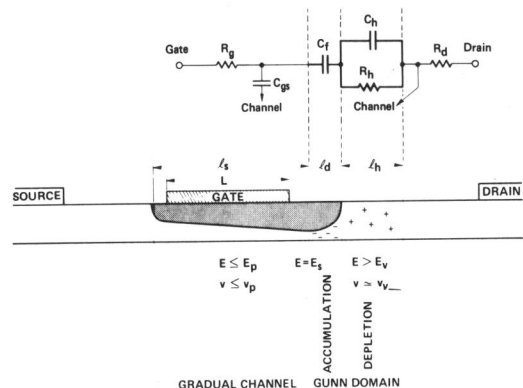

Fig. 6. Location and equivalent circuit of stationary Gunn domain.

lurgical drain contact since, because of the reduced drift velocity, electron depletion due to channel widening need not be large. It is thus expected that most of the additional drain-gate potential applied beyond the MESFET saturation voltage does not drop between the drain edge of the gate depletion layer and the gate contact as in the pinched-off MESFET, but across the high-field region between the metallurgical drain contact and the depletion-layer edge [18]. Dynamically, this high-field region acts as a capacitance, C_h, between the drain and gate in series with the fringing capacitance C_f of the gate depletion layer. The dynamic parallel resistance R_h of the high-field portion is considered to be negligibly high because of velocity saturation. With $C_{dg} = 0.014$ pF ($V_{DS} = 8$ V, $V_{GS} = 0$) and $C_f = 0.086$ pF (taken from $V_{DS} = V_{D\,sat}^{(exp)}$, $V_{GS} = 0$), this capacitance is $C_h = 0.017$ pF, which corresponds to a length of the high-field region of $l_h = 0.66$ μm. Assuming a potential drop across the fringing gate depletion layer of about 2 V (pinch-off voltage, $V_p^{(exp)}$), the average field along l_h becomes larger than 90 kV/cm. This value compares well with even the highest valley velocity fields for GaAs quoted in the literature ($E_v \approx 85$ kV/cm [19], also, cf., [20]). Even though these estimates are fairly rough, they demonstrate that the high-field domain model is consistent with the experimental data.

The bias behavior of the feedback capacitance C_{dg}'' of the InP MESFET's, on the other hand, suggests that the formation of the high-field dipole domain is fairly weak in current saturation (Fig. 4(b) and (c)) and, hence, the capacitance is mainly determined by the gate depletion layer as in the pinch-off case. A qualitative explanation follows from the differences in the drift velocity versus field relationship [20] between GaAs and InP: In InP the negative mobility is typically a factor of two smaller and the fields for valley velocity saturation are substantially larger, viz., by a factor of five. Thus substantially higher drain bias levels than currently possible appear to be necessary for InP FET's to induce similar Gunn domain effects as observed in GaAs MESFET's.

The length l_h of the velocity saturated region (Fig. 6) can also be estimated from the phase shift in the intrinsic transconductance Y_m if the delay in the rest of the device is neglected (cf., [21]). In the GaAs MESFET, the delay time τ_0'' (see Fig. 1) was found to increase with drain bias consistent with the simultaneous decrease of C_{dg}. At V_{DS} = 9 V, we obtained τ_0'' = 5.6 ps which leads to l_h = 0.85 μm assuming a delay analogous to that in a collector depletion layer of a bipolar transistor [28], $l_h \approx 2v_v\tau_0''$, with v_v = 0.75 $\times$ 10^7 cm/s (estimated at 15 percent below the room temperature value due to self heating, cf., [25]). This is quite consistent with the 0.66 μm deduced from C_{dg}. On the other hand, no clear trend for the V_{DS} dependence of τ_0'' evolved with the InP MESFET's [9].

Another fact that might contribute to reduced space charge effects in InP MESFET's is their very soft pinch-off behavior mentioned previously (Figs. 3 and 4). The softer pinch-off implies less constriction of the flow of charge carriers and hence less tendency for space charge formation. Additionally, the high gate leakage currents that flow in InP MESFET's [4] at the larger drain-gate bias (Fig. 2) are expected to affect the field distribution in the region between the depletion-layer edge and drain and may prevent a substantial buildup of field.

In conclusion, the main reason for the larger feedback capacitance C_{dg} in current saturation in InP MESFET's as compared to GaAs ones has been linked to a reduced field buildup and, hence, to a less complete velocity saturation at the "valley" velocity in InP at the bias values employed.

IV. OUTPUT RESISTANCE R_{ds}

The bias dependence of the π circuit output resistance R_{ds}'' (3 GHz) is presented in Fig. 7. The drain bias plots (V_{GS} = 0, Fig. 7(a)) exhibit about an equal relative increase of R_{ds}'' up to the saturation voltage $V_{D\,sat}^{(exp)}$ for both GaAs and InP. However, beyond $V_{D\,sat}^{(exp)}$, the resistance R_{ds}'' increases further for the GaAs MESFET by a much larger factor than for the InP versions. This difference appears to result from the softer pinch-off behavior in InP devices already noted in Figs. 3 and 4, which is clearly evidenced again by the pinch-off behavior of R_{ds}'' in Fig. 7(b). In this connection, it is of interest to note that in InP devices R_{ds}'' tends to increase for units with lower drain current I_D. Fig. 8 shows data taken for a number of fabrication runs covering a large range in epilayer doping and gate-trough depth and employing a variety of different substrates. An inverse proportional relationship between R_{ds} and I_D is expected when the RF current simply scales with the dc current. Such a relationship emerges for higher current InP units (Fig. 8). In GaAs MESFET's, on the other hand, variations in substrate, interface quality, and gate-trough depth play a major role in determining R_{ds} and no correlation to I_D evolved. Here, R_{ds} appears to be strongly influenced by the varying properties of an epi-substrate interface depletion layer resulting from electron trapping. Drain current relaxation effects and gating by substrate bias ("back gating") as observed in GaAs MESFET's have been related to these interface properties [22], [23]. So far, we were not able to produce "back gating" in InP units.

These facts taken together lead to the tentative conclusion that no electron trapping occurs at the epi-sub-

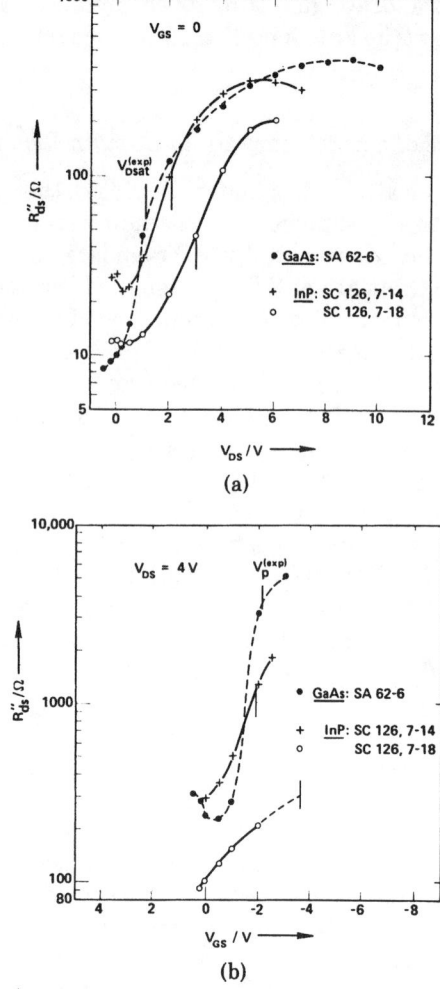

(a)

(b)

Fig. 7. π circuit drain-source resistance R_{ds}'' at 3 GHz: (a) versus drain bias V_{DS} at V_{GS} = 0; (b) versus gate bias V_{GS} at V_{DS} = 4 V.

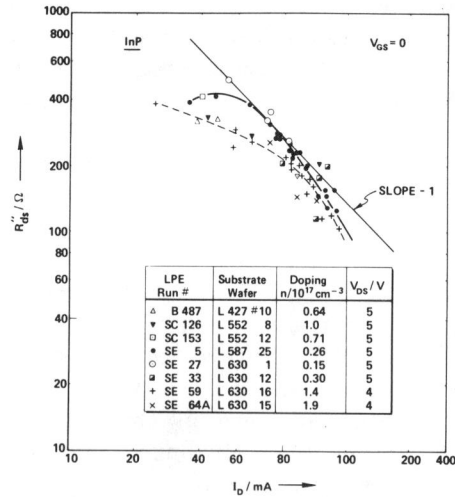

Fig. 8. π circuit drain-source resistance R_{ds}'' at 3 GHz versus drain saturation current I_D at V_{GS} = 0 for different InP MESFET runs.

strate interface in InP. This also explains the softer pinch-off behavior since carriers can be injected into the substrate more easily if no enhanced interface barrier is present. The absence of electron trapping might be related

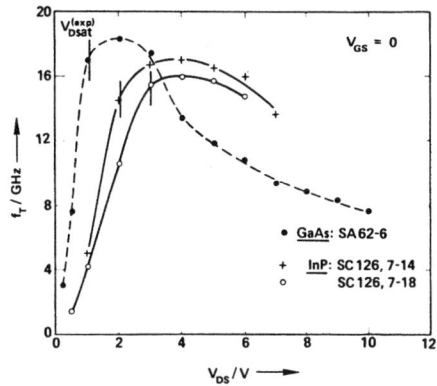

Fig. 9. Current-gain cutoff frequency f_T versus drain bias V_{DS} at $V_{GS} = 0$.

to a different nature of the deep levels in the lower resistivity, Cr-doped InP substrates used (10^3–10^4 $\Omega \cdot$cm, as compared to 10^6–10^8 $\Omega \cdot$cm for GaAs).

V. CURRENT-GAIN CUTOFF FREQUENCY f_T

The current-gain cutoff frequency f_T was directly determined from current-gain versus frequency curves as deduced from the measured s parameters [9]. As a function of drain bias (Fig. 9), we observe a rise toward a peak at about 1–2 V above the saturation voltage. The decrease of f_T beyond this peak is particularly strong in the GaAs MESFET. This suggests that the decrease is related to the Gunn domain formation discussed in Section III. For testing this contention [18], we consider the following relationships for the intrinsic small-signal transconductance G_{mo} (cf., [21], [24]) and for the gate-source capacitance C_{gs}:

$$G_{mo} = \frac{\Delta I_D}{\Delta V_{GS}} \approx \frac{\Delta Q \bar{v}}{l_{|s|} \Delta V_{GS}} = \epsilon w \frac{\bar{v}}{\bar{d}_d} \qquad (4)$$

$$C_{gs} = \frac{\Delta Q}{\Delta V_{GS}} = \epsilon w \frac{l_s}{\bar{d}_d} \qquad (5)$$

and, hence,

$$f_T = \frac{G_{mo}}{2\pi C_{gs}} \approx \frac{\bar{v}}{2\pi l_s}. \qquad (6)$$

Here, $\bar{v}$ is the average drift velocity in the gradual channel region and l_s is its effective length; $\bar{d}_d$ is the average gate depletion-layer penetration; and Q is the total charge in the gradual channel region. It can be shown [18] that $\bar{d}_d$ is hardly affected by the drain bias. Thus the drain bias dependence of G_{mo} and C_{gs} (Figs. 10 and 11) is indicative of that of $\bar{v}$ and l_s, respectively. Fig. 10 then clearly indicates that $\bar{v}$ falls off more strongly in GaAs than in InP with drain bias V_{DS} beyond its maximum value. This reduction of $\bar{v}$ with drain bias can be considered as evidence for the formation of a Gunn domain at the drain end of the channel: Because of current continuity, the lower carrier velocity in the high domain fields requires a simultaneous drop in carrier velocity further upstream in the gradual channel region, as long as the reduced velocity in the domain is not completely compensated by carrier accumu-

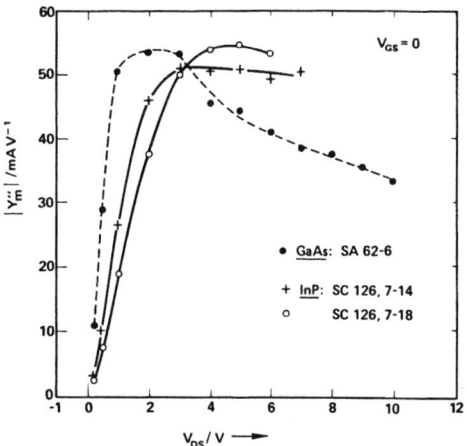

Fig. 10. Intrinsic π circuit transconductance $|Y'_m|$ at 11 GHz versus drain bias V_{DS} at $V_{GS} = 0$.

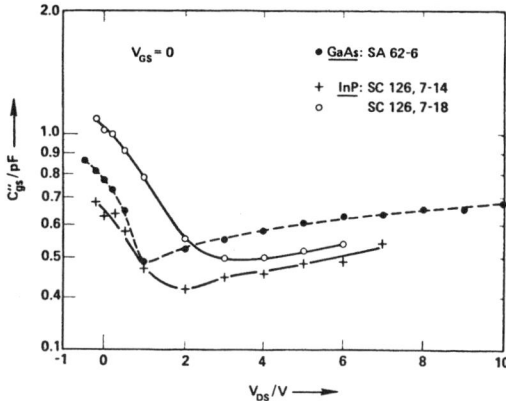

Fig. 11. π circuit gate-source capacitance C'_{gs} at 11 GHz versus drain bias V_{DS} at $V_{GS} = 0$.

lation and/or channel widening [18]. Fig. 10 is thus consistent with the conclusion, reached in Section III-B, that stationary Gunn domain formation is stronger in GaAs than in InP. Fig. 11, on the other hand, suggests an increase of l_s beyond saturation that is similar for both GaAs and InP devices. It has been argued [18] that the increase of l_s is linked to carrier relaxation effects [15] within the short channel (velocity "overshoot"). Hence, it appears that these effects are at least as strong in InP as in GaAs (cf., [17]) even though the stationary Gunn domain that actually has formed is weaker in InP. On the other hand, a field-dependent surface conduction (a surface inversion layer spreading from the gate toward drain) might cause a similar effect.

The stronger decrease of the current-gain cutoff frequency f_T for GaAs MESFET's than for InP equivalents, as demonstrated in Fig. 9, is thus mainly caused by a stronger reduction of the average carrier velocity $\bar{v}$ in the gradual channel from which a stronger stationary Gunn domain growth is inferred.

Fig. 9 reveals that the originally reported improvement of f_T for InP over GaAs MESFET's obtained at $V_{DS} = 5$ V (Table I) [4] is partly due to the strong reduction of f_T already present in GaAs at this bias. In fact, for the units shown, the f_T peak at the optimum drain bias of the GaAs

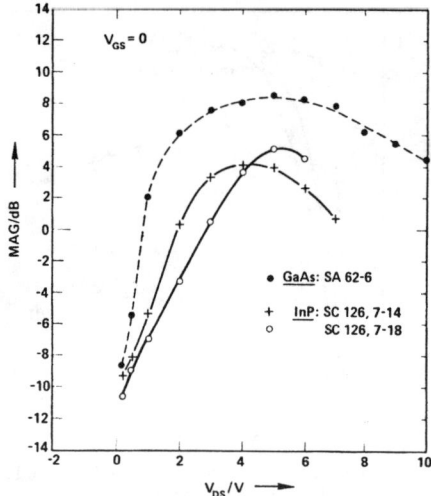

Fig. 12. Maximum available gain (MAG) at 11 GHz versus drain bias V_{DS} for $V_{GS} = 0$.

MESFET happens to be even higher than those of the two InP MESFET's. *Typical* peak values, however, are still somewhat higher in InP devices ranging from 18–24 GHz as compared to 16–20 GHz in GaAs devices.

For completeness, Fig. 12 compares the maximum available power gain of the two kinds of devices at 10 GHz. As already pointed out in [4], the gain is lower in the InP devices due to the combined influence of C_{dg}, R_{ds} and of $R_g + R_s + R_i$ (see Section I).

VI. DISCUSSION AND CONCLUSION

The investigation of the bias dependence of MESFET parameters in GaAs and in InP led to the following conclusions.

1) The roughly five times higher feedback capacitance C_{dg} in current-saturated InP MESFET's as compared to GaAs ones ($V_{DS} = 5$ V, $V_{GS} = 0$, see Table I) is mostly a consequence of a weaker Gunn domain formation in InP between the gate and drain at the drain bias levels employed. This leads to a lesser degree of decoupling of the depletion-layer fringing capacitance from the drain, particularly far beyond the onset of current saturation.

2) The small drain-source resistance R_{ds}, frequently observed in InP MESFET's, was found to be correlated to high drain current. We suggest that a softer pinch-off behavior in InP is responsible for this trend. The lower resistivity of the InP substrates (10^3–10^4 $\Omega \cdot$ cm, Cr-doped, as compared to 10^6–10^8 $\Omega \cdot$ cm for GaAs) that were available for this study and the apparent absence of trapped charges at the interface are most likely the cause for this behavior.

3) The different behavior between GaAs and InP MESFET's in the bias dependence of the current-gain cutoff frequency f_T was linked to the same Gunn domain formation phenomenon that causes the C_{dg} variation. When a Gunn domain forms at the drain side of the channel, the average carrier velocity in the channel is reduced. At the drain bias levels generally employed (V_{DS}

≈ 4–5 V), this effect is stronger in GaAs than in InP and, hence, InP MESFET's appear to have a relatively higher f_T value over GaAs ones than theoretically estimated [4]. However, at the optimum bias level for f_T, the improvement for InP against GaAs in only 10–20 percent as compared to the 30 percent theoretical estimate.

4) The lower *absolute* value of the experimental current-gain cutoff frequency f_T as compared to the theoretical estimates in both materials [4] follows partly from the fact that the average carrier velocity in the channel is significantly below the velocity peak value. At higher drain bias levels, in addition, an effective lengthening of the channel beyond the metallurgical gate length L contributes to the reduction in f_T. For a GaAs MESFET, e.g., at a drain bias of $V_{DS} = 4$ V ($V_{GS} = 0$), the average carrier velocity $\bar{v}$ in the gradual channel was estimated at about 30 percent below the peak value v_p and the effective channel length l_s was estimated at about 35 percent above L [18]. In addition, current saturation measurements on ungated devices led to v_p values that are lower than expected from theoretical estimates [4]. We determined for GaAs at a doping level near 10^{17} cm^{-3} a value of $v_p \approx 1.4 \times 10^7$ cm/s [18] (theoretical value 1.7×10^7 cm/s [25]) and for InP at a doping level near 0.5×10^{17} cm^{-3} a value ranging from 1.3–1.7×10^7 cm/s [9b]. These lower v_p values, in combination with the further reduction for the *average* channel velocity, explain why the theoretical estimates for f_T in Table I are too favorable.

The difference in Gunn domain formation between GaAs and InP is related to the fundamental difference in the carrier velocity/field characteristic of the two materials. Similar Gunn domain effects as in GaAs would be expected in InP at drain bias levels that are at least three times higher than in GaAs, i.e., considerably beyond the currently achievable Schottky gate junction breakdown values.

We have also speculated that the different epilayer–substrate interface properties in GaAs and InP which are believed to cause the differences in R_{ds} might affect the process of Gunn domain formation. This contention would predict stronger Gunn domain formation in InP FET's, with its obvious effects on C_{dg} and f_T, by providing an enhanced interface barrier. This, of course, should also lead to larger R_{ds} values. Higher resistivity Fe-doped substrates [26] or heteroepitaxial interfaces [27] should be investigated in this context. On the other hand, better control of the GaAs epilayer–substrate interface may lead to less electron trapping and, hence, a reduction of the interface barrier causing a weakening of the Gunn in domain in current saturation. This would be particularly desirable for power FET's which require a high drain bias capability.

The overall conclusion of this investigation is then that an f_T advantage over GaAs is present in InP but is marginal. However, the generally higher fields of the InP velocity/field curve and its reduced tendency for Gunn domain formation can be of potential significance for a high power device. In order to sustain the high drain bias levels necessary for such an application, a leakage-free gate is of paramount importance, of course. Replacement of the Schottky gate by a p-n junction or an insulated gate appears to be necessary. An additional advantage for the high-power application would be the higher thermal conductivity of InP.

ACKNOWLEDGMENT

The authors express their gratitude to B. Lizenby for his measurement and computation assistance, to J. Barrera for supplying some of the InP MESFET's and for his stimulating interest in the study, and to R. J. Archer and R. Van Tuyl for many critical comments to the manuscript.

REFERENCES

[1] C. A. Liechti, "Microwave field-effect transistors—1976," *IEEE Trans. Microwave Theory Tech.*, vol. MTT-24, pp. 279–300, June 1976.
[2] P. Wolf, "Microwave properties of Schottky-barrier field-effect transistors," *IBM J. Res. Develop.*, vol. 14, pp. 125–121, 1970.
[3] *IEEE Trans. Microwave Theory Tech. (Special Issue on Microwave Field Effect Transistors)*, vol. MTT-24, June 1976.
[4] J. S. Barrera and R. J. Archer, "InP Schottky-gate field-effect transistor," *IEEE Trans. Electron Devices*, vol. ED-22, pp. 1023–1030, Nov. 1975.
[5] S. Ohkawa, K. Suyama, and H. Ishikawa, "Low noise GaAs field-effect transistors," *Fujitsu Scient. Tech. J.*, vol. 11, pp. 151–173, Mar. 1975.
[6] J. A. Turner and B. L. H. Wilson, "Implications of carrier velocity saturation in a gallium arsenide field-effect transistor," in *1968 Symp. on GaAs Proceedings*, pp. 195–204 (Inst. of Phys., Phys. Soc., London, 1969).
[7] P. L. Hower and N. G. Bechtel, "Current saturation and small-signal characteristics of GaAs field effect transistors," *IEEE Trans. Electron Devices*, vol. ED-20, pp. 213–220, 1973.
[8] C. A. Liechti, E. Gowen, and J. Cohen, "GaAs microwave Schottky gate FET," *1972 ISSCC Dig. Tech. Papers*, pp. 158–159.
[9] R. W. H. Engelmann, "Low noise Schottky barrier FET transistor (InP)," a) Hewlett-Packard Laboratories, Semiannu. Rep. ECOM-75-1300-1, Oct. 1975; b) Hewlett-Packard Laboratories, Final Rep. ECOM-75-1300-F, Apr. 1976.
[10] E. Wasserstron and J. McKenna, "The potential due to a charged metallic strip on a semiconductor surface," *Bell Syst. Tech. J.*, vol. 49, pp. 853–877, May/June 1970.
[11] M. Reiser, "A two dimensional numerical FET model for DC, AC and large signal analysis," *IEEE Trans. Electron Devices*, vol. ED-20, pp. 35–45, Jan. 1973.
[12] D. P. Kennedy and R. R. O'Brien, "Computer aided two-dimensional analysis of the junction field-effect transistor," *IBM J. Res. Develop.*, vol. 14, pp. 95–116, Mar. 1970.
[13] K. Lehovec and R. S. Miller, "Field distribution in junction field-effect transistors at large drain voltages," *IEEE Trans. Electron Devices*, vol. ED-22, pp. 273–281, May 1975.
[14] B. Himsworth, "A two-dimensional analysis of gallium arsenide junction field effect transistors with long and short channels," *Solid State Electron.*, vol. 15, pp. 1353–1361, 1972.
[15] R. W. Hockney, R. A. Warriner, and M. Reiser, "Two-dimensional particle models in semiconductor device analysis," *Electron. Lett.*, vol. 10, p. 484, Nov. 1974.
[16] J. G. Ruch, "Electron dynamics in short channel field effect transistors," *IEEE Trans. Electron Devices*, vol. ED-19, pp. 652–654, May 1972.
[17] T. J. Maloney and J. Frey, "Frequency limits of GaAs and InP field-effect transistors at 300K and 77K with typical active layer doping," *IEEE Trans. Electron Devices*, vol. ED-23, p. 519, May 1976.
[18] R. W. H. Engelmann and C. A. Liechti, "Gunn domain formation in the saturated-current region of GaAs MESFETs," *IEDM Tech. Dig.*, pp. 351–354, Dec. 1976.
[19] P. A. Houston and A. G. R. Evans, "Saturation velocity of electrons in GaAs," *IEEE Trans. Electron Devices*, vol. ED-23, pp. 584–586, June 1976.
[20] W. Fawcett and D. C. Herbert, "High field transport in gallium arsenide and indium phosphide," *J. Phys. C.: Solid State Phys.*, vol. 7, pp. 1641–1654, 1974.
[21] K. E. Drangeid and R. Sommerhalder, "Dynamic performance of Schottky-barrier field effect transistors," *IBM J. Res. Develop.*, vol. 14, pp. 82–94, Mar. 1970.
[22] P. L. Hower, W. W. Hooper, B. R. Cairns, R. D. Fairman, and D. A. Tremere, "The GaAs field-effect transistor," in *Semiconductors and Semimetals*, R. K. Willardson and A. C. Beer, Eds. New York: Academic Press, 1971, vol. 7, part A, pp. 147–200.
[23] P. L. Hower, W. W. Hooper, D. A. Tremere, W. Lehner, and C. A. Bittmann, "The Schottky barrier gallium arsenide field-effect transistor," in *1968 Symp. on GaAs Proceedings*, pp. 187–194 (Inst. Phys., Phys. Soc., London, 1969).
[24] B. R. Pruniaux, J. C. North, and A. V. Payer, "A semi-insulating gate gallium arsenide field effected transistor," *IEEE Trans. Electron Devices*, vol. ED-19, pp. 672–674, May 1972.
[25] J. G. Ruch and W. Fawcett, "Temperature dependence of the transport properties of gallium arsenide determined by a Monte Carlo method," *J. Appl. Phys.*, vol. 41, pp. 3843–3849, Aug. 1970.
[26] O. Mizuno and H. Watanabe, "Semi-insulating properties of Fe doped InP," *Electron Lett.*, vol. 11, pp 118–119, Mar. 6, 1975.
[27] D. R. Decker, R. D. Fairman, and C. K. Nishimoto, "Microwave InGaAs Schottky-barrier gate field effect transistors—Preliminary results," in *Proc. 5th Biennial Cornell Electrical Engineering Conf.*, pp. 305–314 (Cornell Univ., Ithaca, NY, 1975).
[28] E. Stoneham, Hewlett-Packard Company, private communication.

Noise Performance of Gallium Arsenide Field-Effect Transistors

ROBERT A. PUCEL, SENIOR MEMBER, IEEE, DANIEL J. MASSÉ, MEMBER, IEEE,
AND CHARLES F. KRUMM, MEMBER, IEEE

Abstract—The Schottky-barrier gate gallium arsenide field-effect transistor (GaAs FET) is the first three-terminal, solid-state amplifying device to have demonstrated low-noise performance at *X*-band and higher. For example, noise figures approaching 3 dB at 10 GHz have been reported, while theory predicts still lower values.

After a brief review of the noise-generating mechanisms intrinsic to the GaAs FET, an enumeration is given of the various parasitic elements associated with the FET which affect the noise performance. These elements include, among others, the gate metallization and source contact resistances, drain-gate feedback capacitance, and source lead inductance. Numerous graphs are presented to illustrate the effects of these elements and the various design parameters on the noise performance.

A comparison is made between the theoretically predicted and the measured noise performance of microwave GaAs FET's.

The best state-of-the-art noise performance as reported by various laboratories is illustrated graphically for single-stage and multistage FET amplifiers.

Finally, some speculation is attempted in regard to the possible reductions in noise figure to be expected from technological and design improvements of GaAs FET's.

I. INTRODUCTION

THE GALLIUM arsenide Schottky-barrier field-effect transistor (FET) is the first three-terminal solid-state device to exhibit linear power amplification at *X*-band frequencies and higher. Its unique signal-handling capabilities and low-noise properties have been demonstrated by many workers. For example, noise figures approaching 3 dB at 10 GHz have been reported, while theory predicts still lower values.

The GaAs FET is now being used in low-noise amplifiers from low *C*-band and up. As such it nicely supplements the silicon bipolar transistor which still dominates at frequencies below *C*-band. However, with the noise reductions now being achieved with buffered-layer FET's, this frequency range will not long remain the sole province of bipolars. Fig. 1 is a comparison of the state-of-the-art performance of low-noise, narrow-band amplifiers using silicon bipolar transistors and GaAs FET's as of July 1975.

Gallium arsenide field-effect transistors also show potential as low-noise microwave mixers and oscillators [1]–[3]. In this paper we shall restrict ourselves to their performance as small-signal amplifiers.

Manuscript received September 15, 1975. This work was based on an oral presentation given at the 5th Biennial Conference on Active Semiconductor Devices for Microwave and Integrated Optics held at Cornell University, Ithaca, NY, August 19–21, 1975.

The authors are with the Research Division, Raytheon Company, Waltham, MA 02154.

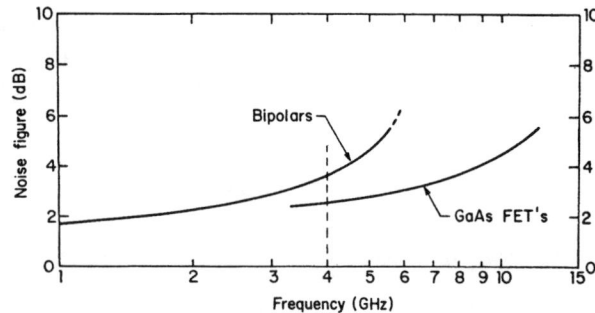

Fig. 1. Noise performance of cascaded (narrow-band) transistor amplifier stages as of July 1975.

As an introduction only a brief review of the present theory of noise of microwave GaAs FET's will be given, since a comprehensive description of the development of this theory has been given [4], [5]. Using this theory we shall assess the relative contributions to the noise performance by sources both intrinsic and extrinsic to the FET. With this as a background, we shall show how these contributions depend on the various material and design parameters at one's disposal. This will allow us to estimate the improvements in noise performance likely to be made in the future with advances in materials and device technology.

Finally, we will compare the theoretical predictions and measured results, and present a summary of the best noise performance obtained with FET devices and multistage amplifiers as of the writing of this paper.

II. SYNOPSIS OF THE NOISE THEORY OF THE GaAs FET

The basic principle of operation of the field-effect transistor was first described by Shockley [6] who assumed a constant mobility throughout the conducting channel region. Van der Ziel, in a series of classic papers, used Shockley's model to derive the small-signal parameters [7] and intrinsic noise properties of the FET [8], [9]. Van der Ziel showed that the intrinsic noise is thermal in origin, and can be represented by two white noise generators, one in the drain circuit, and one in the gate circuit. The gate noise generator, which represents the noise induced on the gate electrode by the passing thermal fluctuations in the drain current, is partially correlated with the drain noise generator.

The constant mobility model of Shockley and van der Ziel, though applicable to long-gate devices, does not apply to microwave devices whose gate lengths are in the micron range. For these devices, when biased in the current saturation

Reprinted from *IEEE J. Solid-State Circuits*, vol. SC-11, no. 2, pp. 243–255, April 1976.

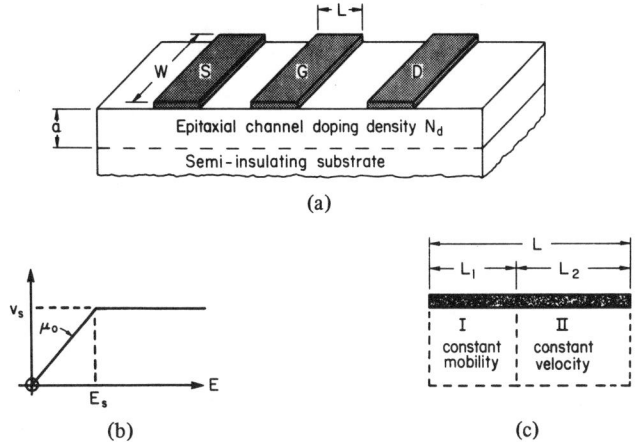

(a)

(b)   (c)

Fig. 2. Perspective sketch and two-section model of FET used in noise analysis. (a) FET model. (b) Assumed velocity-field characteristic. (c) Two-region model of channel.

regime, the average value of the longitudinal dc field in the channel is in the range where the mobility is a decreasing function of field, and indeed, where the carrier velocity is approaching a constant ("saturated") value. Consider, for example, a typical case of a GaAs FET with a 1 μm gate operating with a drain voltage of 3 V. The average longitudinal channel field is 30 kV/cm, approximately ten times the threshold value at which the velocity begins to saturate. Thus, the effects of velocity saturation must be included in any model of a GaAs FET designed for microwave operation.

Velocity saturation within the channel not only modifies the small-signal parameters, but the noise performance as well. Many workers have introduced some aspects of velocity saturation into their FET models, though none of these models include the diffusion noise introduced by electrons experiencing velocity saturation. In the noise and small-signal model developed at the authors' laboratory by Statz *et al.* and Pucel *et al.* [4], [5] this high-field diffusion noise is taken into account. It is the dominant intrinsic noise of microwave GaAs FET's.

A brief description of this model will be given now with the help of Fig. 2. Fig. 2(a) is a perspective sketch of a planar FET consisting of a source electrode (S), gate electrode (G), and drain electrode (D), all of width W. The gate length is denoted by L. The conducting n-type epitaxial channel of thickness a, situated on a semi-insulating substrate, is assumed to be uniformly doped at density N_d cm^{-3} with a low-field mobility μ_0. Typical values for these material parameters are $N_d \sim 10^{17}$ cm^{-3}, $a \sim 0.2 - 0.4$ μm, and $\mu_0 \sim 3000$–4500 cm^2/V · s.

Following Turner and Wilson [10], Statz *et al.* idealized the velocity-field characteristic by a piecewise linear approximation shown in Fig. 2(b). To obtain good agreement with experimental FET data, and reasonable agreement with experimental and theoretical velocity-field data [11], [12], the critical field E_s denoting the onset of velocity saturation was chosen to be 2.9 kV/cm, and the limiting velocity v_s to be 1.3×10^7 cm/s at room temperature.

This piecewise linear approximation to the velocity-field

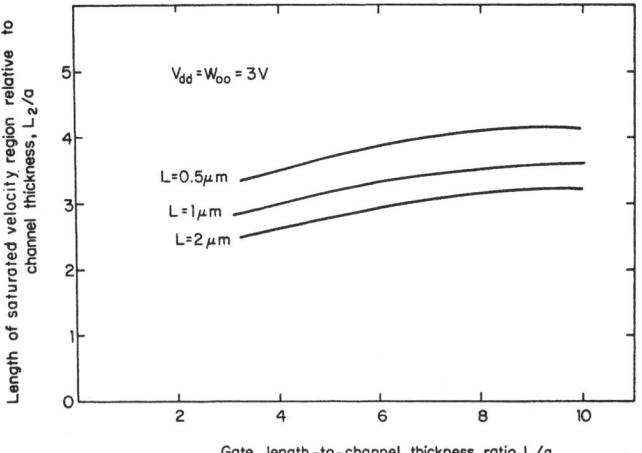

Fig. 3. Length of velocity-saturated zone relative to epi-layer thickness as a function of the gate length and the ratio of the gate length to epi-layer thickness.

characteristic allows one to divide the conducting channel underneath the gate region into two zones as Grebene and Ghandhi [13] have suggested. In this two-zone model, shown in Fig. 2(c), a portion of the channel near the source end is assumed to be in the constant mobility regime, while the remaining portion near the drain end is postulated to be in velocity saturation. The position of the boundary between these two zones, representing the onset of velocity saturation, is a strong function of the source-drain bias, but a weaker function of the gate-source bias. The length of the velocity-saturated zone increases monotonically with source-drain bias.

By a correct application of this model, it can be shown that when the FET is biased in current saturation, that is, above the knee of the drain-voltage-current characteristic, the length of the velocity-saturated zone L_2 is of the order of two to four times the epitaxial layer thickness a [5]. Fig. 3 shows how the length of the velocity-saturated zone, relative to the channel thickness, varies as a function of the geometric ratio L/a for various gate lengths. The drain voltage V_{dd} is assumed to be

59

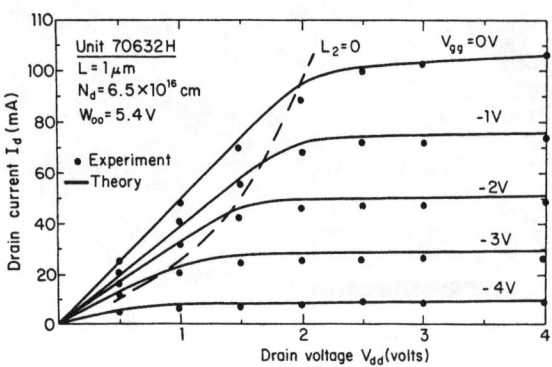

Fig. 4. Comparison between the theoretical and measured drain current-voltage characteristic for an X-band GaAs FET.

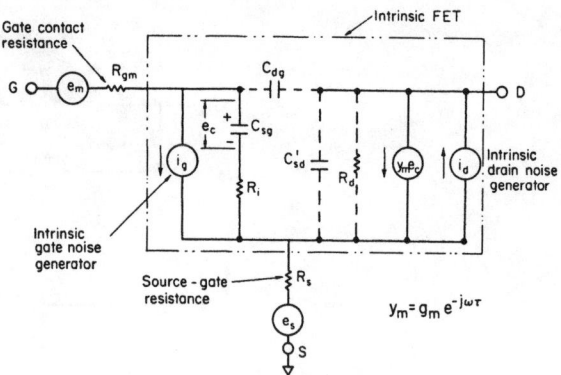

Fig. 6. Noise equivalent circuit of FET showing intrinsic and extrinsic noise sources.

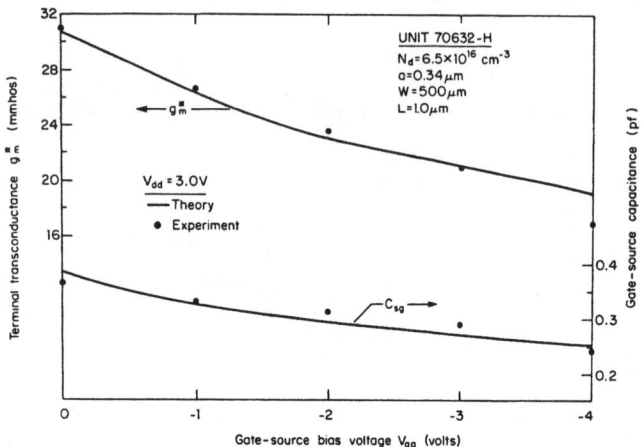

Fig. 5. Comparison between the theoretical and measured values of gate-source capacitance C_{sg} and terminal transconductance g_m^* for an X-band device.

equal to the intrinsic (internal) pinch-off voltage $W_{00} = 3$ V. With contemporary device designs using channel thicknesses of the order of 0.2–0.4 μm, the velocity-saturated zone comprises most of the channel length for gate lengths one micrometer or less. Thus, velocity saturation plays an important role in microwave GaAs FET's.

The piecewise linear approximation, chosen for analytic convenience, is an extreme idealization of the actual $v(E)$ characteristic in that it eliminates any negative resistance regime. In short-channel devices, the assumption of velocity saturation itself may be difficult to justify since the transit time of the electrons in the channel is comparable to the relaxation time of electrons, as Ruch [14] and later Maloney and Frey [15] have pointed out. Despite these recognized limitation, there *does* appear to be an appreciable degree of velocity saturation since this assumption works extremely well for the dc and small-signal characteristics. Fig. 4 demonstrates the agreement between the theoretical model and the measured I-V characteristic. Fig. 5 demonstrates this agreement for the small-signal terminal transconductance g_m^* and source-gate capacitance. The locus $L_2 = 0$ in Fig. 4 denotes the bias conditions for which velocity saturation first begins to manifest itself at the drain end of the gate. To the left of this line, that is, below the "knee" of the I-V characteristic, the channel

is entirely in the constant mobility mode of operation. Thus, in the current "saturation" regime, i.e., to the right of the locus, the channel is always in velocity saturation over a portion of its extent. The FET is normally operated in this current-saturated mode.

We shall show later that using the two-zone model for the noise analysis, the agreement between the predicted and measured noise performance of GaAs FET's is equally as good as it is for the dc and small-signal properties, as demonstrated by Figs. 4 and 5.

Statz *et al*. [4] assume that the noise in zone I is thermal, as in the van der Ziel treatment, but enhanced by hot electron effects as postulated by Baechtold [16], [17]. Zone II, however, cannot be treated as an ohmic conductor. Its noise contribution (which is new in FET theory) is dominant in microwave devices and must be represented as a high-field diffusion noise as Shockley *et al*. [18] and van der Ziel [19] have shown.[1] This diffusion noise is proportional to the high-field diffusion coefficient and is linearly dependent on drain current [4], [5]. On the other hand, the thermal noise of region I decreases with increasing drain current. Although the high-field diffusion noise is high, a strong correlation (approaching unity) exists between the drain noise and the induced gate noise. This correlation leads to a high degree of cancellation in the noise output of the GaAs FET.

Fig. 6 is a noise equivalent circuit of the FET, valid for high frequencies. The noise generator i_g represents the induced gate noise of the intrinsic device (shown in dotted lines). Its mean-square value varies as the square of the frequency, i.e., ω^2. The intrinsic drain noise generator i_d has a flat spectrum. The coupling between these noise generators, represented by the correlation coefficient C

$$jC = \frac{\overline{i_g^* i_d}}{\sqrt{|\overline{i_g^2}| \, |\overline{i_d^2}|}} \qquad (1)$$

[1]Actually, as van der Ziel [19] has shown, the noise of the constant mobility zone also can be represented as diffusion noise. Since the Einstein relation $D_0 = kT\mu_0/q$ holds in this zone, the diffusion noise expression can be transformed into the more familiar thermal or Johnson noise form. This transformation, of course, is invalid when velocity saturation occurs.

where (*) denotes the complex conjugate, and the overbar (⁻) represents a statistical average, approaches unity in magnitude for short-gate devices. (By comparison, in a constant mobility model, $|C| \sim 0.3$–0.4 [9].) In addition to the intrinsic noise sources, the parasitic source–gate resistance R_s and gate metallization resistance R_{gm} introduce thermal noise. This thermal noise is represented, respectively, by the generators labeled e_s and e_m. The resistance R_i represents the resistive charging path for the gate capacitance in the intrinsic FET. The noise associated with R_i is imbedded in the gate noise generator i_g [7].

It is not necessary to include all of the equivalent circuit parameters of the FET since some have a small effect on the noise figure. For example, for simplicity we shall neglect the (small) feedback drain-gate capacitance C_{dg} as well as the source-drain capacitance C_{sd}. The small perturbation of the noise figure produced by these capacitances can be added later if necessary. We may also neglect the small effect of the output drain resistance R_d, and any source lead inductance. We shall show later that inclusion of these parameters, for a well designed device, alters the minimum noise figure by at most a few tenths of a decibel. Thus, as a first approximation C_{dg}, C_{sd}, and R_d^{-1} will be assumed equal to zero. With these approximations, the equivalent circuit used in the noise figure derivation reduces to that shown in Fig. 7. This circuit also includes the signal source impedance Z_g and its associated thermal noise source e_g.

III. NOISE FIGURE

The configuration shown in Fig. 7 with the source terminal common to input and output is often called the grounded-source or common source connection. Although we shall present the expression for the noise figure for this circuit, our results should apply with negligible error to the common-gate and common-drain configurations [20], [21].

The noise figure F can be expressed as the ratio of the sum of the mean-square noise components in the short-circuited drain-source path produced by all of the noise sources in Fig. 7 to the mean-square thermal noise current component produced by the signal source e_g alone.

By a straightforward (but lengthy) circuit analysis the noise figure can be written in the form

$$F = 1 + \frac{1}{R_g}\left(r_n + g_n|Z_g + Z_c|^2\right) \quad (2)$$

where R_g is the real part of the source impedance (assumed to be at the reference temperature $T_0 = 290$ K). The parameters r_n and g_n are the so-called noise resistance and noise conductance, respectively, and Z_c the correlation impedance [22].

In terms of r_n, g_n, and Z_c all the noise properties of the FET with parasitic resistances are embodied in a very simple noisy network shown in Fig. 8, which precedes the FET (now considered noise-free). Thus, r_n represents a thermal noise voltage generator at the reference temperature; g_n, a shunt thermal noise current generator at the same temperature; and Z_c, an impedance at absolute zero (noiseless). The noise figure of

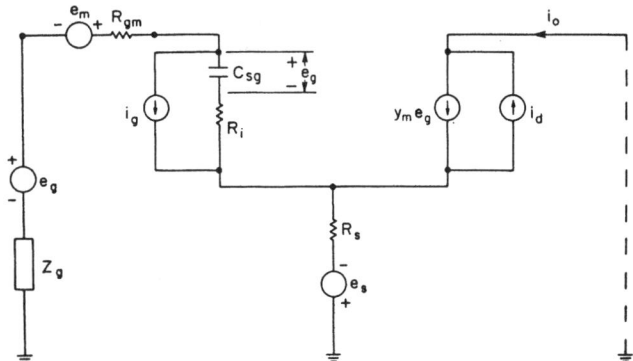

Fig. 7. Simplified equivalent circuit used in noise analysis of FET.

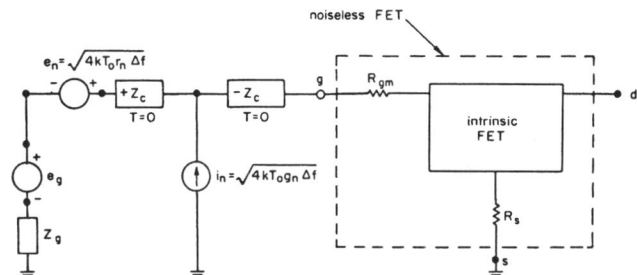

Fig. 8. Representation of noisy FET by a noiseless FET preceded by a noise network.

this combined network is the same as that of the original noisy FET, Fig. 7.

The noise functions are given by the simple expressions [5]

$$r_n = (R_s + R_{gm})\frac{T_d}{T_0} + K_r\left(\frac{1 + \omega^2 C_{sg}^2 R_i^2}{g_m}\right) \quad (3a)$$

$$g_n = K_g\frac{\omega^2 C_{sg}^2}{g_m} \quad (3b)$$

$$Z_c = R_s + R_{gm} + \frac{K_c}{Y_{11}} \quad (3c)$$

where T_d is the temperature of the FET. The parameters K_g, K_r, and K_c are numerical noise coefficients which represent the properties of the intrinsic noise generators i_g, i_d and their correlation (1). For an FET not at room temperature, these noise coefficients, as well as the small-signal parameters g_m, R_i, C_{sg}, the parasitic resistances R_s and R_{gm}, and the input impedance Y_{11}^{-1} of the intrinsic device, Fig. 7, given by

$$Y_{11}^{-1} = R_i + \frac{1}{j\omega C_{sg}} \quad (4)$$

are assumed to be evaluated at the device temperature T_d.

IV. MINIMUM NOISE FIGURE

The first stage of a low-noise amplifier chain is often designed to have a minimum noise figure. The noise figure is optimized by the proper choice of the source impedance $Z_g = R_g + jX_g$. This optimization can be achieved by a suitable lossless matching network between the signal source and the input (gate-source) terminals of the FET. It is easy to show that the minimum noise figure is achieved when the real and imaginary parts of the source impedance are equal to

$$R_g = R_{g0} = \sqrt{R_c^2 + \frac{r_n}{g_n}} \qquad (5a)$$

$$X_g = X_{g0} = -X_c \qquad (5b)$$

where R_c and X_c are the real and imaginary parts of the correlation impedance. The minimum value of F corresponding to this "noise match" can be expressed as

$$F_{\min} = 1 + 2g_n(R_c + R_{g0}). \qquad (6)$$

In decibels, $F_{\min}$ (dB) $= 10 \log_{10} F_{\min}$.

When the source is not optimized for best noise performance, the noise figure is given by

$$F = F_{\min} + \frac{g_n}{R_g} \{(R_g - R_{g0})^2 + (X_g - X_{g0})^2\}. \qquad (7)$$

We shall refer to this equation later when we discuss the experimental procedure for determining $F_{\min}$.

The expression for $F_{\min}$ given by (6) can be written to a very good approximation by the simple three-term power series expansion in frequency

$$F_{\min} = 1 + 2\left(\frac{\omega C_{sg}}{g_m}\right)\sqrt{K_g[K_r + g_m(R_s + R_{gm})]}$$

$$+ 2\left(\frac{\omega C_{sg}}{g_m}\right)^2 [K_g g_m(R_{gm} + R_s + K_c R_i)] + \cdots \qquad (8)$$

valid at room temperature. This simplified form delineates the roles played by the noise sources intrinsic to the device, embodied in the noise coefficients K_g, K_c, and K_r and the noise sources corresponding to the parasitic resistances R_s and R_{gm}.

The frequency dependence of $F_{\min}$ is a consequence of the ω^2 dependence of the induced gate noise. Note that $F_{\min}$ decreases with increasing gain-bandwidth factor g_m/C_{sg} of the FET. The gain-bandwidth factor is a function of gate bias, eventually decreasing as the gate bias approaches the pinch-off condition. In terms of g_m and C_{sg} individually, note that $F_{\min}$ increases with gate capacitance but decreases approximately in proportion to the inverse of the transconductance.

The noise coefficients are frequency-independent numerical factors which are gate-bias dependent, and to a lesser extent, drain-bias dependent. A typical bias dependence of these coefficients is shown in Fig. 9. Note that K_r is an order of magnitude lower than K_g and K_c. The bias dependence is expressed in terms of the drain current I_d normalized to its value I_{dss} at zero gate bias. All three noise coefficients depend in a complicated manner on gate length, channel thickness, and other parameters [5]. Observe that K_g is a strong function of the drain current, increasing at a rate faster than linear with I_d at high currents.

It is evident from (8) that $F_{\min}$ can be lowered by minimizing the parasitic resistances R_s and R_{gm}. As we shall see later, it also can be lowered by a proper choice of gate bias, or equivalently, drain current since the noise coefficients as well as the small signal parameters (mainly g_m) are bias-dependent.

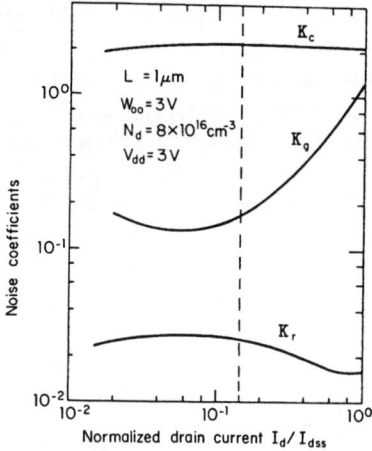

Fig. 9. Drain current dependence of noise coefficients of a GaAs FET for a specific set of design parameters and drain voltage.

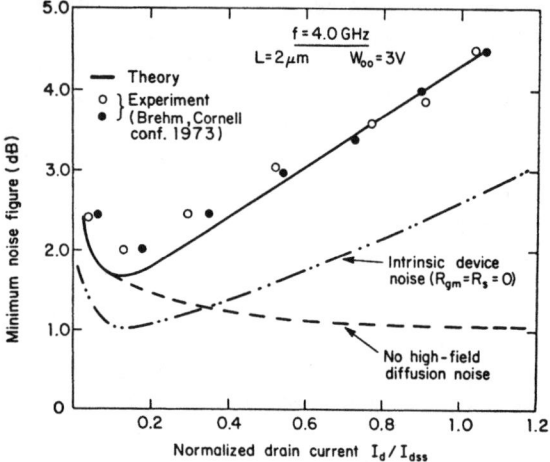

Fig. 10. Theoretical and measured noise figure for a GaAs FET with a 2 μm gate.

V. COMPARISON OF THEORY AND EXPERIMENT

The applications of the two-zone noise model to practical GaAs FET's will be exemplified now.

The solid line in Fig. 10 illustrates the validity of the noise model applied to a device with a 2 μm gate [23]. The nearly linear current dependence of $F_{\min}$ in the high current range demonstrates clearly the contribution of the high-field diffusion noise produced in the velocity-saturated zone. This fact is emphasized further by the lowest dashed line which represents $F_{\min}$ if this diffusion noise were set equal to zero. The increase in $F_{\min}$ at low currents is attributable to the decrease in g_m and the increase in the noise contribution of zone I as pinch-off is approached. The important role played by the parasitic resistances is illustrated by the middle dashed line representing the intrinsic noise obtained by setting R_{gm} and R_s equal to zero. Note that at the minimum the parasitic resistances contribute nearly half of the noise. Fig. 11 illustrates the agreement obtained with the noise data reported for a 1 μm gate device [24]. Again the general features of the bias dependence of $F_{\min}$ are reproduced. If we allow for the

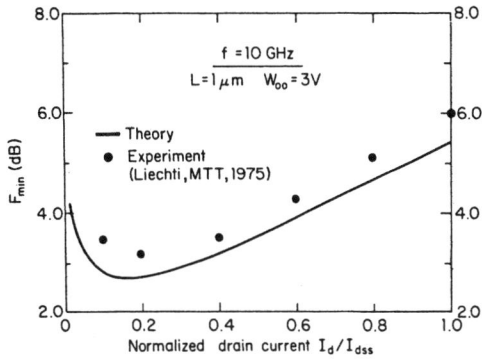

Fig. 11. Theoretical and measured noise figure for a GaAs FET with a 1 μm gate.

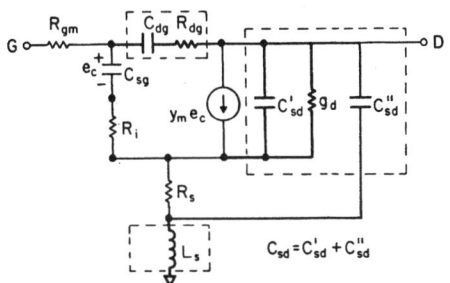

Fig. 12. Inclusion of neglected elements in equivalent circuit for noise figure analysis.

possible measurement uncertainty in the value of the noise figure, which may amount to as much as ±0.4–0.5 dB, the difficulty in accounting for the circuit losses accurately, and the errors introduced by use of the simplified equivalent circuit, the agreement between theory and experiment can be considered satisfactory.

VI. EFFECTS OF NEGLECTED PARASITICS

We shall assess now the effects of including the equivalent circuit elements neglected in the derivation of the noise figure in Section III. Fig. 12 illustrates these parasitics (delineated by dashed lines). They are, principally, the source-drain capacitance C_{sd}, drain conductance $g_d = R_d^{-1}$, source lead inductance L_s, and the drain-gate feedback admittance Y_{12} represented by a resistance R_{dg} in series with the drain-gate capacitance C_{dg}. This resistance (which is assumed to generate thermal noise) represents the resistive charging path for C_{dg} between the drain and gate terminals as suggested by Vendelin [25] and others. It is possible to include the effects of these parasitics in an exact manner; however, the resultant expression for the noise figure is unwieldy. Fortunately, for the small values of these neglected elements, typical of well-designed FET's, the perturbations to the minimum noise figure are linear functions of the element values, and can be added algebraically to the expression for F_{min}, (6).

Furthermore, since these perturbations are small in comparison to F_{min}, as we shall show, the corrections to F_{min}, expressed in decibel form, are also linear. If we denote the perturbation to F_{min} by ΔF, where the latter represents the inclusion of one of the neglected elements, or any combina-

tion of them, then the correction ΔF, expressed in decibels, is given by

$$\Delta F \text{ (dB)} = 10 \log_{10} \frac{F_{min} + \Delta F}{F_{min}} \tag{9a}$$

$$= 4.34 \ln \left(1 + \frac{\Delta F}{F_{min}} \right) \tag{9b}$$

$$\approx 4.34 \frac{\Delta F}{F_{min}} \tag{9c}$$

where the last equation arises from the assumption $|\Delta F| \ll F_{min}$. If this is not true (9a) must be used.

We shall now demonstrate the magnitude and sign of these corrections to F_{min} for the 1 μm gate device discussed earlier. The corrections as evaluated here apply for the bias conditions corresponding to the lowest value of F_{min}, namely 2.75 dB. See Fig. 11.

Unfortunately, we do not have the values of the neglected parasitic elements for the specific 1 μm device discussed earlier. However, since we are discussing small perturbations, we may use the element values obtained by Vendelin [25] for a similar device [26] by a computer optimized fit to the S-parameters. These are $C_{sd} = 0.16$ pF, $R_d = g_d^{-1} \approx 200\ \Omega$, $C_{dg} \approx 0.014$ pF, $R_{dg} \approx 660\ \Omega$, and $L_s \approx 26$ pH.

Consider first the inclusion of the drain-gate feedback, illustrated in Fig. 13(a) for a range of feedback admittances Y_{12}. As one might expect, the resistive feedback increases the noise figure. On the other hand, the capacitive component decreases it, in accordance with the findings of others [27]. For the specific values of the feedback elements, Re $Y_{12} = 0.38$ m℧, Im $Y_{12} = 0.66$ m℧, the corresponding corrections to the noise figure are $\Delta F = 0.45$ dB and $\Delta F = -0.12$ dB, or a net change of +0.33 dB.

We turn next to the inclusion of the output admittance consisting of g_d and C_{sd} in shunt, shown in dotted lines in Fig. 12. Since there is still some uncertainty amongst workers in the field as to the fraction of the source-drain capacitance which should be terminated at the upper end of R_s, as shown in Fig. 12, and the fraction that should tie to the lower end, we shall only consider the perturbation caused by inclusion of the drain output conductance. This is illustrated in Fig. 13(b). As is evident, the correction is negative. For the stated output resistance $g_d = 5$ m℧, $\Delta F = -0.24$ dB.

The effect of source lead inductance is of second-order importance. For values of this inductance in the range below 200 pH, the noise figure decreases slightly. This range exceeds by almost an order of magnitude the values of parasitic source lead inductance in a well designed device. For the specific value of inductance of the device under consideration, $\Delta F \approx -0.01$ dB.

Thus taken together, all of the neglected parasitics considered increase the noise figure by about 0.1 dB. This is a negligible error. Therefore, for a well designed device, use of the simplified model for noise analysis shown in Fig. 7 is justified.

It must be cautioned that the corrections implied by Fig. 13(a) and (b) apply only to the device considered in the text.

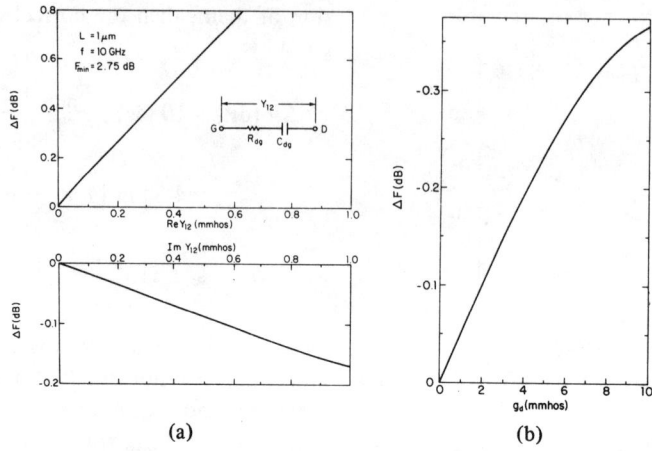

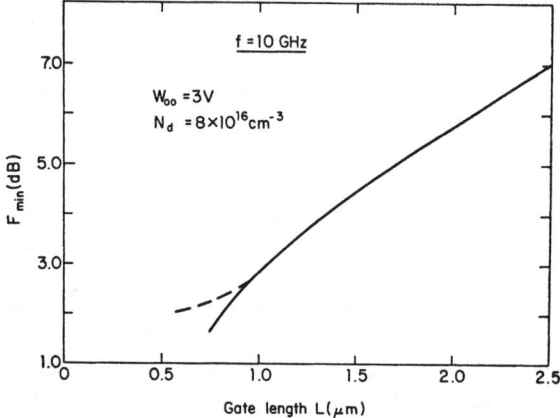

Fig. 13. Corrections to minimum noise figure attributable to neglected equivalent circuit elements. (a) Corrections to F_{min} at 10 GHz attributable to drain-gate feedback. (b) Corrections to F_{min} attributable to drain resistance.

Although corrections for other devices of similar design will be comparable, the quantitative values necessarily will be different.

It should be mentioned in passing that since the noise figure decreases with capacitive feedback, one might use this as a means for improving the noise performance of an FET amplifier by external feedback [27]. We do not believe this to be a satisfactory approach for several reasons. First, increasing feedback in this manner reduces the stability factor of the FET [28]. Second, the available gain decreases. Third, since it is the noise measure, rather than the noise figure that one should minimize, as we shall argue later, no improvement is achieved since noise measure does not change under capacitive feedback, or for that matter, for any lossless feedback scheme, as Haus has shown [29].

VII. THEORETICAL DEPENDENCE OF NOISE FIGURE ON DEVICE GEOMETRY AND PARASITIC RESISTANCES

We shall use the results of the noise theory described earlier to show how the noise sources intrinsic and extrinsic to the FET depend on the various material and geometrical parameters at the disposal of the device designer. With this as a background, we estimate the improvements likely to be made in the future with advances in materials and device technology. Specifically, we shall discuss the dependence of the noise performance on gate length, frequency, and extrinsic parasitic resistances.

We will limit ourselves, for convenience, to a specific design based on a channel doping density $N_d = 8 \times 10^{16}$ cm^{-3} and intrinsic pinch-off voltage $W_{00} = qN_d a^2/2\kappa\epsilon_0 = 3$ V typical of contemporary microwave devices where $\kappa = 12.5$ is the dielectric constant of GaAs. However, the general conclusions to be drawn will also apply to other microwave devices with similar, though not identical, design parameters.

Dependence of Minimum Noise Figure on Gate Length

The dependence of F_{min} on gate length is embodied in the source-gate capacitance and transconductance, and in a more complicated manner in the noise coefficients. The theoretical

Fig. 14. Theoretical minimum noise figure as a function of gate length.

value of F_{min} at $f = 10$ GHz, as a function of gate length, is illustrated in Fig. 14. Notice the rapid rate of decrease of F_{min} as the gate length approaches 1 μm. Gate length reduction is the single most productive means of improving the noise performance of an FET—up to a point! Although our theoretical curve extends down to $L = 0.5$ μm, we show an additional, arbitrarily drawn dashed line, since we believe the validity of our theory becomes questionable below $L = 1$ μm, for the channel thickness ($a \approx 0.225$ μm) corresponding to the assumed value of N_d and W_{00}.

Below $L = 0.5$ μm, there are other, more fundamental reasons why we believe that the rate of decrease in F_{min} will "flatten out" as implied by the dashed line.

First, as the gate length continues to decrease below a micron, the fringing capacitance of the gate (which does not decrease with gate length) [5] puts a lower asymptote on the gate capacitance; and hence on F_{min} [see (8)]. For example, for $L = 0.5$ μm, this fringing capacitance is over 30 percent of the gate capacitance.

Second, unless the channel thickness is reduced correspondingly, in accordance with the gate length, the electric field in the channel will begin to deviate markedly from a longitudinal field configuration, to one conforming more to a cylindrical

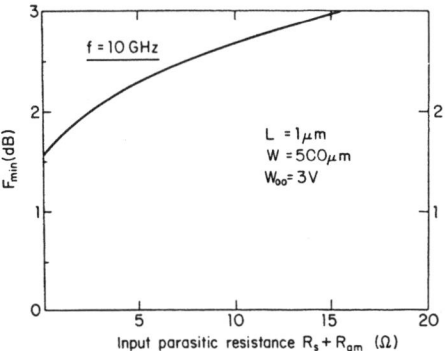

Fig. 15. Theoretical dependence of minimum noise figure on parasitic resistances for a GaAs FET with a 1 μm gate.

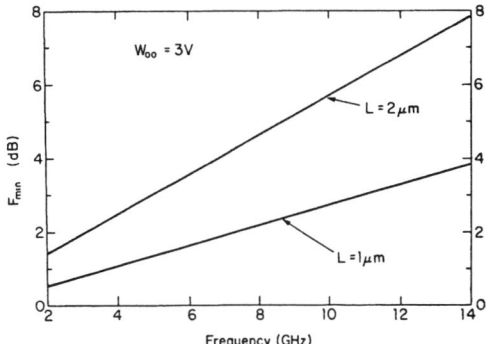

Fig. 16. Predicted frequency dependence of minimum noise figure of a GaAs FET for two values of gate length.

pattern about the gate electrode. The reduced longitudinal component of the electric field leads to a diminished control of the electron flow by the gate potential and to a "softer" drain current-voltage saturation characteristic [30], [32]. Although the noise performance and gain will still improve with decreasing gate length, the rate of improvement should decrease.

If the channel thickness is reduced in proportion to the gate length to reestablish a longitudinal field pattern, this requires use of epitaxial layers 0.1 μm thick, or less. Most of the channel doping profile, in this case, will not be constant, but will be decreasing rapidly toward the substrate. This means that the rate of decrease of transconductance with gate bias will be faster than it would be for an ideal (step) profile. Thus the upturn in F_{min} with decreasing drain current will occur at a higher value of drain current. Hence, the advantages of reducing both gate length and channel thickness, simultaneously, will be partially nullified. Although there appears to be some promise of growing epitaxial layers with a steeper transition zone, eventually one is limited to a lower value of the transition zone fixed by the Debye length [32].

There is another limitation imposed by thinner channel layers, namely, the increase in source gate resistance which must accompany a reduced epitaxial layer.

All of the above considerations must be taken into account in matching the possible advantages of reducing gate lengths much below a micrometer against the additional cost and complexity of producing submicron gate devices with acceptable yield.

Dependence of Minimum Noise Figure on Parasitic Gate and Source Resistance

The theoretical dependence of F_{min} on the parasitic resistances is shown in Fig. 15. Values of $R_s + R_{gm}$ typical of contemporary devices fall in the range from 8-11 Ω for a 500 μm wide gate device. A reduction of the order of perhaps 0.5 dB might be possible with improvements in the design and technology of contacts.

Dependence of Minimum Noise Figure on Frequency

The predicted frequency dependence of F_{min} is illustrated in Fig. 16 for two gate lengths. Note that the curves are nearly linear with frequency. This frequency dependence is exhibited

by experimental data also, as we shall demonstrate. The noise degradation with increasing frequency is attributable to the frequency dependence of the induced gate noise.

The theoretical noise figures displayed in the previous graphs were all computed for channel doping profiles exhibiting a slope near the epi-substrate interface—that is, for a nonrectangular profile, representative of present epitaxial layers. Some further reductions in the noise figure can be expected as "steeper" transition zones are achieved by improvements in the technology of epitaxial growth.

VIII. MEASUREMENT OF NOISE FIGURE

Introduction

In this section we shall discuss some of the practical problems in determining the minimum noise figure peculiar to the FET, and the means for overcoming these difficulties.

Earlier we had demonstrated the strong bias dependence of the minimum noise figure and had mentioned that the gain is also bias-dependent. We also pointed out that not only are the lowest noise figure and the maximum available gain achieved at different gate bias values, but that at a given bias condition, the matching conditions for the best noise figure and highest available gain also differ. This is one problem.

Next, the gain associated with the lowest value of the noise figure of present GaAs FET's is not high enough, at least at the upper end of the microwave band, to permit one to neglect in a noise measurement the correction for postamplifier or mixer noise. Thus it is a very tedious procedure to determine the minimum noise figure of an FET by simply varying the tuning adjustments of the input matching circuit because the correction also varies, nor is it a very precise method.

Equation (7) suggests a more direct approach. Note that this equation contains four unknowns, F_{min}, g_n, R_{g0}, and X_{g0}. Thus, at each bias one may, in principle, ascertain F_{min} and the remaining noise parameters by measuring the noise figure and gain for four selected source impedances.[2] However, unless these impedances are chosen judiciously, so that the resultant noise figures do not all cluster near F_{min}, or conversely, be all far removed from F_{min}, large errors can be

<hr>

[2]It is necessary to measure the gain in order to correct for the postamplifier and/or mixer noise.

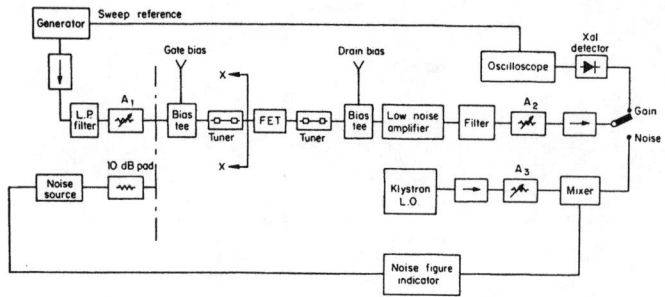

Fig. 17. Experimental set-up for measurement of the microwave gain and noise figure of an FET.

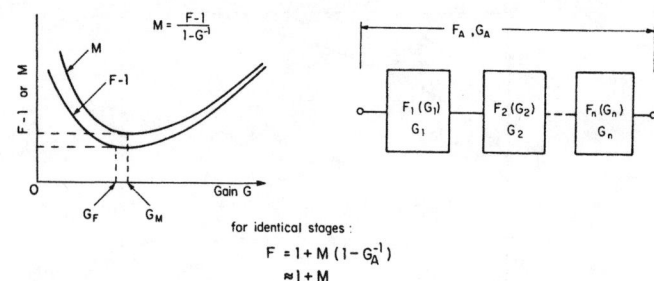

Fig. 18. Relevant to the design of a low-noise multistate, narrow-band FET amplifier.

introduced in the computation of F_{min}. To avoid this pitfall, it is advisable to use more than four measurements.

One such method, widely used [33] is based on seven chosen source impedances and seven noise figure and gain measurements. These seven values, and the corresponding noise figures, are used to obtain the best fit to the four unknown noise parameters (actually four others derived from these) in the minimum mean-square error sense. Naturally, more than seven source impedances may be used, but this lengthens the measurement procedure, and the point of diminishing returns is soon reached.

Noise Measurement Setup

A typical microwave setup to measure the gain and noise figure of FET's is shown in Fig. 17.

The FET is tuned with two coaxial double slug tuners which present very low loss (<0.2 dB) thus reducing the error correction in the noise figure measurement. The low-pass filter at the input eliminates errors in gain measurements due to harmonics; the narrow-band tunable filter (a high-Q cavity) at the output eliminates the image frequencies which would affect the noise measurement.

The gain is measured by a substitution method with a constant level maintained at the output of the crystal detector shown on the oscilloscope. The input level to the FET is adjusted with the attenuator A_1. With the FET and tuners removed, the attenuator A_2 is set at zero and a convenient level set on the oscilloscope. Then the FET is introduced in the circuit and A_2 adjusted to reestablish the original level. The gain is read directly on A_2. With this procedure one must be careful to adjust A_1 such that the saturation level of the low noise amplifier is never approached.

The low-noise amplifier is an essential part of the setup. It facilitates the tuning of the FET for minimum noise and reduces the postamplifier correction. If the noise figure of the postamplifier is F_2 and the noise figure and gain of the FET stage are F_1 and G, respectively, the measured noise figure at the input is given by

$$F = F_1 + \frac{F_2 - 1}{G}. \tag{10}$$

At high microwave frequencies, 10 GHz and higher, the gain G is generally low, less than 6 dB. If F_2 is large, the second term on the right of (10) can be comparable to F_1. In that case, by tuning the FET, one would more likely minimize F by maximizing G rather than minimizing F_1.

The noise figure is measured either with an automatic noise figure indicator such as AILTECH Model 75 or with a receiver and calibrated attenuator, by the so-called Y-factor method. The pad in front of the noise source is necessary only if the source impedance varies with its state (on or off). The attenuation must be taken into account in the noise calculations.

First, the noise figure F_2 of the postamplifier-mixer stages must be determined carefully. Then for a given bias condition of the FET, the output tuner is adjusted for maximum gain and the input tuner for minimum noise. The value of the noise figure measured is recorded together with its associated gain, and F_1 is calculated from (10). It can then be corrected if necessary for input circuit losses. The impedance seen by the FET input is measured (at the plane X–X on Fig. 17) with a network analyzer. This series of measurements is repeated seven times with the input tuner adjusted for slightly different positions each time.

The data are then processed by a computer to obtain F_{min}, g_n, and the optimum source impedance $Z_{g0} = R_{g0} + jX_{g0}$, as described earlier.

This procedure is long and tedious. It can be simplified if many measurements have to be made in the same frequency range. In that case, one can use a set of seven preadjusted tuners which are interchanged for each measurement.

IX. Design Considerations for Cascaded Amplifier

It was mentioned earlier that the lowest noise figure and the highest power gain do not occur at the same bias and tuning conditions. Since the gain at the minimum noise figure condition is not usually high enough to allow one to neglect the noise of the second and succeeding stages, one should not design the first stage of an FET amplifier to have its minimum noise figure if a minimum noise figure for the overall amplifier is to be achieved.

We shall illustrate why this is true with the help of Fig. 18. Shown is a block diagram of a cascaded stage amplifier, assumed to be narrow-band.[3] Since to each value of gain, G, there corresponds a noise figure, F, we have denoted this correspondence as $F(G)$. From the formula for the noise

---

[3]For wide-band amplifiers, other considerations enter besides noise figure.

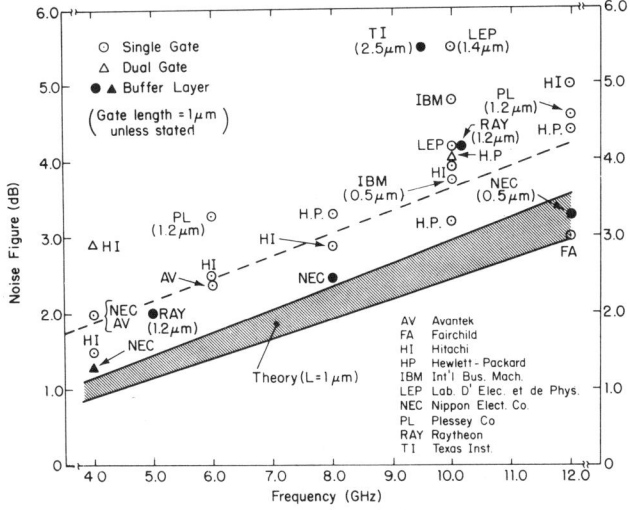

Fig. 19. Noise performance of GaAs FET's obtained at various laboratories (July 1975).

figure of cascaded stages, we find for the overall noise figure of the amplifier

$$F_A = F_1(G_1) + \frac{F_2(G_2) - 1}{G_1} + \frac{F_3(G_3) - 1}{G_1 G_2} +$$

$$\cdots + \frac{F_n(G_n) - 1}{G_1 G_2 G_3 \cdots G_{n-1}} \qquad (11)$$

where the subscript denotes the amplifier stage number. For convenience of our discussion we shall assume identical stages $G_1, G_2, G_3, \cdots, G_n = G; F_1, F_2, F_3, \cdots, F_n = F$. Then (11) becomes

$$F_A = 1 + (F - 1) \left(1 + \frac{1}{G} + \frac{1}{G^2} + \cdots \frac{1}{G^{n-1}} \right) \qquad (12a)$$

$$= 1 + M \left(1 - \frac{1}{G_A} \right) \qquad (12b)$$

where $G_A = G^n$ is the power gain of the amplifier, and $M = M(G)$ is the noise measure of each stage,

$$M(G) = \frac{F(G) - 1}{1 - \frac{1}{G}}. \qquad (13)$$

Equation (12b) is equivalent to the statement that the noise measure of n identical cascaded stages is equal to the noise measure of an individual stage [29].

It is evident that in cascade design, where the overall gain G_A is prescribed, one should minimize the noise measure rather than the noise figure of each stage to minimize the overall amplifier noise figure. When the overall gain is high, $F_A = 1 + M(G_A^{1/n})$. The sketches in Fig. 18 show qualitatively how the noise measure and noise figure vary as a function of stage gain. The minimum noise measure usually occurs at a slightly higher current and gain than the minimum noise figure. Also the value of the minimum noise measure exceeds the minimum value of the excess noise figure of a stage, i.e., $M_{min} > F_{min} - 1$. It follows that the lowest possible value of the amplifier noise figure is greater than the minimum noise

figure of any individual stage, that is $F_{A,min} > F_{min}$. This, of course, is what one would expect. However, when the gain per stage is of the order of 6 dB or more at the bias condition corresponding to minimum stage noise figure, then the difference between $F_{A,min}$ and F_{min} is small. For example, for the 1 μm gate device described by Fig. 11, the lowest measured value of F_{min} = 3.2 dB [24]. The computed value of M_{min} = 3.6 dB. Hence for a three-stage amplifier, with ≈ 7.5 dB gain per stage, the amplifier noise figure F_A = 3.63 dB, only 0.43 dB greater than the single-stage minimum noise figure.

X. SUMMARY OF NOISE PERFORMANCE OBTAINED AT VARIOUS LABORATORIES

We shall present now a compilation of the best noise performance reported by laboratories around the world as of the time of this writing (July 1975). First, the results for single-stage amplifiers (devices) will be given, then cascaded narrow-band amplifiers, and finally, wide-band amplifiers.

Fig. 19 is a graphical presentation of the device performance reported. All devices have 1 μm gates, except where noted. The circles refer to single-gate FET's, the triangles to dual-gate devices. Also shown (by the shaded strip) is the theoretical noise figure for a 1 μm gate for a spread of parasitic resistance values typical of present devices. Note that the buffered-layer device performance is within 0.5 dB of the theoretical.[4] Inclusion of circuit losses and corrections for neglected parasitics will reduce this gap.

Use of buffered layers not only improves the performance of single-gate devices, but also of dual-gate devices as the low noise figure for the NEC device at 4 GHz testifies.

The advantages of buffering are emphasized even more dramatically by the noise performance reported for cascaded narrow-band amplifiers (bandwidth < 20 percent) shown in

---

[4] A buffered-layer device is one that has an epitaxial growth of a high resistivity layer, of the order of 5–10 μm thick, over the substrate prior to channel epitaxial growth.

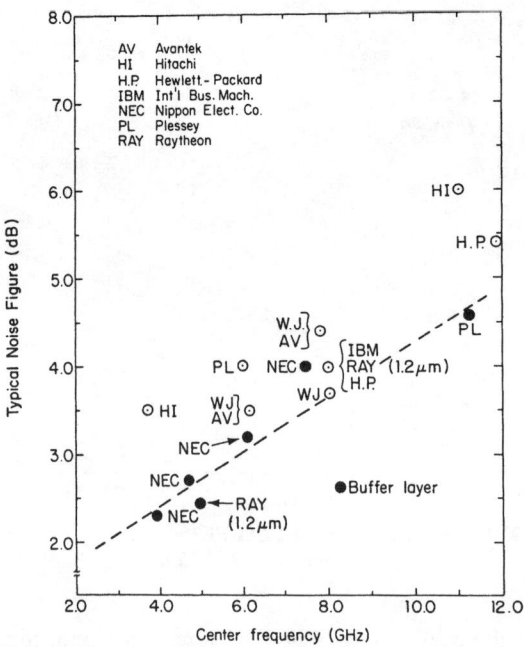

Fig. 20. Noise performance of narrow-band GaAs FET amplifiers as reported by various laboratories (July 1975).

Fig. 20. Notice in particular that the improvement is most pronounced in C-band and in the lower end of X-band.

Buffering appears to improve the noise performance in several ways. First, it covers or "shields" interface traps from the channel. (Present conjecture is that these traps, their nature unknown, may be ionized by the high channel fields and generate a noise spectra extending up to at least the low microwave band.) Second, with a buffer layer, the channel mobility near the substrate side increases substantially above the values with no buffer layer [34]. This not only increases the transconductance of the FET, but also decreases the source-gate parasitic resistance R_s. Reductions in R_s by nearly a factor of two are observed. These latter two improvements also lead to a higher power gain.

It seems reasonable to assume that the noise improvement in C-band and below can be attributed mainly to the reduction in trap noise. On the other hand, in the upper end of C-band and higher, where the trap noise would be expected to have diminished significantly, it is the increase in g_m and the reduction in R_s that is primarily responsible for the improvement in the noise performance. (Note the greater sensitivity of the noise figure to variation in parasitic resistance at the higher end of the frequency band exhibited by the theoretical shaded region in Fig. 19.)

It is interesting to note that the dashed lines through the experimental data in Figs. 19 and 20 both have approximately the same slope, namely 0.3–0.35 dB/GHz, as the theoretical lines in Fig. 19. However, the amplifier noise figures, on the average, exceed the single device values by approximately 0.5–0.6 dB.

Fig. 21 is a sampling of the noise performance reported for wide-band amplifiers. The upper and lower values of F_{min} in each case are not to be construed as the value of F_{min} at the band edges but merely the upper and lower values within the

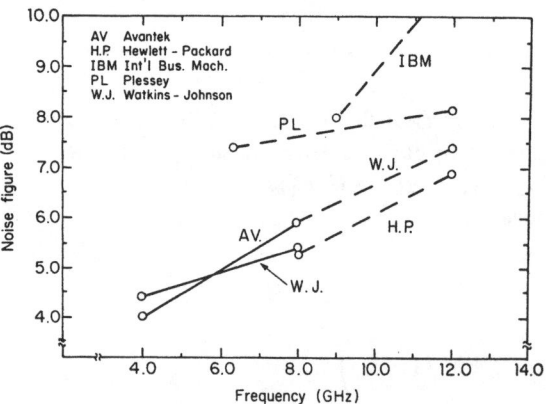

Fig. 21. Noise performance of some wide-band GaAs FET amplifiers (early 1975).

band. Since it is impossible to obtain a good noise match over a wide frequency range, the average noise figures are substantially higher than the narrow-band results.

XI. Conclusions

The measured noise figures of GaAs FET's with buffer layers are approaching the theoretically predicted values based on presently realizable channel doping profiles. With some advances in the design and technology of contacts and the achievement of steeper slopes in the doping profile at the substrate-channel interface, still further improvements in the noise performance should be possible.

It is believed that with the present planar device configuration, gate length reductions substantially below a micron will reach a point of diminishing returns. The reasons are 1) fringing gate capacitance, 2) slower rate of increase in transconductance, 3) increased series resistance of the channel layer, and

4) the need for an extremely narrow doping transition zone at the channel-substrate interface.

ACKNOWLEDGMENT

The authors wish to express their appreciation to Dr. J. Thompson and S. R. Steele who supplied the excellent epitaxial material, to J. Curtis who took the measurements, and to R. W. Bierig for his constant encouragement.

The authors also wish to convey a special note of gratitude to the various laboratories and individuals who were willing to share their best noise figure results. Without these, the last three graphs would not have been possible. Sincere apologies are extended to those laboratories which were inadvertently omitted in the survey.

REFERENCES

[1] R. A. Pucel, D. Massé, and R. Bera, "Integrated GaAs FET mixer performance at X-band," *Electron. Lett.*, vol. 11, pp. 199–200, May 1975.

[2] R. A. Pucel, R. Bera, and D. Massé, "Experiments on integrated gallium-arsenide FET oscillators at X-band," *Electron. Lett.*, vol. 11, pp. 219–220, May 1975.

[3] M. Maeda, K. Kimura, and H. Kodera, "Design and performance of *X*-band oscillators with GaAs Schottky-gate field-effect transistors," *IEEE Trans. Microwave Theory Tech.*, vol. MTT-23, pp. 661–666, Aug. 1975.

[4] H. Statz, H. A. Haus, and R. A. Pucel, "Noise characteristics of gallium arsenide field-effect transistors," *IEEE Trans. Electron Devices*, vol. ED-21, pp. 549–562, Sept. 1974.

[5] R. A. Pucel, H. A. Haus, and H. Statz, "Signal and noise properties of gallium arsenide field-effect transistors," in *Advances in Electronics and Electron Physics*, vol. 38. New York: Academic, 1975.

[6] W. Shockley, "A unipolar 'field-effect' transistor," *Proc. IRE*, vol. 40, pp. 1365–1376, Nov. 1952.

[7] A. van der Ziel and J. W. Ero, "Small signal, high-frequency theory of field-effect transistors," *IEEE Trans. Electron Devices*, vol. ED-11, pp. 128–135, Apr. 1964.

[8] A. van der Ziel, "Thermal noise in field-effect transistors," *Proc. IRE*, vol. 50, pp. 1808–1812, Aug. 1962.

[9] ——, "Gate noise in field-effect transistors at moderately high frequencies," *Proc. IEEE*, vol. 51, pp. 461–467, Mar. 1963.

[10] J. A. Turner and B. L. H. Wilson, "Implications of carrier velocity saturation in a gallium arsenide field effect transistor," in *Proc. 2nd Intl. Symp. Gallium Arsenide*, 1968, pp. 195–204.

[11] J. G. Ruch and G. S. Kino, "Transport properties of gallium arsenide," *Phys. Rev.*, vol. 174, pp. 921–931, Oct. 1968.

[12] J. G. Ruch and W. Fawcett, "Temperature dependence of the transport properties of gallium arsenide determined by a Monte Carlo method," *J. Appl. Phys.*, vol. 41, pp. 3843–3849, Aug. 1970.

[13] A. B. Grebene and S. K. Ghandhi, "General theory for pinched operation of the junction gate FET," *Solid-State Electron.*, vol. 12, pp. 573–589, July 1969.

[14] J. G. Ruch, "Electron dynamics in short channel field-effect transistors," *IEEE Trans. Electron Devices*, vol. ED-19, pp. 652–654, May 1972.

[15] T. J. Maloney and J. Frey, "Effects of nonequilibrium velocity-field characteristics on the performance of GaAs and InP field-effect transistors," in *1974 Dig. Tech. Papers, Int. Electron Devices Meeting*, pp. 296–298.

[16] W. Baechtold, "Noise behavior of Schottky barrier gate field-effect transistors at microwave frequencies," *IEEE Trans. Electron Devices*, vol. ED-18, pp. 97–104, Feb. 1971.

[17] W. Baechtold, "Noise behavior of GaAs field-effect transistors with short gate lengths," *IEEE Trans. Electron Devices*, vol. ED-19, pp. 674–680, May 1972.

[18] W. Shockley, J. A. Copeland, and R. P. James, "The impedance field method of noise calculations in active semiconductor devices," in *Quantum Theory of Atoms, Molecules, and the Solid State*, P.-O. Löwdin, Ed. New York: Academic, 1966.

[19] A. van der Ziel, "Thermal noise in the hot electron regime in FET's," *IEEE Trans. Electron Devices* (Corresp.), vol. ED-18, p. 977, Oct. 1971.

[20] ——, "Equivalence of the noise figures of common source and common gate FET circuits," *Electron. Lett.*, vol. 5, pp. 161–162, Apr. 1969.

[21] R. D. Kässer, "Noise factor contours for field-effect transistors at moderately high frequencies," *IEEE Trans. Electron Devices*, vol. ED-19, pp. 164–171, Feb. 1972.

[22] H. Rothe and W. Dahlke, "Theory of noisy fourpoles," *Proc. IRE*, vol. 44, pp. 811–818, June 1956.

[23] G. E. Brehm, "Variation of microwave gain and noise figure with bias for GaAs FET's," in *Proc. 4th Biennial Cornell Elect. Eng. Conf.*, 1973, pp. 77–85.

[24] C. A. Liechti, "Performance of dual-gate GaAs MESFET's as gain-controlled low-noise amplifiers and high-speed modulators," *IEEE Trans. Microwave Theory and Tech.*, vol. MTT-23, pp. 461–469, June 1975.

[25] G. D. Vendelin, "Circuit model for the GaAs M.E.S.F.E.T. valid to 12 GHz," *Electron. Lett.*, vol. 11, pp. 60–61, Feb. 1975.

[26] C. A. Liechti and R. L. Tillman, "Design and performance of microwave amplifiers with GaAs Schottky-gate field-effect transistors," *IEEE Trans. Microwave Theory Tech.*, vol. MTT-22, pp. 510–517, May 1974.

[27] G. D. Vendelin, "Feedback effects on the noise performance of GaAs FET's," in *1975 Dig. Tech. Papers, IEEE S-MTT Int. Microwave Symp.*, pp. 324–326.

[28] P. Wolf, "Microwave properties of Schottky barrier field effect transistors," *IBM J. Res. Develop.*, vol. 14, pp. 125–141, Mar. 1970.

[29] H. A. Haus and R. B. Adler, *Circuit Theory of Linear Noisy Networks*. New York: Wiley, 1959.

[30] J. R. Hauser, "Characteristics of junction field effect devices with small channel length-to-width ratios," *Solid-State Electron.*, vol. 10, pp. 577–587, June 1967.

[31] T. L. Chiu and H. N. Ghosh, "Characteristics of the junction-gate field effect transistor with short channel length," *Solid-State Electron.*, vol. 14, pp. 1307–1317, Dec. 1971.

[32] C. P. Wu, E. C. Douglas, and C. W. Meuller, "Limitations of the CV technique for ion-implanted profiles," *IEEE Trans. Electron Devices*, vol. ED-22, pp. 319–329, June 1975.

[33] R. Q. Lane, "The determination of device noise parameters," *Proc. IEEE* (Lett.), vol. 57, pp. 1461–1462, Aug. 1969.

[34] T. Nozaki, M. Ogawa, H. Terao, and H. Watanabe, "Multi-layer epitaxial technology for the Schottky barrier GaAs field-effect transistor," in *Proc. 5th Intl. Symp. Gallium Arsenide and Related Compounds*, 1975, pp. 46–54.

Current Saturation and Small-Signal Characteristics of GaAs Field-Effect Transistors

PHILIP L. HOWER AND N. GEORGE BECHTEL

Abstract—The model previously proposed by Turner and Wilson is developed in detail and compared with experiment. Deviations from Shockley's classical theory can be accounted for in terms of a single quantity Γ, which is related to E_m the peak field for GaAs. A discussion of the physical mechanism of current saturation shows that the formation of domains within the channel is hampered in a conventional GaAs FET.

A y-parameter analysis is presented that permits calculation of transconductance and the unity current gain frequency f_T. Measurements of drain current, transconductance, and f_T versus gate voltage all show good agreement with values predicted by the theory. Estimates are given which show that the current saturation mechanism described will be important in the design of GaAs microwave FET's.

LIST OF SYMBOLS NOT DEFINED IN THE TEXT

a Metallurgical film thickness.

$G_0 = q\mu_0 a N_0 Z/L$. Open channel conductance.

$I_0 = G_0 U_0/3$. Normalizing drain current, see (2).

L Length of the gradual portion of the channel.

L_g Metallurgical gate length.

N_0 Channel donor concentration.

$U(x)$ Total electrostatic potential difference, gate to channel.

$U_0 = qN_0 a^2/2\epsilon$. Total electrostatic potential required to fully deplete the channel.

V_D, V_G, V_S Terminal voltages at the intrinsic drain, gate, and source, respectively.

x Space variable, in the direction of current flow.

Z Transverse gate dimension.

μ_0 Low-field mobility.

I. INTRODUCTION

CONSISTENT progress in the technological development of the Schottky barrier GaAs FET has demonstrated that this device holds the promise of being a useful microwave amplifier. To properly exploit this potential, it is necessary to have a device design theory which permits the calculation of both low- and high-frequency terminal characteristics of the FET.

It is generally recognized that any useful analysis of the GaAs FET must properly account for the drift

Manuscript received January 28, 1971; revised August 29, 1972. This research was supported in part by the Air Force Avionics Laboratory, Wright-Patterson Air Force Base, Ohio, under Contract F33615-69-C-1093.

P. L. Hower is with the Research and Development Center, Westinghouse Electric Corporation, Pittsburgh, Pa. 15235.

N. G. Bechtel is with the Microwave and Optoelectronic Division, Fairchild Camera and Instrument Corporation, Palo Alto, Calif. 94304.

velocity versus electric field characteristic of this material. Typically the gate length of a microwave FET is in the 2–5 μm range. As a result, the electric field in the channel can reach E_m ($=3.2$ kV/cm), the value for which the electron drift velocity peaks, at relatively small values of drain–source voltage. Thus any direct application of "classical" or low-field FET analyses will lead to overoptimistic frequency performance predictions.

In this paper the modification of the gradual channel solution proposed by Turner and Wilson [1] is developed in detail and compared with experimental results. Section II is devoted to a discussion of current saturation in the GaAs FET and a calculation of the drain characteristic. In Section III, an approximate small-signal analysis is used to calculate the y-parameters and the current-gain cutoff frequency f_T. Section IV contains a comparison of theoretical and experimental results, with proper accounting for the effects of extrinsic source and drain resistance.

II. CALCULATION OF THE DRAIN CHARACTERISTIC

A. Review

The two analyses of the GaAs FET that are presently available both use Shockley's gradual channel approximation [2], but differ in the approximation used for electron drift velocity v_{dr}. In the Lehovec and Zuleeg [3] analysis v_{dr} is approximated by

$$v_{dr} = \frac{\mu_0 E}{1 + \mu_0 E/v_m} \qquad (1)$$

which predicts that v_{dr} saturates at a value v_m. In the Turner and Wilson analysis, a piecewise linear approximation is used for v_{dr}, which provides a more accurate fit to both measured [4] and predicted [5] v_{dr} data up to fields in the neighborhood of E_m.

In [3], μ_0 is measured and v_m is set equal to 1.5×10^7 cm/s. An iterative procedure is then used to adjust the channel doping so that agreement is obtained between theoretical and experimental g_m versus V_G data. This procedure is restricted to the case where the extrinsic source resistance can be neglected, that is, $R_s \cong 0$.

In [1], saturation is assumed to occur when the drain field reaches E_m, and the saturation current is calculated by imposing the boundary condition that the field at the drain is equal to E_m. There are no adjustable parameters in this analysis and, as we show in Section IV, the effect of R_s can be accounted for quite easily. This solution

Reprinted from *IEEE Trans. Electron Devices*, vol. ED-20, no. 3, pp. 213–220, March 1973.

yields an abrupt saturation and, because of the "rounding" of the v_{dr} characteristic in the vicinity of E_m, it cannot be expected to give accurate results just prior to the saturation point.

It should also be noted that neither approximation accounts for the possibility of electron accumulation near the drain. Accumulation has been predicted for silicon FET's [6] and most likely a similar conclusion would be reached for a GaAs FET. In this case, the presence of additional carriers suggests that the observed drain current should exceed the predicted value, and indeed this is the case, as we describe further in Section IV.

The "excess" current is typically 10 to 15 percent of the predicted value. This relatively small value of excess current is consistent with the view that any large buildup of electron charge within the gradual channel is prevented by the channel electric field, which is in a direction to push negative charge toward the drain contact. For this reason the formation of domains within the channel is unlikely.

We have elected to use the piecewise linear model of Turner and Wilson for the velocity–field characteristic for GaAs, since it closely approximates measured v_{dr} data and, in addition, permits prediction of the drain characteristic without recourse to experimental curve fitting.

B. Drain Current and Saturation Voltage

Below saturation, the gradual channel solution for drain current can be written as

$$I_D = I_0[F(\zeta) - F(\eta)] \tag{2}$$

where F has the functional form

$$F(X) = 3X - 2X^{3/2} \tag{3}$$

and X is a dummy variable. In (2) the quantities η and ζ denote the normalized electrostatic potential difference at the source and drain, respectively. That is,

$$\eta = U_S/U_0 = (V_S - V_G + V_B)/U_0 \tag{4}$$

$$\zeta = U_D/U_0 = (V_D - V_G + V_B)/U_0 \tag{5}$$

where U_0 includes the built-in voltage V_B and is the total electrostatic potential necessary to pinch off the channel. Thus for an FET operating under normal bias conditions, we can say

$$V_B/U_0 \le \eta \le \zeta < 1.$$

The condition for current saturation is given by [1], $dU/dx = E_m$ at $x = L$. This boundary condition on electric field defines an additional equation which can be written as

$$\frac{I_{D \text{ sat}}}{I_0} = \frac{3}{\Gamma}(1 - \sqrt{\zeta_{\text{sat}}}) \tag{6}$$

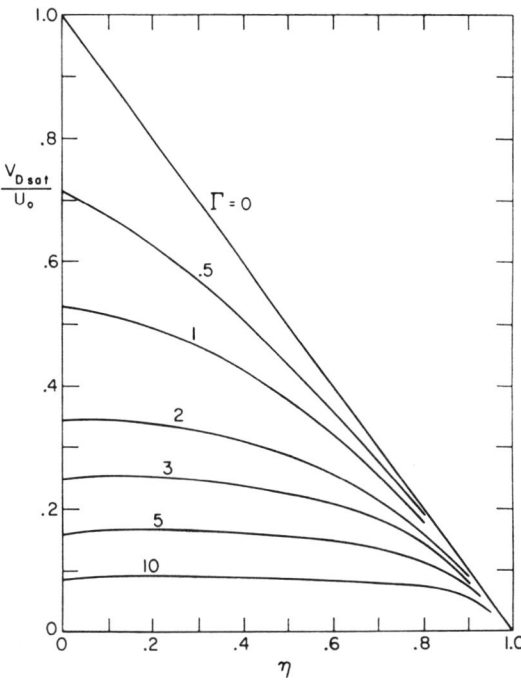

Fig. 1. Normalized saturation voltage versus normalized gate bias η and the field ratio Γ.

where

$$\Gamma = \frac{U_0}{E_m L}. \tag{7}$$

The quantity Γ is a ratio of a "channel field"[1] U_0/L to the field E_m, at which carriers achieve the maximum velocity and is a measure of the degree to which "hot" electron effects dominate the drain characteristic of an FET.[2]

Of course, ζ_{sat} must also satisfy (2). That is, both

$$\frac{I_{D \text{ sat}}}{I_0} = F(\zeta_{\text{sat}}) - F(\eta) \tag{8}$$

and (6) must be true at saturation. For $V_{DS} > V_{D \text{ sat}}$, $I_{D \text{ sat}}$ is then determined only by the value of η. In (6) and (8), ζ_{sat} may then be viewed as an auxiliary variable that links the two equations. By taking the difference of these two equations a cubic in $\zeta_{\text{sat}}^{1/2}$ results. The normalized saturation voltage $V_{D \text{ sat}}/U_0$, which is given by (9), is plotted in Fig. 1 as a function of Γ and η:

$$V_{D \text{ sat}}/U_0 = \zeta_{\text{sat}} - \eta. \tag{9}$$

$I_{D \text{ sat}}/I_0$ is plotted in the same way in Fig. 2. As these figures show, FET's having a channel field U_0/L comparable to E_m will show a significant reduction in both $I_{D \text{ sat}}$ and $V_{D \text{ sat}}$ from the classical ($\Gamma = 0$) case.

Comparison of the results shown in Figs. 1 and 2 with those of [3] indicate that the piecewise linear model

---

[1] U_0/L is equal to the average field over the length of the gradual portion of the channel, when $\eta = \Gamma = 0$.
[2] The field ratio Γ is equivalent to the parameter Z in [3].

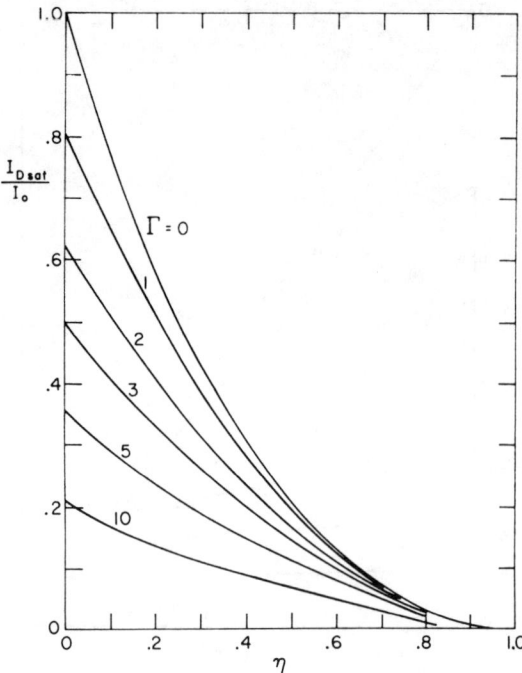

Fig. 2.   Normalized drain current versus η and Γ.

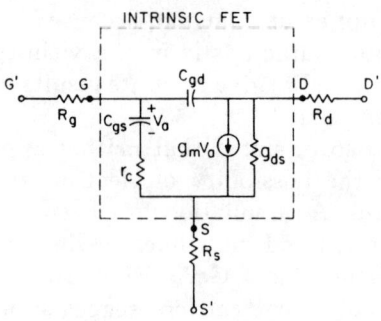

Fig. 3.   Circuit model showing extrinsic resistances R_s, R_d, and R_g.

predicts a larger drain current and a smaller saturation voltage. This result is expected since the "effective mobility," introduced into the gradual channel solution by (1), decreases as the electric field increases, while the piecewise linear model assumes constant mobility for fields up to E_m.

III. Small-Signal Analysis

Common Source y Parameters

A lumped element circuit model [7] that is frequently used to describe the small-signal behavior of the FET is shown in Fig. 3. In the present section we treat only the intrinsic FET and postpone to Section IV inclusion of the extrinsic resistances R_s and R_d. The series combination C_{gs} and r_c is used to approximate the distributed gate–channel network. We have found that this model is adequate up to frequencies in the neighborhood of f_T, the frequency where the current gain of the device reaches unity.

To calculate the network elements g_m, C_{gs}, and r_c we use a modification of the analysis of Hauser [8], which is given in the Appendix. From that analysis we can write y_{21} as

$$y_{21} = \frac{i_d}{v_g} = \frac{g_{m\ sat}}{1 + j\omega/\omega_a} \qquad (10)$$

where we define

$$g_{m\ sat} = \frac{G_s}{1 + 2\Gamma\sqrt{\zeta_{sat}}(1 - \sqrt{\zeta_{sat}})} \qquad (11)$$

$$G_s = G_0(1 - \sqrt{\eta})$$

$$\frac{1}{\omega_a} = \left(\frac{1}{\omega_0} + \frac{\alpha}{\omega_1}\right)\Big/(1 + \alpha) \qquad (12)$$

$$\alpha = 2\Gamma\sqrt{\zeta_{sat}}(1 - \sqrt{\zeta_{sat}}). \qquad (13)$$

(The frequencies ω_0 and ω_1, and also ω_3, which is used in the following, are defined in the Appendix.) The frequency ω_a is the "cutoff frequency" for y_{21} and is equal to ω_0 for the classical case ($\Gamma = 0$). Normalized curves of $g_{m\ sat}$ and ω_a are plotted in Figs. 4 and 5.

If the analysis in the Appendix is strictly interpreted one can show that

$$i_g = v_g G_s \left[\frac{j\omega/\omega_3}{1 + j\omega/\omega_0} + \frac{\alpha}{1 + \alpha}\right.$$
$$\left. \cdot \frac{j\omega/\omega_1}{(1 + j\omega/\omega_a)(1 + j\omega/\omega_0)}\right] \qquad (14)$$

which suggests that a more complicated network is needed in place of the series combination r_c and C_{gs} of Fig. 3. However, it must be remembered that the approximate analysis [8] neglects second-order terms in ω. This means that only one of the two "single-pole" terms in (14) should be retained. As ω is increased, the term involving ω_a will become dominant before the ω_0 term, since $\omega_a < \omega_0$ (we are assuming $\Gamma > 0$). Therefore we retain only the ω_a term (which is equivalent to assuming $\omega_0 = \infty$) and approximate y_{11} as

$$y_{11} = \frac{i_g}{v_g} \simeq \frac{j\omega C_{gs}}{1 + j\omega/\omega_a} \qquad (15)$$

which is the admittance of a series RC network. In (15) C_{gs} is the low-frequency gate capacitance, which is given by

$$C_{gs} = G_s\left(\frac{1}{\omega_3} + \frac{\alpha}{1 + \alpha}\frac{1}{\omega_1}\right). \qquad (16)$$

From (15) the series channel resistance r_c is then

$$r_c = 1/\omega_a C_{gs}. \qquad (17)$$

In terms of the h and y network parameters, the current gain is given by

$$h_{21} = y_{21}/y_{11}. \qquad (18)$$

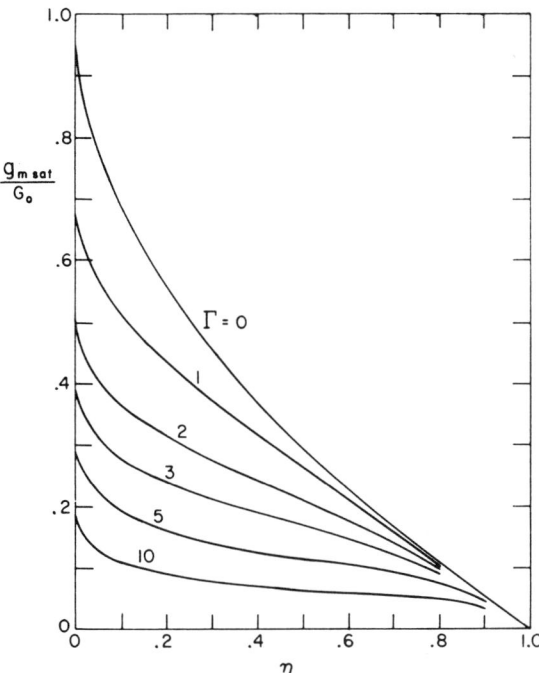

Fig. 4. Normalized saturation transconductance versus η and Γ.

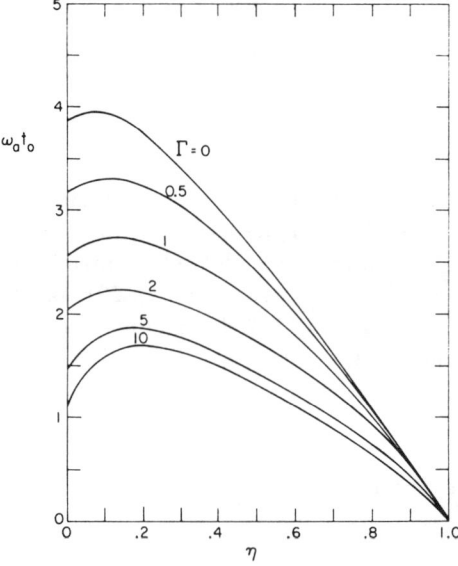

Fig. 5. Normalized y_{21} cutoff frequency versus η and Γ.

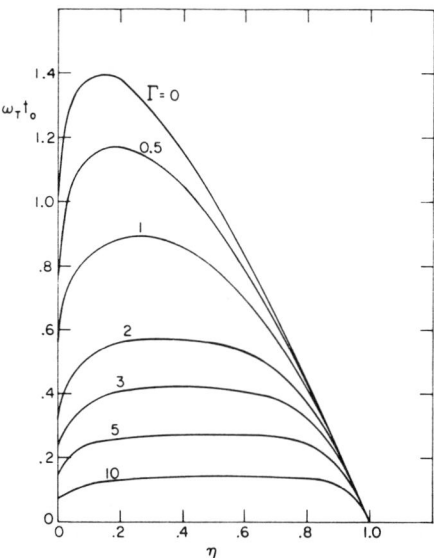

Fig. 6. Normalized unity current gain frequency versus η and Γ.

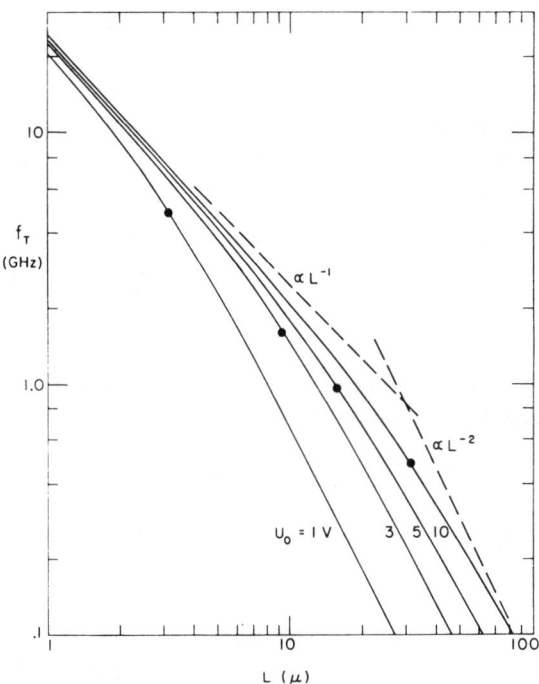

Fig. 7. f_T versus channel length L for various values of the total pinchoff voltage U_0. The dots mark the point where $\Gamma = 1$. ($\eta = 0.2$, $\mu_0 = 5 \times 10^3$ cm²/V·s, $E_m = 3.2$ kV/cm.)

For an FET operating in saturation, $C_{gs} \gg C_{gd}$, and the common-source current gain is then

$$h_{21} \cong \frac{g_{m\,sat}}{\omega C_{gs}} \qquad (19)$$

and the unity gain frequency is

$$f_T = \frac{\omega_T}{2\pi} = \frac{g_{m\,sat}}{2\pi C_{gs}} = \frac{1}{2\pi}\left[\frac{\omega_3 \omega_1}{\alpha\omega_3 + (1+\alpha)\omega_1}\right]. \qquad (20)$$

The normalized radian frequency $\omega_T t_0$ is plotted in Fig. 6.

The product of the channel resistance and transconductance, which is frequently used in FET analysis, is given by

$$r_c g_{m\,sat} = \omega_T/\omega_a. \qquad (21)$$

For the classical FET, r_c is approximately equal to 1 ($3g_{m\,sat}$) over most of the saturated portion of the drain characteristic. As Γ increases, r_c decreases to an even smaller fraction of $1/g_{m\,sat}$. This is to be expected since the channel becomes less tapered as Γ is increased, and consequently, the lumped "charging" resistance r_c is made up of a channel having a larger cross section. Thus r_c will decrease as Γ is increased.

In Fig. 7, we have plotted f_T versus L with U_0 as a parameter, keeping the gate bias fixed. This plot shows that for large enough L, that is, small Γ, f_T is nearly proportional to $1/L^2$, the classical result. As L is decreased, f_T becomes more nearly proportional to $1/L$ and f_T increases less rapidly than predicted by the classical solution.

Present device technology permits the fabrication of FET's with L_g in the mid 10^{-4}-cm range with U_0 values of a few volts. From Fig. 7, it is clear that these devices will lie in the range where f_T is nearly proportional to $1/L$.

IV. COMPARISON BETWEEN THEORY AND EXPERIMENT

In this section, the results derived in the preceding sections are compared with device measurements. For these FET's, evaporated aluminum is used as the Schottky barrier and the ohmic contacts at source and drain consist of a gold–germanium alloy. The channel material is n-type GaAs grown by vapor epitaxy on a semi-insulating substrate. The actual details of device fabrication have been given elsewhere [9] and will be omitted here.

The gate–source and gate–drain contact spacings are both 2 μm and the metallurgical gate length is 4 μm. The transverse gate dimension Z is equal to 510 μm.

From differential capacitance measurements on the wafer (prior to FET fabrication), the film doping N_0 and thickness a are estimated to be

$$N_0 \cong 4 \times 10^{16} \text{ cm}^{-3} \qquad a \cong 0.4 \times 10^{-4} \text{ cm}.$$

Using $\epsilon = 1.06 \times 10^{-12}$ F/cm for GaAs gives $U_0 \simeq 5$ V. (When computing the theoretical curves for an individual FET, it is preferable to use a value for U_0 determined from source–drain conductance measurements on the device of interest.)

A. Measurement of U_0, R_0, R_s, and R_d

The results of Sections II and III are given in terms of the voltages V_D, V_S, and V_G, which exist at the terminals of the intrinsic FET. Measured terminal voltages will differ from the intrinsic values because of unavoidable resistances in series with the terminals.

Although the effect of a series gate resistance R_g is negligible for the FET's described, the voltage difference introduced by the source and drain resistances R_s and R_d can be quite significant. If the sheet resistance of the film is known, R_s and R_d can be very roughly estimated from the contact dimensions; however, the actual resistance will be larger than this value due to an effective resistance contributed by the contact [9]. Uncertainties in this contact resistance, and also the film sheet resistance, make it desirable to have a method for measuring the extrinsic resistance. We describe here the results of a technique [9] that can be used to determine the sum $R_s + R_d$. This is sufficient for our purposes, since $R_s \cong R_d$ can be assumed.

The technique makes use of the fact that when $I_D = 0$, the small-signal source–drain resistance is given by

$$r_{ds0}' = R_s + \frac{1}{G_0(1 - \sqrt{\eta})} + R_d \qquad (22)$$

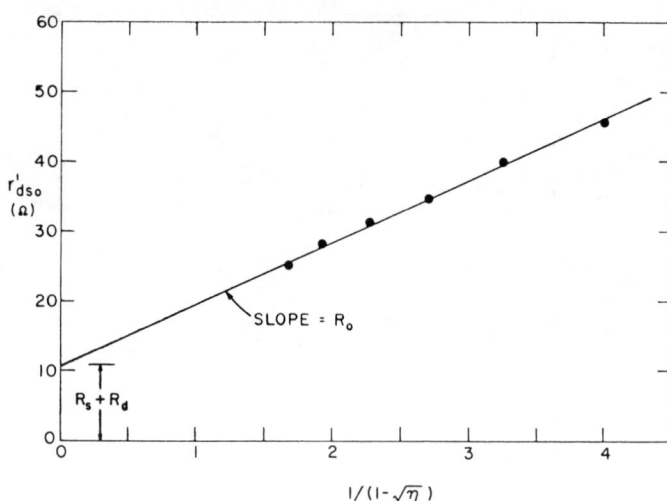

Fig. 8. Illustration of the method used for determining extrinsic resistance.

TABLE I
EXTRINSIC RESISTANCE AND RELATED DATA FOR FET HD-14

$U_0 = 4.82$ V	$V_p = 4.1$ V	$V_B = 0.72$ V
$R_0 = 8.45$ Ω	$R_s = R_d = 5.9$ Ω	
$L = 4.5 \times 10^{-4}$ cm	$\Gamma = 3.35$	

Note: V_p is determined by extrapolating $g_{ds0}'(=1/r_{ds0}')$ versus V_{GS} data to the $g_{ds0}' = 0$ intercept, and V_B is computed using [9, eq. (15)].

where $\eta = (-V_{GS} + V_B)/U_0$, as defined earlier. The prime denotes the fact that the measurement is made at the external terminals of the FET. In the plot of Fig. 8, measured values of r_{ds0}' are plotted against $1/(1 - \sqrt{\eta})$, and a straight line, as predicted by (22), is fit to the data. $R_s + R_d$ and R_0 are then determined from the intercept and slope of this line. The results of this procedure are summarized in Table I.

B. Corrections Due to R_s and R_d

From the circuit of Fig. 3, we can write

$$V_{GS}' = V_{GS} + I_D R_s \qquad (23)$$

and

$$V_{DG}' = V_{DG} + I_D R_d. \qquad (24)$$

In normalized form (23) and (24) are

$$\eta' = \eta + R_s G_0[F(\zeta) - F(\eta)] \qquad (25)$$

$$\zeta' = \zeta + R_d G_0[F(\zeta) - F(\eta)] \qquad (26)$$

where $\eta' = (-V_{GS}' + V_B)/U_0$ and $\zeta' = (V_{DG}' + V_B)/U_0$.

A drain characteristic for an actual FET will then be a plot of I_D versus V_{DS}', with V_{GS}' as a parameter. To find $V'_{D \text{ sat}}$ with η' fixed, it is necessary to simultaneously solve (6), (8), and (25) with $\zeta = \zeta_{\text{sat}}$ and $I_D = I_{D \text{ sat}}$. The solution, which is best done numerically, will give η, ζ_{sat}, and $I_{D \text{ sat}}$. With these results, $V_{D \text{ sat}}'$ can be computed from

$$V_{D \text{ sat}}' = U_0(\zeta_{\text{sat}} - \eta') + I_{D \text{ sat}} R_d. \qquad (27)$$

The effect of R_s is to reduce $I_{D\,\text{sat}}$ below the $R_s = 0$ value, and the effect of R_d is to increase $V_{D\,\text{sat}}'$ above the value for which $R_d = 0$. Since the product $I_{D\,\text{sat}}R_d$ appears in the equation of $V_{D\,\text{sat}}'$, the effects of increasing R_s and R_d counteract each other, and $V_{D\,\text{sat}}'$ tends to be relatively insensitive to extrinsic resistance.

Perhaps the most important effect of extrinsic resistance is the reduction of saturation transconductance. As is well known, the differential form of (23) gives

$$g_{m\,\text{sat}}' = \frac{dI_D}{dV_{GS}'} = \frac{g_{m\,\text{sat}}}{1 + R_s g_{m\,\text{sat}}}. \tag{28}$$

It is important to remember that for fixed V_{GS}', the intrinsic transconductance $g_{m\,\text{sat}}$ will also vary with R_s due to the dc "debiasing" introduced by this resistance.

Although both y_{11} and y_{21} will change with R_s, the ratio y_{21}/y_{11} remains very nearly constant. If the effects of R_d and C_{gd} can be neglected by assuming that

$$R_d g_{m\,\text{sat}} \ll C_{gs}/C_{gd} \tag{29}$$

and

$$(R_s + R_d)g_{ds} \ll 1 \tag{30}$$

and, if it is also assumed that the operating frequency is low enough so that

$$\frac{g_{m\,\text{sat}}}{1 + R_s g_{m\,\text{sat}}} \gg \omega C_{gd}, \tag{31}$$

then it is possible to show that

$$h_{21} = \frac{g_{m\,\text{sat}}}{\omega C_{gs}} = \frac{\omega_T}{\omega} \tag{32}$$

which holds even for the $R_s > 0$ case. Since the above assumptions are valid for the FET's discussed in Section IV-C, f_T can be assumed to be independent of R_s, and no correction to the intrinsic f_T given by (20) is necessary.

C. Measurements

$I_{D\,\text{sat}}$, $g_{m\,\text{sat}}'$, and f_T were all measured for the FET discussed in Table I and Fig. 8. $I_{D\,\text{sat}}$ and $g_{m\,\text{sat}}'$ data are compared with theory in Figs. 9 and 10. The data of Table I and a value for Γ are all that is needed to convert the normalized forms of (8), (6), and (11) when plotting the theoretical curves. For the calculations of this section, we have used $L = 4.5 \times 10^{-4}$ cm and $E_m = 3.2$ kV/cm, which gives $\Gamma = 3.35$ for the FET under discussion. (L has been chosen 0.5×10^{-4} cm larger than L_g to account for the additional channel length contributed by the cylindrical depletion region at the source edge of the gate.)

In general, good agreement between theory and measurement for $I_{D\,\text{sat}}$ and $g_{m\,\text{sat}}'$ is routinely obtained for FET's having different values of U_0, and hence different values of Γ. A comparison is shown in Table II, where

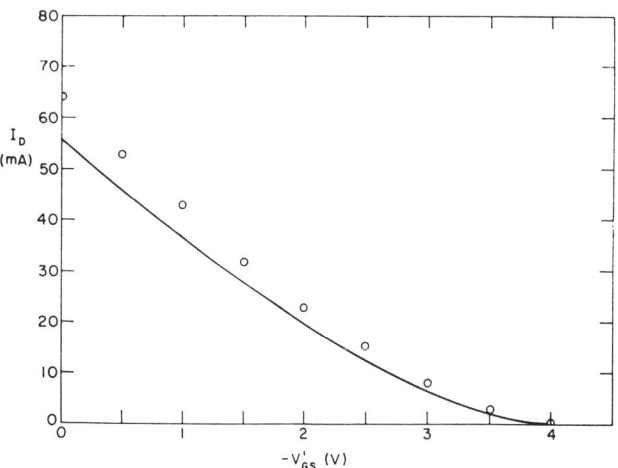

Fig. 9.   Solid line: Calculated drain current versus gate voltage for FET HD-14. Measured values are indicated by the circles. ($V_{DS}' = 5$ V.)

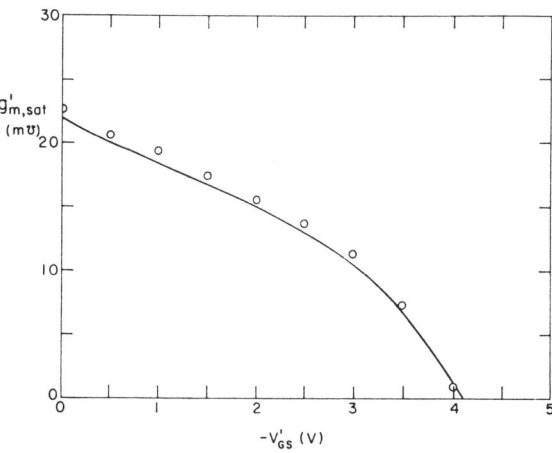

Fig. 10.   Solid line: Calculated transconductance versus gate voltage for FET HD-14. Measured values are indicated by the circles. ($V_{DS}' = 5$ V.)

TABLE II
COMPARISON OF MEASURED AND THEORETICAL DRAIN CURRENT
AND TRANSCONDUCTANCE FOR DIFFERENT GAMMAS

V_p (V)	Γ	Measurement I_D (mA)	$g_{m\,\text{sat}}'$ (mmho)	Calculation I_D (mA)	$g_{m\,\text{sat}}'$ (mmho)
3.62	2.89	57	21.5	52.6	23.4
4.10	3.21	64	22.8	57.0	22.6
4.60	3.55	72	20.0	61.6	23.8
4.90	3.75	70	16.8	61.8	20.0
6.70	4.95	100	18.5	91.3	22.2

measured and calculated values of $I_{D\,\text{sat}}$ and $g_{m\,\text{sat}}'$ are given for several FET's as a function of Γ.

To determine f_T versus V_{GS}', the small-signal current gain h_{21} was measured versus frequency and V_{GS}', using the Hewlett-Packard 8542A network analyzer system. The h_{21} data was observed to be proportional to $1/f$

TABLE III
CIRCUIT ELEMENT VALUES FOR FET HD-14

$g_m = 24$ mmho	$R_s = R_d = 5.9$ Ω
$C_{gs} = 0.90$ pF	$g_{ds} = 1.4$ mmho
$r_c = 7.3$ Ω	$C_{gd} = 0.05$ pF
Bias condition: $\eta' = 0.2$	($\eta = 0.262$)

Note: C_{gd} and g_{ds} are determined from direct device measurements with $V_{DS}' = 5$ V and $V_{GS}' = 0.24$ V. C_{gd} includes the package feedback capacitance of approximately 0.04 pF.

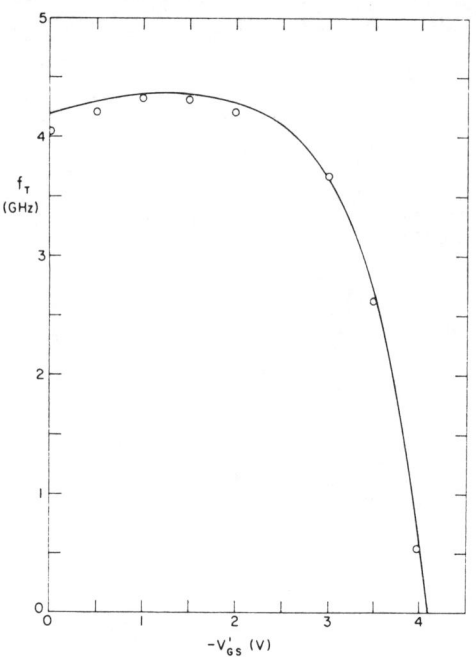

Fig. 11. Solid line: Calculated f_T versus gate voltage for FET HD-14. Measured values are indicated by the circles. ($V_{DS}' = 5$ V.)

in agreement with (32), up to 2 GHz, the limit imposed by the measurement jig. Extrapolating the $h_{21}(f)$ data to $h_{21} = 1$ gave a measured value for f_T', which was corrected for package capacitance C_p using

$$f_T = f_T'(1 + C_p/C_{gs}). \tag{33}$$

$C_p = 0.65$ pF was measured for the package (TO-18) and C_{gs} was determined from y-parameter data given by the network analyzer.

For the theoretical calculation of f_T, it is necessary to know μ_0 the low-field mobility. Although no Hall measurements were made on this particular wafer, a reasonable estimate is $\mu_0 = 5 \times 10^3$ cm²/V·s. As shown in Fig. 11, the calculated f_T shows excellent agreement with the measured data. For this device $\Gamma = 3.35$ and from Fig. 6 it is clear that for small values of gate bias, f_T is approximately one-fourth of the f_T of an equivalent classical FET. Other data are summarized in Table III.

V. CONCLUSION

By employing a simple modification of Shockley's classical FET theory, it is possible to calculate the drain characteristic of a GaAs FET. A similar modification to Hauser's first-order y-parameter analysis permits the

calculation of the input and transfer network parameters y_{11} and y_{21}. Good agreement between experiment and theory is obtained for both low- and high-frequency behavior of FET's having 4-μm gate lengths.[3]

As gate length L_g is decreased to improve the f_T, the corrections to the classical FET theory become quite significant. For small values of L_g, f_T becomes proportional to L_g^{-1}, rather than the classically predicted L_g^{-2}. The fabricating techniques that limit L_g also usually dictate the smallest value of source–gate and drain–gate spacing. In general, substantial corrections are also necessary to account for the resulting extrinsic source and drain resistance. In this paper, these corrections are made using an independent measurement of the extrinsic resistance.

Finally, we note that there are two important problems that remain to be solved before the design theory of the GaAs FET can be considered to be complete. These are: 1) the prediction of the maximum V_{DS} that can be sustained; and 2) the prediction of the source–drain conductance g_{ds} above saturation. We generally observed g_{ds} values that are about an order of magnitude greater than values predicted by conventional theories of channel length modulation [2], [11], [12]. In addition, g_{ds} is observed to increase, that is, I_D comes out of saturation, at V_{DS} values considerably below that required to initiate gate–drain breakdown.

It is our belief that these departures from conventional junction FET behavior are due to an interaction of the channel with the semi-insulating substrate; however, there is presently no satisfactory theory that provides a quantitative prediction of g_{ds} versus V_{DS}.

Such a theory would permit a more accurate prediction of f_{max} [9], the maximum frequency of oscillation, and also a prediction of the maximum Class A output power, which is important in microwave power applications.

APPENDIX

DETAILS OF THE y-PARAMETER ANALYSIS

Although both the circuit elements g_m and C_{gs} of Fig. 3 can be found by differentiating I_D and the depletion layer charge with respect to V_{GS} [1], an alternate approach is taken here. This analysis gives explicit formulas for both these elements and has the added benefit of predicting the series element r_c, which cannot be found by direct differentiation. These results are obtained by modifying the y-parameter analysis of Hauser [8] to account for the fact that current saturation takes place at a fixed field E_m.

Hauser's first-order solution for the small-signal gate and drain currents can be written as

---

[3] Good agreement has also been obtained for FET's having even smaller gate lengths. A detailed comparison, similar to that of Section IV-C, is given in [10] for a 2-μm gate FET.

$$i_d = v_g G_s \frac{1}{1 + j\omega/\omega_0} + (v_d - v_g)G_d \frac{1 + j\omega/\omega_1}{1 + j\omega/\omega_0} \quad (34)$$

$$i_g = v_g G_s \frac{j\omega/\omega_3}{1 + j\omega/\omega_0} + (v_g - v_d)G_d \frac{j\omega/\omega_1}{1 + j\omega/\omega_0} \quad (35)$$

where the source is taken to be the common terminal $(v_s = 0)$, and the conductances are defined as

$$G_s = G_0(1 - \sqrt{\eta}) \quad (36)$$

$$G_d = G_0(1 - \sqrt{\zeta}). \quad (37?)$$

The radian frequencies ω_0, ω_1, and ω_3 are found by integrating appropriate functions of the potential difference $U(x)$ along the gradual portion of the channel. As a result of this procedure, the three frequencies can be expressed as functions of the normalized terminal potentials ζ and η. Using our notation, Hauser's formulas take the form

$$\frac{1}{\omega_0 t_0} = \left[\frac{6}{7} (\zeta^{7/2} - \eta^{7/2}) - 3(\zeta^3 - \eta^3) + \frac{12}{5}(\zeta^{5/2} - \eta^{5/2}) \right.$$
$$- 6(\zeta^2 \eta^{3/2} - \zeta^{3/2}\eta^2) + 9(\zeta^2 \eta - \zeta\eta^2)$$
$$\left. - 12(\zeta^{3/2}\eta - \zeta\eta^{3/2}) \right] / \left[F(\zeta) - F(\eta) \right]^3 \quad (37)$$

$$\frac{1}{\omega_1 t_0} = \frac{[2\zeta^{3/2} - \zeta^2 - 2\zeta^{1/2}(3\eta - 2\eta^{3/2}) + 4\eta^{3/2} - 3\eta^2]}{[F(\zeta) - F(\eta)]^2} \quad (38)$$

$$\frac{1}{\omega_3 t_0} = \frac{[4\zeta^{3/2} - 3\zeta^2 - 2\eta^{1/2}(3\zeta - 2\zeta^{3/2}) + 2\eta^{3/2} - \eta^2]}{[F(\zeta) - F(\eta)]^2} \quad (39)$$

where

$$t_0 = \frac{3}{2} \frac{L^2}{\mu_0 U_0} \quad (40)$$

and has the dimensions of time. (For $\eta = 0$, t_0 is the transit time [7] of a classical FET operating in saturation.)

Equations (34) and (35) were originally derived for the case of a classical FET operating below saturation with v_d and v_g acting as independent variables. In saturation, $v_d = 0$ and $\zeta = 1$ were used to compute the y parameters. Of course, these conditions do not apply for a GaAs FET in saturation. However, by putting the constant field condition (6) into small-signal form, we

can still use the classical y-parameter solutions (34) and (35) with v_d being a variable that is now dependent upon v_g and i_d. The differential form of (6) can be written as

$$dI_{D\,\text{sat}} = -\frac{3I_0}{2\Gamma\sqrt{\zeta_\text{sat}}} d\zeta_\text{sat} \quad (41)$$

or

$$i_d = -\frac{G_0}{2\Gamma\sqrt{\zeta_\text{sat}}}(v_d - v_g) \quad (42)$$

which can be used to eliminate v_d in (34) and (35). This result will then permit the calculation of y_{11} and y_{21} as is done in the text.

ACKNOWLEDGMENT

The authors wish to thank R. Fairman, who provided the epitaxial GaAs, and W. Hooper, who fabricated the devices used in this work.

REFERENCES

[1] J. A. Turner and B. L. H. Wilson, "Implications of carrier velocity saturation in a gallium arsenide field-effect transistor," in *1968 Proc. Symp. GaAs, Inst. Phys. and Phys. Soc.*, 1968, pp. 195–204.

[2] W. Shockley, "A unipolar 'field-effect' transistor," *Proc. IRE (The Transistor Issue)*, vol. 40, pp. 1365–1376, Nov. 1952.

[3] K. Lehovec and R. Zuleeg, "Voltage-current characteristics of GaAs J-FETs in the hot electron range," *Solid-State Electron.*, vol. 13, pp. 1415–1426, 1970.

[4] J. G. Ruch and G. S. Kino, "Measurement of the velocity-field characteristic of gallium arsenide," *Appl. Phys. Lett.*, vol. 10, pp. 40–42, 1967.

[5] J. G. Ruch and W. Fawcett, "Temperature dependence of the transport properties of gallium arsenide determined by a Monte Carlo method," *J. Appl. Phys.*, vol. 41, pp. 3843–3849, 1970.

[6] D. P. Kennedy and R. R. O'Brien, "Computer aided two-dimensional analysis of the junction field-effect transistor," *IBM J. Res. Develop.*, vol. 14, pp. 95–116, 1970.

[7] G. C. Dacey and I. M. Ross, "The field effect transistor," *Bell Syst. Tech. J.*, vol. 34, pp. 1149–1189, 1955.

[8] J. R. Hauser, "Small signal properties of field effect devices," *IEEE Trans. Electron Devices*, vol. ED-12, pp. 605–618, Dec. 1965.

[9] R. K. Willardson and Albert C. Beer, Ed., *Semiconductors and Semimetals*, vol. 7, *Applications and Devices, Part A*. New York: Academic, pp. 147–200, 1971.

[10] N. G. Bechtel, W. W. Hooper, and R. D. Fairman, "High frequency gallium arsenide Schottky barrier gate field effect transistor," Air Force Avionics Lab., Wright-Patterson AFB, Ohio, Tech. Rep. AFAL-7R-71-331, Nov. 1971.

[11] A. B. Grebene and S. K. Ghandi, "General theory for pinched operation of the junction-gate FET," *Solid-State Electron.*, vol. 12, pp. 573–589, 1969.

[12] P. L. Hower, "A theory of saturation and a charge-control analysis of a junction field-effect transistor," Stanford Electron. Lab., Stanford, Calif., Tech. Rep. 4726-1, Apr. 1967.

Control of Gate–Drain Avalanche in GaAs MESFET's

STUART H. WEMPLE, MEMBER, IEEE, WILLIAM C. NIEHAUS, MEMBER, IEEE, HERBERT M. COX,
JAMES V. DILORENZO, AND W. O. SCHLOSSER, MEMBER IEEE

Abstract—The onset of gate–drain avalanche imposes an important fundamental constraint on the drain voltage swing, and hence, on the output power of GaAs FET's. In this paper we show that recognition of the role of surface depletion and proper attention to channel design can yield avalanche voltage factors of 2–3 above bulk values. The appropriate design strategy is *minimization* of the undepleted epitaxial charge per unit area (Q_u) between gate and drain, which, in turn, dictates a gate-notch depth approximately equal to the surface zero-bias depletion depth. A simple lateral spreading model is proposed which predicts that $V_L \sim 50 Q_u^{-1}$ V, where V_L is the gate–drain avalanche voltage and Q_u is measured in units of 10^{12} electrons/cm^2. This prediction is supported by a large body of experimental dc and pulse data, although considerable scatter is observed which we have attributed to epi charge nonuniformities, premature avalanche at the rough edges of Al gates formed by a liftoff process, and surface charging variations associated with dielectric passivation. The observed dependence of V_L on epi charge rather than on doping level, as predicted for bulk avalanche, provides convincing evidence for nonbulk two-dimensional avalanche in the thin-film ($Q_u < 2.3$) FET geometry. In thick films ($Q_u > 2.6$), on the other hand, it is found that the bulk avalanche predictions are reasonably accurate. In terms of saturated epi current I_S, the bulk regime corresponds to $I_S > 450$ mA/mm and the lateral spreading (thin-film) regime to $I_S < 400$ mA/mm. Finally, we have found that gate–drain avalanche is the major cause of output saturation as a function of drain potential in power GaAs FET's.

I. INTRODUCTION

IT IS GENERALLY recognized that the onset of gate avalanche limits the permissible RF excursion of the drain potential in Schottky-barrier GaAs FET's and thereby imposes a fundamental constraint on output power. Although bulk avalanche voltages are commonly employed [1] when addressing this point, we show in the present paper that gate–drain breakdown voltages significantly above bulk values are easily achieved in the planar FET geometry. As we shall show, the appropriate design strategy for maximum output power requires *minimization* of the epitaxial charge per unit area in the region between gate and drain. A corollary requirement is a gate notch approximately equal to the surface zero-bias depletion depth. In contradiction to recent proposals [2], we find essentially no dependence of breakdown on gate–drain spacings above 1 μm nor do we find a significant dependence on doping level over the range 3–19 $\times 10^{16}$ cm^{-3}.

II. ANALYSIS AND MODEL

Any quantitative avalanche model must 1) determine the magnitude of the electric field along the drain edge of the gate and 2) predict the resulting avalanche current. The first step

Manuscript received January 14, 1980; revised February 6, 1980.
The authors are with Bell Laboratories, Murray Hill, NJ 07974.

necessarily demands a realistic two-dimensional solution to the semiconductor transport equations, while the second implies knowledge of both the electric field and the ionization coefficients appropriate to the actual, and possibly rough, gate edge under consideration.

Although a variety of two-dimensional computer models have been reported in the literature [3]–[9] it is difficult to assess their reliability in view of the simplifying assumptions employed. For example, the important role played by surface depletion is generally neglected, ad hoc diffusion coefficient terms are sometimes assumed, and computer mesh-size constraints sometimes limit doping levels to values considerably below those used in real devices. Several simplified analytic models have also been proposed, and these, as well as the numerical models, predict the occurrence of an accumulation-depletion dipole on the drain side of the gate. Most of the source–drain potential is necessarily dropped across this high-field region. Sone and Takayama [5], for example, obtain the straightforward result that the high-field domain moves towards the drain with increasingly positive drain potential, while Shur *et al.* [2] and Shur [8] suggest that the domain extends a large distance away from the gate edge with predicted values near 3 μm for a doping level of 10^{17} cm^{-3}. These authors also propose that expansion of the gate–drain spacing so as to accommodate this domain would improve breakdown voltages and permit design of remarkably high-power transistors.

In the present work, we use a very simple heuristic approach which takes into account two key elements of the avalanche problem, viz., 1) lateral spreading of the depletion region towards the drain with increasing positive drain potential, and 2) surface depletion between gate and drain. The physical situation is shown schematically in Fig. 1. For this simplified model we ignore any accumulation-depletion domains that may exist and assume a uniform positive depletion charge given by the doping level N. The two-dimensional Poisson's equation between gate and drain becomes

$$\frac{\partial E_x}{\partial x} = \frac{Ne}{\epsilon} - \frac{\partial E_y}{\partial y} \qquad (1)$$

where e is the electron charge, and ϵ is the material dielectric constant. For GaAs $Ne/\epsilon = 14N$ V/μm^2, where now N is measured in units of 10^{16} cm^{-3}, the electric field E in volts per micrometer, and both x and y in micrometers. In the one-dimensional lateral case (i.e., $\partial E_y/\partial y$ is negligible compared to $14N$), (1) integrates to the bulk result

$$E = 14Nx \qquad (2)$$

Reprinted from *IEEE Trans. Electron Devices*, vol. ED-27, no. 6, pp. 1013–1018, June 1980.

FET CHANNEL SCHEMATIC

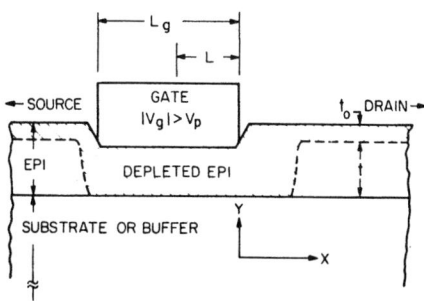

Fig. 1. Schematic drawing of GaAs FET channel geometry showing notched gate; surface depletion depth t_0, depleted epi associated with gate voltage V_g biased beyond pinchoff V_p; undepleted epi thickness t, gate length L_g; and effective gate length L.

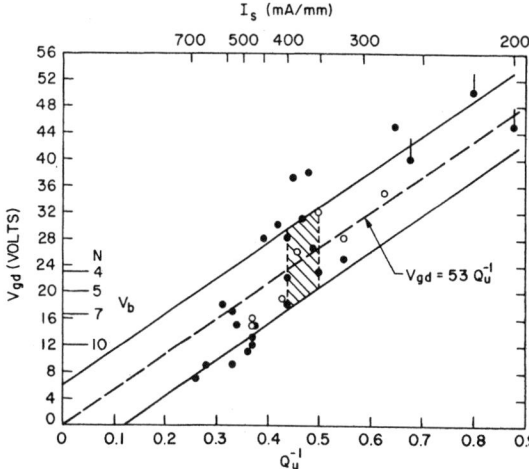

Fig. 2. Experimental dc gate avalanche voltages, defined at a fixed current of 1 mA/mm, as a function of reciprocal undepleted epitaxial charge $Q_u = Nt$, where N is the doping level in units of 10^{16} cm^{-3} and t is the undepleted epi thickness in micrometers. The top scale, which gives the associated saturated epi current prior to gate-notch etching, has been obtained using the relation $I_s = 176 Q_u$. The inserted vertical scale gives the bulk avalanche voltage for the indicated doping levels. Open circles correspond to a nonuniform wafer with $N = 10^{17}$ cm^{-3} (see text).

and

$$V = -7Nx^2. \tag{3}$$

Combining (2) and (3) yields the familiar expression

$$V_b = E_a^2/28N \text{ V} \tag{4}$$

where V_b and E_a refer to bulk avalanche voltage and avalanche electric field, respectively. The latter quantity is approximately 40–50 V/µm.

In the general two-dimensional geometry the final term in (1) cannot be neglected arbitrarily. For sufficiently thick epitaxial layers, however, it seems clear that the one-dimensional "bulk" result given by (4) should hold. For very thin layers, on the other hand, the $\partial E_y/\partial y$ term in (1) must be included, i.e., the problem is inherently two-dimensional. Indeed, the experimental results presented in Section III reveal that $\partial E_y/\partial y > 0$ so that avalanche breakdown voltages larger than bulk values are obtained for thin epitaxial layers.

To address the difficult two-dimensional problem we resort to a simplified model that can be experimentally checked. The basic assumptions are 1) all electric field lines associated with lateral spreading of the depletion region terminate on the gate metallization, and 2) the GaAs surface between gate and drain is depleted to a depth t_0, consistent with midgap pinning of the surface Fermi level. This depth is easily obtained from (3) with $|V| \cong 0.75$ eV. Straightforward application of Gauss' law to the geometry of Fig. 1 yields

$$E = \frac{Ne}{\epsilon} \frac{t}{L} x \tag{5}$$

or, in practical units,

$$E = 14Ntx/L \text{ V/µm}. \tag{6}$$

In (6) t is the *undepleted* thickness of the active layer between gate and drain, and L is some unknown distance along the gate over which the electric field lines terminate. We take L to be the adjustable parameter of the model but note that $L \leqslant L_g$, where L_g is the gate length. Integration of (6) yields the simple result

$$V_L = E_a^2 L/28 Q_u \text{ V} \tag{7}$$

where $Q_u \equiv Nt$ is the *undepleted* epitaxial charge per unit area (in units of 10^{12} cm^{-2}) in the region adjacent to the gate,

and V_L refers to the breakdown voltage predicted by this lateral spreading model. Note that $V_L \propto Q_u^{-1}$ for thin layers (7), whereas $V_b \propto N^{-1}$ for thick layers (4). These predictions provide ideal limiting cases that can be checked experimentally. It is noteworthy that the two limits impose very different constraints on power FET design, i.e., (7) dictates epi charge minimization without regard for doping level, whereas only the doping level enters into (4). A straightforward consequence of the thin epi (or lateral spreading) model is that gates should be placed in notches which are etched to a zero-bias depletion depth below the top surface. Deeper notches, for a given thickness under the gate, result in excess Q_u which does nothing but degrade gate–drain avalanche, whereas shallower notches, also for a fixed thickness under the gate, degrade the current-handling capability and limit power.

A key question related to our epi charge minimization criterion is whether or not improved gate–drain avalanche can be achieved with values of Q_u that are in a suitable range for power FET design. This point is addressed in the following experimental section where we give an affirmative answer to that question and show, furthermore, that the thin-film two-dimensional regime applies for $Q_u < 2.3$ whereas the thick-film bulk case corresponds to $Q_u > 2.6$. These values correspond to "open-gate" channel currents of 400 and 460 mA/mm, respectively.

III. Experimental Results

A. Two-Terminal Measurements

The experimental investigation has encompassed both two-terminal and three-terminal dc gate-current measurements as well as three-terminal pulse-current observations. Two-terminal (drain and source grounded) results are summarized in Fig. 2. The indicated breakdown voltage has been defined at a gate current per unit width of 1 mA/mm. We have calibrated undepleted epi charge Q_u by measuring the saturated source–drain current I_s before etching the gate notch using the relation $I_s = 176\, Q_u$, where I_s is in units of milliamperes per millimeter. The coefficient 176 corresponds to an average electron drift velocity of 1.1×10^7 cm/s. Values of I_s are shown on the upper scale in Fig. 2. The key experimental result is the general trend towards increasing breakdown voltages with decreasing values of Q_u consistent with the lateral spreading model given by (7). Open circles in Fig. 2 apply to a unique wafer which was unusually nonuniform in thickness and consequently provided a check on our model free of wafer-to-wafer processing variations. We note that the doping level for this wafer was 10^{17} cm^{-3} corresponding to bulk breakdown (V_b) near 12 V as shown on the vertical scale insert. This result clearly shows that breakdown voltages up to three times bulk values are achievable provided Q_u is sufficiently small. The solid points in Fig. 2, as well as hundreds of data points not shown, support the indicated trend of improved avalanche with reduced undepleted epi charge. Moreover, we have not observed a significant doping dependence over the range 3–19×10^{16} cm^{-3} for $I_s < 400$ mA/mm, a result inconsistent with bulk avalanche but consistent with the lateral spreading model. The very low avalanche voltages shown in Fig. 2 for $I_s > 500$ mA/mm all apply to wafers with $N > 10^{17}$ cm^{-3} and, therefore, are consistent with bulk values.

The data presented in Fig. 2 apply to devices having optimum gate-notch depths approximately equal to the surface zero-bias depletion depth t_0. More generally, it is possible to hold Q_u constant and reduce the channel height under the gate, and in turn reduce I_{dss}, by etching deeper notches. We have found that this procedure can yield up to 10-V avalanche improvement for very low I_{dss} values ($\lesssim 50$ mA/mm) corresponding to a total charge under the gate of only 10^{12} cm^{-2} for a doping of 4×10^{16} cm^{-3}. This low-current (deep-notch) design is not only unsuitable for power FET's, since we typically require $I_{dss} = 150$–250 mA/mm, but it is also non-optimum for improving avalanche. For example, reducing Q_u to the undergate value of unity would increase V_L to near 50 V which is some 15 V above the deep-notch ($Q_u = 2.0$) value of 35 V (i.e., approximately 10 V above the experimental results shown in Fig. 2). We mention these observations only to point out that the simple lateral spreading model may, under some circumstances, underestimate V_L, a not unsurprising result in view of the complicated two-dimensional nature of the transport problem immediately adjacent to the gate notch.

The rather large scatter shown in Fig. 2 suggests that factors other than Q_u influence gate avalanche as well. Of course, uncertainty in Q_u itself due to wafer edge to center thickness variations may also lead to scatter. For example, we expect $\Delta Q_u \approx 0.2$ leading to several volts uncertainty in V_L from this cause alone. A more important uncontrolled variable is very likely the "effective" avalanche electric field E_a in (7). This quantity is well-defined for an ideal gate geometry having microscopically smooth edges; however, the liftoff process used to define the Al gates yields rough, burred edges. For a fixed applied voltage, the electric field at a nonplanar surface is proportional to the radius of curvature (r) of local metal

defects ($E \sim (V/r)$). Therefore, it is plausible to suggest that burred edges can induce premature and spotty avalanche. Direct evidence for this contention has been obtained by visual observation of the gates during avalanche measurements. White-light emission along the drain-side gate edge under these conditions is sometimes very spotty and localized rather than uniform, a result we attribute to electric-field enhancement at sharp points introduced by the liftoff tearing process. Here we stress that the relevant burrs are not necessarily those easily observed in the optical microscope since only the rough Al edge in physical contact with GaAs would locally enhance the electric field. Indeed, we have found that the gate edges that appear most burred optically generate the more uniform white-light emission. To account for this behavior we speculate that the tearing process that yields rough microscopic burrs also tends to lift the Al edge away from contact with the underlying GaAs. Further experimental support for the rough-gate-edge model has been obtained from avalanche measurements on Pt/Ti/Pt/Au gates formed by ion milling rather than by liftoff. As discussed in more detail below, values of V_L are on the average several volts higher with Pt gates than with our standard Al gates. Moreover, the avalanche white-light emission is more uniform.

A third factor which may lead to scatter in avalanche is the uncertain influence of the passivation dielectric (SiO_2 or Si_3N_4) on Q_u and, in turn, on V_L. We have found that dielectric deposition usually reduces V_L; however, the specific value of the decrease varies from near zero to as much as 10 V. The general trend is for the higher initial values to decrease more than the lower ones. For example, we typically find that 40-V values decrease to near 30 after dielectric deposition, 30-V values to near 25, and 25-V values would remain almost unchanged. Because the data presented in Fig. 2 apply only to passivated devices, it is reasonable to ascribe some fraction of the observed scatter to the possible variable influence of the dielectric. This passivation sensitivity cannot be accounted for using a "bulk" avalanche model, but it is clearly in agreement with our lateral spreading model if we allow surface charge variations to influence Q_u and thereby affect V_L. Thus the contention is that dielectric passivation can change surface band bending by introducing positive charge on the GaAs, thereby decreasing the surface depletion depth and increasing Q_u. This model is also qualitatively consistent with the observation noted above that larger initial values of V_L tend to drop the most, i.e., the smaller the initial value of Q_u the greater the influence of a fixed surface charge $\Delta Q_s = -\Delta Q_u$ on V_L. A value $\Delta Q_s \sim 0.3$ is required to account for most of the observed variations. Note that the "nominal" surface charge associated with 0.75-V surface band bending is 0.7–1.0 (in units of 10^{12} cm^{-2}) for $N = 5$–10×10^{16} cm^{-3}. An additional point worth noting is that the dielectric deposition itself could, in principle, alter gate-edge roughness and influence the "effective" E_a.

In summary, we have found that V_L increases with decreasing values of Q_u and is essentially independent of doping. These results not only show that bulk avalanche does not impose a fundamental limit as is often claimed, but also reveal the two-dimensional surface dominated behavior of avalanche in the planar thin-film geometry. This surface sensitivity, we have suggested, leads to poorly controlled scatter in V_L associated with dielectric passivation. Furthermore, we have also suggested that additional uncontrolled scatter is induced by variations in gate-edge roughness associated with the Al liftoff process. Thus smooth gate edges appear desirable, a claim supported by the Pt/Ti/Pt/Au gate results.

Returning again to Fig. 2, it is of interest to compare the slope of the indicated trend lines with that predicted by (7) (i.e., $dV_L/dQ_u^{-1} = E_a^2 L/28 = 53$, or $E_a = 39L^{-1/2}$ V/μm). Because L is an adjustable parameter of the model, we are not able to make a detailed quantitative comparison, although it should be noted that for $L \approx 1~\mu$m, $E_a \approx 40$ V/μm which is a perfectly reasonable avalanche electric field for GaAs. Thus the observed dependence of V_L on Q_u is consistent with the proposed lateral spreading model.

The cross-hatched region shown in Fig. 2 encompasses our selected design for power GaAs FET's. Typically, we restrict Q_u to the range 2.0–2.3 corresponding to I_s = 350–400 mA/mm so as to maintain the available open-channel current (i.e., the maximum current developed during the positive gate swing) above 350 mA/mm with $V_L > 20$ V. As emphasized earlier, uncontrolled scatter occasionally leads to excursions near 35 V on the high side and somewhat below 20 V on the low side. Further improvements into the 30–40-V range have been achieved on several wafers by reducing I_s to the 300–350-mA/mm range corresponding to $1.7 < Q_u < 2.0$. For $Q_u < 1.7$, we generally obtain $V_L > 40$ V with occasional examples above 50 V. Finally, thick-film bulk avalanche is observed for $Q_u > 2.6$ corresponding to $I_s > 450$ mA/mm.

Implications of the preceding dc results on RF power performance would appear to be straightforward since an optimum notch-depth design implies that the maximum drain current swing is proportional to Q_u, and the maximum drain voltage swing is nearly proportional to Q_u^{-1}. The saturated output power should then be almost independent of Q_u with a value $\frac{1}{8}$ (0.176 × 53) ≈ 1 W/mm. In the following section we show that this "dc" conclusion actually requires substantial modification.

B. Three-Terminal Pulse and DC Measurements

Turning now to the three-terminal results, we show in Fig. 3 a pulse transistor characteristic for a typical 6-mm periphery device, where $2.0 < Q_u < 2.3$. The drain excitation consists of 100-ns pulses applied at a 50-Hz repetition rate and the gate excitation is dc. The key observation is the occurrence of nonpinchable source–drain current for $V_D > 14$ V. This excess current, which flows in the gate–drain circuit, provides a simple measure of pulse avalanche. Note that the avalanche curves not only coalesce at large drain potentials (30 V) but also appear to saturate rather than turn upwards. Such behavior is counter to that expected for bulk avalanche for which sharp thresholds should be observed at decreasing drain bias as the gate voltage becomes more negative, i.e., the avalanche threshold should correspond to a fixed gate–drain potential. Such classic behavior has been observed only for $Q_u > 2.6$ (see Fig. 4), whereas in the lateral spreading regime ($Q_u < 2.3$) we universally obtain pulse characteristics similar to those shown in Fig. 3. As a result, there is no sharply de-

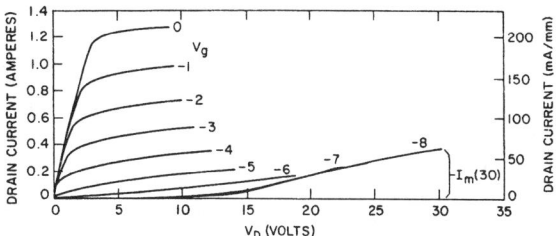

Fig. 3. Pulse source–drain I–V characteristic of typical (Q_u = 2.0–2.3) 6-mm periphery FET showing definition of pulse avalanche current $I_m(30)$. The 100-ns drain voltage pulses have been applied at a 50-Hz repetition rate with dc applied to the gate. The right-hand scale shows the drain current on a milliampere per millimeter basis.

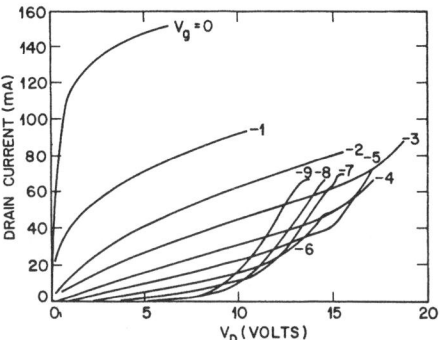

Fig. 4. Pulse source–drain I–V characteristic of thick epi ($Q_u > 2.6$) 1-mm periphery device showing classical avalanche behavior.

fined *maximum* value for the gate–drain potential. Rather, pulse avalanche is best characterized by the *minimum* achievable drain current at some specified drain voltage. This quantity, $I_m(V_D)$, is defined in Fig. 3 at $V_D = 30$ V. For this example, $I_m(30) = 400$ mA corresponding to 67 mA/mm. The choice of 30 V is somewhat arbitrary, although it does correspond roughly to the maximum RF voltage associated with our typical 14-V dc drain bias.

In qualitative agreement with the dc two-terminal observations outlined earlier, we have found that $I_m(30)$ can be adjusted to almost any value between zero and 200 mA/mm simply by controlling Q_u. For example, $I_m(30) = 125$ mA (21 mA/mm) when $Q_u = 1.9$ as shown in Fig. 5. Further reduction to $Q_u = 1.6$ can yield $I_m(30)$ values close to zero as shown in Fig. 6. A particularly poor example of avalanche behavior is shown in Fig. 7. For our standard product ($2 < Q_u < 2.3$) we usually obtain values of $I_m(30)$ between 30 and 80 mA/mm with occasional excursions outside these limits. This scatter in the pulse data can be attributed to the same passivation and liftoff mechanisms suggested earlier to account for the dc scatter, and indeed there is good correlation between $I_m(30)$ and V_L. We also emphasize that $I_m(30)$ does not depend on doping level over the range 3–19×10^{16} cm^{-3}.

To address the performance implications of our three-terminal pulse results, we show in Fig. 8 several plots of $I_m(30)$ versus I_{dss}, the source–drain saturated current at zero gate bias. These data display a clear trend that is universally observed, viz., high values of I_{dss} correlate with high-avalanche currents. Note that this relationship ($[\Delta I_m(30)/\Delta I_{dss}] \approx \frac{1}{2}$) is tight within the same wafer but that wafer-to-wafer variability

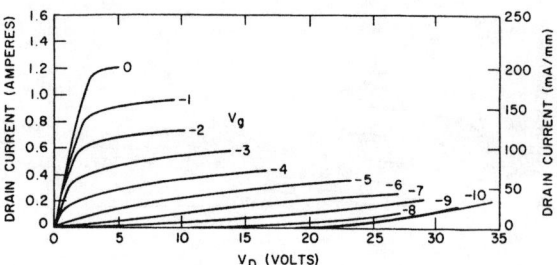

Fig. 5. Pulse source–drain I–V characteristic of thin epi ($Q_u \cong 1.9$) 6-mm periphery FET. The right-hand scale shows the drain current on a milliampere per millimeter basis.

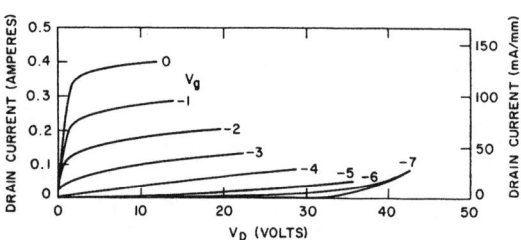

Fig. 6. Pulse source–drain I–V characteristic of a very thin epi ($Q_u \approx 1.6$) 3-mm periphery FET. The right-hand scale shows the drain current on a milliampere per millimeter basis.

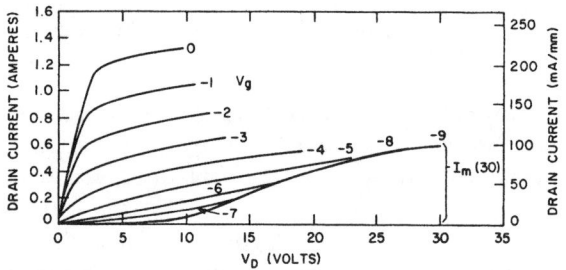

Fig. 7. Pulse source–drain I–V characteristic of a poor example of our standard ($Q_u = 2.0$–2.3) 6-mm periphery FET. The right-hand scale shows the drain current on a milliampere per millimeter basis.

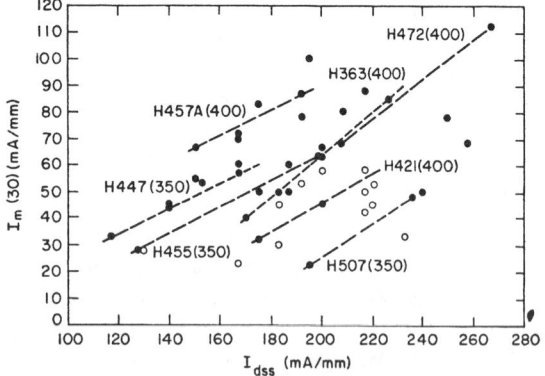

Fig. 8. Pulse avalanche current $I_m(30)$ defined in Fig. 4 as a function I_{dss} on a milliampere per millimeter basis for various devices having notched gates etched to a depth t_0 defined in Fig. 1. The dashed lines show the trend of increasing $I_m(30)$ with increasing I_{dss} within the specified wafers. The space between H457A and H421 encompasses nearly all of our standard devices (i.e., $Q_u \cong 2.0$–2.3 or $I_s < 400$ mA/mm). The lower separation between H455 and H507 depicts the improved avalanche range associated with thinner epi-layers (i.e., $Q_u \cong 1.8$–2.0 or $I_s < 350$ mA/mm). The open circles denote devices with Pt/Ti/Pt/Au gates and $I_s < 400$ mA/mm.

TABLE I
OBSERVED RELATIONSHIP BETWEEN PULSE AVALANCHE CURRENT I_m (30) AND INCREMENTAL POWER INCREASE ΔP_s AT 4 GHz ASSOCIATED WITH A 14–20-V DRAIN POTENTIAL VARIATION

I_m(30) mA/mm	ΔP_s W/mm
>100	<0
60	~ 0.1
30	~ 0.2
<10	~ .25

is rather large. The detrimental effects of avalanche are obviously most pronounced at elevated drain bias so that the compressed output power P_s tends to saturate and can even decrease with increasing drain potential. In Table I, the experimentally observed relationship between I_m(30) and the incremental saturated power increase between 14- and 20-V drain bias is summarized. Our general result is that gate–drain avalanche is the major cause of power saturation as a function of drain potential in GaAs FET's. A final important point revealed by the open-circle data of Fig. 8 is the apparent beneficial effects of Pt/Ti/Pt/Au gates formed by ion milling rather than by the liftoff process associated with our standard Al gates. Although a full statistical analysis must await more extensive data, the general impression at this point is that the smooth Pt/Ti/Pt/Au gate yields roughly 20-mA/mm reduction in I_m(30). We presume that other gate-forming processes yielding smooth gate edges, such as chemical etching, would be equally advantageous.

We may ask, based on the foregoing pulse observations, what limit avalanche imposes on the saturated power of GaAs MESFET's. Using the optimum gate-notch geometry described previously, $Q_u = 2.3$ ($I_s = 400$ mA/mm) appropriate to lateral spreading, I_m(30) = 30 mA/mm corresponding to the very best Pt gate results shown in Fig. 8, and $V_D = 16$ V, we estimate (using V(knee) = 2 V) that $P_{\text{sat}} \approx 14$ (0.4 - 0.03)/4 ≈ 1.3 W/mm. With a higher drain potential and even better gate-edge characteristics, a value closer to 1.5 W/mm might be possible, but such achievements may be very difficult except for isolated and nonreproducible examples.

In addition to the pulse data just outlined, we have also measured dc transistor characteristics in the three-terminal configuration. These measurements yield dc avalanche currents that are substantially below pulse values for equivalent bias conditions. Taking as an example the transistor of Fig. 6, we find that I_D(dc) < 1 mA for $V_D = 37$ V and $V_g = -8$ V which compares with a minimum value of I_d(pulse) = 60 mA at $V_D = 37$ V. This is a general result, although the magnitude of the difference varies from device to device. For example, we have observed I(pulse)/I(dc) ratios at 30 V of 500/290, 270/3, and 120/2. These data were all obtained in the dark since photo effects are also generally observed. Another common observation is the presence of a gradual downward drift in I_D(dc) near pinchoff with time constants of several minutes to perhaps 30 min. Both the dc drift observations and the enhanced avalanche current during pulse excitation point to the presence of interfacial charging effects. Although this conjecture has not been studied in detail, it would be reasonable to presume that a finite dc surface electron current from gate to drain could induce a negative surface charge which would, in turn, further deplete the epitaxial layer and thereby reduce Q_u. Lower avalanche currents for dc excitation would be accountable, according to this model, in terms of enhanced lateral spreading associated with a lower effective Q_u. Although the model is speculative, the observation that pulse avalanche currents significantly exceed dc values leads to the important conclusion that performance degradation associated with avalanche cannot be assessed from dc measurements alone.

IV. SUMMARY AND CONCLUSIONS

An extensive series of dc and pulse avalanche measurements on GaAs FET's has revealed the key role played by the undepleted epitaxial charge Q_u in the region between gate and drain. The appropriate design strategy is minimization of this charge, which implies a gate notch about equal to the surface depletion depth. A further important result is that for thin films (i.e., $Q_u < 2.3 \times 10^{12}$ cm^{-2}) bulk breakdown values are no longer applicable. We find instead that the observed dc avalanche data are consistent with a simple two-dimensional lateral spreading model for which $V_L \approx 53\, Q_u^{-1}$ V. Although the expected trend is clearly observed, there is considerable scatter in the data, a result we have attributed to epi charge nonuniformities, premature avalanche at the rough edge of Al gates formed by a liftoff process, and surface charging phenomena associated with dielectric passivation. Normal bulk breakdown is observed when $Q_u > 2.6$ corresponding to saturated epi currents above 450 mA/mm.

Pulse three-terminal avalanche measurements, which are more applicable to RF performance questions, also reveal nonclassical behavior in the thin-film ($Q_u < 2.3$) regime with currents considerably above dc values. Moreover, these data show that gate–drain avalanche imposes a minimum value on drain current rather than a maximum value on drain voltage

as for classical avalanche. This result has important implications for both power-handling capability and power rolloff as a function of dc gain bias. Again we find that avalanche, as defined in terms of minimum drain current, is a simple function of Q_u, although the scatter is large and presumably due to gate edge roughness and dielectric passivation. Finally, our experiments do not reveal a dependence on doping or on gate–drain spacing, at least over the 1–3-μm range.

ACKNOWLEDGMENT

The authors wish to thank all members of the GaAs FET project for numerous helpful discussions and continued experimental support. In particular, we appreciate the processing support of G. E. Mahoney and P. F. Sciortino.

REFERENCES

[1] M. Fukuta, K. Suyama, H. Suzuki, Y. Nakayama, and H. Ishikawa, "Power GaAs MESFET with a high drain-source breakdown voltage," *IEEE Trans. Microwave Theory Tech.*, vol. MTT-24, p. 312, 1976.
[2] M. S. Shur, L. F. Eastman, S. Judraprawira, J. Gammel, and S. Tiwari, "Design criteria for GaAs MESFETs related to stationary high field domains," in *Int. Electron Device Meet., Dig. Tech. Papers* (Washington, DC, 1978), pp. 381–384.
[3] B. Himsworth, "A two-dimensional analysis of gallium arsenide junction field effect transistors with long and short channels," *Solid-State Electron.*, vol. 15, pp. 1353–1361, 1972.
[4] M. Reiser, "Two-dimensional analysis of substrate effects in junction f.e.t.'s," *Electron Lett.*, vol. 6, pp. 493–494, 1970.
[5] J. Sone and Y. Takayama, "Analysis of field distributions in a GaAs MESFET at large drain voltages," *Electron Lett.*, vol. 12, pp. 622–624, 1976.
[6] K. Yamaguchi, S. Asai, and H. Kodera, "Two-dimensional numerical analysis of stability criteria of GaAs FET's," *IEEE Trans. Electron Devices*, vol. ED-23, pp. 1283–1289, 1976.
[7] R. A. Warriner, "Computer simulation of gallium arsenide field-effect transistors using Monte-Carlo methods," *Solid-State and Electron Devices*, vol. 1, pp. 105–110, 1977.
[8] M. Shur, "Maximum electric field in high-field domain," *Electron Lett.*, vol. 14, pp. 521–522, 1978.
[9] J. Frey and T. Wada, "Mobility, transit time and transconductance in submicrometric-gate-length MESFETs," *Electron Lett.*, vol. 15, pp. 26–28, 1978.

Proximity Effect of Dislocations on GaAs MESFET Threshold Voltage

SHINTARO MIYAZAWA AND FUMIAKI HYUGA

Abstract--Spatial inhomogeneity of semi-insulating LEC-grown GaAs is currently a subject of considerable attention, and there have been contradictory reports that dislocations affect FET threshold voltage either directly or indirectly.

Careful measurements of the threshold voltage are carried out in the vicinity of cellular dislocation networks, lineages, and the densely dislocated wafer periphery, which are dislocation distributions peculiar to LEC-grown GaAs crystals. The measurements were made in an attempt to verify the proximity effect of dislocations on the threshold voltage. It is concluded that, even though there is still ambiguity at very close gate-to-pit distances, the proximity effect is apparently observed only when the gate-to-pit distance is less than about 50 μm. A threshold voltage difference between dislocation-free and dislocated areas is also demonstrated.

I. INTRODUCTION

SEMI-INSULATING GaAs crystals, either undoped or Cr doped, grown by the liquid encapsulated Czochralski (LEC) pulling technique have recently received practical application as a substrate for GaAs integrated circuits with direct ion implantation. Although device performance is controlled predominantly by the implant and post-implantation annealing conditions, it is well known to be affected by inherent properties of the substrate. The uniformity of electrical behavior across the substrate is of current interest for the practical realization of large-scale high-performance integrated circuits. Particular interest in uniformity evaluation of the LEC-grown GaAs is devoted to the effect of dislocation density variations ranging from 10^4 to more than 10^5 cm^{-2} across a (100) wafer and almost all results reported earlier [1]–[8] on material parameters showed a macroscopic variation, with either a "W" or "M" shape, related to the well-pronounced W-shaped variation of dislocation density across the wafer.

Spatial inhomogeneity of semi-insulating LEC-grown GaAs has been the subject of considerable attention, because of the great interest in enhancement-mode FET integrated circuits. In particular, all of these device parameters have been shown to be spatially correlated, either in a positive or negative way, with dislocation distributions in the substrate. One of the authors [9], [10] was the first to clearly demonstrate that dislocations exert a direct influence on the threshold voltage for GaAs MESFET, and to also point out [11] the effect of isothermal annealing on

Manuscript received July 12, 1985; revised September 10, 1985.

The authors are with NTT Atsugi Electrical Communication Laboratories, 3-1, Morinosato-Wakamiya, Atsugi-shi, Kanagawa 243-01 Japan.

IEEE Log Number 8406436.

the improvement of threshold voltage uniformity from the viewpoint of the proximity effect of dislocations on FET threshold voltage. To the contrary, Winston *et al.* [12] and Lee *et al.* [13] have recently shown experimentally with their own process on a dislocated substrate that there exists no correlation between dislocation proximity and MESFET threshold voltage, although the threshold voltage for MESFET's fabricated on an area with high dislocation density was more negative than that on an area with low dislocation density.

This paper provides a definitive result on the proximity effect of dislocations on MESFET threshold voltage on a dislocated substrate grown by the conventional LEC technique, and sheds light on the puzzle. The authors have paid particular attention to dislocation distributions peculiar to LEC-grown crystals, i.e., lineage structure, cellular network structure, and highly dislocated peripheral regions, in order to reveal and verify the proximity effect. It is shown that dislocation in LEC-grown crystals influence virtually the FET threshold voltage, and that the dislocation proximity effect on the threshold voltage is observed at close gate-to-pit distances.

II. EXPERIMENTS

The substrates used were prepared from conventional LEC grown, semi-insulating, 2-in-diameter crystals. FET's with a gate length/width of 1/6- and 5-μm source–drain distance were fabricated with mesa etching by means of conventional photolithography, followed by ^{28}Si implantation with a dose of 2×10^{12} cm^{-2} at 60 keV and annealing at 800°C for 20 min with a SiN cap. FET's were located at intervals of 200 μm over the entire substrate area. The averaged threshold voltage over the entire wafer was measured to be -0.012 V with a standard deviation of 97 mV. Details of the fabrication process and the threshold voltage measurements have already been published elsewhere [14]. After threshold voltage measurements, the wafer was etched with molten KOH to reveal dislocation pits. The KOH also removed FET metallization but the location of the FET gate portion was still distinguishable, as was shown in the previous papers [9], [10].

In order to confirm the effect of dislocations on FET performances, a measurement of sheet carrier concentration in ^{28}Si-implanted and annealed layers was carried out. Micro Hall chips with a 40×40-μm measuring area were

Reprinted from *IEEE Trans. Electron Devices*, vol. ED-33, no. 2, pp. 227–233, Feb. 1986.

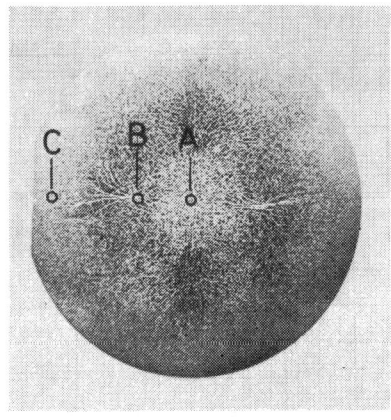

(a)

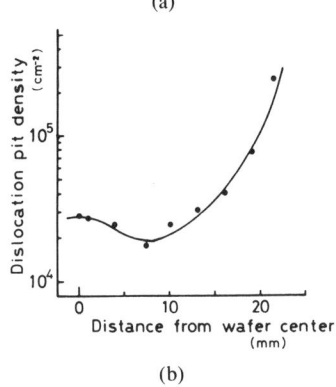

(b)

Fig. 1. (a) A photograph of KOH-etched wafer adjacent to the wafer examined. A, B, and C are 1 × 2-mm areas where FET threshold voltage was measured. (b) The variation in dislocation pit density along the <100> direction from the wafer center to the periphery.

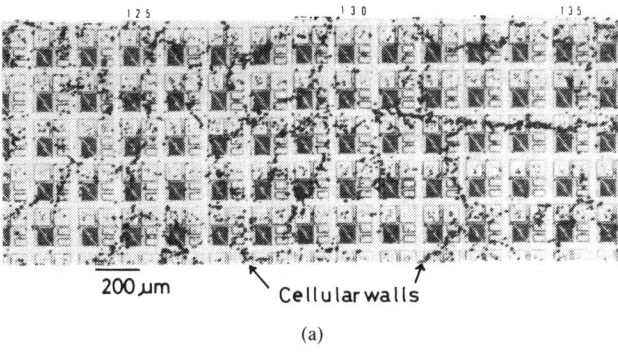

(a)

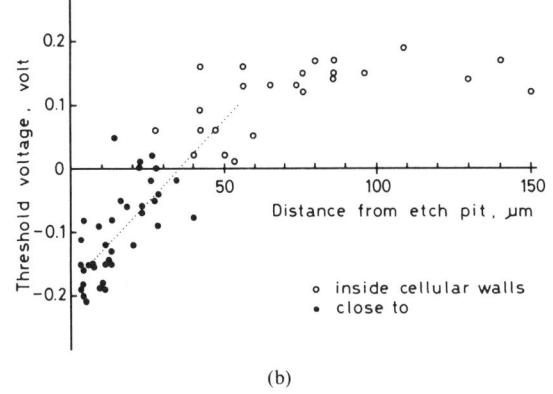

(b)

Fig. 2. (a) KOH-etched pattern for area A. Cellular dislocation networks are visible in the background of the FET metalization. (b) The threshold voltage variation as a function of the gate-to-pit distance. The slope of the dotted line is roughly 5 mV/μm.

fabricated on a Si-implanted wafer at intervals of 400 μm by means of conventional photolithography followed by annealing with a SiN cap. The detailed fabrication process has been published elsewhere [15].

III. EXPERIMENTAL RESULTS

Fig. 1(a) shows a KOH-etched wafer adjacent to the wafer examined for threshold voltage measurements where lineages running along the ⟨110⟩ direction and cellular dislocation pit networks are clearly visible. The figure is typical of conventional LEC-grown materials. The dislocation density variation across the ⟨110⟩ direction on half of the wafer is shown in Fig. 1(b). The variation shows a well-pronounced W shape across the wafer. Threshold voltage was measured automatically by means of a MAP system [14] for, typically, 50 FET's occupying a 1 × 2-mm^2 area. The measured areas are indicated by A, B, and C in Fig. 1(a). Area A, in the central part of the wafer, includes cellular networks with 100–300 μm in diameter, which is peculiar to conventional LEC-grown GaAs. This dislocation configuration was widely studied by imaging using cathodoluminescence (CL), chemical etching and image transmission microscopy. Within the cellular walls, there are essentially no dislocations. Area B includes a few arrays consisting of dislocation pits, which are dislocation-polygonized arrays often referred to as a lineage structure, leaving narrow dislocation-free regions among adjacent lineages. This dislocation-free area was also well pronounced [16]. Area C is part of the wafer periphery, where dislocation density was greater than 8×10^4 cm^{-2}, and there are neither lineages nor cellular networks.

Hereafter, dislocation proximity effect on FET threshold voltage is investigated for the areas A, B, and C.

A. Cellular Networks Structure

Fig. 2(a) shows a KOH-etched pattern of area A, where cellular networks consisting of dislocation pits are visible. Since dislocations are localized by forming cellular walls, "dislocation density" can not be defined in a strict sense. Apparent dislocation density was calculated to be as greater as 3×10^4 cm^{-2} on the average. The background patterns are KOH-etched FET metalizations. There are very few dislocations in the region inside cellular walls, which are basicaly dislocation-free. Earlier studies [9], [17] of the cellular structure on FET threshold voltage demonstrated that the threshold voltage for FET's located in the region inside the cellular wall shifted remarkably toward a normally-off side compared with that close to and/ or on the walls.

Then, the proximity effect measurements were performed for two parts; one is for FET's located inside the cellular wall and another is for FET's close to the cellular walls. The results are shown in Fig. 2(b), where the open and closed circles correspond to the values for FET's inside and close to the cellular walls, respectively. The mean threshold voltage was −0.015 V for 62 FET's. Obviously, it is apparent that a value of approximately 30–50 μm is critical, beyond which the threshold voltage exhibits a normally-off value. This means that FET's inside the walls, i.e., in dislocation-free regions, indicate a normally-off state, while those close to the walls, i.e., less than about 30 μm, indicate a normally-on state. Since the FET pitch is 200 μm and cell size varies from 100 to 300 μm, this experiment provides a wide range of FET pit distances. Up to about 50 μm, the threshold voltage appears to increase almost linearly by 5 ± 2 mV per 1 μm with the distance, as indicated by the dotted line. This tendency was also recognized in another region including the cellular structure. On the other hand, the threshold voltage for FET's located inside the walls (dislocation-free) is essentially independent of the distance at greater than about 50 μm.

From Fig. 3(b), it could still be argued that dislocations should influence FET threshold voltage at close gate-to-pit distances, especially less than about 50 μm. There have been many reports concerning the cellular structure. In particular, recent studies with CL revealed that CL intensity around cellular walls is higher than that inside the walls, and that the high intensity width is approximately 50–100 μm [16], [18]. The high CL intensity width corresponds closely to the range of the gate-pit proximity effect.

B. Around Lineage Structure

Fig. 3(a) shows an etched figure of area B, where lineages are visible as an array of dislocation pits. Lineage $L3$ is very wavy, as compared with lineages $L1$ and $L2$. Threshold voltage varies with the gate-to-pit distance as shown in Fig. 3(b), (c), and (d) for FET arrays a, c, and b and d, respectively. Notice here that lineage $L1$ runs less parallel to the FET gate array a and intersects the FET array at a low angle. This feature is very favorable for exhibiting the dislocation proximity effect on FET threshold voltage, because the gate-to-pit distance varies with each FET location. The threshold voltage variation for FET array a is shown in Fig. 3(a). The distance between lineage $L1$ and the FET channel varied from several micrometers to about 50 μm. From this result, it can be emphasized that the proximity effect of dislocations on threshold voltage is evident, and it also can be seen that a threshold voltage increases by roughly 5 ± 2 mV/μm with the distance, identical to the effect of the cellular structure described earlier (see Fig. 2(b)). The direct evidence that lineage dislocations influence the area extending several tens of micrometers on both sides was clearly demonstrated by mapping FET performance [1]. Lineage $L2$ runs almost parallel and close to FET array c. Almost all FET's on array c exhibited a normally-on threshold voltage at a proximity of less than about 30 μm, as shown in Fig. 3(c), and the dislocation proximity effect is very difficult to distinguish because of scattering.

Fig. 3(d) shows the threshold voltage for FET arrays b and d, where the open circles and closed triangles are for FET's on line b and line d, respectively. It appears that,

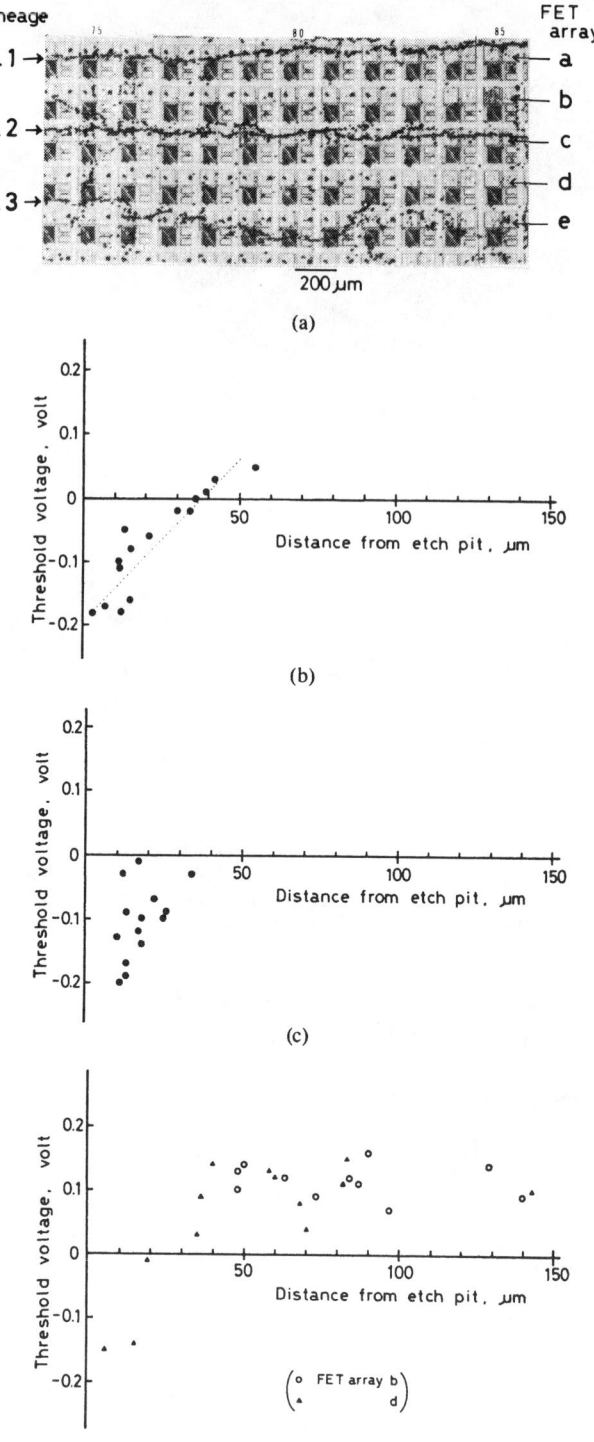

(a)

(b)

(c)

(d)

Fig. 3. (a) KOH-etched pattern for area B. Two distinct lineages ($L1$ and $L2$) and a very wavy lineage $L3$ are observed. The position indicated by the arrows for FET arrays is an FET gate. (b), (c) Threshold voltages for FET's of arrays a and c, respectively. (d) Threshold voltage for FET's of arrays b and d (see in text).

between two adjacent lineages $L1$ and $L2$, essentially no dislocation pits are observed, i.e., dislocation-free. Dislocation-free area adjacent to lineages has been imaged by CL, as was demonstrated by Chin et $al.$ [16], [18]. Contrary to FET array c, FET's of line b are more than about 50 μm away from lineages $L1$ and $L2$ in the dislocation-

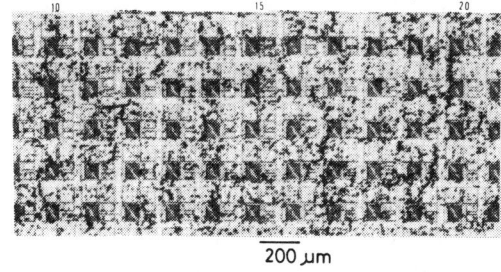

(a)

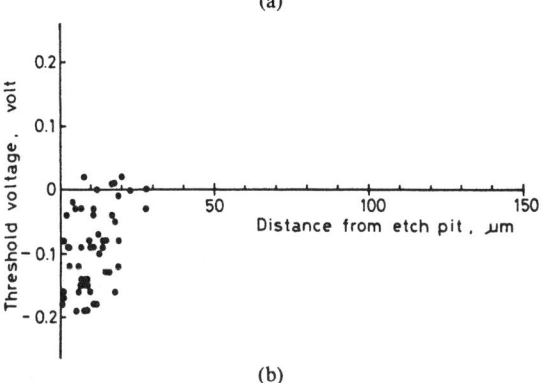

(b)

Fig. 4. (a) KOH-etched pattern for area C. Dislocations are densely distributed. (b) Threshold voltage distribution at close gate-to-pit distances.

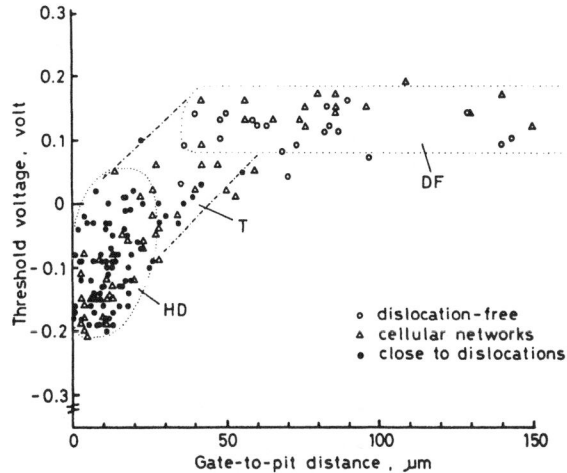

Fig. 5. The composite plot of all measured threshold voltages as a function of the distance to dislocations. This can be divided into three envelopes *DF*, *T*, and *HD* according to the FET channel locations at three types of dislocation distributions.

free area. It is noticed that the threshold voltages lie around +0.12 V with a standard deviation of 32 mV and are essentially independent of the gate-to-pit distance. This implies that dislocation-free substrates provide high uniformity in FET threshold voltage. A careful observation of FET array *d* revealed a few isolated dislocations branched out from the wavy lineage *L3*, and the threshold voltage was scattered to some extent, as shown by the closed triangles.

The envelope which aggregates the data in Fig. 3(b), (c), and (d) is very similar to that of Fig. 2(b), and consequently it can be emphasized that FET threshold voltage is strongly influenced by the dislocations according to gate-to-pit distances of less than about 50 μm, although the correspondence is less clear at very small distances.

C. High-Density Area

At the peripheral region of the wafer, dislocations distribute randomly, as shown in Fig. 4(a), and the dislocation density can reach as high as about 8×10^4 cm^{-2}. Threshold voltage variations are shown in Fig. 4(b). It is apparent that the distance between a FET channel and its nearest-neighbor dislocation pit does not exceed about 20 μm, and that almost all FET's exhibit a normally-on value. This feature is quite similar to the results plotted in Fig. 3(c) for FET's very close to lineage dislocations. In Fig. 4(b), it can be stressed that the proximity effect on the threshold voltage is not clearly seen, because the threshold voltage scatters within 200 mV, independent of the distance. This scattering is roughly equal to that for FET's located adjacent to cellular dislocation pit networks and close to the lineages, as shown in Fig. 2(b) and Fig. 3(c), respectively. The standard deviation of the threshold volt-

age for 55 FET's on 1×2.2-mm^2 area was estimated to be as small as 50 mV. A less distinct proximity effect observed here is almost consistent with that for FET's located inside cellular walls (see Fig. 2(b)) and close to lineages (see Fig. 3(c)), where the gate-to-pit distance was less than about 30 μm.

IV. DISCUSSION

All results shown in Figs. 2(b), 3(b)–(d), and 4(b) are plotted as shown in Fig. 5. This composite plot is quite consistent with the result first reported [9], where threshold voltage measurements were carried out at random in a stripe area of 1×24 mm^2 running along the $\langle 110 \rangle$ direction from the center of the wafer to the periphery. Substantially, the composite plot can be divided into three regions, i.e., dislocation-free (DF), transitional (T), and highly dislocated (HD). The measured areas corresponding to envelope DF were essentially dislocation-free, i.e., inside cellular walls and far from lineages. Threshold voltages lie around $+0.13 \pm 0.04$ V and are independent of the gate-to-pit distance. In envelope HD, which represents distances of less than about 30 μm, threshold voltages scatter within about 200 mV, but threshold voltages are lower than those for FET's about 50 μm away from dislocations. The measured areas corresponding to envelope HD were highly dislocated, i.e., close to cellular walls and lineages, and the dislocated periphery. Between the two envelopes, the threshold voltage varies with an increase of about 5 mV/μm, from normally-on to normally-off through 0 V, depending on the gate-to-pit distances. The measured areas were close to cellular dislocation networks and adjacent to wavy lineages, as described in Section III. This increase ratio is not definitive because it should be dependent of device process and/or device structure. Although there is an ambiguity at very close gate-to-pit distance, the dislocation proximity effect on FET threshold voltage can be seen up to about 50 μm.

Fig. 6 demonstrates the variation of sheet carrier con-

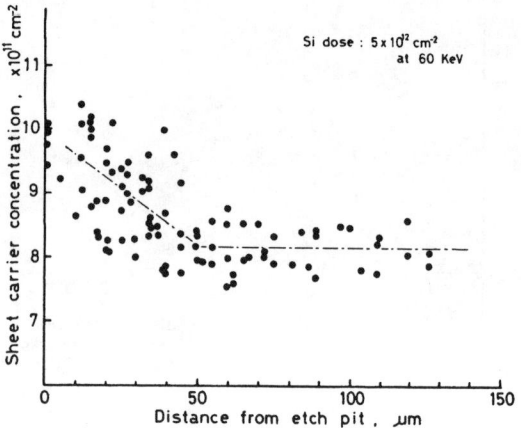

Fig. 6. Variation in sheet carrier concentration as a function of the distance between a micro Hall chip and dislocation pit. Si dose was 5×10^{12} cm^{-2} at 60 keV, and activation annealing was at 800°C for 20 min with a SiN cap. The dotted line shows a slope of 0.04×10^{11} cm^{-2}/μm.

centration in Si-implanted layer with the distance from the nearest dislocation pit. The broken line shows that the sheet carrier concentration decreases by roughly $0.04 \pm 0.02 \times 10^{11}$ cm^{-2} μm with a distance of up to about 50 μm. Since FET threshold voltage is a function of carrier concentration for the first approximation, the higher the carrier concentration, the lower the threshold voltage. The composite plot shown in Fig. 5 can be explained by the variations in sheet carrier concentration. The decrease ratio of $0.04 \pm 0.02 \times 10^{11}$ cm^{-2}/μm in carrier concentration is in agreement with the increase in threshold voltage of about 5 mV/μm estimated from Figs. 2(b) and 3(c). Therefore, it can be stressed that FET performance is strongly influenced by spatially inhomogeneous dislocations in the substrate.

Winston et al. [12] and Lee et al. [13] reported that there was no distinct relation between the gate-to-pit distance and threshold voltage. However, they described that the mean threshold voltage is dependent on localized dislocation density; the average threshold voltage was about -0.73 and -0.85 V for the areas with 3.7×10^4 and 1.3×10^5 cm^{-2} of dislocation density, respectively. This correlation has been upheld in the previous works [1], [2], [18], [19], where the threshold voltage varies in an "M" shape across the wafer corresponding to a "W" shape in the variation of dislocation density. In Fig. 5, threshold voltage scatters within about 200 mV at less than about 30 μm, resulting in no distinct proximity effect, whereas at greater than about 50 μm, threshold voltage appears to be independent of the distance and is almost constant. However, between 30–50 μm, a rough linear correlation between threshold voltage and distance can be seen. Assuming that one dislocation pit exists in an area 30–50 μm in radius, the dislocation density can be calculated to be about 3.5×10^4–1.3×10^4 cm^{-2}, respectively. Considering the present results, it is worth concluding that the threshold voltage for FET's located less than 30 μm from a dislocation, i.e., dislocation density greater than 3.6×10^4 cm^{-2} of dislocation density, does not vary noticeably

with the distance from the nearest neighbor dislocation pit, since there is a scattering within 200 mV as shown in Figs. 2(b) and 4(b). This implies less of a distinct proximity effect for dislocations on the absence of a proximity effect, argued for in [12] and [13], understandably became to be as a matter of course, because the dislocation density in the area they measured the proximity effect was 3.7×10^4 and 1.3×10^5 cm^{-2} and their scattering was about 200 mV. When we replot the Hughes data [12], [13] on Fig. 5, they fall within the envelope covered by HD, although the threshold voltages were more negative.

Additionally, the most conspicuous feature is that the Hughes' data in [12] and [13] for threshold voltages were for depletion-mode FET's namely -0.8 V or less. One of the authors [15] has recently studied the effect of a single dislocation on sheet carrier concentration in a Si-implanted and annealed layer, and estimated that the carrier concentration around a dislocation increases by 3–4 $\times$ 10^{15} cm^{-3} more than that far from the dislocation. This increase explained the composite plot shown in Fig. 5. Therefore, it is assumed here that a few crowded or closely spaced dislocations should exert a large effect on threshold voltage for enhancement-mode FET's of which the peak carrier concentration in a channel is usually around 1×10^{17} cm^{-3} and the effective channel depth is conventionally defined as the depth at which the carrier concentration is about 1×10^{16} cm^{-3}, according to the LSS theoretical carrier profile. In fact, it has been demonstrated [20] experimentally that a few crowded dislocations exert a greater effect on threshold voltage than do isolated dislocations. It can then be said that enhancement-mode FET's exhibiting approximately 0-V threshold voltage are more sensitive to dislocations than depletion-mode FET's.

Another interesting point that can be understood from Figs. 2(b), 3(c)–(d) and 4(b) is that the activation efficiency for implanted ions in dislocation-free regions is lower that in dislocated regions, resulting in a greater normally-off threshold voltage in the dislocation-free region. This was clearly confirmed by measuring sheet carrier concentrations in ion-implanted layers of dislocated and dislocation-free undoped crystals. The author's group has succesfully grown a small-diameter (as large as 14 mm) undoped semi-insulating dislocation-free crystal [21]. Fig. 7 shows the sheet carrier concentration variations across wafers of dislocated and dislocation-free crystals. The dislocated wafer shows a so-called "W"-shaped variation of dislocation density and a corresponding "W"-shaped variation in sheet carrier concentration. On the other hand, dislocation-free wafer shows a relatively uniform sheet carrier concentration, and it can be remarked that the concentration is lower than the minimum value for the dislocated wafer. This is direct evidence that threshold voltage in dislocation-free areas exhibits a more normally-off voltage than those close to dislocations.

In a photoluminescence study, Heinke and Queisser [22] reported the existance of a cylindrical area several tens of micrometers in diameter around dislocations where acceptors are "depleted." It is generally accepted that some

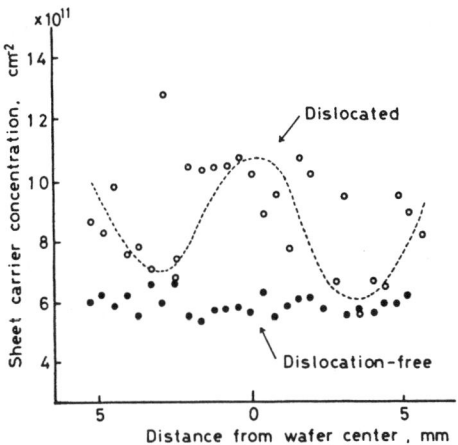

Fig. 7. Variations in sheet carrier concentration in Si-implanted layer across wafers for dislocated and dislocation-free, undoped wafers. Si dose was 5×10^{12} cm^{-2}.

impurities and/or point defects are gettered at a dislocation core due to the Cottrell atmosphere, leaving a so-called "denuded" zone 50–100 μm in radious around a dislocation. However, it has not been definitely established whether the impurities and/or point defects acting as carrier activation killers for implanted ions are gettered. The CL image at the highly dislocated periphery exhibited overlapping in the "denuded" zones with each other, resulting in an image with spatially irregular CL intensity. This may contribute to the relatively large scattering of threshold voltage. The scattering shown in Fig. 4(b) has been analyzed in terms of the stress field around a dislocation [10]. If it is true that dislocations act as sinks for some point defects or shallow acceptors with the Cottrell atmosphere during the growth process, the effect around dislocations on FET performance can be explained, but not much information about species being gettered at dislocations has been reported [23], [24].

V. CONCLUSIONS

By measuring in full the effect of dislocations on the threshold voltage for MESFET's fabricated in a conventional liquid encapsulated Czochralski (LEC) grown semi-insulating GaAs wafer, a distinct correlation was seen between the distance from the dislocation pit to FET channel and the corresponding FET threshold voltage. The examined areas were typically 1×2 mm in dimension at the wafer periphery, around lineages and cellular dislocation networks on a wafer. Mean threshold voltage over a 2-in-diameter wafer examined was -0.012 V. No dislocation pits were observed inside cellular walls and intralineages, and the threshold voltages for FET's located on these areas was normally-off. Each FET was found to be beyond about 50 μm from dislocations, and the threshold voltage lay within $+0.13 \pm 0.04$ V, independent of the distance from dislocations. On the other hand, it was recognized that threshold voltage for FET's less than about 30 μm were scattered to some extent, typically within 200 mV, and exhibited a normally-on threshold voltage. Additionally, it was roughly estimated that the threshold volt-

age shifts toward normally-off by about 5 mV/μm at distances less than about 50 μm. This was recognized by measuring the variation in sheet carrier concentration with the distance from the dislocation pits, showing a carrier concentration of roughly $0.04 + 0.01 \times 10^{11}$ cm^{-2}/μm.

Several reports on the proximity effect of dislocations on FET threshold voltage have concluded that no significant correlation exists between dislocations and threshold voltage shift. However, the data [12], [13] was given for areas with relatively high dislocation density, greater than 3.7×10^4 cm^{-2}. From the present experimental analyses, areas with greater than about 3.6×10^4 cm^{-2} of dislocation density shows a less distinct shift of threshold voltage with change in proximity, because of a larger scattering at very close gate-to-pit spacings less than about 30 μm. Even though there is still some ambiguity about the dislocation proximity effect at very close gate-to-pit distances, it can be emphasized that the proximity effect apparently exists up to a distance of about 50 μm.

The present result leads to the conclusion that materials which are perfectly dislocation-free are required for substrates for GaAs LSI's with direct ion implantation. More recently, the author's group has succeeded in growing perfectly dislocation-free, semi-insulating 2-in-diameter GaAs [25], and has also confirmed the high uniformity of electrical behavior for both As-grown and Si-implanted materials [15]. Another approach to obtain homogeneous substrates is to perform post-growth high-temperature annealing [11]. However, more detailed experiments and analyses including dislocation effects are needed for practical uses.

ACKNOWLEDGMENT

The authors would like to express their thanks to Y. Ishii for measuring the FET performance, and are also indebted to K. Hoshikawa for his thoughtful discussions and to Dr. M. Fujimoto and Dr. T. Ikegami for their constant encouragement. Thanks are also due to K. Yamada for growing a dislocation-free GaAs used in this study.

REFERENCES

[1] Y. Nanishi, S. Ishida, and S. Miyazawa "Correlation between dislocation distribution and FET performance observed in low Cr-doped LEC GaAs," *Japan J. Appl. Phys.*, vol. 22, no. 1, pp. L54–L56, 1983.
[2] S. Miyazawa and Y. Nanishi, "Characterization of semi-insulating GaAs substrate for GaAs IC's," in *Proc. 1982 Int. Conf. Solid-State Devices*, (Tokyo), 1982, also *Japan J. Appl. Phys.*, vol. 22, no. 22-1, pp. 419–425, 1983.
[3] R. T. Blunt, S. Clark, and D. J. Stirland, "Dislocation density and sheet resistance variations across semi-insulating GaAs wafers," *IEEE Trans. Electron Devices*, vol. ED-29, no. 7, pp. 1039–1045, 1982.
[4] S. Miyazawa, T. Mizutani, and H. Yamazaki, "Leakage current I_L variation correlated with dislocation density in undoped, semi-insulating LEC-GaAs," *Japan J. Appl. Phys.*, vol. 21, no. 9, pp. L542–L544, 1982.
[5] I. Grant, D. Rumsby, R. M. Ware, M. R. Brozel, and B. Tuck, "Etch pit density, resistivity, and chromium distribution in chromium-doped LEC GaAs," in *Semi-Insulating III-V Materials*, S. Makram-Ebeid and B. Tuck, Eds. Cheshire, U.K.: Shiva Publishing, 1982, pp. 98–106.
[6] K. Kitahara, K. Nakai, and S. Shibatomi, "One-dimensional photoluminescence distribution in semi-insulating GaAs grown by CZ and

HB methods," *J. Electrochem. Soc.*, vol. 129, no. 4, pp. 880–883, 1982.

[7] M. Tajima and Y. Okada, "Characterization of deep levels in LEC GaAs crystals by photoluminescence technique," *Physica*, vol. 116B, pp. 404–408, 1983.

[8] D. E. Holmes, R. T. Chen, and J. Yang, "EL2 distributions in doped and undoped liquid encapsulated Czochralski GaAs," *Appl. Phys. Lett.*, vol. 45, no. 5, pp. 419–421, 1983.

[9] S. Miyazawa, Y. Ishii, S. Ishida, and Y. Nanishi, "Direct observation of dislocation effects on threshold voltage of a GaAs field-effect transistor," *Appl. Phys. Lett.*, vol. 43, no. 9, pp. 853–855, 1983.

[10] S. Miyazawa and Y. Ishii, "Dislocations as the origin of threshold voltage scattering for GaAs MESFET on LEC-grown, semi-insulating GaAs substrate," *IEEE Trans. Electron Devices*, vol. ED-31, no. 8, pp. 1057–1062, 1984.

[11] S. Miyazawa, T. Honda, Y. Ishii, and S. Ishida, "Improvement of crystal homogeneities in liquid-encapsulated Czochralski grown, semi-insulating GaAs by heat treatment," *Appl. Phys. Lett.*, vol. 44, no. 4, pp. 410–412, 1984.

[12] H. V. Winston, A. T. Hunter, H. M. Olsen, R. P. Bryan, and R. E. Lee, "Substrate effects on the threshold voltage of GaAs field-effect transistors," *Appl. Phys. Lett.*, vol. 45, no. 4, pp. 447–449, 1984.

[13] R. E. Lee, A. T. Hunter, R. P. Bryan, H. M. Olsen, H. V. Winston, and R. S. Beaubien, "Threshold voltage uniformity of MESFET's fabricated on GaAs and In-alloyed GaAs substrates," *IEEE Tech. Dig. 1984 GaAs IC Symp.* (Boston), pp. 45–48, 1984.

[14] Y. Ishii, S. Miyazawa, and S. Ishida, "Characterization of thin active layer on semi-insulating GaAs by mapping of FET array performance," *IEEE Trans. Electron Devices*, vol. ED-31, no. 8, pp. 1051–1056, 1984.

[15] F. Hyuga, H. Kohda, H. Nakanishi, T. Kobayashi, and K. Hoshikawa, "Electrical uniformity for Si-implanted layer of completely dislocation-free and striation-free GaAs," *Appl. Phys. Lett.*, Sept. 15, 1985.

[16] A. K. Chin, A. R. VonNeida, and R. Caruso, "Spatially resolved cathodoluminscence study of semi-insulating GaAs substrates," *J. Electrochem. Soc.*, vol. 129, no. 10, pp. 2386–2388, 1982.

[17] Y. Ishii, S. Miyazawa, and S. Ishida, "Threshold voltage scattering of GaAs MESFET's fabricated on LEC-grown semi-insulating substrate," *IEEE Trans. Electron Devices*, vol. ED-31, no. 6, pp. 800–804, 1984.

[18] A. K. Chin, R. Caruso, M. S. S. Young, and A. R. VonNeida, "Uniformity characterization of semi-insulating GaAs by cathodoluminescence imaging," *Appl. Phys. Lett.*, vol. 45, no. 5, pp. 552–554, 1984.

[19] T. Egawa, Y. Sano, H. Nakamura, T. Ishida, and K. Kaminishi, "The dependence of threshold voltage scattering of GaAs MESFET on annealing method," *Japan J. Appl. Phys.*, vol. 24, no. 1, pp. L35–L38, 1985.

[20] Y. Matsuoka, K. Ohwada, and M. Hirayama, "Uniformity evaluation of MESFET's for GaAs LSI fabrication," *IEEE Trans. Electron Devices*, vol. ED-31, no. 8, pp. 1062–1067, 1984.

[21] K. Yamada, unpublished.

[22] W. Heinke and H. J. Queisser, "Photoluminescence at dislocations in GaAs," *Phys. Rev. Lett.*, vol. 33, no. 18, pp. 1082–1084, 1974.

[23] B. Wakefield, P. A. Leigh, M. H. Lyons, and C. R. Elliot, "Characterization of semi-insulating liquid encapsulated Czochralski GaAs by cathodoluminescence," *Appl. Phys. Lett.*, vol. 45, no. 1, pp. 66–68, 1984.

[24] K. Watanabe, H. Nakanishi, K. Yamada, and K. Hoshikawa, "Spatially resolved electrical and spectroscopic studies around dislocations in GaAs single crystals," *Appl. Phys. Lett.*, vol. 45, no. 6, pp. 643–645, 1984.

[25] H. Kohda, K. Yamada, H. Nakanishi, T. Kobayashi, J. Osaka, and K. Hoshikawa, "Crystal growth of completely dislocation-free and striation-free GaAs," *J. Cryst. Growth*, vol. 71, no. 3, pp. 813–816, 1985.

Mechanisms for Low-Frequency Oscillations in GaAs FET's

DANIEL J. MILLER AND MARINA BUJATTI, MEMBER, IEEE

Abstract—Low-frequency oscillations in GaAs MESFET's were observed under back-gating conditions. The FET oscillations are directly related to oscillations in leakage currents in the semi-insulating GaAs substrate. The occurrence of these oscillations in the substrate is strongly dependent upon GaAs material. It is proposed that oscillating substrate leakage currents modulate the FET current in two ways; first, by modulating the active channel-substrate junction and second, by inducing periodic voltage fluctuations on the gate via gate pad contacts on the semi-insulating substrate. The latter mechanism is dominant and dependent upon gate bias and gate impedance.

I. INTRODUCTION

LOW-FREQUENCY current oscillations in semi-insulating GaAs have been extensively modeled [1]–[4]. Although the exact mechanism is not fully understood, it is generally accepted that low-frequency oscillations result from the field-enhanced capture of electrons by deep traps. The charge domains established by trapping move with a velocity dependent upon the current density and the time constant of the trap involved [1], [2]. The frequency of the current oscillation is determined by the transit time of the high field domains traveling between an electron-injecting cathode and an anode on the semi-insulating GaAs.

Low-frequency oscillations in various GaAs FET structures have also recently been reported [5]–[7]. In one case, oscillations observed in MESFET's were attributed to the high-field tunneling of electrons into traps in the gate depletion region [5]. Inconsistent with a tunneling mechanism, however, the same phenomenon was recently observed in GaAs MISFET's under forward gate bias [6]. Low-frequency oscillations in GaAs FET's have also been observed under side-gating conditions and related in a general way to oscillations occurring in semi-insulating substrates [7].

The purpose of this paper is to identify the mechanisms whereby oscillations in the leakage current in a semi-insulating GaAs substrate directly result in oscillations in GaAs MESFET's. The work originates from the observation of unusually strong and anomalous phase noise in

several GaAs circuits. Spurious sidebands, which were often found to depend upon the assembly technique, were observed in low-noise GaAs FET amplifiers fabricated with hybrid technology. However, the problem was found to be more severe in monolithic circuits. Anomalous and substantial peaks, superimposed on the expected $1/f$ and phase noise, were found to severely limit the noise performance of several GaAs MMIC's. It was found that these noise characteristics were typically the result of coherent oscillations in the FET's mixed by nonlinear circuit elements. One specific example has been previously reported [8].

Two mechanisms for the modulation of a MESFET current by an oscillating substrate leakage current are presented. It is felt that an understanding of these mechanisms is of particular pertinence to the design of low-noise GaAs IC's. Measurements and observations of oscillations in various semi-insulating GaAs materials are also presented. These measurements should be of practical significance to circuit designers and, it is hoped, be of value to those concerned with elucidating the fundamental nature of oscillations in GaAs.

II. OSCILLATIONS IN SEMI-INSULATING GaAs

Oscillations in semi-insulating GaAs were measured in the configuration shown in Fig. 1(a). Ohmic contacts on semi-insulating GaAs were formed by standard ion implantation, thin-film deposition, and wet etching techniques. The rectangular ohmic contacts were separated by various distances (10, 25, and 100 μm) of mesa-isolated semi-insulating GaAs. These ohmic contacts were biased in series with a 10-MΩ resistor. An HP 3582A low-frequency spectrum analyzer was used to measure the voltage functions across the resistor as the leakage current oscillated. In a similar fashion, oscillations in leakage current between a topside ohmic contact and a grounded backside metallization were measured (Fig. 1(b)). These measurements were made on GaAs wafers lapped to thicknesses ranging from approximately 450 to 40 μm. In these arrangements, the -125-dB $\cdot$ V sensitivity of the HP 3582A enabled the resolution of oscillations of magnitude as small as 1.0 pA. Low-frequency oscillations are sensitive to illumination. All measurements reported here were taken under dark conditions. The thermal activation of oscillations was determined by measurements taken between 0 and 125°C with the sample mounted in a cryo-

Manuscript received September 23, 1986; revised February 2, 1987.

D. J. Miller was with the Microwave Technology Division, Hewlett-Packard Company, Santa Rosa, CA 95401. He is now with the Advanced Packaging Laboratory, Hewlett-Packard Laboratory, Palo Alto, CA 94304.

M. Bujatti was with the Microwave Technology Division, Hewlett-Packard Company, Santa Rosa, CA 95401. She is now with Microwave Power Research, Santa Clara, CA 95054.

IEEE Log Number 8714141.

Reprinted from *IEEE Trans. Electron Devices*, vol. ED-34, no. 6, pp. 1239–1244, June 1987.

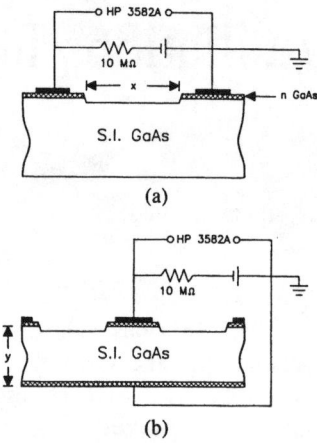

Fig. 1. Configurations for the measurement of low-frequency oscillations (a) between isolated surface contacts and (b) through the substrate thickness.

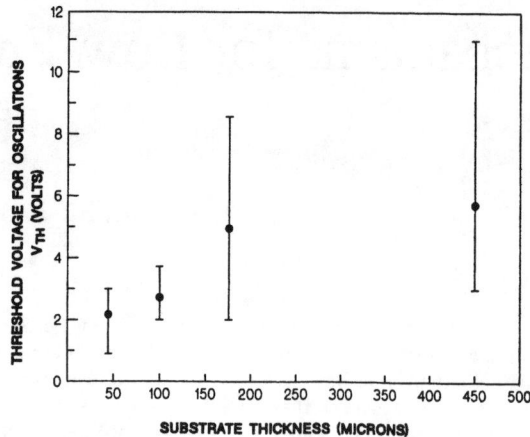

Fig. 2. Dependence of threshold voltage for low-frequency oscillations upon substrate thickness.

system with temperature control of $\pm 0.2°$C. During measurements of oscillation activation, a constant bias was maintained across the semi-insulating GaAs sample. This was accomplished by adjusting the voltage across the series connection of the 10-MΩ resistor and GaAs sample in such a way as to account for the temperature dependence of the GaAs resistance. The accuracy of these measurements is estimated to be ± 0.1 eV based on the reproducibility of the results.

For both geometries current oscillations arose at a distinct threshold voltage (V_{TH}) consistent with a mechanism of field-enhanced capture of electrons by traps. Often more than one distinct oscillation frequency was observed with the different oscillations exhibiting slightly different threshold voltages. At room temperature, oscillation frequencies were typically less than 100 Hz. Threshold voltages for current oscillations between a front side ohmic contact and the backside ground are plotted as a function of substrate thickness in Fig. 2. A fairly wide range in V_{TH} is observed indicating the effects of nonuniform electric fields and substrate inhomogeneities. For the thinnest GaAs substrate, in which adjacent contacts can be expected to perturb the electric field the least, threshold fields of 200 to 600 V/cm are calculated. These values are in close agreement with previously reported values calculated under the same assumption of uniform field distribution [2], [9], [10].

The threshold voltage for oscillations should not be confused with the well-known trap-filled-limit voltage associated with the space-charge limited current in S. I. GaAs. A representative $I–V$ curve for semi-insulating GaAs is seen in the log-log plot of Fig. 3. As electrons are injected into traps, the current–voltage characteristics undergo a transition from an ohmic to an $I \alpha V^2$ dependence. At the trap-filled-limit voltage V_{TFL}, a dramatic increase in current occurs. This sharp rise in current results when the quasi-Fermi level crosses a trap level in the semi-insulating GaAs substrate. The threshold voltage for oscillations was always reached before the trap-filled-limit voltage was attained. Oscillation amplitudes were approx-

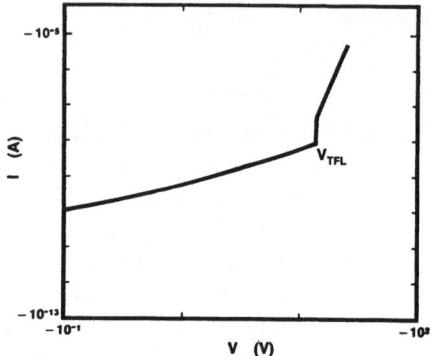

Fig. 3. Typical $I–V$ curve for space-charge-limited current in semi-insulating GaAs with indication of trap-filled-limit voltage.

imately three orders of magnitude lower than the dc leakage current. At the threshold voltage, oscillation amplitudes were typically of the order of 1.0–3.0 pA in a dc current of a few nanoamperes. As the applied voltage was increased beyond the threshold voltage, the oscillation amplitudes also increased, reaching values as great as a few nanoamperes. Beyond the trap-filled-limit, however, the noise background would typically increase to the extent that coherent oscillations were masked.

Since it is well known that semi-insulating GaAs materials grown by different technologies or under different conditions will exhibit distinctly different trap densities, it is not surprising that the susceptibility of semi-insulating GaAs to low-frequency oscillations is strongly material dependent. Large-amplitude oscillations were observed in undoped, low Cr(< 0.5 wt-ppm)-doped, In-doped (2–3 wt-ppm), and ingot annealed LEC materials. A thick buffer (10 μm) on undoped LEC GaAs caused a reduction in oscillation amplitudes. Highly Cr-doped LEC and low Cr-doped horizontal Bridgman material showed significantly lower oscillatory behavior. Semi-insulating Bridgman material with no chromium doping, however, was susceptible to large amplitude oscillations. This dependence of oscillation susceptibility upon chromium content is also in evidence in a previously reported study of oscillations in MESFET's [5].

LOW FREQUENCY OSCILLATION TEMPERATURE DEPENDENCE

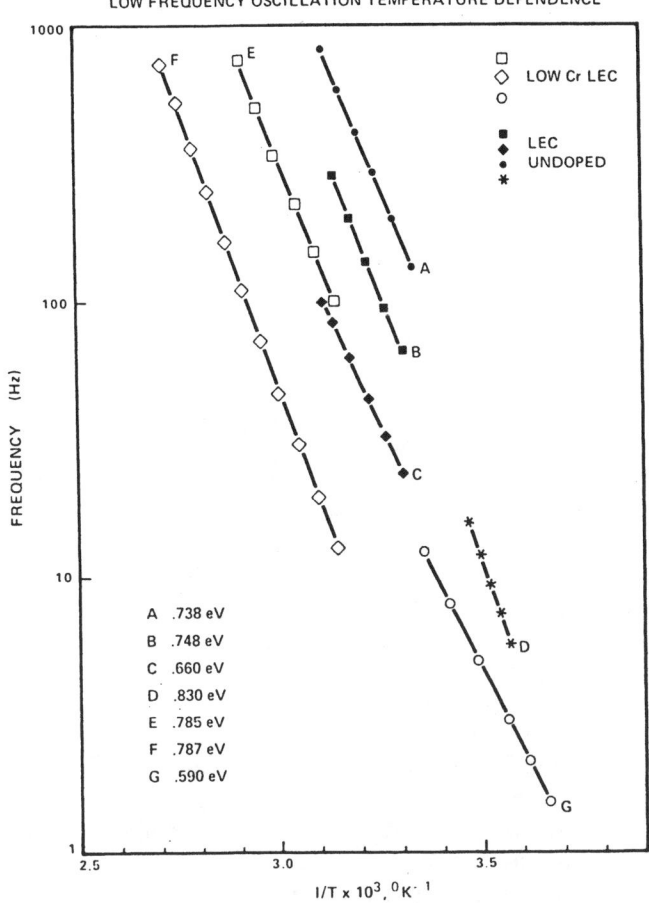

Fig. 4. Thermal activation of low-frequency oscillations in undoped and low-Cr-doped LEC GaAs.

The thermal activations of oscillations in two representative semi-insulating GaAs materials (undoped and low Cr-doped LEC) are shown in Fig. 4. At a fixed bias condition, the oscillations were observed to be stable within a specific temperature range. As the temperature increased, the oscillations would eventually decrease in magnitude and become less coherent until they disappeared into the background noise. The two samples shown in Fig. 4 were measured under a constant bias condition of approximately 1 kV/cm. Thus, the field dependence of oscillation activation energies [10] is not a factor in these measurements. The thermal activation of oscillations will depend upon the thermal activation of the current density in semi-insulating GaAs as well as the thermal activation of the time constant of responsible traps [1], [2]. The activation energies presented in Fig. 4 should not, therefore, be interpreted as the energy levels of traps in semi-insulating GaAs. The significant feature of Fig. 4, however, is the coexistence of oscillations such as those marked "A" and "C," i.e., oscillations of significantly different activation energies. In some samples this behavior was even more conspicuous as spectral lines of oscillations passed through each other with change in temperature. These results, along with previous studies in which multiple activation energies for oscillations have been measured [5], [9], indicate that the recombination mechanisms associated with low-frequency oscillations involve more than one trap.

III. OSCILLATIONS IN GaAs FET'S

Because of our interest in extending the results of this investigation to the case of integrated circuits, most of the work was carried out on FET's fabricated with a process very similar to our standard IC process [11]. In particular, direct ion implantation was used in most of this work, with a dose of $5 \times 10^{12}/cm^2$ of Si at both 60 and 200 keV. Annealing was carried out at 850°C for 30 min in hydrogen with a SiO_2 cap. In some cases, implantation was into an epitaxial buffer grown by organometallic vapor phase epitaxy.

In order to reduce process-induced effects, the FET processing was as simple as possible and consisted of Au/Ge ohmic contacts, chemical mesa etching for isolation, a constant gate recess by chemical etching, and Ti/Pt/Au gates deposited by electron-beam evaporation and defined by direct photoresist lifting. The FET had a gate length of 1 μm and a gate width of 350 μm.

Although the main focus was on ion-implanted devices, several epitaxial layers were also processed and measured in exactly the same manner. In particular, several MBE layers, grown on different substrate materials, were included in this investigation. The active layer, 0.3 μm thick and doped with a constant Si concentration of $2 \times 10^{17}/cm^3$, was grown on a 0.5-μm undoped buffer. All the measurements performed on epitaxial layers indicated that the low-frequency oscillation characteristics of the devices are dominated by the substrate used. In other words, devices fabricated on epitaxial layers grown on a given type of substrate exhibited the same characteristics as devices formed by direct ion implantation into the same substrate.

Oscillations in GaAs FET's were observed under sidegating and back-gating conditions. The condition most comparable to that found in the operation of a GaAs IC is a sidegate condition. It is that geometry, in which a negative bias is applied to an adjacent ohmic contact, that we will discuss in detail. The test configuration is shown in Fig. 5 in which relevant currents have been indicated. The adjacent ohmic contact was isolated from the FET by 100 μm of semi-insulating GaAs. The FET was typically biased to saturation ($V_{DSS} \sim$ 1.5–2 V, $I_{DSS} \sim$ 80–120 mA) although FET oscillations were evident at all values of V_{DS}. Using the HP3582A spectrum analyzer, oscillations in the substrate leakage current I_{BG} and the FET current I_{DS} were detected as voltage fluctuations across the 10-MΩ and 100-Ω resistors, respectively. The total leakage current I_{BG} consists of currents from the FET active region I_A; from the gate pad located on the semi-insulating substrate I_{GP}; and, should the connection be made, from the grounded backside I_B. The measurable gate current is denoted I_{GT} and consists of the gate-pad current I_{GP} and the gate leakage current I_G. It was essential to determine the relative magnitudes of these currents in order to understand their contributions to FET oscillations. The lim-

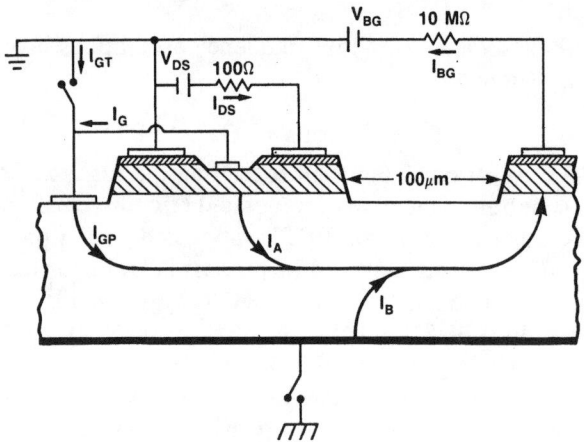

Fig. 5. Configuration for the measurement of low-frequency oscillations in FET current I_{DS} and back-gating current I_{BG}.

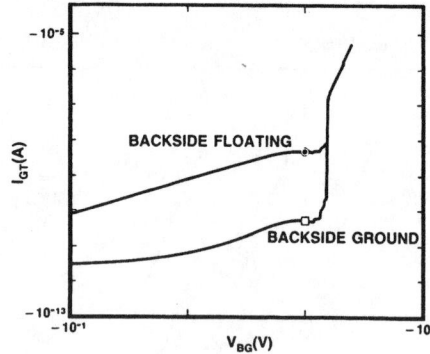

Fig. 6. Dependence of total gate-leakage current upon backside ground condition.

itations imposed by the dimensions of the active-gate (1-μm length) made the isolated measurement of I_G impossible. The relative magnitude of I_G was determined from a diode structure with a Schottky metal contact approximately equivalent in area to the active gate. The diode was located in the vicinity of the FET under test and located on n-type GaAs. An adjacent isolated ohmic contact (100 μm distant) was negatively biased to determine the effect on the diode current.

The magnitudes of the various currents shown in Fig. 5 were determined with an HP 4145 semiconductor parameter analyzer. The relative magnitudes of I_{GP} and I_A can be determined from measurements of I_{GT} and I_{BG} if the contribution of I_G to I_{GT} is known. Let us first consider the case in which the backside of the substrate is floating and the gate pad is grounded. In this case, $I_{BG} = I_{GP} + I_A$, and $I_{GT} = I_{GP} - I_G$, where the currents are defined as positive as drawn in Fig. 5. Inherent in this logic is the assumption that I_A does not contribute significantly to I_G. This was confirmed from measurements of the influence of side-gating on our diode current. A diode (or the active-gate region of a FET) will tend to be forward biased by a negative voltage on an adjacent ohmic contact. The diode current resulting from this side-gate voltage will, however, be extremely small due to voltage drop in the semi-insulating GaAs and also due to the potential gradient in the depletion region of the Schottky contact. Actual measurements show that, for side-gate voltages less than the trap-filled-limit (V_{TFL}), the component of I_A contributing to I_G will be less than 1.0 pA. The active-gate leakage current I_G for a current-saturated FET was typically of the order of 0.1–1.0 nA. This is the same order of magnitude as the leakage current through the gate pad I_{GP}. It was, in fact, possible to observe a change in the sign of I_{GT} depending upon the bias condition of the FET and the applied side-gate voltage. As is expected from the dominance of the resistance of the semi-insulating GaAs, the leakage currents through the FET structure, I_{GP} and I_A, are of the same order of magnitude, i.e., 0.1–10 nA at voltage less than the trap-filled limit.

Next, let us consider the condition in which the backside of the substrate is grounded. The substrate leakage current in this case will be $I_{BG} = I_{GP} + I_A + I_B$. The parallel current path to the backside will tend to shunt current from the FET structure. An example of this is shown in Fig. 6 in which the total gate current I_{GT} is shown as a function of backside ground condition. These measurements were made with the source, gate, and drain of the FET at ground potential so that $I_{GT} \sim I_{GP}$ as previously explained. A large decrease in leakage current through the gate pad is evident due to the shunting of current by the grounded backside. The current through the active area I_A will be affected in the same manner.

Measurements of oscillations in GaAs FET current revealed them to be directly related to oscillations in the semi-insulating substrate leakage current. This correlation is shown in Fig. 7(a), in which the spectra of oscillations in both FET current I_{DS} and substrate leakage current I_{BG} are presented. Note the gate pad and the substrate backside are grounded for this measurement. A complex spectrum of oscillations in I_{BG} is seen (lower trace). The spectrum consists of several major peaks. Oscillations in I_{DS} (upper trace) can also be seen above the background noise. These FET oscillations are matched in frequency to the oscillations in the substrate leakage current. Whereas the oscillations in the back-gate current are typically of the order of 1–100 pA, the oscillations in the FET are typically of the order of 0.1–5.0 μA. The effect of disconnecting the gate from ground is shown in Fig. 7(b). New oscillations in I_{BG} become evident and the oscillations in the FET current I_{DS} increase in magnitude. Most typically disconnecting the gate from ground would cause the FET oscillations to increase in size by an order of magnitude. In order to explain this dependence of FET oscillation upon gate connection, oscillations were measured in FET structures which, while having gate-pad contacts, contained no active gate metallization. Oscillations in these "gateless" FET's were observed under back-gating conditions as with normal FET's. Likewise, the frequencies of FET oscillations were matched to oscillations occurring in the substrate leakage current. In contrast to FET oscillation measurements, however, the oscillations in gateless FET structures were unaffected by gate-pad ground connection.

In both normal and gateless FET's, low-frequency oscillations in I_{BG} and I_{DS} were altered by disconnecting the

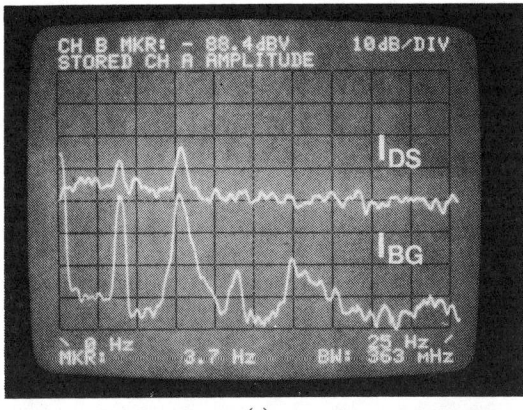

(a)

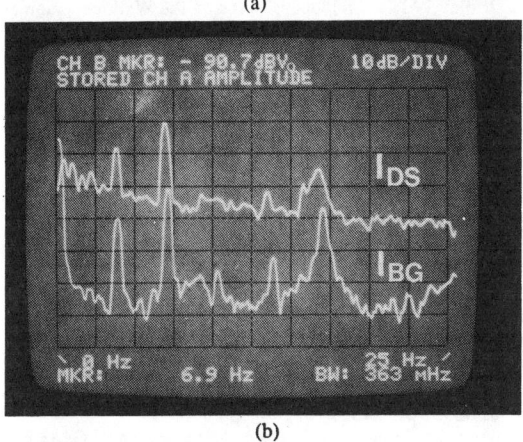

(b)

Fig. 7. Correlation of low-frequency oscillations in FET current I_{DS} and back-gate current I_{BG} with (a) gate pad ground and (b) gate pad floating. Substrate backside is grounded in both measurements.

backside ground. The changes in the oscillation spectrum were pronounced and dependent upon bias conditions. FET oscillations were largest when both the FET gate and backside were disconnected from ground. Table I presents representative results for the oscillations in a current-saturated FET under these different conditions.

An attempt was made to measure low-frequency oscillations in the reverse-biased current of our diode. No oscillations in this current were observed.

Based on these results, we propose that low-frequency oscillations in FET's are a direct result of oscillations in the leakage current in the semi-insulating GaAs substrate. The total leakage current between an FET and an adjacent circuit element will consist of current through gate pads on the semi-insulating GaAs substrate and current through the active FET channel. These two current paths result in two mechanisms for the modulation of the FET by the oscillating leakage current; first, by modulation of the active gate region through interaction with the gate pad, and, second, by modulation of the active layer–substrate interface. In Fig. 5 these currents are denoted as I_{GP} and I_A.

Oscillations in I_{GP}, the gate-pad component of the leakage current, can result in modulation of the voltage at the active gate. The severity of this modulation, and thus the severity of FET oscillation, will depend upon the low-frequency impedance to ground for the gate contact and the magnitude of the leakage-current oscillation. If the impedance to ground is large, oscillations in the leakage

TABLE I
TYPICAL DEPENDENCE OF LOW-FREQUENCY FET OSCILLATION MAGNITUDE UPON GROUND CONDITION OF GATE PAD AND BACKSIDE
(Conditions of measurement: $V_{DS} = 1.5$ V, $I_{DS} = 87$ mA, $V_{BG} = 8.5$ V.)

BACKSIDE \ GATE PAD	GROUND	FLOATING
GROUND	~ 2 μA	~ 10 μA
FLOATING	~ 5 μA	~ 100 μA

current will result in voltage oscillations at the active gate causing large FET oscillations. If the gate is well grounded, the oscillating leakage current will be shunted and will not result in FET oscillations. This argument explains the observed dependence of oscillations upon gate connection as shown in Fig. 7.

Oscillations in the FET current also result from the modulation of the space-charge region at the active layer–substrate interface by an oscillating leakage current. This mechanism of FET oscillation is identical to the proposed mechanism for back-gating in GaAs FET's whereby electrons injected into the semi-insulating substrate are balanced by an increase in the positive donors at the active layer–substrate interface [12]. This basic mechanism can be extended to the periodic modulation of the active layer–substrate interface by the traveling charge domains responsible for oscillations in the semi-insulating substrate. The oscillations in gateless FET structures can be attributed to this mechanism. In a normal FET, the two mechanisms will act in conjunction.

The dependence of FET oscillations on backside ground condition is a result of the influence of ground condition on the leakage current through the FET. When the backside of a FET is grounded, the parallel current path reduces the magnitude of leakage current through the FET structure and the magnitude of oscillations is reduced.

Based on this understanding of the mechanisms by which oscillating leakage currents modulate the FET current, FET's were fabricated with gate pads isolated from the semi-insulating substrate with a layer of polyimide. In order to appreciably reduce the leakage current through the pad, a 4.0-μm layer of polyimide was required. The difficulty of producing a continuous step coverage of the gate metal over such a relief was overcome by ion milling a cavity into the semi-insulating substrate before the polyimide deposition. In comparison to FET's with gate pads located directly on the semi-insulating GaAs, the isolated FET's exhibited a gate-pad leakage current reduced by an order of magnitude. Consistent with the mechanisms proposed here, the isolated FET's also exhibited oscillations that were unaffected by the impedance to ground of the gate.

IV. CONCLUSIONS

In agreement with earlier work, low-frequency oscillations were observed to be induced in semi-insulating GaAs when a threshold voltage, related to a threshold field for electron capture by deep traps, was applied. The low values of threshold fields are of practical significance; considering typical geometries and voltages used in GaAs IC's, there is a high probability that the leakage current between adjacent circuit elements will be oscillatory.

The tendency of various semi-insulating GaAs materials to exhibit oscillations is strongly dependent upon crystal growth technology. The coexistence of oscillations with distinctly different activation energies indicates that more than one trap is involved in the phenomenon. Chromium-doped material is much less prone to exhibit oscillations. Certainly, more detailed work is required to understand the relation between trap concentrations and phenomenon of low-frequency oscillations.

The spectral lines of oscillations observed in an active device were correlated one to one with spectral lines of oscillations in the substrate leakage current. This was true both for devices fabricated by direct ion implantation into the substrate and for devices fabricated on epitaxial layers grown on the substrate.

Two mechanisms were identified for the transfer of oscillations from the substrate to the active devices, corresponding to the two major leakage paths through the FET structure: across the substrate-active layer and through the gate-pad metallization located on the semi-insulating substrate. The leakage current through the substrate–active layer interface modulates the FET current by a back-gating effect. As such this mechanism has already received some attention in the literature. On the other hand, the oscillating leakage currents collected by the gate pad induced voltage fluctuations on the gate and are amplified in the same manner. This mechanism has not been pointed out before and is strongly dependent upon the impedance of the gate to ground. If the impedance is large, this mechanism is predominant. Large effective gate pads, including any metallizations directly connected to the gate in GaAs IC's, exacerbate this modulation mechanism and can lead to severe noise problems.

Both modulation mechanisms are significantly reduced when the leakage currents present in the substrate are shunted away from the active device by thinning down the substrate and providing effective backside ground.

ACKNOWLEDGMENT

This work was accomplished at the Microwave Technology Division of Hewlett-Packard in Santa Rosa, California. The authors would like to acknowledge helpful suggestions from D. Estreich and C. Stolte, who are located at that division. Likewise, we are very grateful to S. Harris for her careful processing and meticulous measurements. We also wish to extend our thanks to F. C. Wang, now with Morgan Semiconductor Division, Ethyl Corp., for his stimulating discussions.

REFERENCES

[1] B. K. Ridley, "The influence of traps on the Watkins–Gunn effect," *Brit. J. Appl. Phys.*, vol. 17, pp. 595–602, 1966.

[2] H. K. Sacks and A. G. Milnes, "Low frequency oscillations in semi-insulating gallium arsenide," *Int. J. Electron*, vol. 28, no. 6, pp. 565–583, 1970.

[3] H. K. Sacks and A. G. Milnes, "An experimental study of high-field moving domains produced by deep centres in semi-insulating GaAs," *Int. J. Electron*, vol. 30, no. 1, pp. 49–54, 1971.

[4] B. K. Ridley, "A mechanism for negative differential resistance in III–V and II–VI semi-conductors associated with the enhanced radiative capture of hot electrons," *J. Phys. C: Solid State Phys.*, vol. 7, pp. 1169–1173, 1974.

[5] M. B. Das and P. K. Ghosh, "Low frequency emissions from deep levels in GaAs MESFETs," *Electron. Lett.*, vol. 18, no. 5, Mar. 1982.

[6] P. Canfield and L. Forbes, "Gate-bias-dependent low frequency oscillations in GaAs MISFET's," *IEEE Electron Device Lett.*, vol. EDL-6, no. 5, pp. 227–228, May 1985.

[7] S. Makram-Ebeid and P. Minondo, "Side-gating in GaAs integrated circuits: surface and bulk related phenomena," in *GaAs IC Symp. Tech. Dig.*, 1983.

[8] D. Miller, M. Bujatti, and D. Estreich, "Low-frequency oscillations in GaAs ICs," in *GaAs IC Symp. Tech. Dig.*, pp. 31–34, 1985.

[9] G. N. Maracas, D. A. Johnson, and H. Goronkin, "Experimental evaluation of low-frequency oscillations in undoped GaAs to probe deep level parameters," *Appl. Phys. Lett.*, vol. 46, no. 3, pp. 305–307, 1985.

[10] M. Kaminska, J. M. Parsey, J. Lagowksi, and H. C. Gatos, "Current oscillations in semi-insulating GaAs associated with field enhanced capture of electrons by major deep donor EL2," *Appl. Phys. Lett.*, vol. 41, no. 10, pp. 989–991, Nov. 15, 1982.

[11] R. L. Van Tuyl *et al.*, "A manufacturing process for analog and digital gallium arsenide integrated circuits," *IEEE Trans. Electron Devices*, vol. ED-29, pp. 1031–1037, July 1982.

[12] C. Kocot and C. Stolte, "Backgating in GaAs MESFET's," *IEEE Trans. Electron Devices*, vol. ED-29, pp. 1059–1064, July 1982.

Frequency-Dependent Electrical Characteristics of GaAs MESFET's

J. MICHAEL GOLIO, SENIOR MEMBER, IEEE, MONTE G. MILLER,
GEORGE N. MARACAS, MEMBER, IEEE, AND DAVID A. JOHNSON

Abstract—Output resistance and transconductance of GaAs MES-FET's have been observed to change significantly at very low frequencies. Extensive measurements of these characteristics as a function of device bias are reported in this work. Direct measurements of the dispersive behavior between dc and 100 kHz and over a broad temperature range have been made on ion-implanted MMIC devices. Conductance DLTS and microwave *s*-parameter measurements have also been made to investigate this behavior. These measurements reveal that surface or channel–substrate interface traps in the material are most likely to be responsible for the observed behavior. A new equivalent circuit model is developed which accounts for many of the observed characteristics. Unlike previously proposed equivalent circuits, the new model does not rely on physically unrealistic circuit element values in order to obtain accurate performance predictions. The bias dependence of circuit element values are computed for one device. Effects not described by the model are also discussed.

I. INTRODUCTION

MANY OF THE electrical characteristics of GaAs MESFET's shift dramatically in value at extremely low frequencies (below about 1 MHz). Device characteristics which have been observed to shift include output resistance [1]–[5], [7], [8], [14], transconductance [4]–[6], [9], and device capacitances [4]. As frequency is increased above dc, measured device output resistance has been observed to drop by as much as 85% [1]. The characteristic frequencies at which these decreases occur vary from less than 100 Hz [1] to approximately 100 kHz [3]. Shifts in transconductance values are more typically on the order of 5% to 30% of the dc value, and also occur at widely varying frequencies [4], [6], [9].

The low-frequency dispersions of device output resistance and transconductance are important phenomena to understand because of the implications they have for large-signal device modeling. Microwave circuit designers have been able to ignore these frequency dispersion effects for small-signal applications because the dispersion occurs well below the frequency bands typically of interest. The increasing importance of large-signal modeling, however,

has focused attention on these characteristics. Large- signal device models must describe both the dc and the RF characteristics of the device in order to produce accurate performance predictions. The existence of low-frequency dispersion effects also has significant implications related to the parameter extraction process utilized with large signal models. The use of only dc characterization data, for example, has been shown to be inadequate for microwave applications involving the determination of large-signal model parameter values [1], [13]. In order to obtain accurate predictions, designers have had to develop more tedious characterization techniques involving specialized equipment [1] or involving large numbers of *s*-parameter measurements [13].

An understanding of the causes of the observed low-frequency dispersion effects will lead to greatly simplified large-signal device characterization and parameter extraction processes. Only after the cause of this anomalous behavior is understood, can processing techniques be developed to control or eliminate the effect. Similarly, physically based models which are capable of predicting the effect can be developed and confirmed only if it has been adequately characterized. The development of these capabilities has the potential to allow the RF current–voltage relationships of GaAs MESFET's (including transconductance and output resistance) to be determined solely from dc data—thus simplifying the large-signal modeling process.

The low-frequency dispersion of GaAs MESFET characteristics has been observed by several researchers using a variety of different measurement techniques. Smith *et al.* have reported on an elaborate measurement apparatus which utilizes a 1-MHz signal source, power divider, phase shifters, and computer-controlled voltage and current probes [1]. Similar information can also be obtained using a pulsed current–voltage measurement technique. A technique which requires less sophisticated equipment has been reported by Canfield *et al.* [2]. The observation by Scheinberg *et al.* of "drain lag" in linear amplifier circuitry has also been correlated to output resistance dispersion [5].

The observed dispersion effects have been attributed to a wide range of physical phenomena. Electron trapping at the channel–substrate interface has been indicated as the mechanism responsible for output resistance dispersion by several researchers [2], [3], [8]. Canfield *et al.* [8] have

Manuscript received October 12, 1989; revised December 20, 1989. This work was supported by Internal Research and Development Funding of the Motorola Government Electronics Group, Chandler, AZ. The review of this paper was arranged by Associate Editor S. Tiwari.

J. M. Golio and M. G. Miller are with Motorola Government Electronics Group, Chandler, AZ 85248-2899.

G. N. Maracas and D. A. Johnson are with the Center for Solid State Electronics, Arizona State University, Tempe, AZ 85287.

IEEE Log Number 9034611.

Reprinted from *IEEE Trans. Electron Devices*, vol. 37, no. 5, pp. 1217–1227, May 1990.

indicated that surface-state occupation also plays a role in the observed behavior. This surface state explanation has also been employed by other researchers [4], [9], [14]. Other explanations involve hole traps [6] or hole injection from the channel–substrate interface [7]; and backgating phenomena [5]. Although each of the above mentioned theories succeeds in describing some of the observed behavior, none of these explanations have been developed in a rigorous fashion from first physical principles. Comparisons of theoretical results to measured behavior has also been presented for only a limited number of bias levels and other electrical conditions. The complexity of the low-frequency dispersion problem and the scarcity of comprehensive characterization data makes a formal detailed theoretical development extremely difficult. Determination of the validity of existing theories also requires that more comprehensive experiments be performed.

Several different models have been employed to describe the observed dispersion characteristics. Each of these models has been found to be extremely valuable at furthering our understanding of this phenomena. One of the earliest attempts to develop an equivalent circuit model of the dispersion of output resistance uses a simple series resistor/capacitor network in parallel with the standard output resistance of the device [8]. Although this approach successfully describes both dc and microwave output resistance behavior at one bias level, it does not predict the device characteristics in the transition region from high to low resistance. This model also fails to predict characteristics at more than one bias level or the corresponding observed transconductance dispersion. This model has been recently modified to include some bias effects [14], but the improved model still does not account for transconductance dispersion. Another unsettling aspect of this model is that an extremely high value for the capacitor element (on the order of microfarads) is required to obtain reasonable predictions. Some researchers [4], [8] have noted high measured output capacitance of the device at low frequencies; however, a single-capacitor element value on the order of microfarads is difficult to justify on a physical basis. A model which is capable of predicting the observed apparent high output capacitance without resorting to capacitor elements of excessive value would be preferred. Some researchers have chosen to ignore the dc characteristics of the device and use only RF data in the parameter extraction processes utilized with their models [1], [13]. This technique can be exploited effectively for some microwave circuit applications. The resulting model fails, however, to accurately predict dc device characteristics. This failure can be extremely important for many applications (e.g., when a device is used as a current source for MMIC circuitry). The research efforts of Scheinberg et al. have produced a valuable model which utilizes an additional resistor, capacitor, and dependent current source. This model successfully predicts the existence of frequency dispersion not only in device output resistance, but also in transconductance. Furthermore, the model does not require capacitance values to be

larger than is physically realistic while predicting an observed high output capacitance at low frequency. The predicted dispersion of the model, however, depends on the existence of a voltage difference and large resistance value between substrate and source contacts. Measurements presented in this paper clearly indicate that the dispersion exists without such a voltage drop or resistance. This model also predicts a transition of the output resistance over frequency which is much more abrupt than that actually measured. Bias dependance of the observed device behavior is also inadequately addressed by this model. Larson [3] has suggested use of a standard large-signal GaAs MESFET model in conjunction with an added resistor/capacitor network and two MOSFET devices. The model development does devote some effort to describing bias dependance of the observed behavior. Such a model offers considerable versatility in matching device characteristics, but the complexity of the model makes it difficult to gain insight into the cause of the problem.

II. Measurements and Measurement Techniques

The data presented here have been obtained from measurements of similar 0.8×400 μm MMIC devices fabricated at one foundry. The low-frequency measurements presented represent average values taken from five nearly identical devices. In addition to these measurements, microwave s-parameter data have been taken between 1 and 15 GHz. These data were used to determine equivalent circuit element values for the device at 18 different bias levels. Conductance DLTS data have also been utilized in order to obtain an indication of the material properties of the FET device.

Output resistance measurements were made using the test configuration of Fig. 1. Coaxial transmission lines were utilized for bias and signal paths to avoid unwanted device oscillations. The circuit was monitored for such oscillations using a spectrum analyzer and no measurements were made with unwanted signals present.

Both the RF and the dc output resistance and transconductance measurements were made over many bias points (approximately 20 to 50 for most devices). The low-frequency RF measurements ranged in frequency from 20 Hz to 100 kHz. Signal voltage amplitudes were set to a minimal value (100 to 250 mV typically) to ensure the device was not overdriven.

Referring to the quantities labeled in Fig. 1, the RF output resistance at the terminals of the device can be expressed as

$$r_{ds(t)} = V_{ds} \frac{R}{V_r} \quad (\Omega).$$

For the transconductance measurements, a similar equipment configuration to that shown in Fig. 1 was utilized. For these measurements, an RF signal V_{gg} was applied to the gate of the device instead of the signal V_{dd} applied to the drain. With such an equipment configuration, trans-

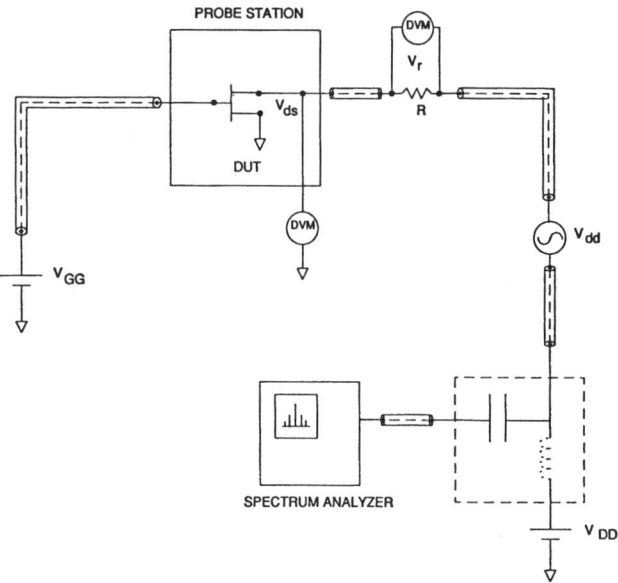

Fig. 1. Measurement configuration used to make dc and low-frequency output resistance measurements. A similar configuration was also used for transconductance measurements.

conductance can be expressed as

$$g_{m(t)} = \frac{V_r}{R \, V_{gs}} \quad (S)$$

where V_{gs} is the measured gate–source signal level.

The dc output resistance was determined by incrementing the drain–source bias voltage in 0.1-V steps and then measuring the current difference through the resistor R. This process was done using a delay of several seconds between each bias point measurement. This delay is essential to obtaining accurate dc data since device characteristics can drift for times on the order of several milliseconds. DC transconductance was determined in a similar manner.

Microwave *s*-parameter measurements were made using an automated network analyzer and probe station configuration. *s*-parameters were measured from 1 to 15 GHz at 18 different bias voltages. DC current–voltage data were gathered concurrently with the *s*-parameters. Equivalent circuit model element values were extracted from measured *s*-parameters using commercially available optimization software.

Conductance Deep Level Transient Spectroscopy (DLTS) measurements were made on a 0.8×400 μm ion-implanted device identical to those used for the previously described measurements. The DLTS spectra were obtained by digitally recording conductance transients resulting from a 2-ms filling pulse. The pulsed signal applied to the gate terminal of the device varied from one specific reverse-bias level to a second bias level closer to zero volts. A small 100-mV drain–source voltage was used to make the measurements. The measurements were made over a temperature range between approximately 90 and 410 K.

III. Device

All of the devices used in these measurements were fabricated with geometries and doping density profiles which were intended to be identical. The devices were wafer probable test FET's for use in microwave monolithic circuit applications. The source electrode and substrate of each of the devices were physically connected to each other by metalized via holes through the 4-mil-thick GaAs substrates. This via connection insures that the substrate and source contacts are at identical voltage levels, and that the resistance between regions of the active channel and substrate is small. The gates were comprised of four separate 0.8×100 μm fingers for a total gate width of 400 μm. Both the gate-to-source and gate-to-drain electrode spacings were 1.5 μm. Processing of each of the devices was done at the same foundry using identical procedures. The devices were defined by an optical lithographic technique. Ohmic contacts at the source and drain were formed using a selective n^+ implant. The channel of the devices was formed by a shallow n-type implant, and this process was followed by an anneal with arsenic overpressure. A sintered Au/Ge–Ni–Au metallization system was used for ohmic metal while electron-beam evaporation of Ti–Pt–Au was used for the gate electrodes. Plasma-deposited Si_3N_4 approximately 5000 Å thick was used as surface passivation for the devices. Finally, a high-energy boron implant was employed to provide device isolation.

IV. Measurement Results

Fig. 2(a) and (b) shows measured current-voltage characteristics for one of the 0.8×400 μm devices described in Section III. The characteristics are fairly typical of ion-implanted microwave MESFET's with similar gate dimensions and thereshold voltage. No negative differential resistance was observed in these devices.

Small-signal output resistance as a function of frequency is shown in Fig. 3(a) and (b). The data presented in Fig. 3(a) are obtained using the technique described in Section II. The device is biased with a gate–source voltage of $V_{GS} = -1.0$ V. This corresponds to the depletion region extending approximately midway between the Schottky-gate metallization and the channel–substrate interface. The drain–source voltage is varied from $V_{DS} = 1.0$ to 5.0 V as indicated on the curves. The observed shift in output resistance is insignificant for V_{DS} values below the drain–source saturation voltage, but increases monotonically with increased drain–source voltage. Fig. 3(b) shows measurements made with the device biased at $V_{DS} = 3.0$ V. The percentage shift in output resistance is seen to be greatest for low gate–source bias values (i.e., V_{GS} approximately 0 V). At these bias levels, the depletion region under the gate is shallow (i.e., is located near the semiconductor surface). As the reverse gate bias is increased, the depletion region is forced to extend deeper into the device and the observed dispersion in RF output resistance is decreased. Also of importance is the observed characteristic frequency of the shift in resistance.

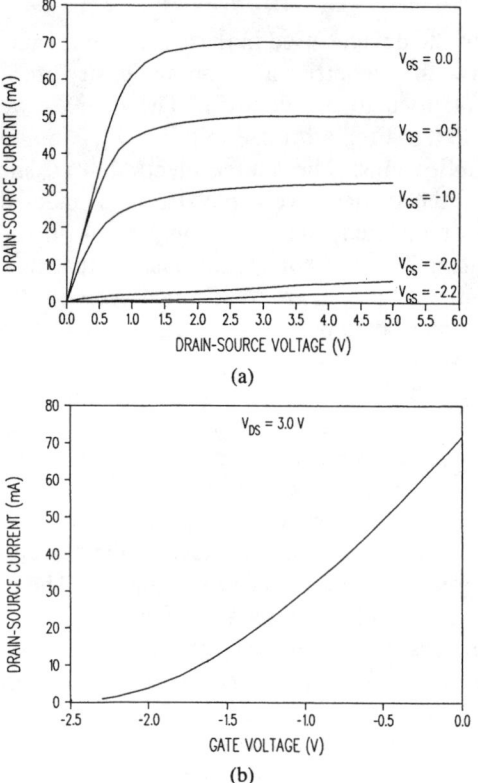

Fig. 2. DC drain-source current for the 0.8 × 400 μm device used in this study (a) as a function of drain-source voltage, and (b) as a function of gate-source voltage.

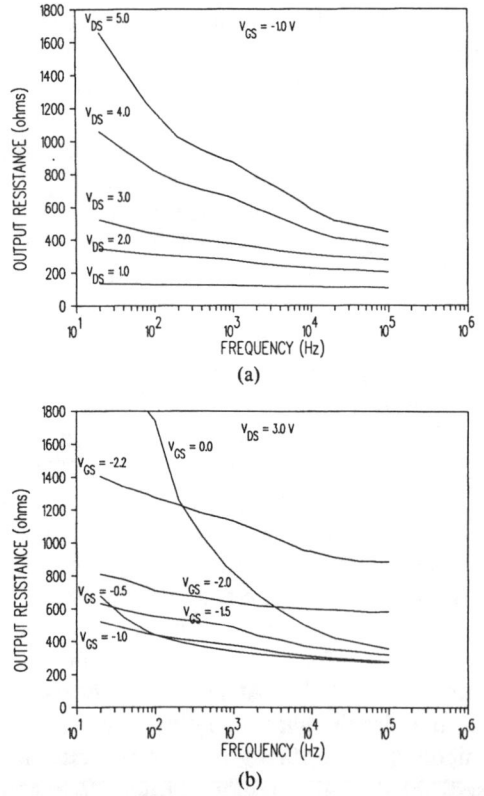

Fig. 3. Measured output resistance as a function of frequency for the 0.8 × 400 μm device used in this study. The frequency range is 20 Hz to 100 kHz. (a) Gate-source voltage level of −1.0 V. (b) Drain-source voltage level of 3.0 V.

For low gate-source bias values, the characteristic frequency is below the 20-Hz minimum measurement frequency capabilities of the equipment. As the device is biased nearer pinchoff, however, a second, less pronounced shift in resistance is also observed to occur at a much higher frequency (closer to 1 kHz). Referring to Fig. 3(a), it is seen that a less significant higher frequency shift in resistance is also observed for drain-source bias levels greater than about 4 or 5 V. The occurrence of this higher frequency shift is at bias levels which correspond to a depletion region at the drain end of the gate extending deep into the active channel of the device.

The 1-kHz shift becomes more pronounced at lower operating temperatures. This fact is illustrated in Fig. 4(a) where the measured output resistance as a function of frequency is plotted for three different temperatures. As the temperature is decreased from 325 to 275 K, the dominant low-frequency shift moves lower in frequency while the 1-kHz shift becomes more prominent. To obtain the data presented in Fig. 4(a), the device is biased very near pinchoff. When the device is biased well away from pinchoff, the 1-kHz output resistance shift is not observed even at 275 K (as shown in Fig. 4(b). For this case, the dominant low-frequency shift in output resistance is again observed to move to lower frequencies as the temperature is reduced from 325 to 275 K.

Transconductance dispersion for the same device is presented in Fig. 5. The observed shift in this value is seen to be much smaller than the corresponding shift in output resistance. The dc to microwave drop in transconductance for this particular device was found to be approximately 16% of the dc value for a wide range of bias values.

Results of the conductance DLTS measurements are presented in Fig. 6. To obtain the data of the curves in Fig. 6, an initial gate-source bias voltage of $V_{GS} = -1.00$ V was used. This corresponds to the depletion region extending to approximately mid-channel. The DLTS spectra show three distinct electron peaks—at temperatures of about 200, 325, and 400 K. On the plots of the ratio of change-in-conductance to conductance, these electron traps cause spikes in the spectrum which are negative in direction. Only the 400 K peak is well resolved. The data indicate that this deep level is located approximately 0.7 eV below the conduction band with an electron cross section of about 4×10^{-15} cm^{-2}. This peak could be the same as the H peak identified by Dhar et al. [11] to be due to implantation damage. There is no discernable EL2 peak in this sample.

The lower curve of Fig. 6 was produced by pulsing the gate bias from $V_{GS} = -1.0$ to 0.0 V, while the upper curve was produced by pulsing the gate bias from $V_{GS} = -1.0$ to -0.25 V. Although a zero-volt gate bias corresponds to a much shallower depletion region than a -0.25-V bias, the details of the two spectra are nearly identical. Both traces reveal identical trap signals at the same temperatures. This similarity in the relative magnitudes of the DLTS spectral peaks indicates that there is

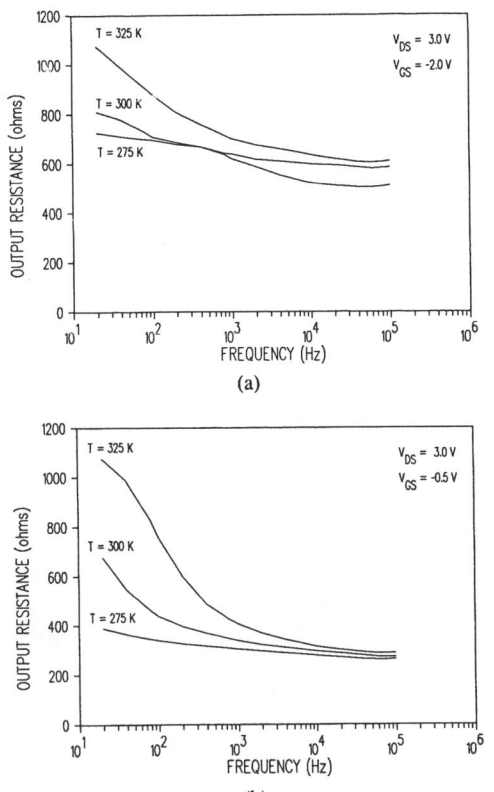

(a)

(b)

Fig. 4. Measured output resistance as a function of frequency for various temperatures. The frequency range is 20 Hz to 100 kHz. (a) Gate–source voltage level of −2.0 V. (b) Gate–source voltage level of −0.5 V.

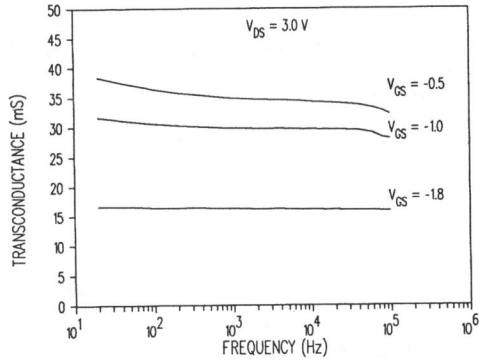

Fig. 5. Measured transconductance as a function of frequency for the 0.8 × 400 μm device used in this study.

CONDUCTANCE DLTS SPECTRA

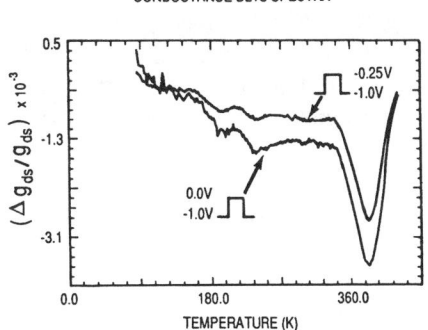

Fig. 6. Measured conductance deep-level transient spectroscopy spectrum for the 0.8 × 400 μm device used in this study.

no significant difference in the bulk material traps located close to the surface of the device. Additional DLTS data were taken for gate pulses from −1.25 to 0.0 V and from −0.5 to 0.0 V. No other deep levels were discernable from this additional data.

The lack of correlation between the bulk trapping state profile and the observed bias dependence of the dispersion phenomena complicates a theoretical explanation which exploits bulk trapping phenomena. Such a theory might still be utilized provided an explanation for bias-dependent carrier injection into these traps could be provided. It should also be noted that the DLTS data do not provide information about the deep levels very close to the channel–substrate interface. Such data are very difficult to obtain using conventional conductance DLTS measurement techniques.

V. FREQUENCY DISPERSION MODEL

Most of the current theories regarding low-frequency dispersion of transconductance or output resistance in GaAs MESFET's have recognized the role of deep levels in the observed phenomena. Electrons trapped in these deep levels will remain captured for times on the order of microseconds to seconds. These long time constants correspond to frequencies in the appropriate hertz to megahertz range of observed low-frequency dispersion. Because fields at microwave frequencies change polarity at rates much faster than these time constants, trapped electrons cannot participate in the RF current-carrying process. This trapped charge does, however, affect RF device performance since the total net charge (trapped or mobile) must be considered in electric field calculations. In contrast, dc current can be carried by charge which is alternately trapped and released at time intervals typical of the deep level time constants.

Measurements presented in this paper suggest that the primary trapping effect is most significant for bias levels corresponding to the shallowest depletion region extent. A second trapping mechanism is also observed to occur for extremely deep depletion region extent, but this effect is much less significant. Because the shallow depletion phenomenon is dominant for this FET, modeling efforts will focus on a description with only one mechanism present. It must be noted that FET's which do not have source-to-substrate via connections may exhibit dominant properties which differ from those discussed here.

As shown in Fig. 3(a), frequency dispersion of the output resistance is insignificant for low drain–source bias levels. This fact may be explained in terms of the GaAs carrier transport properties. The transport mechanisms which dominate electron velocity at low values of electric field are significantly different than those which dominate high-field behavior. Low drain–source voltage values produce low electric fields in the GaAs material. At these low-field levels, electron velocities V are proportional to the electric field value E and given by

$$v = \mu_0 E$$

where μ_0 is the low-field mobility of electrons in the semiconductor. For low drain–source bias levels, therefore, perturbations in voltage result in perturbations in current caused primarily by changes in electron velocities. This is equivalent to stating that for low drain–source bias levels, output resistance is dominated by these velocity changes. Because carrier velocity is primarily a bulk material property, this mechanism is essentially unaffected by the occupancy condition of surface-state or channel–substrate interface layer energy levels. Thus no frequency dispersion is observed for low drain–source biases.

In contrast to the low-field situation, carrier velocities in GaAs saturate for electric field values above a critical field level. The drain–source voltage associated with this critical field level is typically on the order of 1 V. Since carrier velocity is saturated, the output resistance under these conditions arises primarily because of the redistribution of depletion region charge in the device. As charge is redistributed, the undepleted channel dimensions are reshaped and the current is changed. The redistribution of depletion charge is directly affected by the occupation state of any traps in the material. The traps act as a mechanism which limits the rate at which the redistribtion of charge can take place. Therefore, a frequency dispersion of the output resistance is observed for high drain–source voltage levels. This description of why dispersion is only observed at high drain–source bias levels applies for deep trapping centers on the semiconductor surface or at the channel–substrate interface.

It should be noted that Larson [3] has observed similar drain–source bias dependence in the output resistance dispersion. He has used the observation to support an argument that the traps involved in the process are located at the channel–substrate interface. If either of these two arguments is correct, it must also be capable of describing the gate–source bias dependance of the characteristics noted in Section IV.

The observed dispersive characteristics can be described in terms of a bias-dependant injection of free carriers into trapping states, and the capacitive coupling of this charge-carrier movement into the device channel. Because the injected charge produces an electric field which affects the current-carrying channel, the injected carriers directly affect the output resistance of the device. Although this process will occur at both the source and drain sides of the gate terminal, the injection on the drain side is assumed to have the most pronounced effect on output resistance. This assumption is based on the fact that the total output resistance of the device is dominated by the resistance of this portion of the device. Also, the electric fields within the channel are much higher in this region of the device.

The process described in the preceding paragraph can be modeled using the equivalent circuit of Fig. 7. The equivalent circuit elements C_{gs}, g_{m1}, and R_{ds} represent a standard simplified equivalent circuit model for a MESFET. This simplified model is useful in the analysis of the frequency dispersion model, but a more sophisticated

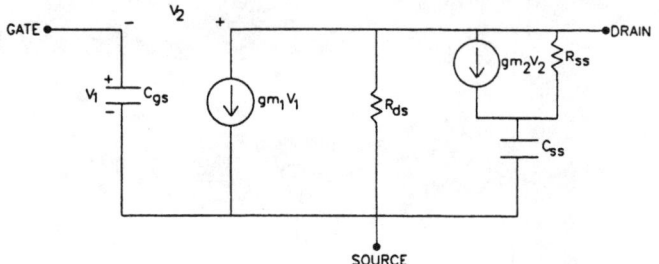

Fig. 7. A simplified equivalent circuit of a MESFET including frequency dispersion elements: g_{m2}, R_{ss}, and C_{ss}.

representation of the device may also be easily fitted with the three additional elements of the dispersion model. The injection mechanism into the trapping states in modeled with the dependent current source g_{m2}. The injected current value is dependant on the signal level between the gate and drain terminals V_{gd}. The injection mechanism is a high-impedance mechanism (i.e., the injected current levels are nearly independent of device load, but are instead dependent only on device deep-level properties). The high but finite injection impedance of the current source is modeled by the resistor R_{ss} in the equivalent circuit model. Finally, the injected current is coupled to the drain–source current through the capacitance C_{ss}.

Analysis of the equivalent circuit of Fig. 7 produces analytical expressions for both output resistance and transconductance which agree well with the observed phenomena. The output impedance of the circuit may be expressed as

$$Z_{ds} = \frac{R_{ds}\left[1 + j\omega C_{ss} R_{ss}\right]}{1 + j\omega C_{ss}\left[R_{ss} + R_{ds} + g_{m2} R_{ss} R_{ds}\right]} \quad (1)$$

where ω is the radian frequency. The device output resistance observed at the terminals of the device $r_{ds(t)}$ is the real part of (1). For dc conditions ($\omega = 0$) this is given by

$$r_{ds(t)}(\omega = 0) = R_{ds} \quad (2)$$

The value of output resistance at microwave frequencies for the circuit of Fig. 7 can similarly be obtained from (1) by taking the limit of the real part of (1) as ω approaches infinity

$$r_{ds(t)}(\omega \to \infty) = \frac{R_{ds}}{1 + R_{ds}\left[1/R_{ss} + g_{m2}\right]}. \quad (3)$$

Another important performance parameter related to the output resistance of the circuit of Fig. 7 is the frequency where $r_{ds(t)}$ has fallen half way to its final value. This characteristic frequency of the output resistance can be expressed as

$$f_{0(rds)} = \frac{1}{2\pi R_{ss} C_{ss}}\left[\frac{R_{del}}{K^2\left(R_{ds} + r\right) + 2Kr - R_{del}}\right]^{1/2} \quad (4)$$

where

$$R_{del} = R_{ds} - r$$
$$r = r_{ds(t)}(\omega \to \infty)$$

and

$$K = R_{ds}\left[g_{m2} + 1/R_{ss}\right].$$

In a similar manner, the terminal transconductance $g_{m(t)}$ of the equivalent circuit of Fig. 7 can be expressed as

$$g_{m(t)} = g_{m1} - g_{m2}\left[\frac{j\omega C_{ss} R_{ss}}{1 + j\omega C_{ss} R_{ss}}\right]. \qquad (5)$$

Under dc conditions, (5) reduces to

$$g_{m(t)}\left(\omega = 0\right) = g_{m1} \qquad (6)$$

while the observed high-frequency transconductance is

$$g_{m(t)}\left(\omega \rightarrow \infty\right) = g_{m1} - g_{m2}. \qquad (7)$$

The characteristic frequency where the real part of $g_{m(t)}$ has fallen half way to its final value is given by

$$f_{0(gm)} = \frac{1}{2\pi C_{ss} R_{ss}}. \qquad (8)$$

Equations (1)–(8) illustrate that the circuit of Fig. 7 possesses characteristics in agreement with the measured behavior presented in Section IV. Both output resistance and transconductance of the circuit decrease as a function of frequency as expected.

VI. MODEL RESULTS

Measured output resistance at dc and from 20 Hz to 100 kHz was used along with (1)–(8) to determine the behavior of the equivalent circuit element values over bias. Data considered in determining the element values include: 1) dc output resistance, 2) microwave output resistance determined from measured s-parameters, 3) the characteristic frequency of the output resistance as defined by (4), and 4) the dc transconductance value.

The characteristic frequency of the transconductance and the microwave transconductance value of the device were not used to determine equivalent circuit element values. Actual measurement of the characteristic frequency of the transconductance as defined by (8) is difficult because the shift in transconductance is typically small and the shift from the dc to RF value occurs over a wide frequency range as illustrated in Fig. 5. The microwave transconductance can be easily measured and the value of this quantity predicted from the resulting model is compared to the measured value after the other equivalent circuit element values have been established.

In order to establish equivalent circuit element values, the gate–source capacitance element C_{gs} of Fig. 7 is first determined using standard s-parameter measurements. The value for trapping-state capacitance C_{ss} is next set equal to this value. This choice of values for C_{ss} is somewhat arbitrary, but represents a value of capacitance which is not unrealistically large for the device structure. This choice may be justified if the traps responsible for the observed behavior are located on the device surface and the trapped charge movement is then coupled through the depletion-layer capacitance into the channel. Transconductance g_{m1} is set equal to the measured dc transconductance

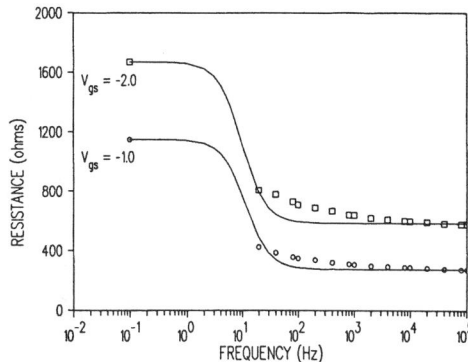

Fig. 8. A comparison of measured and modeled output resistance as a function of frequency. $V_{ds} = 3.0$ V.

to satisfy (6). Similarly, output resistor R_{ds} is set equal to the measured dc output resistance to satisfy (2). Finally, the values of the second transconductance g_{m2} and trapping resistance R_{ss} are chosen to match the output resistance characteristic frequency of (4) and the microwave output resistance value of (3). When a more complex equivalent circuit is utilized, a similar process can be used to serve as a starting point for equivalent circuit element value optimization.

Fig. 8 shows the agreement obtained between measured and modeled characteristics for the device described in the previous sections, and for a drain–source voltage of 3.0 V. Similar agreement is obtained for all bias cases. In the figure, the measured dc resistance value is assumed to hold for a frequency of 0.1 Hz. The procedure described in the preceding paragraph to determine the equivalent circuit element values guarantees exact agreement between measured and modeled output resistance values at both dc and microwave frequencies. Although the low transition frequency of these particular devices does not allow an adequate comparison between measured and modeled output resistance during transition, the modeled characteristics do appear to change more abruptly than those measured.

Table I presents the equivalent circuit values determined using the technique described above. The circuit element parameters are defined by the equivalent circuit of Fig. 7. The gate–source voltage dependence of the second transconductance g_{m2} and trapping resistance R_{ss} is also presented in Fig. 9. The trapping-state transconductance decreases monotonically as gate–source voltage is decreased. The trapping-state injection resistance plotted in Fig. 9 is an increasing function of reverse gate bias. The values required of this resistance are extremely high (in the 10^9- to 10^{10}-Ω range). Such values, however, are very close to the conductance per contact area values reported by other researchers for p^+-n step junctions compensated with deep acceptors [12]. The physical phenomena investigated in this previous research have many similarities to the theory presented in Section V. Such structures are expected to, and do exhibit electrical behavior very similar to the characteristics described in Section IV. Both the trapping resistance and second transconductance curves are observed to change abruptly at a

TABLE I
EQUIVALENT CIRCUIT ELEMENT VALUES CORRESPONDING TO THE CIRCUIT OF
FIG. 7 FOR THE DEVICE MEASURED

V_{DS}	V_{GS}	$c_{gs}=c_{ss}$	g_{m1}	R_{ds}	g_{m2}	R_{ss}
(volts)	(volts)	(pF)	(mS)	(ohms)	(mS)	(10^9ohms)
3.0	0.0	0.57	55.0	5000	3.45	0.8
3.0	-0.5	0.49	40.3	1600	3.07	5.6
3.0	-0.75	0.44	36.0	1270	2.89	7.6
3.0	-1.0	0.40	35.2	1150	2.75	8.4
3.0	-1.5	0.32	25.0	1170	2.27	9.2
3.0	-1.8	0.26	17.8	1400	1.22	14.7
3.0	-2.0	0.23	17.8	1670	1.11	24.9
0.5	-1.0	0.37	23.4	44	3.45	37.2
1.0	-1.0	0.40	31.0	190	4.00	15.7
2.0	-1.0	0.42	36.0	580	3.18	8.6
3.0	-1.0	0.40	35.2	1150	2.75	8.4
4.0	-1.0	0.42	35.0	2000	2.19	3.9
5.0	-1.0	0.43	35.0	2400	1.82	1.3

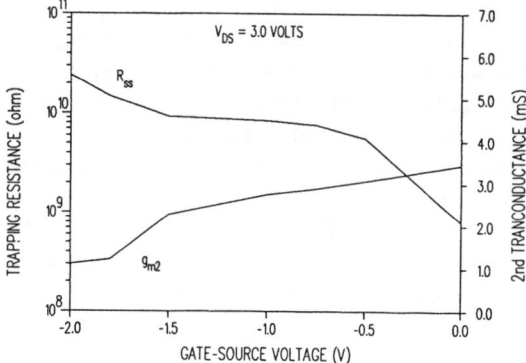

Fig. 9. Trapping-state injection resistance R_{ss} and second transconductance g_{m2} as a function of gate bias voltage.

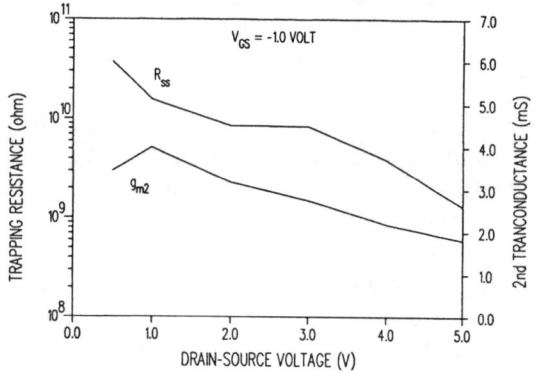

Fig. 10. Trapping state injection resistance R_{ss} and second transconductance g_{m2} as a function of drain bias voltage.

bias of about $V_{gs} = -1.5$ V. This corresponds to the bias where a second (higher frequency) shift in the dispersion characteristics of the output resistance became noticeable (see Fig. 3(b)).

Plots of the second transconductance and output injection resistance as a function of drain–source bias are presented in Fig. 10. Again an abrupt change in the curves is observed for drain–source bias levels below about 1 V. This bias level corresponds to the onset of current saturation in the device.

More detailed modeling of the measured device behavior was accomplished by using the more complex equivalent circuit model of Fig. 11. Table II presents the equiv-

alent circuit element values determined to describe the device behavior at one bias level (typical of the bias used for amplifier applications). To obtain the element values listed in Table II, the values listed in Table I for the appropriate bias level were used as a starting point and optimization was performed. Trapping-state capacitance C_{ss} was set equal to the sum of the gate–source and gate–drain capacitances. Other aspects of the optimization process are as described for the previous simple circuit modeling case.

Excellent agreement between measured and modeled s-parameters was obtained for the device at this bias level. The agreement obtained for all of the microwave s-parameters is as good as the agreement which can be obtained without the added trapping-state injection elements g_{m2}, R_{ss}, and C_{ss}. The circuit of Fig. 11, however, also successfully predicts the dc and low-frequency characteristics of the device.

Fig. 12 compares measured versus modeled microwave transconductance for the device. Although this parameter was not considered in matching modeled device performance to measured characteristics, agreement is seen to be excellent.

It should also be pointed out that (1) does predict a high value of device output capacitance at low frequencies. Although our measurements do not allow us to compare measured to modeled output capacitance of the device in the 20-Hz to 100-kHz range, these predictions are in at least qualitative agreement with observations of others [4], [8].

Although the proposed trapping state injection mechanism does provide an adequate explanation for many of the observed nonideal characteristics of GaAs MESFET's, it does not successfully explain some of the observed dc characteristics of the device. Fig. 13 presents the measured dc output resistance as a function of gate–source voltage for the device used in this work. For small values of drain–source voltage, the output resistance increases monotonically for gate bias decreasing toward pinchoff. This behavior is consistent with standard theories of device operation.

For higher values of drain–source voltage, however, the

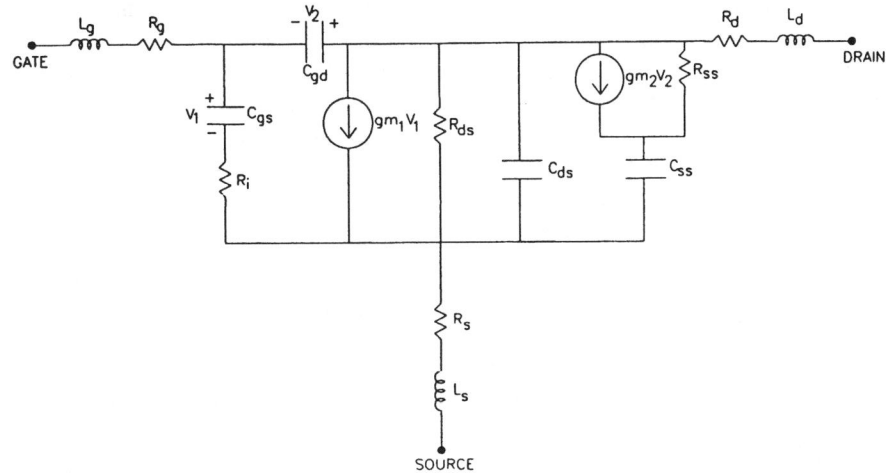

Fig. 11. Equivalent circuit model of a MESFET including trapping-state frequency dispersion elements: g_{m2}, R_{ss}, and C_{ss}.

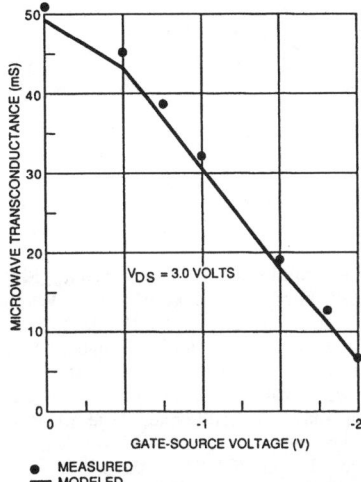

Fig. 12. A comparison of measured and modeled microwave transconductance as a function of gate–source voltage.

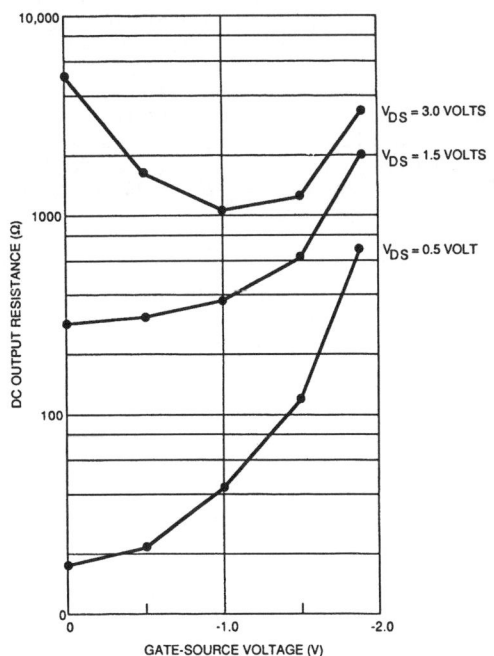

Fig. 13. DC output resistance as a function of gate–source voltage.

TABLE II
EQUIVALENT CIRCUIT ELEMENT VALUES CORRESPONDING TO THE CIRCUIT OF
FIG. 11 FOR THE DEVICE MEASURED
(Element values are for a bias level of $V_{DS} = 3.0$ V and $V_{GS} = -0.75$ V.)

EQUIVALENT CIRCUIT ELEMENT	ELEMENT VALUE
g_{m1}	39.3 (mS)
g_{m2}	2.87 (mS)
R_{ds}	1270 (ohms)
R_{ss}	7.6×10^9 (ohms)
C_{gs}	0.433 (pF)
C_{gd}	0.045 (pF)
C_{ss}	0.477 (pF)
C_{ds}	0.126 (pF)
R_i	1.11 (ohms)
R_g	1.77 (ohms)
R_s	1.66 (ohms)
R_d	2.83 (ohms)
L_g	0.055 (nH)
L_s	0.027 (nH)
L_d	0.307 (nH)

device output resistance is seen to be as large for small gate–source voltage values as it is for values near pinchoff—but to be smaller for intermediate values. This behavior is not expected from traditional MESFET theory and is not explained by the trapping-state injection arguments alone.

VII. CONCLUSIONS

Most researchers who have investigated the frequency-dispersive characteristics of MESFET's have recognized the importance of the role of traps in this behavior. It is not clear, however, where the traps which contribute to this behavior are located in the device structure and what mechanism causes electrons (or holes) to be injected into these traps. The present measurements have not com-

IEEE TRANSACTIONS ON ELECTRON DEVICES, VOL. 37, NO. 5, MAY 1990

pletely resolved these issues, but have determined the attributes of low-frequency dispersion in much greater detail than previously reported work.

Important attributes determined by this and previous investigations include:

1) Canfield et al. [2] found that both a p-type surface and p-type buffer were required to produce a buried-channel device which did not exhibit output resistance dispersion.

2) Output resistance dispersion is not significant for drain–source bias levels below current saturation, but becomes increasingly important as drain–source bias levels are increased (see Fig. 3(a) and [3]).

3) The frequency dispersion of output resistance is most significant for low gate–source bias values (see Fig. 3(b)). These bias conditions correspond to depletion layers nearest the surface of the device.

4) Two distinct output resistance transition frequencies can be observed (see Figs. 3 and 4(a)). In the measurements presented here, a low-frequency (approximately 10–20 Hz) transition dominates, while a higher frequency (approximately 1 kHz) transition appears for gate biases near pinchoff or large drain-bias values. The bias levels where the second, high-frequency transition occurs correspond to depletion layers which extend deep into the device near the channel–substrate interface. This second transition may correspond to a second type of trap becoming important.

5) Transconductance dispersion is much less significant than output resistance dispersion (see Fig. 5).

6) The bias dependence of the dispersion behavior is not accounted for simply by a nonhomogeneity in the bulk deep-level concentrations (see Fig. 6).

To account for the above listed properties, a trapping-state injection mechanism has been described using three equivalent circuit elements added to the standard MESFET equivalent circuit model. This new model has been shown to successfully describe the low-frequency dispersion characteristics of both the output resistance and transconductance.

This new modeling work can serve as a starting point for rigorously derived physically based models. Such models will be able to contribute significantly to the sim-

plification of the large-signal parameter extraction process and to improving device performance. Device processing research directed toward eliminating undesirable deep-level effects (especially at the surface and channel-substrate interface) can also have a significant impact on the large-signal parameter extraction process.

REFERENCES

[1] M. A. Smith, T. S. Howard, K. J. Anderson, and A. M. Pavio, "RF nonlinear device characterization yields improved modeling accuracy," in IEEE Microwave Theory and Techniques Symp. Dig., pp. 381–384, 1986.

[2] P. Canfield, J. Medinger, and L. Forbes, "buried-channel GaAs MESFET's with frequency-independent output conductance," IEEE Electron Device Lett., vol. EDL-8, pp. 88–89, Mar. 1987.

[3] L. E. Larson, "An improved GaAs MESFET equivalent circuit model for analog integrated circuit applications," IEEE J. Solid-State Circuits, vol. SC-22, pp. 567–574, Aug. 1987.

[4] P. H. Ladbrooke and S. R. Blight, "Low-field low-frequency dispersion of transconductance in GaAs MESFET's with implications for other rate-dependent anomolies," IEEE TRANS. Electron Devices, vol 35, pp. 257–267, Mar. 1988.

[5] N. Scheinberg, R. Bayruns, and R. Goyal, "A low-frequency GaAs MESFET circuit model," IEEE J. Solid-State Circuits, vol. 23, pp. 605–608, Apr. 1988.

[6] A. Zylberstejn, G. Bert, and G. Nuzillat, "Hole traps and their effects in GaAs MESFETs," in Inst. Phys. Conf. Ser. No. 45, 1979, ch. 1, pp. 315–325.

[7] J. F. Wager and A. J. McCamant, "GaAs MESFET interface considerations," IEEE Trans. Electron Devices, vol. ED-34, pp. 1001–1004, May 1987.

[8] C. Camacho-Penalosa and C. S. Aitchison, "Modeling frequency dependence of output impedance of a microwave MESFET at low frequencies," Electron. Lett., vol. 21, pp. 528–529, June 6, 1985.

[9] S. R. Blight, R. H. Wallis, and H. Thomas, "Surface influence on the conductance DLTS spectra of GaAs MESFETs," IEEE Trans. Electron Devices, vol. ED-33, pp. 1447–1453, Oct. 1986.

[10] T. M. Barton and P. H. Ladbrooke, "The role of the device surface in the high voltage behavior of the GaAs MESFET," Solid-State Electron., vol. 29, pp. 807–813, Aug. 1986.

[11] S. Dhar, P. K. Bhattacharya, F-Y. Juang, W-P. Hong, and R. A. Sadler, "Dependence of deep-level parameters in ion-implanted GaAs MESFET's on material preparation," IEEE Trans. Electron Devices, vol. ED-33, pp. 111–118, Jan. 1986.

[12.] L. Forbes and C. T. Sah, "Application of the distributed equilibrium equivalent circuit model to semiconductor junctions," IEEE Trans. Electron Devices, vol. ED-16. pp. 1036–1041, Dec. 1969.

[13] M. Weiss and D. Pavlidis, "Power optimization of GaAs implanted FET's based on large-signal modeling," IEEE Trans. Microwave Theory Tech., vol. MTT-35, pp. 175–188, Feb. 1987.

[14] N. Ishihara, H. Kikuchi, and M. Ohara, "Gigahertz-band high-gain GaAs monolithic amplifiers using parallel feedback techniques," IEEE J. Solid-State Circuits, vol. 24, pp. 962–968, Aug. 1989.

Backgating in GaAs MESFET's

CHRISTOPHER KOCOT AND CHARLES A. STOLTE, MEMBER, IEEE

Abstract—The phenomenon of backgating in GaAs depletion mode MESFET devices is investigated. The origin of this effect is electron trapping on the Cr^{2+} and EL(2) levels at the semi-insulating substrate-channel region interface. A model describing backgating, based on DLTS and spectral measurements, is presented. Calculations based on this model predict that closely compensated substrate material will minimize backgating. Preliminary experimental data support this prediction.

I. INTRODUCTION

THE characteristics of GaAs metal-semiconductor field effect transistors (MESFET's) depend strongly on the properties of the interface between the n-type active region and the semi-insulating substrate [1]. GaAs MESFET's can exhibit phenomena such as a drift in the drain current with time and a change in the drain current as a result of a change in the substrate bias. The decrease in the drain current when a negative voltage is applied to the substrate is termed backgating [2], [3]. Backgating is a detrimental effect in complex GaAs integrated circuits due to the interaction between closely spaced devices. This effect is caused by the relatively large capacitance of the substrate-active channel interface due to negative charge accumulated on deep traps in the interface region. The application of a bias to the substrate modulates this space charge region. This results in a change in the active channel region width and, therefore, a change in the drain current. In this paper, we will present the results of investigations into the physical nature of the deep traps responsible for backgating. Based on these investigations, we propose a model which explains backgating and demonstrate one solution, namely the use of closely compensated substrate material to minimize the back-side channel capacitance.

II. EXPERIMENTAL PROCEDURE

A. Substrate Material

GaAs MESFET's fabricated in four different types of substrate materials were investigated. The active region for the devices is produced by ion implantation into: 1) Cr-doped semi-insulating substrates (with Cr concentrations between 5×10^{15} and 1×10^{17} cm^{-3}); 2) high purity semi-insulating substrates (grown with no intentionally added dopants); 3) buffer layers on Cr-doped substrates; and 4) buffer layers on high purity substrates.

Manuscript received January 4, 1982; revised February 1, 1982.
The authors are with Hewlett-Packard Laboratories, Palo Alto, CA 94304.

The substrate material, both Cr-doped and high purity, is grown by the two-atmosphere liquid encapsulated Czochralski (LEC) technique [4]. Chromium incorporates into the GaAs lattice on Ga sites and gives up three electrons to the bonds. The neutral state of Cr with respect to the lattice is Cr^{3+} with the electron configuration $3d^3$. Capture of electrons leads successively to the core states $3d^4$, Cr^{2+} (a singly, negatively charged acceptor), and $3d^5$, Cr^{1+} (a doubly, negatively charged acceptor). Chromium which is neutral, Cr^{3+}, is a double acceptor. The Cr^{2+} and Cr^{3+} levels are located 0.70 eV [5] and 0.45 eV [6] above the valence band, respectively, as shown in the energy level diagram of Fig. 1. There is uncertainty concerning the position of the Cr^{4+} and Cr^{1+} levels. According to the literature, the Cr^{4+} level is 0.15 eV [6] above the valence band and the Cr^{1+} level is degenerate with the conduction band [7]. The Cr^{2+} and Cr^{3+} levels are the most important to the understanding of the behavior of Cr-doped semi-insulating GaAs.

The other commonly observed level in GaAs is the EL(2) donor level which is responsible for the high resistivity of nonintentionally doped, high-purity GaAs [8]. This level is believed to be due to an anti-site defect, Ga on an As site [9]. It is located 0.75 eV below the conduction band (thermal activation energy); the optical ionization energy is 0.82 eV [10]. The necessary conditions for the production of high-resistivity material is that the dopants satisfy the following relations:

$$\text{if } N_d > N_a \text{ then } (N_{da} - N_{dd}) > (N_d - N_a), [11] \quad (1)$$

or

$$\text{if } N_a > N_d \text{ then } (N_{dd} - N_{da}) > (N_a - N_d), [12] \quad (2)$$

where N_d and N_a are the concentrations of shallow donors and acceptors, respectively, and N_{dd} and N_{da} are the concentrations of deep donors, EL(2), and acceptors, Cr, respectively. The recent paper by Martin describes the compensation mechanisms in detail [13].

The high-purity buffer layers are grown on the Cr-doped or high-purity semi-insulating substrates by the liquid-phase epitaxy (LPE) technique. These layers are grown in a horizontal graphite slider system at 700°C. This epitaxial material, characterized using Hall measurements, is n-type with mobilities measured at room temperature of approximately 8000 cm^2 V^{-1} s^{-1} and mobilities measured at 77 K are greater than 120 000 cm^2 V^{-1} s^{-1}. The measured free-carrier concentration at room temperature is less than 1×10^{14} cm^{-3} [14]. The buffer layers used in this investiga-

Reprinted from *IEEE Trans. Electron Devices*, vol. ED-29, no. 7, pp. 1059–1064, July 1982.

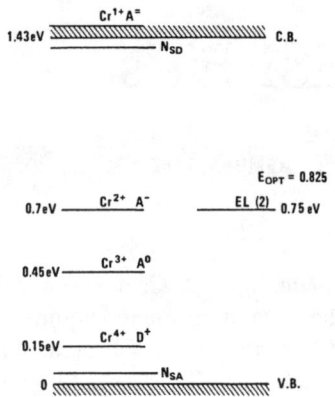

Fig. 1. Energy level diagram of Cr-doped GaAs.

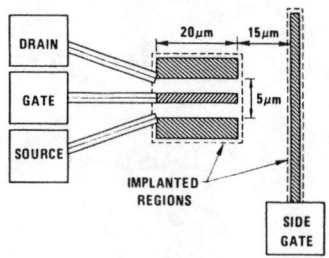

Fig. 2. The MESFET test structure used in the backgating, DLTS, and spectral response experiments.

tion are approximately 3 μm thick and are fully depleted by the surface and semi-insulator to buffer layer interface space charge regions.

The depletion mode MESFET test devices, shown in Fig. 2, are fabricated using selective region ion implantation. The MESFET channel regions are formed by localized ion implantation of 500-keV Se ions to a dose of 6×10^{12} cm^{-2}. A second localized Si implant at 500 keV to a dose of 1×10^{13} cm^{-2} is used in the ohmic contact regions to lower the sheet resistance. The implants are simultaneously annealed at 850°C for 15 min using a Si$_3$N$_4$ cap. The MESFET fabrication includes a recessed gate technology which reduces the modulation of the drain current due to changing depletion layer widths at the free surface during switching transients and also allows the adjustment of the gate cutoff voltage during the device processing [15]. The side-gate electrode is an ohmic contact to an implanted region. This region is isolated from the active region of the MESFET by the semi-insulating substrate or the fully depleted buffer layer.

III. EXPERIMENTAL RESULTS

The magnitude of the backgating effect is determined by measuring the change in the saturated drain current as a result of the application of a negative voltage to the side-gate electrode. In these experiments, the source–drain bias is 4 V, the gate is shorted to the source, and the side-gate bias is -4 V. Results obtained from these devices fabricated in wafers of the types listed above are presented in Fig. 3. These data represent the mean and one sigma variation of the drain-current change with the application of -4-V side-gate voltage for approximately 30 devices on a one-inch-square wafer. These data illustrate the wide spread in the magnitude of backgating on a single wafer and the large variation of the effect from wafer to wafer. In general, the effect is less for buffer layers on Cr-doped substrates than for high-purity substrates, with or without a buffer. The buffer layers used in this experiment had little influence on backgating. This is not understood since the buffer layer should decrease the capacitance between the substrate and the active layer and therefore decrease the magnitude of backgating. Experiments to investigate the effect of buffer layers are in progress. The sample on

the Cr-doped substrate was fabricated in closely compensated material. This will be discussed in detail later.

The time dependence of the drain current following the application of a side-gate potential is shown in Fig. 4. There is a rapid decrease in current when the side-gate bias is applied followed by an additional slow decrease for Cr-doped substrates and a slow increase for high-purity substrates. The general form of the drain-current transients for buffer layers on the different types of substrates is the same as observed without the buffer layer. The magnitude of the drain-current change depends on the particular substrate as illustrated in Fig. 3. These current transients can be understood by considering the band diagram shown in Fig. 5. Negative charge is accumulated on deep levels on the substrate side of the substrate-channel region interface. This negative charge is balanced by the positive space charge region in the channel region and produces the relatively large capacitance of the substrate-channel region interface. A negative potential applied to the substrate increases the width of the depletion region at the back side of the MESFET channel which results in the rapid decrease of the drain current. This change in the depletion layer width is followed by the emission of electrons or holes from the deep levels as equilibrium is reestablished. In the case of Cr-doped substrates, holes are emitted from the deep Cr level which results in a further decrease in the drain current because of the increase of the negative charge on the substrate side of the interface. In the high-purity substrates, electrons are emitted to the conduction band from the EL(2) level and the drain current increases to an equilibrium value.

These drain-current transients have been analyzed using an automated conductance DLTS system [16]. In this technique, the transients are analyzed at a number of temperatures to derive the activation energies and the capture cross sections of the levels responsible for the change in the drain current following the application of the side-gate voltage. The results obtained by this technique for Cr-doped and high-purity substrates are summarized in Table I. In Cr-doped substrates, the major DLTS peak is due to the emission of holes from the Cr level to the valence band. The major DLTS peak in high-purity substrates is due to the emission of electrons from the EL(2) level to the conduction band. The activation energies and the capture cross sections determined in this investigation agree with those reported by Martin [17].

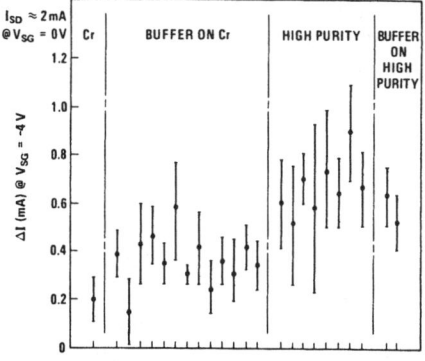

Fig. 3. Backgating results measured on devices fabricated in different substrate materials.

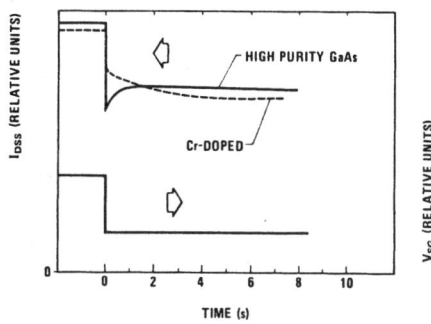

Fig. 4. Typical time dependence of the drain current for Cr-doped and high-purity substrates after the application of a side-gate voltage.

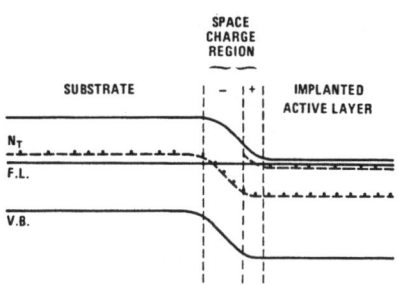

Fig. 5. The energy band diagram of the substrate-active region interface.

TABLE I
DEEP LEVELS DETERMINED BY DLTS MEASUREMENTS OF THE
BACKGATING TRANSIENTS

	Cr-DOPED			HIGH-PURITY	
LEVEL	ACTIVATION ENERGY (eV)	CAPTURE CROSS-SECTION (cm²)	LEVEL	ACTIVATION ENERGY (eV)	CAPTURE CROSS-SECTION (cm²)
Cr	0.781	7 e − 15	EL (2)	0.82	2.7 e − 13
HL	0.644	1.1 e − 13	EL	0.692	2.0 e − 13
HL	0.516	1.4 e − 13	HL	0.535	4.1 e − 15

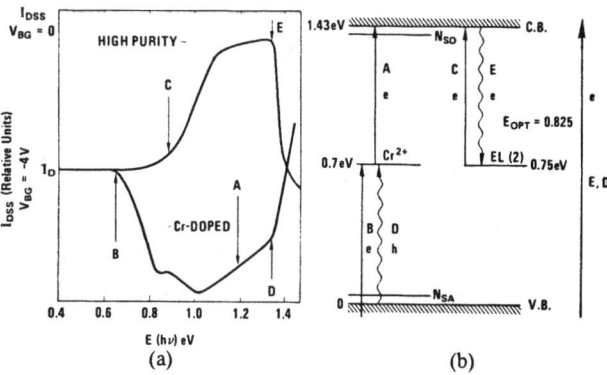

Fig. 6. (a) Spectral dependence of backgating for Cr-doped and high-purity substrates. (b) Energy level diagram showing the transitions responsible for the spectral response.

the MESFET channel, as indicated in Fig. 5, and therefore the drain current decreases. The increase of the current at 1.0 eV is caused by electron transitions from the Cr^{2+} level to the conduction band which reduces the negative charge in the junction, decreases the depletion width, and increases the drain current. Since the Cr^{2+} level is close to the center of the band gap both transitions, valence band to the Cr^{2+} level and Cr^{2+} level to the conduction band, are possible. In the photon energy range from 0.65 eV to 1.0 eV, the optical cross section for electron transitions from the valence band to the Cr level σ_p^0 is greater than the optical cross section for electron transitions from the Cr level to the conduction band σ_n^0, and therefore the transition producing electrons in the Cr^{2+} state is dominant. In the range from 1.0 eV to 1.4 eV, $\sigma_n^0 > \sigma_n^0$, and therefore the transition of electrons from the Cr^{2+} level to the conduction band is dominant [10]. A steep increase in current is seen at the band-gap photon energy because electron-hole pairs are generated and the holes are trapped by the Cr^{2+} level while the electrons are swept away by the space charge field. This results in a decrease in the negative charge at the interface and an increase in the current.

The spectral dependence of backgating is quite different in the case of high-purity substrates. An increase of the drain current is observed with a threshold of 0.8 eV which agrees with the optical ionization energy of the EL(2) deep level [10]. These transitions decrease the concentration of the negative charge on the substrate side of the interface which produces a wider channel region and the increase in the drain current. The drain current decreases as the photon energy approaches the band-gap energy. In this region,

The spectral dependence of backgating has been measured to obtain additional information regarding the levels responsible for the effect and to understand ambient light effects on the MESFET characteristics. Typical spectral dependence of the changes in the saturated drain current, measured with a side-gate voltage of −4 V, for Cr-doped and high-purity substrates are shown in Fig. 6(a). The transitions responsible for the spectral dependence of the drain current are indicated on the energy level diagram of Fig. 6(b). In the case of the Cr-doped substrates, the decrease of the current in the range of 0.65 eV to 1.0 eV is caused by the increase of backgating due to electron transitions from the valence band to the Cr^{3+} level. These transitions produce chromium in the Cr^{2+} charge state. The holes created in the valence band by this transition are swept away by the electric field in the space charge region at the substrate-channel region interface. This increase in the charge on the Cr^{2+} level increases depletion depth into electron-hole pairs are generated and the electrons are captured by the EL(2) level. This increase in the negative charge reduces the drain current as discussed above. It should be noted that there is almost no backgating effect when the high-purity substrate is illuminated with 1.2-eV light. In this condition, no change in the drain current is seen when a side-gate potential is applied.

IV. Backgating Model

A model which is consistent with the experimental results obtained on structures without buffer layers is postulated. The experiments described above indicate that backgating in Cr-doped substrates is caused by the accumulation of negative charge on the Cr^{2+} level (fast transient) and by the emission of holes from the Cr level (slow transient). For high-purity substrates, backgating is caused by the accumulation of negative charge on the EL(2) level (fast transient) and by the emission of electrons from the EL(2) level (slow transient). The net concentration of negative charge, at equilibrium, on the substrate side of the substrate-channel region interface N_{eff} depends on the occupancy of the deep traps in the bulk according to the following relationship:

$$N_{eff} = N_{Cr}\left\{e_p^{Cr}/\left(e_p^{Cr} + e_n^{Cr}\right) - 1/\left[1 + \exp\left[\left(E_{Cr^{2+}} - E_f\right)/kT\right]\right]\right\}$$
$$+ N_{EL(2)}\left\{e_p^{EL(2)}/\left[e_p^{EL(2)} + e_n^{EL(2)}\right] - 1/\left[1 + \exp\left[\left(E_{EL(2)} - E_f\right)/kT\right]\right]\right\} \quad (3)$$

where

N_{Cr}	total chromium concentration in the substrate,
$N_{EL(2)}$	total EL(2) level concentration in the substrate,
e_p^{Cr}	emission rate for holes from the Cr^{3+} level,
e_n^{Cr}	emission rate for electrons from the Cr^{2+} level,
$e_p^{EL(2)}$ and $e_n^{EL(2)}$	emission rates for holes and electrons from the EL(2) level, respectively,
$E_{Cr^{2+}}$ and $E_{EL(2)}$	thermal activation energies with respect to the valence band for the Cr^{2+} and EL(2) levels, respectively, and
E_f	energy of the Fermi level with respect to the valence band in the bulk of the substrate.

In this equation, the Fermi-function terms give the concentration of negative charge on the deep levels in the bulk. The emission-rate terms give the concentration of negative charge in the space charge region. The emission rates are derived from the capture cross sections given in the literature [17]. The difference between the emission-rate terms and the Fermi-function terms is the excess concentration of negative charge on the deep levels on the substrate side of the substrate-channel region interface.

Since the relative occupancy of the deep levels in the space charge region is fixed by the appropriate emission rates, the concentration of negative charge at the interface is controlled by the occupancy of the deep traps in the bulk of the substrate. For a low N_{eff}, the space charge region at the interface between the substrate and the active layer will be very diffuse and backgating will be minimal.

Calculations of the magnitude of backgating as a function of the substrate compensation using this model are discussed below. In these calculations, the change of the drain current ΔI for a -2-V and -4-V bias applied to the substrate side of the MESFET channel is determined. The relationship between the depletion layer width at the substrate-channel region interface and the excess charge N_{eff} in the space charge region is derived using the standard application of Poisson's equation and the neutrality condition which in this case is

$$\int_0^d N_D(x)\,dx = N_{eff}W \quad (4)$$

where d is the depletion width on the channel-region side ($N_D(x)$ is the charge distribution in this region) and W is the depletion width on the substrate side (N_{eff} is the constant negative charge in this region). Using these conditions, the general form for the voltage appearing across the space charge region is derived to be

$$V = \frac{q}{\epsilon}\left\{d\int_0^d N_D(x)\,dx - \int_0^d\left(\int_0^x N_D(x)\,dx\right)dx\right.$$
$$\left. + \frac{1}{2N_{eff}}\left(\int_0^d N_D(x)\,dx\right)^2\right\} \quad (5)$$

where V is the sum of the built-in voltage and the applied backgate bias. In these calculations, it is assumed that the active region donor concentration can be represented by

$$N_D(x) = N_D[1 - \exp(-x/t)] \quad (6)$$

where N_D and t are determined by a fit to the free-carrier concentration profile used in the experimental devices. This approximation, made as a matter of computational convenience, is a good representation for the recessed gate devices investigated. Using (5) and (6), we obtain

$$V = N_D q/\epsilon\left\{d^2/2 + t[d\exp(-d/t) + t\exp(-d/t) - t]\right.$$
$$\left. + N_D/2N_{eff}[d - t[1 - \exp(d/t)]]^2\right\}. \quad (7)$$

This relationship allows the determination of d for given values of N_D, t, N_{eff}, and V. The value of N_{eff} is given by (3) and the values of N_D and t are given by the fit to the experimental concentration profile in the channel region. Therefore, for different values of V, the depletion depth d in the back-channel region is determined.

The saturated drain current at zero gate bias is calculated using

$$I(d) = \int_d^T q v_s W N_D(x)\, dx \qquad (8)$$

where v_s is the saturated velocity, W is the gate width, and T is the total channel region thickness minus the Schottky-gate depletion width. Using the assumed doping profile (6), the general expression for the drain current is

$$I_D = q v_s W N_D \{ (T - d) + t[\exp(-T/t) - \exp(-d/t)] \}. \qquad (9)$$

Finally, the results of these calculations are shown in Fig. 7 where the drain current with 0, -2-V, and -4-V backgate bias are plotted as a function of N_{eff}. The data shown in Fig. 7 were calculated using the following parameters: $N_D = 2.25 \times 10^{17}$ cm^{-3}, $t = 0.1$ μm, $T = 0.13$ μm, $W = 20$ μm, $v_s = 10^7$ cm s^{-1}, and $V_{bi} = 0.7$ V.

This figure graphically demonstrates the role of the substrate compensation on the drain current for zero, -2-V, and -4-V backgate bias. For example, if N_{eff} equals 1×10^{16} cm^{-3}, the application of a -2-V backgate bias reduces the drain current by 0.88 mA, 22 percent, and a -4-V backgate bias reduces the drain current by 1.55 mA, 39 percent. If N_{eff} equals 1×10^{15} cm^{-3}, the decrease in the drain current is 0.35 mA, 9 percent, and 0.55 mA, 14 percent, at -2-V and -4-V backgate bias, respectively. The shallow- and deep-level concentrations corresponding to these values of N_{eff} can be determined from (3). For example, if the EL(2) concentration is 1×10^{16} cm^{-3} and the difference in the shallow-level concentration $(N_d - N_a)$ is 5×10^{15} cm^{-3} then a Cr concentration of 1.8×10^{16} cm^{-3} yields N_{eff} equal to 1×10^{16} cm^{-3} and a Cr concentration of 6×10^{15} cm^{-3} yields N_{eff} equal to 1×10^{15} cm^{-3}. Therefore, using these calculations, the magnitude of backgating can be determined if the properties of the substrate and the active region doping concentration profile are known. In general, the backgating effect is minimized if N_{eff} is small.

This backgating model has been evaluated in a preliminary experiment in which MESFET test devices were fabricated in low Cr-doped material, approximately 5×10^{15} cm^{-3} Cr concentration. The measured backgating characteristics, represented by the first data point, Cr (shown in Fig. 3), are in qualitative agreement with the model and demonstrate reduced backgating for closely compensated material. The calculations described above assume that the entire backgate voltage is applied at the substrate-channel region interface. A model for the voltage communication between the side-gate and the back-channel region will be developed to permit a more quantitative correlation between the calculated and experimental results. Experiments using proton bombardment isolation to reduce the voltage communication between the side-gate and the back-channel region have demonstrated a decrease in the magnitude of backgating for a given side-gate bias [19].

The model predicts that the use of an n^+ layer, separated from the active region by a buffer layer, would eliminate backgating. In this structure, the excess negative charge N_{eff} in the substrate would produce a narrow space charge region entirely in the n^+ region. Therefore, the

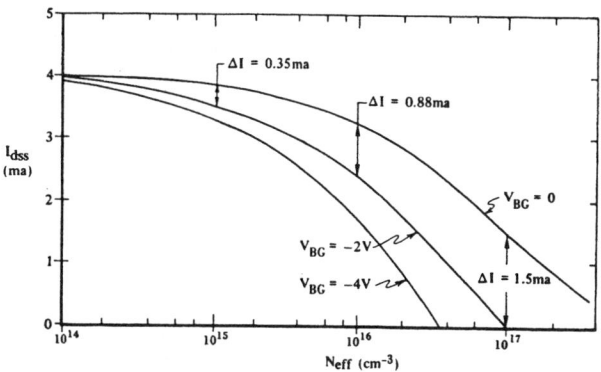

Fig. 7. The calculated dependence of the drain current I_{dss} on the degree of substrate compensation N_{eff} for different backgate bias voltages V_{BG}.

space charge will not reach the channel region and no change in the channel width would be produced by a back-gate bias. This postulate, based on the backgating model, will be evaluated experimentally.

V. CONCLUSIONS

The results of this experimental investigation indicate that the phenomenon of backgating is the result of an accumulation of excess charge at the substrate-channel region interface. This charge resides on deep traps in the substrate material, either on Cr and EL(2) levels in Cr-doped substrates or on EL(2) levels in high-purity semi-insulating substrates. The deep levels responsible for backgating were determined from spectral response and DLTS measurements. Calculations based on this model predict that the magnitude of backgating is dependent on the degree of compensation in the substrate material. Closely compensated substrates have less backgating than substrates with a large excess of deep traps which are unoccupied in the bulk. Preliminary experimental results on lightly Cr-doped material support this model.

Other solutions to the backgate problem include the use of thick buffer layers or n^+ layers as described above. These solutions assure that the space charge region at the semi-insulating to n layer interface does not extend to the MESFET channel region. These solutions and their effect on integrated circuit performance are under investigation.

ACKNOWLEDGMENT

The authors thank the members of the Hewlett–Packard Laboratory who contributed to the work reported. The DLTS measurements were made in conjunction with D. Mars. The encouragement of and the stimulating discussions with R. Archer, C. Bittmann, and C. Liechti contributed to the results reported.

REFERENCES

[1] P. L. Hower, W. W. Hooper, D. A. Tremere, W. Lehrer, and C. A. Bittmann, "The Schottky barrier galium arsenide field-effect transistor," in *Proc. 1968 Inf. Symp. Gallium Arsenide and Related Compounds*, pp. 187–195.
[2] T. Itoh and H. Yanai, "Stability and performance and interfacial problems in GaAs MESFET's," *IEEE Trans. Electron Devices*, vol. ED-27, no. 6, pp. 1037–1045, 1980.

[3] M. Tanimoto, K. Suzuki, T. Itoh, H. Yanai, L. M. F. Kaufmann, W. Nievendick, and K. Heime, "Anomalous phenomena of current–voltage characteristics observed in Gunn-effect digital devices under dc bias conditions," *Electron. Commun. Japan*, vol. 60-C, no. 11, pp. 102–110, 1977.

[4] W. M. Ford and T. L. Larsen, "LEC growth of large GaAs single crystals," in *Proc. Electro Chem. Soc.*, vol. 75-1, p. 517, 1975.

[5] H. R. Szawelska and J. W. Allen, "Photocapacitance measurements of the two acceptor levels of chromium in GaAs," *J. Phys. C.*, vol. 12, pp. 3359–3367, 1979.

[6] U. Kaufmann and J. Schneider, "Chromium as a hole trap in GaP and GaAs," *Appl. Phys. Lett.*, vol. 36, pp. 747–748, 1980.

[7] A. H. Hennel, W. Szuszkiewicz, G. Martinez, B. Clerjoud, A. M. Huber, G. Mouillot, and P. Merenda, "Activation of Cr^{1+} ($3d^5$) level in GaAs: Cr induced by hydrostatic pressure," in *Proc. Semi-insulating III–V Materials, Nottingham, 1980*, G. J. Rees, Ed., 1980, pp. 228–232.

[8] G. M. Martin, A. Mitonneau, and A. Mircea, "Electron traps in bulk and epitaxial GaAs crystals," *Electron. Lett.*, vol. 13, pp. 191–193, 1977.

[9] J. Lagowski, H. C. Gatos, J. M. Parsey, K. Wada, M. Kaminska, and W. Walukiewicz, "Origin of the 0.82-eV electron trap in GaAs and its annihilation by shallow donors," *J. Appl. Phys.*, to be published.

[10] G. M. Martin, G. Jacob, G. Poilblaud, A. Goltzene, and C. Schwab, "Identification and analysis of near-infrared absorption bands in undoped and Cr-doped semi-insulating GaAs crystals," in *11th Int. Conf. on Defects and Radiation Effects in Semiconductors*, (Oiso, Tokyo), Sept. 1980.

[11] P. F. Linquist, "A model relating electrical properties and impurity concentrations in semi-insulating GaAs," *J. Appl. Phys.*, vol. 48, pp. 1262–1267, 1977.

[12] E. M. Swiggard, S. H. Lee, and F. W. Batchelder, "Electrical properties of PBN-LEC GaAs crystals," in *Inst. of Phys. Conf. Ser.* no. 45, pp. 125–133, 1979.

[13] G. M. Martin, J. P. Forges, G. Jacob, and J. P. Hollars, "Compensation mechanisms in GaAs," *J. Appl. Phys.*, vol. 51, pp. 2840–2852, 1980.

[14] C. A. Stolte, "The influence of substrate properties on the electrical characteristics of ion implanted GaAs," in *Proc. Semi-insulating III–V Materials, Nottingham, 1980*, G. J. Rees, Ed., 1980, pp. 93–99.

[15] C. A. Liechti, C. A. Stolte, M. Namjoo, and R. Joly, "GaAs Schottky-gate field effect transistor medium scale integration," Wright–Patterson AFB, Final Report AFAL-TR-81-1082, OH, Jan. 1981.

[16] E. E. Wagner, D. Hiller, and D. Mars, "Fast digital apparatus for capacitance transients analysis," *Rev. Sci. Instrum.*, vol. 51, pp. 1205–1211, 1980.

[17] G. M. Martin, "Key electrical parameters in semi-insulating materials; the methods to determine them in GaAs" in *Proc. Semi-insulating III–V Materials, Nottingham 1980*, R. J. Rees, Ed., 1980, pp. 13–28.

[18] A. Mitonnou, A. Mircea, G. M. Martin, and D. Pons, "Electron and hole capture cross sections at deep centers in Gallium Arsenide," *Rev. Phys. Appl.*, vol. 14, pp. 853–861, 1979.

[19] D. D'Avanzo, "Proton isolation for GaAs integrated circuits," this issue, **pp. 1051–1059.**

Two-Dimensional Numerical Simulation of Side-Gating Effect in GaAs MESFET's

NORIO GOTO, YASUO OHNO, AND HITOSHI YANO

Abstract—Two-dimensional device simulations confirm that side-gating effect of GaAs MESFET occurs on semi-insulating substrates containing hole traps. A negative voltage applied on a side gate, a separate n-type doped region, causes an increase in the thickness of the negatively charged layer at the FET channel interface in the substrate, through hole emission from hole traps. The FET drain current is modulated by the emerged negative charges as an electrical field effect. The magnitude of the drain current reduction is determined by the total acceptor concentration in the substrate and the donor concentration of the channel. However, the magnitude will be independent of the side-gate distances.

I. INTRODUCTION

SIDE-GATING effect, unintentional drain current modulation by a distant terminal voltage, is a serious problem in GaAs MESFET IC's made on semi-insulating substrates (i-substrates). The distant terminal is called a side gate. Extensive research has been performed in order to understand the mechanisms of the effect, such as the relation to substrate leakage current [1], the surface-state-induced mechanism [2], or negative charge formations at FET channel–substrate interface [3]–[6]. However, the complete mechanism and the relation to the nature of the semi-insulating substrates have not yet been clarified.

Hole traps as a cause of side-gating effect were already pointed out experimentally by examining the transient responses of FET drain current to side-gate voltages [3], [6]. Previously, the electrostatic potential profiles in n-i-n structures (n-region/i-substrate/n-region) have been investigated. When i-substrates contain hole traps, the potential profiles become a peculiar form just like the step-like potential profiles in n-p-n structures, under the application of external voltages between the two n-type terminals [7], [8]. Most of the applied voltage appears at the n-i interface of the anode, as if the cathode voltage propagated through the i-substrate to the vicinity of the anode. The potential profile indicates the formation of an electric dipole layer at the interface between the substrate and the anode n-region. The change in the electric dipole layer causes a change in the depleted layer thickness for the anode n-region. When this anode n-region is an n-type

FET channel, the FET drain current will be modulated and side-gating effect would occur.

The purpose of this work is to confirm that hole traps in substrates are responsible for the side-gating effect in GaAs MESFET's, by using a two-dimensional numerical simulation which includes deep-level models. The magnitude of drain current reduction in relation to device structures will also be investigated.

This paper is restricted to side-gating effect without threshold behavior, which is caused by hole traps. Side-gating effects observed in GaAs MESFET's often accompany threshold behavior for side-gate voltages. Threshold behavior is understood to be related to hole injection into substrates from the Schottky-gate metal or p-type electrodes [7], [14]. However, the effects with threshold behavior will be reported elsewhere.

II. MODELS FOR SIMULATION

For numerical simulations, a two-dimensional device simulator BIUNAP-C (BIpolar and UNipolar Analysis Program for Compound semiconductor) was used [9]. BIUNAP-C is a two-dimensional, two-carrier device simulator comprising a deep-trap model. Transports of free carriers are calculated from the commonly used current continuity equations and the Poisson's equation. The carrier drift mobility model used for electrons is assumed to be a simple field-dependent model, in order to avoid the complexity due to negative resistance.

The deep-level model used in the simulator follows SRH (Shockley–Read–Hall) statistics. Electron occupancy ratio for a deep-level f_T and carrier recombination or generation rate through the deep-level R are described as

$$f_T = \frac{C_n n + C_p p_1}{C_n(n + n_1) + C_p(p + p_1)} \quad (1)$$

$$R = \frac{n_1 p_1 - np}{\tau_p(n + n_1) + \tau_n(p + p_1)} \quad (2)$$

where

$$C_n = \sigma_n (3kT/m_n)^{1/2} \qquad C_p = \sigma_p (3kT/m_p)^{1/2}.$$

$$n_1 = 2(2\pi m_n kT/h^2)^{3/2} \exp\left\{-(E_C - E_T)/kT\right\} \quad (3)$$

$$p_1 = 2(2\pi m_p kT/h^2)^{3/2} \exp\left\{-(E_T - E_V)/kT\right\} \quad (4)$$

$$\tau_n = (C_n N_T)^{-1} \qquad \tau_p = (C_p N_T)^{-1}.$$

Manuscript received October 18, 1989; revised March 22, 1990. The review of this paper was arranged by Associate Editor S. E. Laux.

The authors are with Microelectronics Research Laboratories, NEC Corporation, 4-1-1, Miyazaka, Miyamae-ku, Kawasaki 213, Japan.

IEEE Log Number 9036413.

Reprinted from *IEEE Trans. Electron Devices*, vol. ED-37, no. 8, pp. 1821–1827, Aug. 1990.

TABLE I
PHYSICAL PARAMETERS FOR GaAs SUBSTRATE AT 293 K

Energy gap		1.426eV
Electron effective mass	m_n/m_0	0.067
Hole effective mass	m_p/m_0	0.473
Relative dielectric constant		13.1
Electron mobility		$\mu_{0n}/(1+(\mu_{0n}E/v_{sn}))$ cm^2/Vs
		$\mu_{0n}=7300$cm^2/Vs, $v_{sn}=8.0\times10^6$cm/s
Hole mobility		$\mu_{0p}/(1+(\mu_{0p}E/v_{sp}))$ cm^2/Vs
		$\mu_{0p}=425$cm^2/Vs, $v_{sp}=9.5\times10^6$cm/s
		E is electric field in V/cm
Schottky barrier height for gate metal		0.84eV (from conduction band)

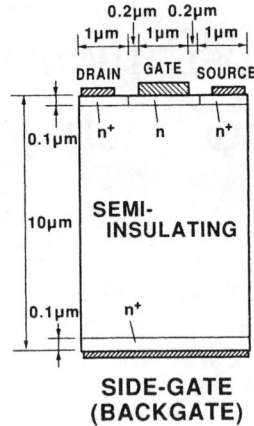

Fig. 1. Device structures used for numerical calculations. Side-gate terminal is placed at the backsurface of substrate. Donor concentrations for n and n$^+$ regions are 2×10^{17} and 8×10^{17} cm^{-3}, respectively. The semi-insulating substrate properties are specified in Table II.

Here, σ_n and σ_p are capture cross sections, m_n and m_p are effective masses, for electrons and holes, respectively. E_T, E_C, and E_V are energies for the deep level, the conduction band bottom, and the valence band top, respectively. N_T is deep-level concentration, n is electron concentration, and p is hole concentration. The degeneracy factor (g factor) is omitted in (3) and (4). However, the difference can be adjusted by a slight shift in deep-level energies. The conventional Poisson's equation and current continuity equations are modified to include deep-level effects, such as carrier occupancy f_T and the generation or recombination rate R. Material parameters used for the simulation are summarized in Table I.

III. SIMULATION OF SIDE-GATING EFFECT

The device structure for the numerical simulations is shown in Fig. 1. In order to reduce the number of mesh points and calculation time, we used a backgate configuration, where the side-gate terminal is placed at the back surface of the substrate. This structure cannot simulate surface-specific mechanisms for side-gating effects, if any of them exist, but it will be sufficient to demonstrate mechanisms caused by properties of bulk substrates.

Calculations were made for three kinds of semi-insulating substrates, Substrates I, II, and III, each containing a deep level and a shallow level listed in Table II. Substrate I assumes an undoped LEC substrate, which contains an EL2-like electron trap of deep donor compensating for shallow acceptors [10]. Substrate III assumes a Cr-doped substrate which contains a hole trap of deep acceptor compensating for shallow donors [11]. In order to emphasize the relation between trap properties and device characteristics, the deep levels are assumed to be located at the middle of the energy gap and the capture cross-section values are not necessarily the same as those reported in references. The capture cross sections for minor trapping carriers are tentatively set at 3/1000 of that for major trapping carriers. Substrate II is a reference substrate to the two substrates. It contains a hole trap of deep donor compensating for shallow acceptors. Using this substrate, it will be shown that the side-gating effect is related to the trap property of hole trap or electron trap,

TABLE II
DEEP AND SHALLOW LEVELS IN THE SUBSTRATES ASSUMED IN THE SIMULATION

	Deep Level					Shallow Level	
	Type	σ_n (cm^2)	σ_p (cm^2)	E_C-E_T (eV)	N_T (cm^{-3})	Type	N (cm^{-3})
Substrate I	donor	1×10^{-13}	3×10^{-16}	0.715	2×10^{16}	acceptor	5×10^{15}
Substrate II	donor	3×10^{-16}	1×10^{-13}	0.715	2×10^{16}	acceptor	5×10^{15}
Substrate III	acceptor	3×10^{-16}	1×10^{-13}	0.715	2×10^{16}	donor	5×10^{15}

not the property of donor or acceptor (see the Appendix). In these substrates, traps are classified into electron-trap type or hole-trap type, only by the capture cross-section values for electrons and holes, since their energy levels are located at the center of the energy gap. Calculated drain current I_{dss} and side-gate current are shown in Fig. 2, as functions of side-gate voltage. In Substrate I, which is compensated for by electron traps, drain current is maintained at an almost constant value, when the side-gate voltage is changed from positive to negative values. In Substrates II and III, which are compensated for by hole traps, drain current decreases from zero to negative side-gate voltages. These show that the side-gating effect occurs when substrates are compensated for by hole traps, as described in [7] and [8], whether the hole traps are donor or acceptor.

Fig. 3(a) and (b) shows two-dimensional electrostatic potential profiles for Substrate I and Substrate II, respectively. The side-gate voltage is −1.5 V. When the substrate contains electron traps, the electrostatic potential in the substrate has a linearly graded shape in the i-substrate region, as shown in Fig. 3(a). In this substrate, drain-current reduction does not occur. On the contrary, as shown in Fig. 3(b), when the substrate contains hole traps, the electrostatic potential becomes flat in the i-substrate region with the value following the side-gate voltage. It looks as if the negative side-gate voltage propagated throughout the substrate. In this substrate, drain current reduction occurs, presumably due to a field effect.

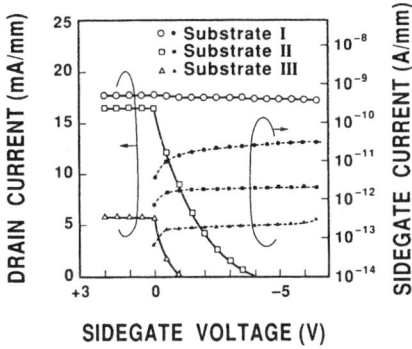

Fig. 2. Calculated FET drain current and side-gate current for three substrates. Voltages applied on the FET source, gate, and drain are 0, 0, and 1 V, respectively.

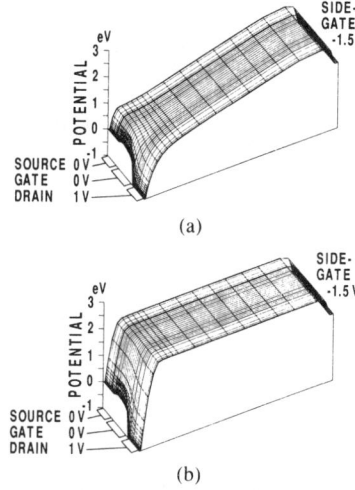

Fig. 3. Two-dimensional electrostatic potential energy profiles for electrons. (a) In Substrate I containing electron traps. (b) In Substrate II containing hole traps.

In order to examine the cause for the potential profile difference, electron and hole concentrations, the occupancy ratio of electrons in traps, and the electrostatic potential profile are plotted along a path from the FET gate to the back gate (side gate). Fig. 4 shows results for Substrate I, and Fig. 5 shows results for Substrate II. The side-gate voltages are -1.5 V, the same as those used in Fig. 3.

The characteristic electron and hole concentrations, n_1 and p_1, defined by (3) and (4), are indicated by arrows at the right side axis in Figs. 4(a) and 5(a). Electron and hole concentration values in the thermal equilibrium are close to n_1 and p_1, respectively. When the side-gate voltage is applied, holes are depleted at the FET side (anode) from the thermal equilibrium value p_1, while holes accumulate at the side-gate side (cathode) in both of the two substrates.

In those substrates, occupancy ratio $f_T = 0.75$ corresponds to the electrical neutrality of the substrate, and $f_T = 1.0$ corresponds to the fully negatively charged state of the substrate, caused by electron occupation at the deep levels together with the ionization of shallow acceptors. Only the substrate compensated for by hole traps, shown

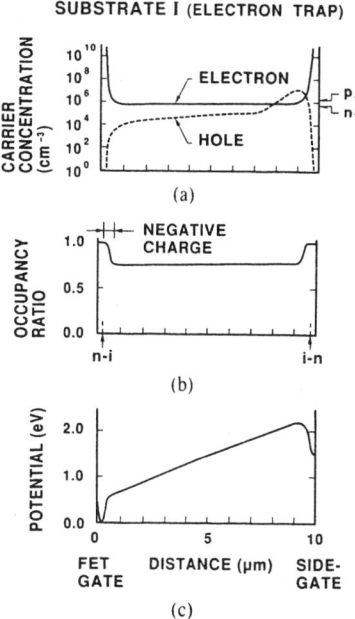

Fig. 4. Calculated quantities inside the substrate along a path from the gate to the side gate for Substrate I. (a) Electron and hole concentrations. (b) Electron occupancy ratio for the deep level. The n-i interfaces positions are indicated on the horizontal axis. (c) Electrostatic potential energy for electrons.

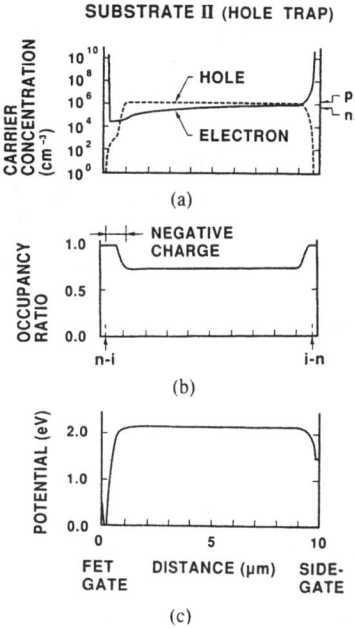

Fig. 5. Calculated quantities inside the substrate along a path from the gate to the side gate for Substrate II. (a) Electron and hole concentrations. (b) Electron occupancy ratio for the deep level. The n-i interfaces positions are indicated on the horizontal axis. (c) Electrostatic potential energy for electrons.

in Fig. 5(b), has an apparent negative space-charge layer at the FET side. As can be seen in (1), the electron occupancy ratio f_T mainly responds to free hole concentration, when $C_p \cdot p_1 \gg C_n \cdot n_1$ is satisfied, i.e., when the deep level is a hole trap. On the contrary, f_T mainly responds to free electron concentration when the deep level is an electron trap.

In hole trap compensating Substrate II, the electron occupancy ratio for the deep level is sensitive to the depletion of holes caused by the application of side-gate bias voltage. The deep levels are negatively charged up by emitting holes, in the layer indicated as "negative charge" in Fig. 5(b). The negatively charged layer extends until the potential profile in the charge-neutral region of the i-substrate becomes flat, as shown in Fig. 5(c). The hole depleted region disappears, except for the negatively charged layer around the anode. The extension of the negatively charged layer modulates the FET channel charges. This is the direct cause of drain current reduction, which will be shown in the next section.

In electron trap compensating Substrate I, however, the deep levels are not sensitive to the depletion of holes. The thickness of the negatively charged layer changes little, only enough to adjust the built-in potential difference for the n-i interface at both the FET and the side gate, as indicated in Fig. 4(b). Then, the potential profile has a linearly graded shape in the charge-neutral region of the i-substrate, as shown in Fig. 4(c).

Another hole trap compensating Substrate III gives results similar to those for Substrate II. This indicates that the charged states for deep levels and the potential profiles in substrates may depend on whether the deep levels are electron traps or hole traps, not on whether they are donors or acceptors. In both substrates, the total bulk charge can be negative by emitting holes from hole traps.

IV. SIDE-GATING EFFECT MAGNITUDE

Fig. 6 shows a band diagram inside the substrate along a path from the FET gate to the side gate. Two n-i junctions are formed, at the FET side and at the side-gate side. Uniform doping is assumed for the FET channel. As the negatively charged layer extends into the i-substrate by applying negative side-gate voltage, for the substrate containing hole traps, the depletion layer in the n-type layer of the FET channel also extends, in order to maintain the charge balance in the n-i junction. The situation is similar to the reverse-biased condition for p-n junctions. In p-n junctions, the p-region is negatively charged by depletion of free holes. For n-i junctions with the i-substrate containing hole traps, the i-region is negatively charged by emitting holes from hole traps. We define the negative charge values as $-qN_A^{\text{eff}}$, where N_A^{eff} is the effective acceptor concentration in the i-substrate, as shown in Fig. 6. This is expressed as

$$N_A^{\text{eff}} = N_{DA} + N_{SA} - N_{SD} \qquad (5)$$

where N_{DA} is deep acceptor concentration, N_{SA} is shallow acceptor concentration, and N_{SD} is shallow donor concentration in the i-substrate. As a matter of fact, qN_A^{eff} is the maximum charge density for the substrate achieved under free hole depletion.

The depleted layer width in the channel n-type region is calculated by the same equations as those for calculating depletion width for the n-region in an abrupt p-n junction [12]. Instead of acceptor concentration for the

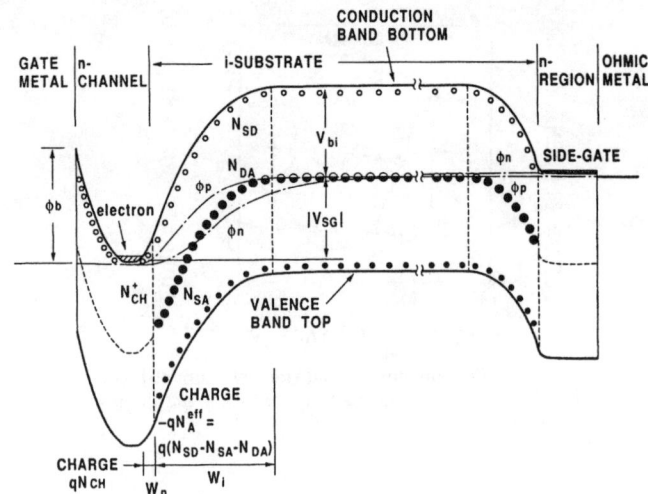

Fig. 6. Band diagram from n-i junctions at FET to side gate. The bulk charge densities at the FET n-i interface are $-qN_A^{\text{eff}} = -q(N_{DA} + N_{SA} - N_{SD})$ and qN_{CH} for i-substrate and FET n-channel, respectively. Here, N_{DA}, N_{SA}, and N_{SD} are deep donor concentration, shallow acceptor concentration, and shallow donor concentration, respectively. N_{CH} is the donor concentration in the channel.

p-region, N_A^{eff} can be used. Then, depletion width for channel n-region W_n is given by

$$W_n = \left(\frac{2\epsilon(|V_{SG}| + V_{bi})}{qN_{CH}} \cdot \frac{N_A^{\text{eff}}}{N_A^{\text{eff}} + N_{CH}} \right)^{1/2} \qquad (6)$$

where N_{CH} is donor concentration for the channel, V_{bi} is built-in potential for the n-i junction, and V_{SG} is side-gate voltage.

Assuming $N_{CH} \gg N_A^{\text{eff}}$, (6) becomes

$$W_n \simeq \frac{1}{N_{CH}} \left(\frac{2\epsilon(|V_{SG}| + V_{bi})N_A^{\text{eff}}}{q} \right)^{1/2}. \qquad (7)$$

Differentiation by V_{SG} gives

$$\frac{dW_n}{dV_{SG}} \simeq \frac{1}{N_{CH}} \left(\frac{\epsilon N_A^{\text{eff}}}{q \cdot (|V_{SG}| + V_{bi})} \right)^{1/2}. \qquad (8)$$

This indicates that a large N_A^{eff} value gives a large change rate of depletion width with respect to V_{SG}.

Drain current I_{dss} is roughly estimated to be proportional to the active channel thickness in the FET. Then, the drain current can be written as

$$I_{dss} = k \cdot N_{CH} \cdot (a - h - W_n)$$

$$\simeq k \cdot N_{CH}$$

$$\cdot \left(a - h - \frac{1}{N_{CH}} \left(\frac{2\epsilon(|V_{SG}| + V_{bi})N_A^{\text{eff}}}{q} \right)^{1/2} \right) \qquad (9)$$

where a is the thickness of the FET n-type doped region, h is the depletion width by gate voltage. Coefficient k is $q\mu(V_{ds}/L)$ for low drain voltage cases, and is $q \cdot v_s$ for high drain voltage cases. Here, μ is the low-field electron drift mobility, V_{ds} is the drain voltage, L is the channel length, and v_s is the electron saturation velocity. Equation (9) indicates that a higher N_A^{eff} value results in smaller I_{dss}, even at no side-gate bias application [13]. Differentiating (9)

$$\frac{dI_{dss}}{dV_{SG}} = -k \cdot N_{CH} \frac{dW_n}{dV_{SG}} \simeq k \cdot \left(\frac{\epsilon N_A^{\text{eff}}}{|V_{SG}| + V_{bi}} \right)^{1/2}.$$

(10)

This shows the higher N_A^{eff} value causes a greater I_{dss} reduction rate with respect to V_{SG}.

Drain current reductions are numerically simulated changing N_A^{eff} values. Here, a substrate similar to Substrate III is assumed and N_{DA} is varied. Fig. 7(a) shows that greater N_A^{eff} values give smaller drain currents at zero side-gate voltage, and greater reduction rates for negative side-gate voltage applications. In other words, a greater N_A^{eff} value gives steeper drain current reduction, compared with the initial current value. Fig. 7(b) indicates this fact clearly.

Thus the magnitude of drain current reduction is determined by doping profiles around the FET channel. It should be noted that, in the above expressions, the distance between the FET and the side gate does not appear. In actual FET's, however, the flat potential region in the i-substrate can have a finite gradient, in order to maintain the leakage current through the n-i-n structure. Therefore, the voltage drop at the n-i junction beneath the FET channel can be less than $|V_{SG}|$, which will be dependent on the side-gate distance. To be precise, the drain current reductions will also weakly depend on the side-gate distance.

V. DISCUSSION

In this paper, we only simulated for substrates with a single deep level. For more complicated deep-level configurations, such as substrates with both hole traps and electron traps, the simple classification by hole trap and electron trap is not adequate. A more detailed classification for the occurrence of the side-gating effect is needed.

The exact conditions for the occurrence of the side-gating effect in n-type FET's are that substrate space charge is sensitive to free hole concentration, or hole Fermi level. In an n-i-n structure, the space charge for the i-region can only be zero or negative, since electrons easily diffuse from the n-regions to a positively charged region, if it appears. To form negatively charged region with hole traps, hole emission from them will be crucial.

The side-gating effect occurs when the hole trap energy is located above, or just on, the Fermi level in thermal equilibrium condition. In the substrate containing a single deep level of the hole trap, the substrate compensation is accomplished by balancing the charges in the hole trap itself. The Fermi level in thermal equilibrium is always located near the hole trap energy level. The hole trap is partially occupied with electrons and holes. Thus the hole trap can always emit holes, responding to the depletion of

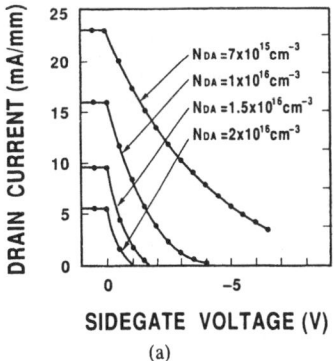

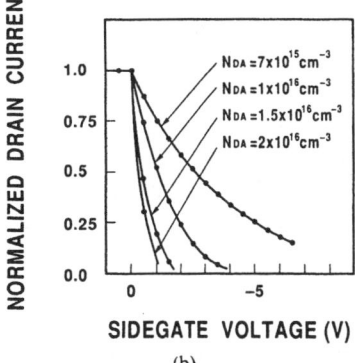

Fig. 7. Calculated FET drain current for substrates with different effective acceptor concentrations. (a) Substrates contain shallow donors with concentration of 5×10^{15} cm^{-3}, and hole trap type deep acceptors with concentrations from 7×10^{15} to 2×10^{16} cm^{-3}. The resulting effective acceptor concentration $N_A^{\text{eff}} = N_{DA} + N_{SA} - N_{SD}$ ranges from 2×10^{15} to 1.5×10^{16} cm^{-3}. (b) Drain current normalized by the value at zero side-gate voltage for each curve.

the free holes. However, if the substrate contains more than two deep levels, the situation becomes a little complicated. Even if the substrate contains a hole trap, it does not necessarily contribute to compensating for the substrate. The Fermi level at the thermal equilibrium is not necessarily located at the hole trap energy, depending on the other trap energy level. If the hole trap is located under the Fermi level, i.e., close to the valence band, the hole trap will be completely depleted of holes, and can no longer emit any holes. Thus in this case, the side-gating effect would not occur, in spite of the existence of hole traps. The situation is realized when the other trap is an electron trap and its energy level is higher than that for the hole trap. If the hole trap is located above the Fermi level, i.e., close to the conduction band, the hole trap is completely occupied by holes. Then, the hole trap can emit holes, when free holes are depleted. Under such conditions, side-gating effects would occur.

By numerical simulations, it has been shown that hole traps are the origin of the side-gating effect. The reported experiments also suggest that hole traps play roles in side-gating effects [3]. It is well known that LEC GaAs substrates contain an electron trap EL2 as a main deep level to compensate for shallow acceptors in the substrates and to maintain their high resistivity. It is generally believed that an as-grown bulk GaAs substrate does not contain a substantial amount of hole traps. Although hole traps in bulk GaAs substrates are not presently well determined, hole traps responsible for side-gating effects would have been introduced during the FET fabrication processes. According to numerical simulations, the formation of few micrometer thick surface layers containing hole traps will be sufficient to cause the side-gating effect [7].

This paper reports on only the side-gating effects beginning from zero side-gate voltage. The simulation indicated that side-gating effect will not appear in Substrate I, which is assumed to be similar to the existing substrates compensated for by EL2. However, side-gating effects with threshold have also been reported for the substrates [1]. Their drain current reduction does not occur from zero side-gating voltage, but it does not occur when side-gate voltage exceeds a certain threshold value. TFL (Trap Fill Limited) mechanism has been thought to be an origin of side-gating effect due to this threshold behavior. If a large voltage is applied at the side gate in Substrate I, the TFL mechanism would occur. However, the voltage for TFL mechanism is as high as 350 V when the acceptor concentration is 5×10^{15} cm^{-3} and side-gate distance is 10 μm. Therefore, we do not believe the threshold is due to the TFL mechanism. In order to explain such threshold in side-gating effect, another mechanism, such as injection of holes into the substrate, will be needed [14].

VI. CONCLUSION

The side-gating effect, drain current reduction by a distant side-gate voltage, in GaAs MESFET's, was numerically investigated by a two-dimensional device simulator, which employs SRH statistics for deep levels. It was confirmed that the drain current reduction occurs when the substrate contained deep levels acting as hole traps, regardless of whether the traps were donor or acceptor. FET channel current was modulated by the electron depletion of the n-type channel, which resulted from the compensation for the extension of the negatively charged layer at the n-i interface into the i-substrate containing hole traps. The negatively charged layer was extended by emitting holes from hole traps in the anode region of the i-substrate, where holes were depleted under negative side-gate voltage application. It is also shown that the magnitude of the side-gating effect is determined by the maximum negative charge density, which can be achieved in the i-substrate under free hole depletion. The maximum negative charge density is given by the total acceptor concentration, less shallow donor concentrations in the substrate. Therefore, the higher acceptor concentration causes larger drain current reduction, whether the acceptors are shallow or deep in the energy band diagram.

From the similarity in the potential profiles between those for n-i-n structures under side-gating effects, and those for n-p-n structures [8], the effect can be approximated to be the substrate bias effect in MOSFET's [15]. Then, the i-substrate containing hole traps corresponds to a p-type substrate. The difference lies in the response time of the reaction. The conductive p-type substrate in MOS-

FET's allows a p-type back gate to cause drain current reduction for the entire IC chip with almost no time delay. Therefore, the effect can be explicitly included in the circuit designs by introducing a substrate bias electrode. On the other hand, the side-gating effect in GaAs MESFET is as slow as the trap response time. It will be difficult to count in circuit designing. So, to realize a GaAs IC, it should be eliminated by avoiding formation of hole traps in substrates, or by implementing other measures in the device structures.

APPENDIX
ELECTRICAL EQUIVALENCE BETWEEN SUBSTRATES

In the text we mentioned that whether the deep trap level is a donor or an acceptor is not crucial for the origin of side-gating effect. In addition to the substrate containing a hole trap of deep acceptor and a shallow donor, we introduced another substrate containing a hole trap of deep donor and a shallow acceptor. Also, a substrate containing only deep levels can be assumed. These three substrates, all containing hole traps, can be mathematically equivalent, if their parameters are suitably chosen.

For instance, it is assumed that substrate A-I contains N_{DA1} deep acceptors and N_{SD1} shallow donors, substrate A-II contains N_{DD2} deep donors and N_{SA2} shallow acceptors, and substrate A-III contains N_{DA3} deep acceptors and N_{DD3} deep donors. It is also assumed that all the deep levels have the same energy level, the same electron capture cross section, and the same hole capture cross section, then the capture rate for electrons C_n and the capture rate for holes C_p are equal among these deep levels, respectively. The values n_1 and p_1, defined by (3) and (4), are also equal among these deep levels.

In steady state, then, occupancy ratio f_T values, given by (1), are also equal among the deep levels. At room temperature, all the shallow levels can be assumed to be completely ionized. Then, the space charges for these substrates are

$$qN_1 = -f_T q N_{DA1} + q N_{SD1} = q N_{SD1} - f_T q N_{DA1}$$

for substrate A-I,

$$qN_2 = (1 - f_T) q N_{DD2} - q N_{SA2}$$
$$= q(N_{DD2} - N_{SA2}) - f_T q N_{DD2}$$

for substrate A-II, and

$$qN_3 = (1 - f_T) q N_{DD3} - f_T q N_{DA3}$$
$$= q N_{DD3} - f_T q (N_{DD3} + N_{DA3})$$

for substrate A-III. If $N_{DD2} = N_{DA1}$, and $N_{SA2} = N_{SD1} + N_{DA1}$, then substrates A-I and A-II have the same space charges at any f_T value. Also, if $N_{DD3} = N_{SD1}$, and $N_{DA3} = N_{DA1} - N_{SD1}$, then substrate A-III has the same space charges as substrate A-I. In the transient states, capture and emission of electrons and holes are given by

$$\frac{dn}{dt} = c_n N_{DA1} \left\{ n_1 f_T - n(1 - f_T) \right\}$$

$$\frac{dp}{dt} = c_p N_{DA1} \left\{ p_1(1 - f_T) - p f_T \right\}$$

for substrate A-I,

$$\frac{dn}{dt} = c_n N_{DD2} \left\{ n_1 f_T - n(1 - f_T) \right\}$$

$$\frac{dp}{dt} = c_p N_{DD2} \left\{ p_1(1 - f_T) - p f_T \right\}$$

for substrate A-II, and

$$\frac{dn}{dt} = c_n (N_{DA3} + N_{DD3}) \left\{ n_1 f_T - n(1 - f_T) \right\}$$

$$\frac{dp}{dt} = c_p (N_{DA3} + N_{DD3}) \left\{ p_1(1 - f_T) - p f_T \right\}$$

for substrate A-III. Generation and recombination processes through shallow levels are always so small that they can be ignored. Occupancy ratio f_T values may not be equal among these substrates in transient, but they should be equal if the preceding free carrier concentration variations are equal. Then, if the same deep and shallow level concentrations as those used in the space charge are chosen, capture and emission rates have the same values among these substrates. Thus on those substrates, equations used in device simulation are all equivalent.

ACKNOWLEDGMENT

The authors would like to thank S. Kumashiro for the use of the device simulator BIUNAP-C, Dr. T. Itoh and Dr. S. Asada for their valuable suggestions and discussion. The authors also wish to thank Dr. H. Sakuma for giving them the opportunity to do this work.

REFERENCES

[1] C. P. Lee, S. J. Lee, and B. M. Welch, "Carrier injection and backgating effect in GaAs MESFET's," *IEEE Electron Device Lett.*, vol. EDL-3, pp. 97–98, 1982.

[2] H. Hasegawa, T. Kitagawa, T. Sawada, and H. Ohno, "A new side-gating model for GaAs MESFETs based on surface avalanche breakdown," in *Gallium Arsenide and Related Compounds, 1984* (Inst. Phys. Conf. Ser. no. 74, Institute of Physics, 1985), pp. 521–524.

[3] T. Itoh and H. Yanai, "Stability of performance and interfacial problems in GaAs MESFET's," *IEEE Trans. Electron Devices*, vol. ED-27, pp. 1037–1045, 1980.

[4] ——, "Experimental investigation of interfacial problems in GaAs MESFETs," in *Gallium Arsenide and Related Compounds, 1980* (Inst. Phys. Conf. Ser. no. 56, Institute of Physics, 1981), pp. 537–542.

[5] C. Kocot and C. A. Stolte, "Backgating in GaAs MESFET's," *IEEE Trans. Electron Devices*, vol. ED-29, pp. 1059–1064, 1982.

[6] M. Ogawa, "Enhanced temperature dependence of MESFET characteristics by backgate and s.degate biasing," *Trans. IEICE*, vol. E70, pp. 847–856, 1987.

[7] Y. Ohno and N. Goto, "GaAs MESFET-IC side-gating effect model," in *IEDM Dig. Tech. Papers*, pp. 252–256, 1987.

[8] ——, "Mechanism of electrostatic potential conduction in semi-insulating substrates," *J. Appl. Phys.*, vol. 66, pp. 1217–1221, 1989.

[9] S. Asada, S. Sugou, K. Kasahara, Y. Kato, and S. Kumashiro, "Semi-insulating current blocking property simulations for buried heterostructure laser diodes," *Appl. Phys. Lett.*, vol. 52, pp. 703–705, 1988.

[10] G. M. Martin, A. Mitonneau, and A. Mircea, "Electron traps in bulk and epitaxial GaAs crystals," *Electron. Lett.*, vol. 13, pp. 191–193, 1977.

[11] A. Mitonneau, G. M. Martin, and A. Mircea, "Hole traps in bulk and epitaxial GaAs crystals," *Electron. Lett.*, vol. 13, pp. 666–668, 1977.

[12] S. M. Sze, *Physics of Semiconductor Devices*, 2nd ed. New York, NY: Wiley, 1981, p. 74.

[13] K. Horio, K. Asada, and H. Yanai, "Numerical analysis of GaAs MESFET with deep acceptors Cr in the Semi-insulating substrate," *Trans. of IEICE*, vol. E72, pp. 303–306.

[14] N. Goto and Y. Ohno, "Two-dimensional simulation of GaAs MESFET side-gating effect" in *Proc. 5th Conf. on Semi-Insulating III-V Materials* (Malmö, Sweden). Bristol, UK: Hilger, 1988, pp. 253–258.

[15] M. B. Das, "Dependence of the characteristics of MOS-transistors on the substrate resistivity," *Solid-State Electron.*, vol. 11, pp. 305–322, 1968.

Theoretical Study of the Piezoelectric Effect on GaAs MESFET's on (100), (011), and (1̄ 1̄ 1̄)Ga, and (111)As Substrates

TSUKASA ONODERA AND HIDETOSHI NISHI

Abstract—This paper discusses the influence of piezoelectric charge on GaAs MESFET's fabricated on (100), (011), and (1̄1̄1̄)Ga, and (111)As substrates. We use a two-dimensional device simulation, including a piezoelectric charge model with an edge-force approximation. We found that piezoelectric charge has very little influence on GaAs MESFET's fabricated on (011) substrates. Furthermore, the threshold voltage shift of MESFET's on (1̄1̄1̄)Ga has the opposite sign as those on a (111)As plane substrate. We also found that orientation effects are much smaller for substrates other than the (100) substrate. This indicates the potential of orientation-effect free alignment of FET's in high-speed GaAs LSI's with (011) or (111) substrates.

I. INTRODUCTION

THE electrical characteristics of GaAs MESFET's fabricated on the (100) substrate depend on the orientation of the gate finger [1]. The stress-enhanced preferential diffusion model [1]–[3] and the piezoelectric effect model [4], [5] were proposed to explain this. We confirmed that the piezoelectric effect was the main cause of orientation effects by observing the gate-orientation dependence of GaAs MESFET parameters, which is related to the sign of the internal stress in the dielectric overlayer [6]. Because a high degree of threshold voltage control for GaAs LSI fabrication is important, there has been much interest in the piezoelectric effect. Two-dimensional device analysis shows that the change in MESFET characteristics is due to the change in the effective thickness of the channel, which is caused by a potential barrier of piezoelectric charge [7]. External stress also affects GaAs MESFET characteristics [8]. Recently, measurements of external stress are compared with calculations, based on two-dimensional finite-element analysis for mechanical stress and one-dimensional device analysis [9].

With gate lengths of 2.0 μm or less, the piezoelectric effect improves device parameters for a certain combination of gate orientation and dielectric overlayer. It reduces the gate length dependence and improves the K-value and subthreshold characteristics [6], [10]. However, the orientation effect still limits GaAs IC layout.

This paper shows that (011) and (1̄1̄1̄)Ga or (111)As plane substrates can achieve orientation-independent GaAs MESFET characteristics.

II. THEORY

We used a two-dimensional device simulator [7] to analyze the piezoelectric effect on MESFET's fabricated on various GaAs substrates. We used the concentrated edge-force approximation [11] to calculate the stress distribution in the substrate. We have generalized the model equations derived by Asbeck *et al.* to calculate the piezoelectric charge distribution for [011] and [01̄1]-oriented FET's on a (100) GaAs substrate, for the calculation of those for FET's along with arbitrary orientations on various planes of a GaAs substrate. The simulator consists of four major parts.

The first part calculates the two-dimensional distribution of the stress tensor σ, or in vector-like notation T, in the vicinity of the MESFET on a GaAs substrate. The cause of the stress in the substrate is assumed to be the stress in both the dielectric overlayer [4]–[7] and the gate electrode [8]. When a stressed overlayer is deposited on the surface of a substrate, the stress on the substrate is intensified near the discontinuities in overlayer thickness [12]. The concentrated edge force f_E represents the intensified stress at the edge of discontinuities. For the edge of the Schottky-gate electrode of the WSi$_x$-gate self-aligned GaAs MESFET, shown in Fig. 1, f_E is modeled by (1), the sum of intensified stress due to the dielectric overlayer σ_f, and the gate metal σ_g.

$$f_E = \sigma_f d_f + \sigma_g d_g. \qquad (1)$$

Here, d_f represents the thickness of the dielectric overlayer and d_g the thickness of the gate metal. Because of the negligibly small stress in WSi$_x$ with 60-percent Si content [13], we only accounted for the stress in the dielectric overlayer. We approximate that the gate width W_g is much larger than gate length L_g, and assume that the GaAs substrate is elastically isotropic; the six elements of the stress

Manuscript received October 7, 1988; revised April 17, 1989. The review of this paper was arranged by Associate Editor S. Tiwari.

The authors are with Fujitsu, Ltd., 1015 Kamikodanaka, Nakahara-ku, Kawasaki 211, Japan.

IEEE Log Number 8928089.

Reprinted from *IEEE Trans. Electron Devices*, vol. ED-36, no. 9, pp. 1580–1585, Sept. 1989.

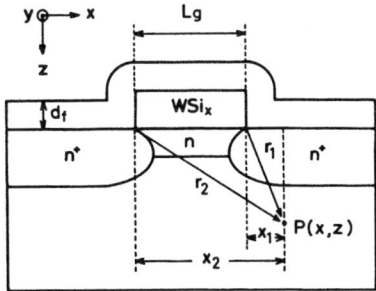

Fig. 1. Cross section of self-aligned MESFET, with coordinate system for calculating piezoelectric charge distribution.

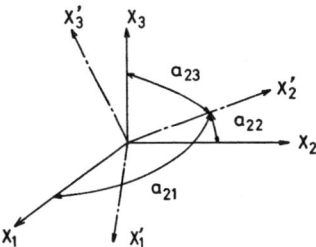

Fig. 2. Relation between primary and rotated coordinate system for calculating rotated piezoelectric tensor.

vector are derived as follows [4], [11]:

$$T_1 = \sigma_{xx} = -\frac{2f_E}{\pi}\left(\frac{X_1^3}{r_1^4} - \frac{X_2^3}{r_2^4}\right)$$

$$T_2 = \sigma_{yy} = -\frac{2\nu f_E}{\pi}\left(\frac{X_1}{r_1^2} - \frac{X_2}{r_2^2}\right)$$

$$T_3 = \sigma_{zz} = -\frac{2f_E}{\pi}\left(\frac{X_1 Z^2}{r_1^4} - \frac{X_2 Z^2}{r_2^4}\right)$$

$$T_4 = \sigma_{yz} = 0$$

$$T_5 = \sigma_{xz} = -\frac{2f_E}{\pi}\left(\frac{X_1^2 Z}{r_1^4} - \frac{X_2^2 Z}{r_2^4}\right)$$

$$T_6 = \sigma_{xy} = 0 \qquad (2)$$

where ν is 0.23, Poisson's ratio for GaAs [11]. The second-order stress caused by the electric field was neglected. Stress components due to the discontinuities in overlayer thickness at the ohmic electrode were calculated in a similar way and added to each of the components of the stress vector. Though (2) is convenient for calculation and for discussing the orientation effect, a more accurate analysis, such as the finite-element method [9], is necessary for quantitative discussion of the elastic and electrical properties of GaAs MESFET's.

The second part of the simulator calculates the 18 elements of the piezoelectric tensor for a GaAs crystal with a coordinate system rotated for the crystallographic plane and gate orientation of interest. The relation between the primary and rotated coordinate system is shown in Fig. 2. The axis subscripts, 1, 2, and 3, denote the X, Y, and Z axes. The rotated coordinate X_i' ($i = 1$ to 3) is expressed by the primary coordinate X_j ($j = 1$ to 3) and the orientation cosines cos (a_{ij}) as follows:

$$X_i' = \sum_{j=1}^{3} X_j \cos(a_{ij}). \qquad (3)$$

The elements of the transformed piezoelectric tensor d_{ijk}' are expressed as [14]

$$d_{ijk}' = \sum_{l=1}^{3}\sum_{m=1}^{3}\sum_{n=1}^{3} d_{lmn} \cos(a_{il}) \cos(a_{jm}) \cos(a_{kn}) \qquad (4)$$

where d_{lmn} represents the elements of the piezoelectric tensor in the primary coordinate system. Because of the zinc-blende structure of the GaAs crystal, all but six of the elements of the piezoelectric tensor are zero in the primary coordinate system. All of the nonzero elements are constant at 2.7×10^{-17} C/dyn [15]. The transformed piezoelectric tensor can then be written as a matrix by reducing the number of elements with equivalent suffices.

The third part calculates the two-dimensional distribution of the components of the piezoelectric polarization vector and the charge. The piezoelectric polarization vector $\boldsymbol{P}_{pz}$ is obtained as a product of the stress vector and piezoelectric tensor and is expressed as

$$\boldsymbol{P}_{pz} = \boldsymbol{d}\boldsymbol{T}. \qquad (5)$$

The piezoelectric charge density ρ_{pz} in the substrate is given by

$$\rho_{pz} = \text{div}\,(\boldsymbol{P}_{pz}). \qquad (6)$$

The fourth part of the simulator calculates the electrical characteristics by solving Poisson's equation and the current continuity equation, which are expressed as follows:

$$\text{div}\,(\epsilon_s \,\text{grad}\, \psi) = -q(N_d - N_a - n + \rho_{pz}) \qquad (7)$$

$$\text{div}\,(\boldsymbol{J}_n) = 0 \qquad (8)$$

where ψ is the potential distribution, ϵ_s the dielectric constant of the substrate, q the electron charge, and N_d, N_a, and n are the ionized donor, acceptor, and mobile electron densities, respectively. Piezoelectric charge is modeled as additional space charge in the substrate. In a nondegenerate condition, the electron current $\boldsymbol{J}_n$ can be expressed as

$$\boldsymbol{J}_n = k_B T \mu_n n_i \exp(q\psi/k_B T) \,\text{grad}\,[\exp(-q\phi_n/k_B T)] \qquad (9)$$

where k_B is Boltzmann's constant, T is the temperature, μ_n is the electron mobility, n_i is the intrinsic carrier density, and ϕ_n is the quasi-Fermi potential for electrons. We modeled the electric field and doping density dependence

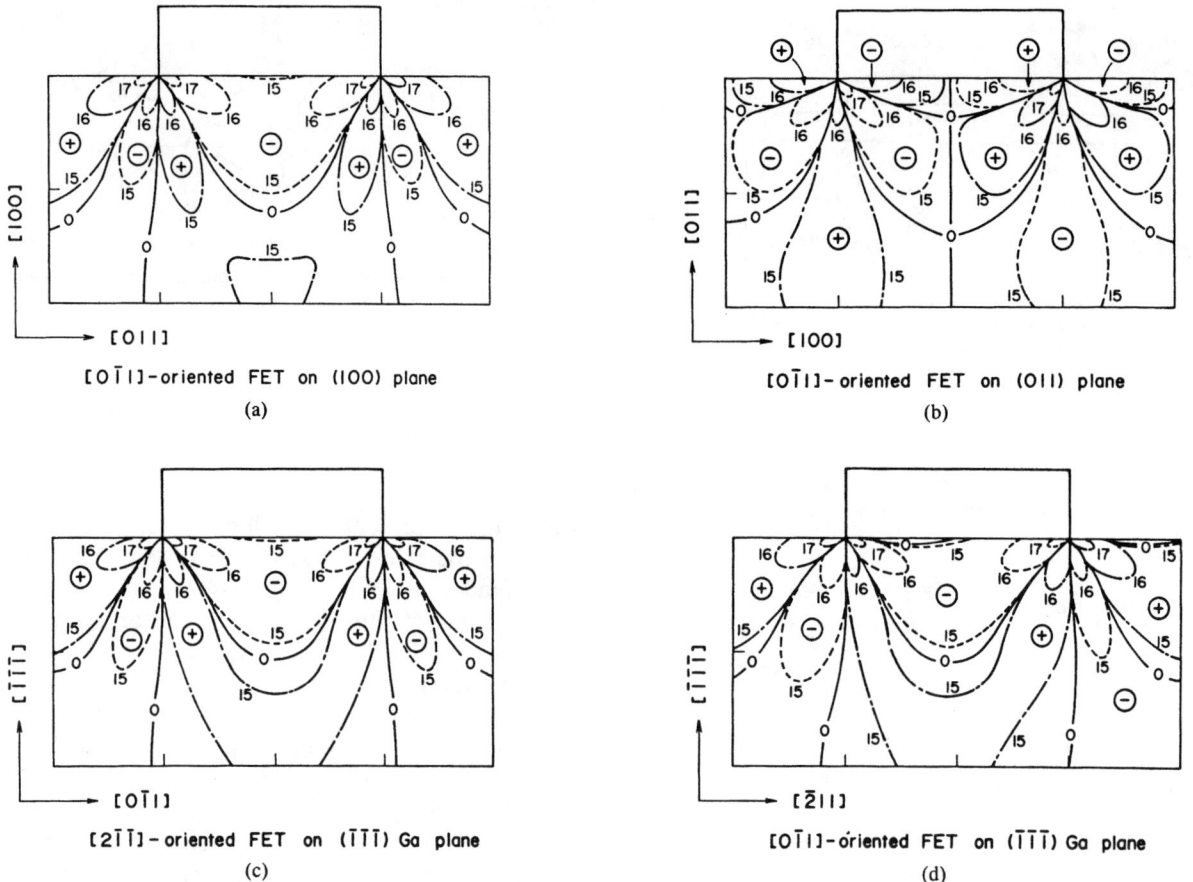

Fig. 3. Two-dimensional distribution of piezoelectric charge in GaAs MESFET's for (a) a [$0\bar{1}1$]-oriented gate on a (100) substrate, (b) a [$0\bar{1}1$]-oriented gate on a (011) substrate, (c) a [$2\bar{1}\bar{1}$]-oriented gate on a ($\bar{1}\bar{1}\bar{1}$)Ga substrate, and (d) a [$0\bar{1}1$]-oriented gate on a ($\bar{1}\bar{1}\bar{1}$)Ga substrate. The contours 0, 15, 16, and 17 correspond to zero, 10^{15}, 10^{16}, and 10^{17} electrons. The concentrated edge force is 9×10^4 dyn/cm. The rectangles indicate the 1.0-μm-long gate-electrodes, and give both horizontal and vertical magnitude of the region.

for μ_n [16], but ignored the influence of piezoelectric polarization.

The boundary conditions were derived from the surface and contact properties [17]. The absence of current through the surface and boundaries in the substrate was modeled by zero normal derivatives of potential and carrier density. The ohmic contacts were modeled with a uniform applied voltage and contact equilibrium carrier density. The operation of the Schottky-barrier gate was modeled using a diffusion theory. The equations were solved using finite difference [18] with a maximum potential error of 0.1 mV. The electrode current was calculated by integrating the current density along the plane surrounding each electrode. Threshold voltage was obtained as the square root of the drain current plotted against the gate voltage.

III. RESULTS AND DISCUSSION

Fig. 3(a)–(d) shows the two-dimensional distribution of piezoelectric charge near the channel for MESFET's with representative gate orientations on (100), (011), and ($\bar{1}\bar{1}\bar{1}$)Ga, and (111)As plane GaAs substrates. The gate length is 1.0 μm. A concentrated edge force of 9.0×10^4

dyn/cm is used. This simulates the 0.9-μm-thick dielectric overlayer under 1.0×10^{10} dyn/cm^2 compressive stress. Orientation is defined in [1].

It would be worthwhile to review the piezoelectric effect on the GaAs MESFET on a (100) substrate for the following discussion on (011) and ($\bar{1}\bar{1}\bar{1}$)Ga, and (111)As substrates. The piezoelectric charge distribution in a [$0\bar{1}1$]-oriented FET on a (100) substrate (see Fig. 3(a)) is similar to that reported in [4]. The sign of the piezoelectric charge changes in all regions when a negative concentrated edge force, which simulates the deposition of dielectric layer under tensile stress, is used in the calculations [4]–[6], [10]. Furthermore, the sign of the charge changes if the gate orientation is rotated 90°, i.e., a [011]-oriented FET. For the combination of gate orientation and the sign of concentrated edge force to induce the negative piezoelectric charge near the channel–substrate interface, for example, [$0\bar{1}1$]-oriented FET's with a CVD-SiO$_2$ overlayer, the effective channel thickness becomes thinner, resulting in a positive threshold voltage shift [7].

For the combination of gate orientation and the sign of concentrated edge force as positive piezoelectric charge is induced near the channel–substrate interface, for example, [011]-oriented FET's with a plasma CVD-Si₃N₄ overlayer [6], the sign of the threshold voltage shift is negative because of the increase in current below the channel–substrate interface [7].

The distribution of piezoelectric charge in a [0$\bar{1}$1]-oriented FET on a (011) substrate is shown in Fig. 3(b). The distribution is not symmetric about the vertical center line. The contour line of the zero piezoelectric charge lies along with the center line from the surface to the back of the substrate. Moreover, the sign of the charge changes twice with depth in the regions under the gate between the edge and the midpoint of the electrode. The sign of the charge on the left is the opposite of that on the right. Because of the rather complicated charge distribution, the two-dimensional nature of charge and potential distribution must be considered when analyzing the characteristics of the [0$\bar{1}$1]-oriented FET on the (011) substrate and will be discussed later. This FET's characteristics depend on the direction of current in the channel because of the nearly anti-symmetric distribution of charge about the center of the device.

We also calculated the distribution of piezoelectric charge in the [100]-oriented FET on a (011) substrate. However, the charge is zero for all parts of the FET cross section. This indicates that this FET's electrical characteristics are not influenced by the piezoelectricity of the GaAs crystal.

The distribution of piezoelectric charge in [2$\bar{1}$$\bar{1}$] and [0$\bar{1}$1]-oriented FET's on the ($\bar{1}$$\bar{1}$$\bar{1}$)Ga substrate is shown in Fig. 3(c) and (d). The angle between the two orientations is 90° and 30° for the equivalent orientations. The sign of the charge distribution is nearly the same for these two orientations. The charge at the depth of the channel–substrate interface, 0.1 μm for a Si⁺-implanted channel at 60 keV, is negative for those two orientations. Therefore, the sign of the threshold voltage shift due to piezoelectric charge for GaAs MESFET's on a ($\bar{1}$$\bar{1}$$\bar{1}$)Ga substrate is positive for all, or at least for every 30°, of the orientations of gate electrode. The distribution of piezoelectric charge in a [0$\bar{1}$1]-oriented FET is slightly asymmetric, as shown in Fig. 3(d). This indicates that the electrical characteristics of a [0$\bar{1}$1]-oriented FET depends on the direction of current in the channel.

We also calculated piezoelectric charge distributions in [2$\bar{1}$$\bar{1}$] and [0$\bar{1}$1]-oriented FET's on a (111)As substrate. The results with a positive concentrated edge force are similar to those obtained for ($\bar{1}$$\bar{1}$$\bar{1}$)Ga substrates and can be represented by changing the sign of the charge density in Fig. 3(c) and (d) for the equivalent gate orientations. Therefore, the sign of the threshold voltage shift due to piezoelectric charge, which is induced by the positive value of the concentrated edge force (i.e., deposition of a dielectric overlayer with compressive stress), for GaAs MESFET's on a (111)As substrate is negative for all orientations of the gate finger. We concluded that the sign

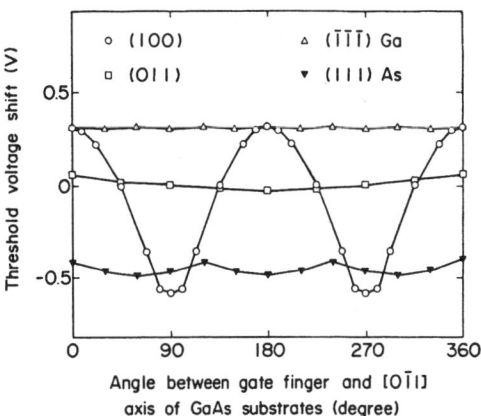

Fig. 4. Simulated gate-orientation dependence of threshold voltage shift for 1.0-μm gate GaAs MESFET's on (100), (011), ($\bar{1}$$\bar{1}$$\bar{1}$)Ga, and (111)As substrates. Concentrated edge force is 9×10^4 dyn/cm.

of the piezoelectric charge induced in a (111)GaAs substrate changes with both the direction of stress in the dielectric overlayer and the atomic plane of the crystal used for FET fabrication.

Fig. 4 shows the relationship between the gate orientation and the simulated threshold voltage shift of 1.0-μm gate MESFET's on the (100), (011), and ($\bar{1}$$\bar{1}$$\bar{1}$)Ga, and (111)As planes of GaAs substrates. We assumed Si⁺ implantations at 60 keV with a dose of 0.9×10^{12} cm⁻² for the n-channel layer and at 175 keV with a dose of 1.7×10^{13} cm⁻² for the self-aligned n⁺-layer. We used a concentrated edge force of 9.0×10^4 dyn/cm. The threshold voltage calculated without piezoelectric charge is −0.15 V.

The threshold voltage shift for MESFET's on a (100) substrate, indicated by circles in Fig. 4, varies with gate orientation by a half-period of 90° from +320 to −690 mV. The threshold voltage shift is zero at 45°, 135°, 225°, and 315°. These results are consistent with Fig. 3(a) and with previous results [5].

The threshold voltage shift for a MESFET on a (011) substrate, shown by rectangles, varies with gate orientation between +80 and −38 mV, which is much smaller than that observed on the (100) substrate. At 90° and 270°, the threshold voltage shift due to piezoelectric effect is zero. It is interesting that a positive threshold voltage shift is observed at 0°, but a negative one is observed at 180°. At both of the two orientations, the gate fingers are parallel to the [0$\bar{1}$1] axis, but the directions of the current in the channel are opposite. Therefore, the current direction dependence of the characteristics of [0$\bar{1}$1]-oriented FET's on a (011) GaAs substrate, predicted by Fig. 3(d), is confirmed. With negative concentrated edge force, which simulates the dielectric overlayer in tensile stress, the result is similar to that shown in Fig. 4, shifting the horizontal axis by 180° because of the opposite sign of the piezoelectric charge.

Fig. 4 also shows that the magnitude of the threshold voltage shift of MESFET's on a ($\bar{1}$$\bar{1}$$\bar{1}$)Ga substrate, indicated by open triangles, varies from 320 to 300 mV with gate orientation. The shift is positive for all the orientations. The variation with gate orientation has a period of 60°, which corresponds to the angle between the equivalent axes of symmetry in the (111) crystal plane. The magnitude of the shift is similar to that of a [0$\bar{1}$1]-oriented FET on a (100) substrate. Despite the asymmetric distribution of piezoelectric charge in a [0$\bar{1}$1]-oriented FET, mentioned in Fig. 3(d), the threshold voltage shift did not depend on current direction. The reason will be discussed later, with a comparison to the (111)As substrate.

The threshold voltage shift of MESFET's on a (111)As substrate, shown by the solid triangles in Fig. 4, varies from −480 to −410 mV with gate orientation. The orientation did not affect the sign of the threshold voltage shift. The variation with orientation has a period of 120°, which is the angle between the axes of threefold symmetry, in a (111)-crystal plane. This indicates the current direction dependence of the threshold voltage shift of a [0$\bar{1}$1]-oriented FET, which is not observed in the ($\bar{1}\bar{1}\bar{1}$)Ga substrate. This discrepancy is caused by the difference in the sensitivity of the effective channel thickness of asymmetric piezoelectric charge distribution near the channel–substrate interface. The effective channel thickness is modified by the channel–substrate built-in potential, which has a logarithmic dependence on the space-charge density on both sides of the interface. Usually, the channel has a much higher charge density, up to mid-10^{17} cm^{-3} for WSi$_x$-gate self-aligned MESFET's [10], than piezoelectric charge (about some 10^{15} cm^{-3}), as shown in Fig. 3(d). When the piezoelectric charge is negative, there is a one-sided p-n-junction-like interface formed on the back of the channel. In this case, the effective thickness of the channel is insensitive to the magnitude of piezoelectric charge because the depletion layer spreads mainly toward the substrate. The channel-substrate interface is n$^-$-n-junction-like, with positive piezoelectric charge. In this case, an additional carrier distribution, which enhances the substrate current, is formed below the channel, due to the positive charge in the substrate. The deeper carrier density depends on the positive piezoelectric charge. The magnitude of substrate current, and hence, the effective thickness of the channel is more sensitive to the change in piezoelectric charge distribution than the p-n-junction-like case.

When the concentrated edge force is negative, a result similar to that shown in Fig. 4 is obtained. The solid and open triangles change planes because of the inverted piezoelectric charge. It has been reported that the K-value and the subthreshold characteristics of GaAs MESFET's on a (100) substrate can be improved by the piezoelectric effect [7], [10]. Other favorable effects, such as a higher K-value and a lower gate-length dependence of threshold voltage, can be expected for (111)GaAs substrates with specific combinations of stress in the dielectric overlayer

and the atomic plane used for FET fabrication. Moreover, the drastic reduction of the gate-orientation dependence of MESFET characteristics will aid in the development of high-speed GaAs LSI's. The results presented in this paper have been confirmed by experiment, and will be presented in the separate paper [19].

IV. CONCLUSIONS

We have demonstrated how the piezoelectric effect influences the electrical characteristics of GaAs MESFET's on (011) and ($\bar{1}\bar{1}\bar{1}$)Ga, and (111)As substrates. We found that the threshold voltage shift and its orientation dependence in MESFET's on a (011) substrate are much smaller than those on (100) substrates. The characteristics of a [0$\bar{1}$1]-oriented FET on a (011) substrate depend slightly on the current direction. We also found that the sign of the threshold voltage shift of MESFET's on a (111) substrate depends on both the sign of the stress in the dielectric overlayer and the atomic plane used. However, we found that the gate-orientation dependence of electrical characteristics is smaller than with (100) substrates. These findings indicate the potential use of (111)GaAs substrates in LSI's with MESFET's with piezoelectrically improved and gate-orientation-independent characteristics.

ACKNOWLEDGMENT

We wish to thank N. Yokoyama and T. Ohnishi for valuable discussions and T. Misugi, M. Kobayashi, A. Shibatomi, M. Fukuta, and A. Takahashi for their encouragement.

REFERENCES

[1] C. P. Lee, R. Zucca, and B. M. Welch, *Appl. Phys. Lett.*, vol. 37, p. 311, 1980.
[2] N. Yokoyama, H. Onodera, T. Ohnishi, and A. Shibatomi, *Appl. Phys. Lett.*, vol. 42, p. 270, 1983.
[3] R. A. Sadler and L. F. Eastman, *Appl. Phys. Lett.*, vol. 43, p. 865, 1983.
[4] P. M. Asbeck, C. P. Lee, and M. F. Chang, *IEEE Trans. Electron Devices*, vol. ED-31, p. 1377, 1984.
[5] M. F. Chang, C. P. Lee, P. M. Asbeck, R. P. Vahrenkamp, and C. G. Kirkpatrik, *Appl. Phys. Lett.*, vol. 45, p. 279, 1984.
[6] T. Ohnishi, T. Onodera, N. Yokoyama, and H. Nishi, *IEEE Electron Device Lett.*, vol. EDL-6, p. 172, 1985.
[7] T. Onodera, T. Ohnishi, and H. Nishi, in *Proc. 12th Int. Symp. GaAs and Related Compounds* (Karuizawa, Japan), 1985, p. 517.
[8] M. Kanamori, H. Ono, T. Furutsuka, and J. Matsui, *IEEE Electron Device Lett.*, vol. EDL-8, p. 228, 1987.
[9] J. C. Ramirez et al., *IEEE Trans. Electron Devices*, vol. 35, p. 1232, 1988.
[10] T. Onodera, T. Ohnishi, N. Yokoyama, and H. Nishi, *IEEE Trans. Electron Devices*, vol. ED-32, p. 2314, 1985.
[11] P. A. Kirkby, P. R. Selway, and L. D. Westbrook, *J. Appl. Phys.*, vol. 50, p. 4567, 1979.
[12] R. J. Jaccodine and W. A. Schlegel, *J. Appl. Phys.*, vol. 38, p. 2913, 1967.
[13] T. Ohnishi, N. Yokoyama, H. Onodera, S. Suzuki, A. Shibatomi, *Appl. Phys. Lett.*, vol. 43, p. 600, 1983.
[14] T. Ogawa, *Applied Physics of Crystals.* Tokyo: Shokabo, 1971, p. 38 (in Japanese).

[15] A. J. Slobodnik, Jr., E. D. Conway, and R. T. Delmonico, *Microwave Acoustics Handbook*, vol. 1A, *Surface Wave Velocities*, AFCRL-TR-73-9597, 1973, p. 27.

[16] C. M. Snowden, M. J. Howes, and D. V. Morgan, *IEEE Trans. Electron Devices*, vol. ED-30, p. 1817, 1983.

[17] K. Yamaguchi, S. Asai, and H. Kodera, *IEEE Trans. Electron Devices*, vol. ED-23, p. 1283, 1976.

[18] M. Kurata, *Numerical Analysis for Semiconductor Devices*. Lexington, MA: Lexington Books, 1982.

[19] T. Onodera, H. Kawata, H. Nishi, T. Futatsugi, and N. Yokoyama, *IEEE Trans. Electron Devices*, this issue, pp. 1586–1590.

Experimental Study of the Orientation Effect of GaAs MESFET's Fabricated on (100), (011), and (1̄1̄1̄)Ga, and (111)As Substrates

TSUKASA ONODERA, HARUO KAWATA, HIDETOSHI NISHI, TOSHIRO FUTATSUGI, AND NAOKI YOKOYAMA, MEMBER, IEEE

Abstract—We investigated the electrical characteristics of GaAs MESFET's fabricated on (100)-, (011)-, and (1̄1̄1̄)-Ga, and (111)-As oriented substrates. About 4600 cm^2/Vs of electron mobility in the Si$^+$-implanted n-layer was obtained from Hall measurement with no distinct difference in results observed for the different substrate orientations. We fabricated WSi$_x$-gate self-aligned MESFET's in all orientations used in the experiment. We observed a remarkable difference in the gate-length dependence of FET parameters, especially at gate lengths shorter than 2 μm. FET's fabricated on (011) substrates had a much smaller dependence of threshold voltage shift on the dielectric overlayer thickness and gate orientation than FET's fabricated on (100) substrates. The threshold voltage shift of FET's fabricated on (111)-cut GaAs substrates depends on the dielectric overlayer thickness and the atomic plane of the substrate on which the devices are fabricated. Gate orientation did not affect FET characteristics much on (111)-cut substrates. We explained this in terms of the piezoelectric effect in a GaAs crystal. Our findings indicate that (011)- and/or (111)-cut substrates are useful in fabricating high-speed GaAs IC's. They allow flexible FET layout, and are free from the gate-orientation effect.

I. INTRODUCTION

THE physical properties of semiconductors depend on the crystal orientation of the wafer. This dependence has been investigated for both material research and industrial applications. For Si devices, such as MOSFET's, the surface-state density N_{ss} of the Si–SiO$_2$ structure depends on the crystal orientation [1]. A smaller N_{ss} is obtained with a (100) substrate than with a (011) or (111) substrate. This correlates with the orientation effect of the oxidation rate, which is proportional to the number of Si bonds per unit area available to reach with oxygen atoms [2], [3]. A correlation between N_{ss} and the orientation-dependent Young's modulus of Si has been explained by the stress effect at the Si–SiO$_2$ interface due to the thermal expansion mismatch [4]. The extremely stable surface properties obtainable for Si MOS transistors on (100) substrates are the backbone of Si device technology.

So far only the (100) substrate has been used for manufacturing GaAs MESFET's [5], [6]. There are many in-

teresting papers on crystal orientation effects. It was reported that dislocation-free growth by LEC in the ⟨111⟩ direction is easier than in the ⟨011⟩ or ⟨100⟩ directions [7]. However, the (100) surface is more stable for epitaxial layers [8]–[10]. The crystal orientation affects the donor-density dependence of the avalanche breakdown voltage V_B of the Schottky-barrier diode [11]. The crystal plane with the highest V_B was (111) with a 10^{15} cm^{-3} donor density and (100) with 10^{17} cm^{-3}. A Schottky-barrier height of 0.9 eV has been reported on n-type (111)B or an As plane substrate with Ni contacts [12]. Because of the nearly parabolic structure of the conduction band near the Brillouin zone center, a low-field electron mobility is expected for random planes of the GaAs substrate [13]. However, analysis of the ballistic transport of electrons in GaAs showed that the maximum electron velocity increases from ⟨100⟩, ⟨011⟩, and ⟨111⟩ [14].

Due to the partially ionic bond nature of GaAs crystals, the MESFET characteristics depend on the orientation of the gate finger [5], [6]. This was explained theoretically by the piezoelectric effect model [15], [16]. There has been much interest in it because it is important to obtain highly controlled threshold voltages in fabricating LSI's. Both experimental [17]–[19] and theoretical [20], [21] studies on the piezoelectric effect in GaAs MESFET's have been published.

Recently, we explained theoretically how the piezoelectric effect affects MESFET characteristics on (011), (1̄1̄1̄)Ga, and (111)As substrates, in another paper [22]. This paper discusses the electrical characteristics of WSi$_x$-gate self-aligned MESFET's fabricated on (100), (011), (1̄1̄1̄)Ga, and (111)As substrates.

II. EXPERIMENT

We fabricated WSi$_x$-gate self-aligned MESFET's [23] on semi-insulating GaAs substrates. The gate orientations were [011] and [01̄1] on a (100) substrate, [100] and [01̄1] on a (011) substrate, and [01̄1] and [21̄1̄] on both (1̄1̄1̄)Ga and (111)As substrates. The substrates were cut from ⟨100⟩-grown In-doped LEC dislocation-free crystal. The gate orientations are shown in Fig. 1. The Gate width was 20 μm. Si$^+$ for the FET n-channel was implanted at 59 keV at a dose of 0.9×10^{12} cm^{-2}, and

Manuscript received October 7, 1988; revised April 17, 1989. The review of this paper was arranged by Associate Editor S. Tiwari.

T. Onodera, H. Kawata, and H. Nishi are with Fujitsu, Ltd., 1015 Kamikodanaka, Nakahara-ku, Kawasaki 211, Japan.

T. Futatsugi and N. Yokoyama are with Fujitsu Laboratories, Ltd., 10-1 Morinosato-Wakamiya, Atsugi 243-01, Japan.

IEEE Log Number 8929093.

Reprinted from *IEEE Trans. Electron Devices*, vol. 36, no. 9, pp. 1586–1590, Sept. 1989.

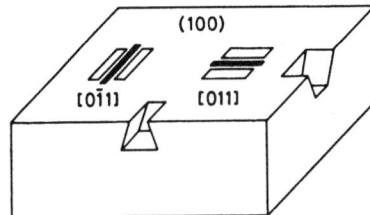

Fig. 1. Schematic drawing for definition of FET orientation.

furnace annealing at 800°C was done for 15 min. The angle of the ion implantation to the substrate surface was determined carefully to avoid an abnormally deep carrier profile due to the channeling effect [24]. Self-aligned Si^+ implantation at 175 keV with a dose of 1.7×10^{13} cm^{-2} for the n$^+$-source/drain was done with 400-nm-thick WSi$_x$ gates as masks and annealed at 750°C for 15 min. The annealing cap for both the n-channel and n$^+$-source/drain was 100-nm-thick CVD SiO$_2$. The ohmic metal for the source/drain electrode was AuGe/Au (20 nm/280 nm), and the condition for making the alloy was 450°C for 1.0 min. CVD SiO$_2$ and plasma-CVD Si$_3$N$_4$ formed the dielectric overlayer. X-ray diffraction showed that the SiO$_2$ and the Si$_3$N$_4$ films used in the experiment cause opposite stress on GaAs [17]. The SiO$_2$ film is compressed by about 10^9 dyn/cm^2, and the Si$_3$N$_4$ film is stretched at about 10^9 dyn/cm^2. After MESFET fabrication, the sequence of threshold voltage measurement and chemical etching of the dielectric overlayer was repeated to investigate the relation between dielectric overlayer thickness and threshold voltage. The measured threshold voltages were averaged for 40 or more FET's in 2-in-diameter wafers.

III. RESULTS AND DISCUSSIONS

A. Electrical Properties of Implanted Layer

Before fabricating the MESFET, we investigated the electrical properties of the Si$^+$-implanted layer in the substrates. The samples were prepared by Si$^+$ implantation at 175 keV with a dose of 2.5×10^{12} cm^{-2} and annealing at 850°C for 10 min, with a 100-nm-thick CVD-SiO$_2$ annealing cap. The data obtained from Hall measurement are summarized in Table I. The electron mobility is 4600 cm^2/Vs at a sheet carrier density of about 8×10^{11} cm^{-2} for (011) and ($\bar{1}\bar{1}\bar{1}$)Ga, and (111)As substrates. For the (100) substrate, a slightly smaller electron mobility was obtained with a slightly higher carrier density. The carrier activation efficiency depends slightly on the crystal plane, and ranges from about 30 to 40 percent. However, no significant difference in implanted carrier profile, as obtained from C–V measurement, was observed for any crystal plane used in the experiment, except near the surface. This indicates that the difference in the sheet carrier density may also be attributed to the orientation dependence in the depletion layer at the crystal surface due to different pinning potential, depending on the orientation. Further detailed investigations on the orientation effect in the electrical properties, such as surface potential and Schottky-barrier height, are necessary. Moreover, higher

TABLE I
ELECTRICAL PROPERTIES OF Si$^+$-IMPLANTED LAYERS AT 175 keV WITH A
DOSE OF 2.5×10^{12} cm^{-2}

CRYSTAL ORIENTATION	ELECTRON MOBILITY (cm^2/Vs)	SHEET CARRIER DENSITY (cm^{-2})
(100)	4300	1.05×10^{12}
(011)	4608	8.17×10^{11}
($\bar{1}\bar{1}\bar{1}$)Ga	4680	8.49×10^{11}
(111)As	4600	7.76×10^{11}

activation can be attained with more optimized annealing conditions.

B. Device Characteristics

After we fabricated the WSi$_x$-gate self-aligned MESFET [7], we confirmed by measuring I_{DS}–V_{DS} characteristics that enhancement-mode FET's had been successfully fabricated on (100), (011), and ($\bar{1}\bar{1}\bar{1}$)Ga, and (111)As substrates. The gate-length dependence of the threshold voltage for the MESFET's on (100), (011), and ($\bar{1}\bar{1}\bar{1}$)Ga, and (111)As substrates is shown in Fig. 2. The FET's with gate fingers parallel to the [$0\bar{1}1$] orientation are compared in the figure because this orientation was used in previously published works [5], [6], [15]. The data were obtained for FET's with a 1.2-μm-thick CVD SiO$_2$ overlayer. The relation between the gate length and the threshold voltage depends on the crystal plane. The threshold voltage for the FET's on the (100) substrate is almost independent of gate length [6], [17]. The slightly deeper threshold voltage for the FET on the (100) substrate, at about a 4-μm gate length, is caused by the difference in sheet carrier density in the n-layer, due to the difference in activation or Schottky-barrier height. Although nearly equal threshold voltages are obtained for the FET's on (011) and ($\bar{1}\bar{1}\bar{1}$)Ga, and (111)As substrates for about a 4-μm gate, the threshold voltage shift changes with reduced gate lengths. A similar phenomenon was observed with [011]- and [$0\bar{1}1$]-oriented FET's fabricated on a (100) substrate and explained by piezoelectric effect [15], [16]. The effect is dependent on the dielectric overlayer thickness [17], [20], and we investigated the relationship between them in detail.

Fig. 3 shows the threshold voltage shift for MESFET's fabricated on a (100) GaAs substrate as a function of the CVD-SiO$_2$ thickness. The gate length was 1.2 μm for [$0\bar{1}1$]- and [011]-oriented FET's. The angle between the gate fingers was 90°. The threshold voltage shift of the [$0\bar{1}1$]-oriented FET was positive, and that of the [011]-oriented FET was negative. The results are consistent with those reported previously [6], [17], [20]. The difference in the threshold voltatge shift between FET's of these two orientations is as much as 2.9 V, with a 1.2-μm-thick SiO$_2$ overlayer. The difference in the sign of the threshold volt-

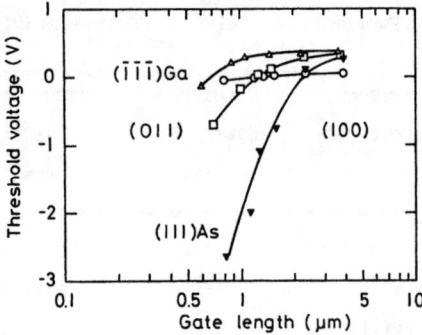

Fig. 2. Relation between the threshold voltage and gate length for a GaAs MESFET with [0$\bar{1}$1]-oriented gate finger on (100), (011), (111)As, and ($\bar{1}\bar{1}\bar{1}$)Ga substrates, with a 1.2-μm-thick SiO$_2$ overlayer.

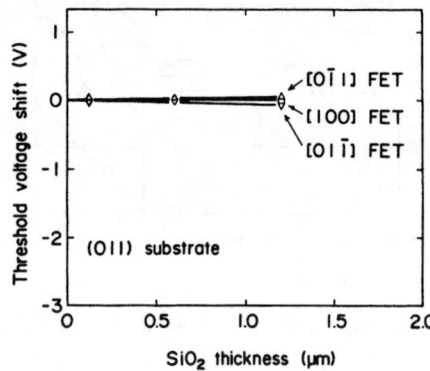

Fig. 4. The relationship between threshold voltage shift and SiO$_2$ thickness for GaAs MESFET's with [0$\bar{1}$1]-, [01$\bar{1}$]-, and [100]-oriented gate fingers fabricated on a (011) substrate.

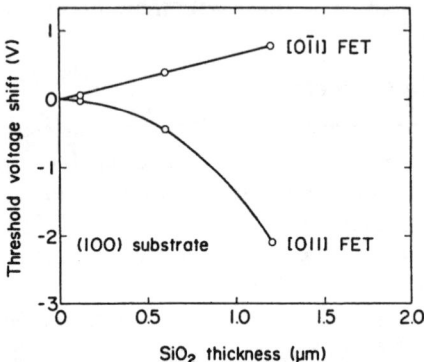

Fig. 3. The relationship between threshold voltage shift and SiO$_2$ thickness of GaAs MESFET's with [0$\bar{1}$1]- and [011]-oriented gate fingers on a (100) substrate.

age shift is caused by the change in sign of the piezoelectric charge [15], [16].

Fig. 4 shows the threshold voltage shift as a function of CVD SiO$_2$ overlayer thickness for [100]-, [0$\bar{1}$1]-, and [01$\bar{1}$]-oriented FET's fabricated on a (011) GaAs substrate. The gate length was 1.0 μm. The angle between the gate fingers of the [100]- and the [0$\bar{1}$1]-oriented FET's was 90°. The [0$\bar{1}$1] and [01$\bar{1}$]-oriented FET's were identical, except for the change in the source/drain electrode (the direction of current flow in the channel). A positive shift is observed for the [0$\bar{1}$1]-oriented FET, and a negative shift is observed for the [01$\bar{1}$]-oriented FET. The difference in the threshold voltage shift was 60 mV with a 1.2-μm-thick SiO$_2$ overlayer. The asymmetric distribution of piezoelectric charge [22] and the process-induced asymmetry in the device structure may be the cause. Unfortunately, we have not been able to confirm that the implanted n$^+$-source/drain and ohmic electrodes are completely symmetric in the devices we used. However, the symmetry of the device structure probably does not vary with the SiO$_2$ overlayer thickness. Therefore, the SiO$_2$ thickness dependence of the threshold voltage shift indicates the existence of an asymmetric piezoelectric charge in GaAs MESFET's with [0$\bar{1}$1]-oriented gate fingers.

Fig. 4 also shows that the threshold voltage shift in the [100]-oriented FET is zero for SiO$_2$ overlayer thicknesses up to 1.2 μm. This is consistent with the theoretical prediction that there is no piezoelectric charge in MESFET's with this gate orientation [22]. Consequently, both orientation effects and the dependence on the SiO$_2$ overlayer thickness of MESFET's fabricated on the (011) GaAs substrate are substantially smaller than those observed with the (100) substrate. For GaAs LSI applications requiring good threshold voltage control, there would be an advantage in using the (011) substrate to provide stress- and orientation-independent MESFET characteristics.

Fig. 5 shows the threshold voltage shift of [0$\bar{1}$1]- and [2$\bar{1}\bar{1}$]-oriented FET's on ($\bar{1}\bar{1}\bar{1}$)Ga, and [01$\bar{1}$]- and [$\bar{2}$11]-oriented FET's on (111)As GaAs substrates. The gate length was 1.1 μm, and the angle between these orientations on each atomic plane was 90°, and 30° for their equivalent orientations. The threshold voltage shift is positive for [0$\bar{1}$1]- and [2$\bar{1}\bar{1}$]-oriented FET's on ($\bar{1}\bar{1}\bar{1}$)Ga, and it increases linearly with SiO$_2$ thickness. The difference between the shifts was about 100 mV at 1.2-μm-thick SiO$_2$. This is much smaller than that observed on the (100) substrate. This indicates that the sign of the piezoelectric charge near the channel–substrate interface of SiO$_2$-encapsulated FET's on the ($\bar{1}\bar{1}\bar{1}$)Ga substrate is negative in any orientation, in 30° increments of the gate finger.

The sign of the threshold voltage shift changes for both [01$\bar{1}$]- and [$\bar{2}$11]-oriented FET's on the (111)As GaAs substrate, although the condition of the CVD-SiO$_2$ overlayer deposition and, hence, the stress in the film is the same as that in the ($\bar{1}\bar{1}\bar{1}$)Ga substrate. Thus, for the (111)-cut GaAs substrate, the direction of piezoelectric polarization and the sign of the piezoelectric charge depends on the atomic plane used for fabrication. It is easy to understand why the direction of piezoelectric polarization in the FET's on the (111)As substrate is opposite that induced in the devices on the ($\bar{1}\bar{1}\bar{1}$)Ga substrate since the Ga and As atoms in the crystal are inverted. The sign of the shift is independent of gate orientation. This was also observed in the ($\bar{1}\bar{1}\bar{1}$)Ga substrate. These observations are consistent with the results obtained from theoretical analysis [22].

Fig. 6 shows the threshold voltage shift of a [01$\bar{1}$]-oriented FET on a ($\bar{1}\bar{1}\bar{1}$)Ga substrate, as a function of

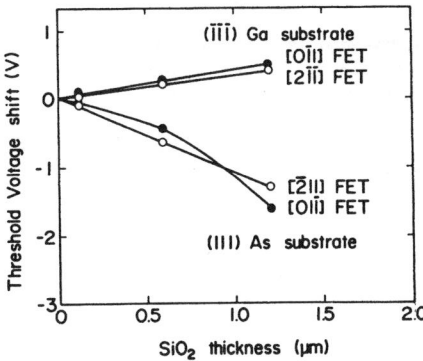

Fig. 5. The relationship between threshold voltage shift and SiO₂ thickness for MESFET's with [2̄1̄1]- and [01̄1]-oriented gate fingers fabricated on (1̄1̄1)Ga and (111)As substrates.

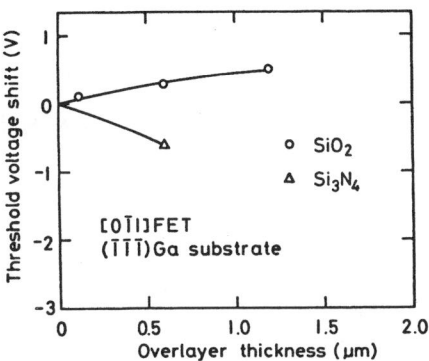

Fig. 6. The relationship between threshold voltage shift and dielectric overlayer thickness for MESFET's with [2̄1̄1]-oriented gate finger fabricated on a (1̄1̄1)Ga substrate.

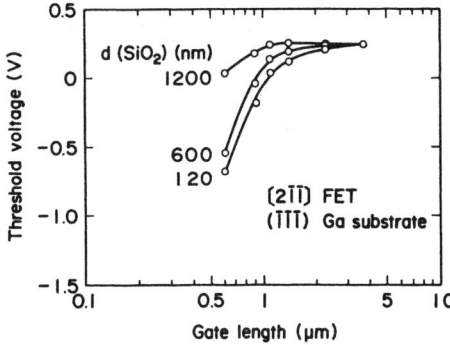

Fig. 7. Relationship between the threshold voltage and gate length for GaAs MESFET with a [2̄1̄1]-oriented gate finger fabricated on a (1̄1̄1)Ga substrate, with various thicknesses of the SiO₂ overlayer.

the CVD-SiO₂ and plasma-CVD Si₃N₄ overlayer thicknesses. The data for Si₃N₄-encapsulated FET were obtained with the same devices as the SiO₂-encapsulated FET, after removing the SiO₂ and depositing the Si₃N₄. A negative threshold voltage shift is observed with 0.6-μm-thick Si₃N₄. The change in the threshold voltage shift can be explained by the stress in the dielectric overlayer and the resultant change in piezoelectric charge, as reported previously with the (100) substrate [17].

Piezoelectric charge also improves the FET characteristics for a certain combination of gate orientation and dielectric overlayers [17]. The increased *K*-value, the reduced gate-length dependence, and the improved subthreshold characteristics are explained by the reduced effective channel thickness [20], [25]. Therefore, similar effects can be expected for MESFET's fabricated on a (1̄1̄1)Ga substrate.

Fig. 7 shows the gate-length dependence of threshold voltage measured for a [2̄1̄1]-oriented FET fabricated on a (1̄1̄1)Ga substrate for various SiO₂ thicknesses. The threshold voltage slowly decreases with gate length, and the greater the SiO₂ thickness, the less the threshold voltage depends on the gate length. The negative piezoelectric charge near the channel–substrate interface reduces the substrate current, which tends to increase at smaller

gate lengths. Fig. 7 shows that the preferable overlayer for MESFET's fabricated on a (1̄1̄1)Ga substrate is SiO₂, or a dielectric in compressive stress. In an analogy based on Figs. 5 and 6, the dielectric overlayer for the FET's fabricated on a (111)As substrate should be in tension, such as in plasma-CVD Si₃N₄ film. These results indicate that the piezoelectric charge could be used to achieve submicrometer-gate GaAs MESFET's with highly controllable and/or adjustable threshold voltage and orientation-independent characteristics, with a certain combination of dielectric overlayer and atomic plane used for fabrication.

IV. Conclusions

We investigated the electrical characteristics of GaAs MESFET's fabricated on (100), (011), and (1̄1̄1)Ga, and (111)As substrates with SiO₂ and Si₃N₄ overlayers. The theoretical predictions of the orientation effect on GaAs MESFET's, obtained with the piezoelectric effect model [22], were verified by experiments.

For MESFET's fabricated on the (011) substrate, the threshold voltage of the [01̄1]-oriented FET depends slightly on the direction of current, and the threshold voltage of the [100]-oriented FET is not influenced by the dielectric overlayer thickness. The threshold voltage shift of MESFET's fabricated on the (011) GaAs substrate is substantially smaller than that of the MESFET's on the (100) substrate. These results indicate the advantage of using the (011) substrate for IC fabrication. Characteristics are independent of gate orientation and dielectric overlayer thickness.

For MESFET's fabricated on the (111)-cut GaAs substrate, the sign of the threshold voltage shift depends on the atomic plane used for device fabrication as well as on the stress in the dielectric overlayer. The dependence of MESFET characteristics on gate orientation is much smaller than that observed for a (100) GaAs substrate for both Ga and As planes. Therefore, (111)-cut GaAs substrates are also useful for fabricating GaAs IC's, and for obtaining highly controllable and adjustable threshold voltage and improved characteristics of MESFET's for high-speed LSI's with specific combinations of the dielectric overlayer and the atomic plane.

IEEE TRANSACTIONS ON ELECTRON DEVICES, VOL. 36, NO. 9, SEPTEMBER 1989

Acknowledgment

The authors thank H. Onodera, M. Nogami, and T. Ohnishi for valuable discussions, Y. Yamaguchi, and S. Suzuki for their contribution during device fabrication, S. Komiya for his assistance in X-ray diffraction measurements, and T. Misugi, M. Kobayashi, A. Shibatomi, M. Fukuta, and A. Takahashi for their encouragement.

References

[1] P. V. Gray and D. M. Brown, *Appl. Phys. Lett.*, vol. 8, p. 31, 1966.
[2] B. E. Deal, M. Skar, A. S. Grove and E. H. Snow, *J. Electrochem. Soc.*, vol. 114, no. 3, p. 266, 1968.
[3] J. R. Ligenza, *J. Phys. Chem.*, vol. 65, p. 2011, 1961.
[4] G. Abowitz, E. Arnold, and J. Ladell, *Phys. Rev. Lett.*, vol. 18, p. 543, 1967.
[5] C. P. Lee, R. Zucca, and B. M. Welch, *Appl. Phys. Lett.*, vol. 37, p. 311, 1980.
[6] N. Yokoyama, H. Onodera, T. Ohnishi, and A. Shibatomi, *Appl. Phys. Lett.*, vol. 42, p. 270, 1983.
[7] A. Steinemann and V. Zimmerli, *J. Phys. Chem. Sol. Suppl.*, p. 81, 1967.
[8] D. W. Shaw, *J. Crystal Growth*, vol. 31, p. 130, 1975.
[9] A. Y. Cho, *J. Appl. Phys.*, vol. 41, p. 2780, 1970.
[10] L. Esaki, *Japan. J. Appl. Phys.*, suppl. 2, part I, p. 821, 1974.
[11] M. H. Lee and S. M. Sze, *Solid-State Electron.*, vol. 23, p. 1007, 1980.
[12] T. Ikoma, *Handbook on Compound Semiconductors.* Tokyo: Science Forum, 1983, p. 164 (in Japanese).
[13] S. M. Sze, *Physics of Semiconductor Devices.* New York: Wiley, 1969, p. 58.
[14] H. Shichijo *et al.*, *Electron. Lett.*, vol. 16, p. 208, 1980.
[15] P. M. Asbeck, C. P. Lee, and M. F. Chang, *IEEE Trans. Electron Devices*, vol. ED-31, p. 1377, 1984.
[16] M. F. Chang, C. P. Lee, P. M. Asbeck, R. P. Vahrenkamp, and C. G. Kirkpatrik, *Appl. Phys. Lett.*, vol. 45, no. 3, p. 279, 1984.
[17] T. Ohnishi, T. Onodera, N. Yokoyama, and H. Nishi, *IEEE Electron Device Lett.*, vol. EDL-6, p. 172, 1985.
[18] C. H. Chen, M. Shur, and A. Peczalski, *IEEE Trans. Electron Devices*, vol. ED-33, p. 792, 1986.
[19] M. Kanamori, H. Ono, T. Furutsuka, and J. Matsui, *IEEE Electron Device Lett.*, vol. EDL-8, p. 228, 1987.
[20] T. Onodera, T. Ohnishi, and H. Nishi, in *Proc. 12th Int. Symp. GaAs and Related Compounds* (Karuizawa, Japan), 1985, p. 517.
[21] J. C. Ramirez *et al.*, *IEEE Trans. Electron Devices*, vol. ED-35, p. 1232, 1988.
[22] T. Onodera and H. Nishi, *IEEE Trans. Electron Devices*, this issue, pp. 1580–1585.
[23] N. Yokoyama, T. Ohnishi, K. Odani, H. Onodera, and M. Abe, *IEEE Trans. Electron Devices*, vol. ED-29, p. 1541, 1982.
[24] P. G. Wilson and V. R. Deline, *Appl. Phys. Lett.*, vol. 37, p. 793, 1980.
[25] T. Onodera, T. Ohnishi, N. Yokoyama, and H. Nishi, *IEEE Trans. Electron Devices*, vol. ED-32, p. 2314, 1985.

A High-Speed and Highly Uniform Submicrometer-Gate BPLDD GaAs MESFET for GaAs LSI's

Minoru Noda, Kenji Hosogi, Tomoki Oku, Kazuo Nishitani, and Mutsuyuki Otsubo

Abstract—We present forming conditions of ion-implanted regions of a GaAs Buried p-layer Lightly Doped Drain (BPLDD) MESFET that can improve short-channel effect, V_{th} uniformity, and FET operating speed, simultaneously. For 0.7-μm gates, we find that a Mg^+ dose of 2×10^{12} cm^{-2} at 300 keV and a Si^+ dose of 2×10^{12} cm^{-2} at 50 keV are suitable for the p and n' layer, respectively. The σV_{th} of 7 mV is realized. To our best knowledge, it is the smallest value measured on an area as large as 3 mm × 3 mm and with the number of measured FET's as large as 2500. Gate-edge capacitance of the 0.7-μm-gate BPLDD that consists of both overlap capacitance and fringing capacitance is successfully reduced to 0.5 fF/μm, which is about 50% of that of a non-LDD Buried p-layer (BP) FET. We also find that another parasitic capacitance due to the p-layer has less effect on the speed than the gate-edge one. Consequently, the gate propagation delay time of the BPLDD can be reduced to 15 ps at power dissipation of 1 mW/ gate, which is about 65% of that of a BP. Applying the 0.7-μm-gate BPLDD to 16-kb SRAM's, we have obtained the maximum access time of less than 5 ns with a Galloping test pattern.

NOMENCLATURE

V_{th}	Threshold voltage.
σV_{th}	Standard deviation of V_{th}.
g_m	Mutual conductance.
K	K value.
I_{sub}	Subthreshold current.
N_g	Subthreshold factor.
ΔV_{th}	V_{th} negative shift.
ΔV_{thp}	ΔV_{th} by piezoelectric charge.
R_s	Source resistance.
α	Channel thickness.
$\Delta \alpha$	Deviation of α by piezoelectric charge.
$\rho_{piezo}(z)$	Piezoelectric charge density.
$\rho_{ch}(z)$	Intrinsic channel charge density.
C_{gs}	Gate–source capacitance.
$C_{gs}(V_{gs})$	Intrinsic gate–source capacitance.
C_{gd}	Gate–drain capacitance.
C_{gp}	Parasitic gate capacitance.
C_{gpe}	Gate-edge capacitance.
C_{gpp}	Parasitic capacitance related to partially depleted p-layer.
C_{gov}	Overlap capacitance.
C_{gfo}	Outer fringing capacitance.
C_{gfi}	Inner fringing capacitance.
C_{gon}	C_{gs} with V_{gs} of 0.3 V.
C_{st}	Stray capacitance.
ϵ_0	Permittivity of air.
ϵ_s	Permittivity of GaAs.
H	Gate thickness.
d	Thickness of depletion layer.
ΔL_g	Length of lateral diffusion of n' region into n-channel.
V_{bi}	Built-in voltage.
f_T	Cutoff frequency.
t_{pd}	Gate propagation delay time.
t_{acc}	Address access time.

Manuscript received April 18, 1991; revised September 3, 1991. The review of this paper was arranged by Associate Editor M. Shur.

The authors are with the Optoelectronic and Microwave Devices R&D Laboratory, Mitsubishi Electric Corporation, 4-1 Mizuhara, Itami, Hyogo 664, Japan.

IEEE Log Number 9105892.

I. INTRODUCTION

A GaAs MESFET with a partially depleted p-layer (Buried p-layer FET) has recently been studied for SRAM applications by Noda *et al.* [1]. In their FET structure, short-channel effects are well suppressed for gate lengths down to 0.5 μm by a rather dense p-layer buried under the channel, which corresponds to a partially depleted condition. At the same time, a V_{th} standard deviation (σV_{th}) as low as 9 mV in an area of 3 mm × 3 mm is attained in 60-μm-pitch microscopic measurement, together with high current drivability of g_m (590 mS/mm) and K value (490 mS/V · mm). Consequently, fully functional 7-ns 4-kb SRAM's operative at 75°C are successfully developed by using the FET with 1-μm gate [2]. And a high chip yield of 22% is obtained in a 3-in wafer [1].

It is recognized, however, that in the fully functional SRAM's, their access times are larger than those of Si counterparts [3]. From the above results of both access time and g_m, we believe that the gate capacitance of a BP FET is quite large. It is generally agreed that this large gate capacitance is mainly caused by parasitic capacitance due to a not completely depleted p-layer under the channel, or by overlap capacitance between the gate edge and the n^+ region [4]. Also, the V_{th} uniformity of component FET's is not adequate enough to make GaAs SRAM's faster than their Si counterparts.

In order to make GaAs LSI's, especially SRAM's, faster, adopting the Lightly Doped Drain (LDD) structure is very promising [5]–[7]. One major reason is that the

Reprinted from *IEEE Trans. Electron Devices*, vol. 39, no. 4, pp. 757–766, April 1992.

131

overlap capacitance can be reduced because the n^+ region is separated from the gate edge by an n' intermediate region, whose carrier concentration is much lower than that in the n^+ region. Another major reason is that the V_{th} uniformity is improved because the distance between the source n^+ region and the drain one, namely, the effective gate length, is increased, and hence the short-channel effect is more suppressed than that in the non-LDD FET. For the above reasons, we have decided to combine a conventional BP FET with the LDD structure. We call this FET a Buried p-layer Lightly Doped Drain (BPLDD) FET, which has recently become a common name for this FET structure. We improve this FET so as to realize three things simultaneously: 1) to suppress the short-channel effect, 2) to improve the V_{th} distribution, and 3) to make the FET operation faster. We also apply the BPLDD to 16-kb SRAM's.

In this paper, we first introduce a newly improved BPLDD structure, in which the p-layer is designed to be partially depleted. A carrier concentration of the n' region is selected so as to keep a reasonable value of current drivability. The n^+ regions are completely surrounded by the p-layer. We then describe a FET fabrication process. We proceed to compare FET characteristics of the BPLDD and the conventional FET's. We find that both the subthreshold leak (I_{sub}) and V_{th} negative shift (ΔV_{th}) are improved by adopting the LDD structure. We then determine such ion implantation conditions of a BPLDD that improve the V_{th} uniformity. We find that an optimal value of dose for the p-layer exists, and that a tradeoff between the V_{th} uniformity and current drivability exists for a dose of the n' region. Furthermore, we examine the relationship between gate capacitance and gate voltage of a BPLDD and identify a FET-forming condition that realizes faster operation. The gate-edge capacitance of a BPLDD is as small as 0.5 fF/μm for a 0.7-μm gate, which is about 50% of that of the BP. Gate propagation delay time of a BPLDD is 15 ps at power dissipation of 1 mW/gate for a 0.7-μm gate, which is about 65% of that of the BP. Finally, we show an application of the FET to 16-kb SRAM's and its results. We confirm their fast, fully functional, and stable operations with a reasonable wafer yield.

II. FET STRUCTURE AND FABRICATION PROCESS

Fig. 1(a) shows a schematic cross-sectional view of the BPLDD FET. We shall describe now the major features of the FET structure. The p-layer under the channel is partially depleted to suppress the short-channel effect. We select a forming condition of n' intermediate regions so that both the current drivability (K value) and uniformity of V_{th} are not degraded. The carrier concentration of the n' region is required to be low so as to decrease the gate capacitance. Therefore, we increase the depth of the n' region to keep its resistivity low. Consequently, the n' region is thicker than the n-channel and the maximum carrier concentration of the n' region is 3–4 $\times$ 10^{17} cm^{-3},

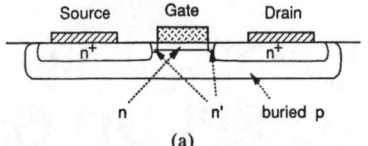

(a)

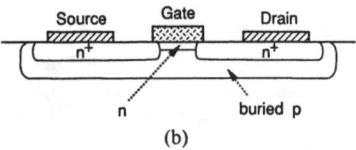

(b)

Fig. 1. Schematic cross-sectional view of a BPLDD FET (a) and BP FET (b).

which is larger by only about 50% than that of the n channel. We then form dense n^+ regions that are completely surrounded by the buried p-layer. From the atomic profile by Pearson IV model, the depth of the n, n', n^+, and p regions are calculated to be about 1500, 2000, 3500, and 7000 Å, respectively. This structure reduces the leakage currents between the source and drain n^+ regions and also suppresses the leakage currents among the n^+ regions of adjacent FET's, which suppressses the sidegating effect well [8], [9]. We can also improve soft-error immunity, which is extremely important in SRAM use [10], [11].

We now introduce the BPLDD fabrication process. After a surface treatment of In-doped GaAs LEC wafer with a wet chemical, a thin p-CVD SiO film is deposited with a thickness of several hundreds angstroms for a through-film implantation. We implant ^{25}Mg$^+$ ions into the channel region to form a buried p-layer under an n channel at an energy of 300 keV. We then implant ^{29}Si$^+$ ions into the channel region to make the n-channel at an energy of 50 keV. Cap annealing with SiO film is performed at 800°C in AsH$_3$ atmosphere for 30 min.

After removing the encapsulant film, WSi$_x$ refractory gate metal is sputter-deposited and patterned by RIE with a mixed gas of CF$_4$ and O$_2$. Then ^{29}Si$^+$ ions are implanted to form the n' intermediate layer using the WSi$_x$ gate as a mask. Compared to other works [5]–[7], we do not use a sidewall technique to separate the n^+ region from the gate. After depositing a SiON film of about 2300 Å on the whole substrate surface, we implant Si$^+$ ions through the film under the condition of 165 keV and 4 $\times$ 10^{13} cm^{-2}. Capless annealing is performed at 800°C, in AsH$_3$ atmosphere for 30 min to activate the ions for both the n' and n^+ regions. We believe that the gap between the n^+ region and gate is about 1000 Å by considering lateral diffusions in the annealing.

We deposit an insulator (SiON) as a spacer for liftoff of source–drain ohmic metals. Openings are made to contact the ohmic metals with the n^+ region. AuGe, Ni, and Au are evaporated successively and lifted off. Finally, they are alloyed. This completes the newly improved BPLDD.

III. COMPARISON OF SHORT-CHANNEL EFFECT IN BPLDD FET'S AND CONVENTIONAL FET'S

Typical I_{ds}–V_{gs} characteristics of both BPLDD FET and conventional FET's for gate length of 0.7 μm are shown in Fig. 2, where three types of FET structure are compared; that is, the BPLDD, conventional BP as seen in Fig. 1(b), and LDD with no p-layer (simple LDD). Here, all the drain–source saturation currents (I_{dss}) of the FET's are almost the same. It is clear that the characteristics differ from each other in the subthreshold regions. Although the subthreshold currents (I_{sub}'s) for both the simple LDD and the BPLDD are almost constant over the entire reverse gate bias region, the absolute value for the simple LDD is higher by about one order of magnitude than that for the BPLDD. This means that the buried p-layer effectively suppresses the subthreshold leakage current. On the other hand, although the minimum I_{sub}'s for both the BP and the BPLDD are nearly the same, the I_{sub} of the BP increases monotonically as the reverse gate bias is increased. We believe that this increase is due to leakage in the reversely biased Schottky junction between the gate and drain [12]. This is because the electric field of the BP is larger than that of the BPLDD, which is due to a smaller separation width between the gate and drain for the BP than that for the BPLDD.

The gate length dependence of V_{th} for the three types of FET's is shown in Fig. 3. In the figure, it should be noted that the V_{th} negative shift (ΔV_{th}) of the BP is larger than that of the simple LDD, though the I_{sub} of the BP is smaller than that of the LDD. This is explained by the difference between the BP and LDD. From our previous experiments, the effective gate length of BP is shorter by 0.3 μm than that of the LDD due to the lateral diffusion of the n^+ dopant.

In summary, from the viewpoint of both I_{sub} and the V_{th} negative shift (ΔV_{th}), the BPLDD is the best of the three types of FET structures.

IV. OPTIMIZATION OF BPLDD FET FOR IMPROVING V_{th} UNIFORMITY

We evaluated various forming conditions of the BPLDD, especially of the implanted regions (n, p, n', n^+) in order to improve the V_{th} uniformity. First of all, we examine the condition of the p-layer. The dose of Mg^+ for the p-layer is varied from 0 to 1.5×10^{13} cm^{-2} at a constant implantation energy of 300 keV. We increased the dose of Si^+ for the n-channel as the dose of Mg^+ was increased in order to keep the V_{th} nearly constant (in this work, -150–-200 mV) for valid comparisons among the FET's with various doses for the p-layers. Fig. 4 shows the Mg^+ ion dose dependence of I_{sub}, N_g, ΔV_{th}, and σV_{th} measured on a whole 3-in wafer, where the N_g is the subthreshold factor that is defined in the equation ($I_{ds} = I_{ds 0} \exp{(qV_{gs}/kTN_g)}$, $I_{ds 0}$: constant) and indicates the degree of the short-channel effect. The ΔV_{th} is defined as $\Delta V_{th} = V_{th}(L_g = 0.7\ \mu m) - V_{th}(L_g = 0.5\ \mu m)$.

In Fig. 4(a), we find that the I_{sub} decreases monotoni-

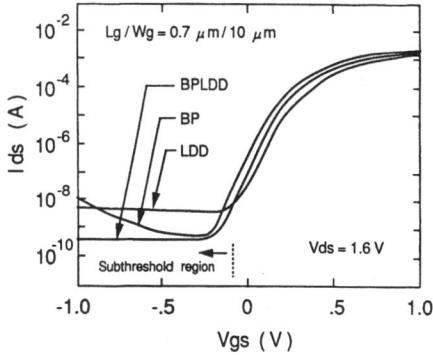

Fig. 2. Typical I_{ds}–V_{gs} characteristics.

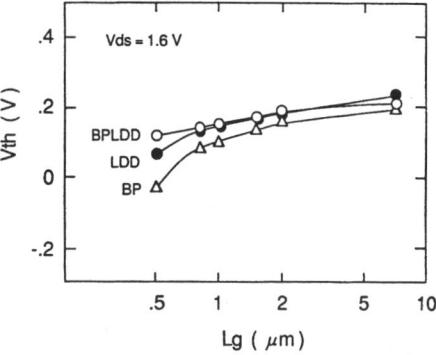

Fig. 3. Gate length dependence of threshold voltage (V_{th}).

cally as the Mg^+ dose is increased up to about 2×10^{12} cm^{-2}, and then the I_{sub} is kept almost constant at less than 1 nA for the dose of more than 2×10^{12} cm^{-2}. Although the I_{sub} for a 0.5-μm-gate FET is slightly larger than that for a 0.7-μm-gate FET, they are of the same order and are nearly the same as that for a long-gate FET, such as the 7-μm-gate FET. Therefore, we find that the I_{sub} is sufficiently suppressed by the rather dense p-layer for both short gate lengths.

We can see from Fig. 4(b) that the N_g's for both the 0.7- and 0.5-μm-gate FET's are minimized for the Mg^+ dose of 2–5 $\times 10^{12}$ cm^{-2}. Then the N_g' start to increase as the Mg^+ dose is increased to more than 2–5 $\times 10^{12}$ cm^{-2}. We believe that this result is explained as follows: as the Mg^+ dose is increased, the resistance of the n^+ region increases. This causes the source resistance (R_s) to increase and hence I_{ds} is reduced. This leads to an increase in N_g because the I_{ds} is proportional to $\exp{(1/N_g)}$. As a result, the Mg^+ dose dependence of both N_g and I_{sub} disagree with each other.

From Fig. 4(c), we find that the ΔV_{th} decreases monotonically as the Mg^+ dose is increased from 0 to more than 15×10^{12} cm^{-2}. We then recognize that the Mg^+ dose dependence of the ΔV_{th} is different from that of the I_{sub} shown in Fig. 4(a). This means that the ΔV_{th} may be increased not only by subthreshold leakage but also by other factors. Two such factors are considered as follows. One is the thinning of the n-channel by the p-layer. The n-channel becomes thinner as the Mg^+ dose is increased, because the depletion layer between the n-channel and the

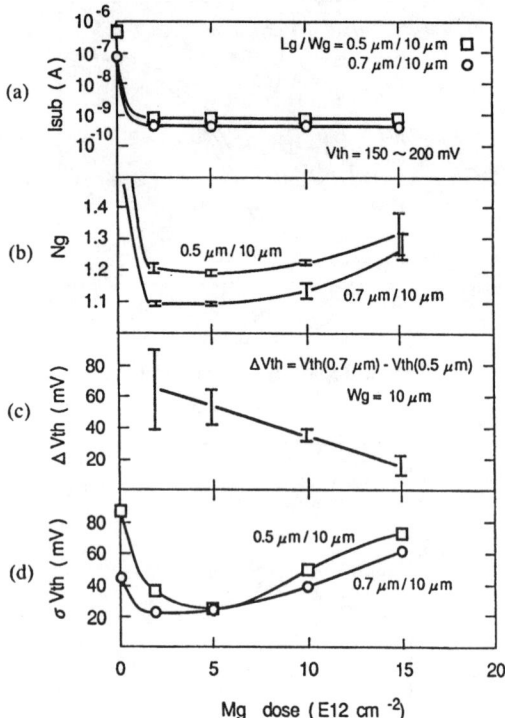

Fig. 4. (a) Mg^+ ion implantation dose dependence of subthreshold current (I_{sub}). (b) Mg^+ ion implantation dose dependence of subthreshold factor (N_g). (c) Mg^+ ion implantation dose dependence of V_{th} negative shift (ΔV_{th}). (d) Mg^+ ion implantation dose dependence of V_{th} standard deviation (σV_{th}).

p-layer spreads more toward the n-channel. This suppresses the two-dimensional effect of the electric field by a gate [13].

The other factor is the suppression of the piezoelectric effect by a denser p-layer. The change in V_{th} by the piezoelectric charge (ΔV_{th_p}) is approximately expressed as follows [14]:

$$\Delta V_{th_p} = (1/\epsilon_s) \left(\int_{Z=0}^{Z=\alpha+\Delta\alpha} z\rho_{piezo}(z)\, dz \right.$$
$$\left. + \int_{Z=\alpha}^{Z=\alpha+\Delta\alpha} z\rho_{ch}(z)\, dz \right) \quad (1)$$

where ϵ_s is the permittivity of GaAs, α is the n-channel thickness, $\Delta\alpha$ is the deviation of α by the piezoelectric charge, $\rho_{piezo}(z)$ is piezoelectric charge density, and $\rho_{ch}(z)$ is channel charge density. As the p-layer is made denser, α becomes smaller. To keep the V_{th} constant, we also make the n-channel denser. This leads to a reduction in $\Delta\alpha$ because the density ratio of the n-channel to the piezoelectric charge increases. Hence, the first term in (1) is decreased. The second term in (1) is also reduced because $\rho_{ch}(z = \alpha$ or $\alpha + \Delta\alpha)$ is a carrier density at the channel edge in the channel depth direction. Therefore, $\rho_{ch}(z = \alpha$ or $\alpha + \Delta\alpha)$ is nearly the same in the case with the p-layer and that without the p-layer. It is as small as 1–10×10^{15} cm^{-3}. As a result, the ΔV_{th_p} is reduced as the p-layer becomes denser.

Fig. 4(d) shows that the Mg^+ dose for the minimum σV_{th} for a 0.5-μm gate is larger than that for a 0.7-μm

gate. That is, the σV_{th} for a 0.5-μm gate is minimized at the Mg^+ dose of 5×10^{12} cm^{-2}, though the σV_{th} for a 0.7-μm gate is minimized at the Mg^+ dose of 2×10^{12} cm^{-2}. This agrees with the scaling law, in which denser layers are needed for shrinking the FET's dimensions. The increase in V_{th} scattering for the Mg^+ dose higher than 2–5×10^{12} cm^{-2} is considered to be due to the nonuniformity of the activation ratio of carriers enhanced by implantation damage at a high dose implantation. As most of the values of σV_{th}'s are more than 20 mV, they seem to be larger than those cited in other publications. The reason for this is unknown at the present stage of our research. We measured V_{th}'s on a whole 3-in wafer, including some extraordinary V_{th} values at peripheral positions. The V_{th}'s of 56 FET's were measured on a wafer.

We also examined the forming conditions for both the n' intermediate region and the n$^+$ region. Before determining optimum Si^+ doses for both regions, we checked both the upper and the lower limits of the concentration of the n' region from the viewpoint of current drivability. Fig. 5 shows the K value versus the Si^+ dose for the n' region. From this figure, the dose of 2×10^{12} cm^{-2} is found to be optimum, at which we can obtain almost the maximum K value. Then we examine the dependence of V_{th} uniformity on the dose for the n' region in the range of more than 2×10^{12} cm^{-2}.

The σV_{th} is plotted against the Si^+ ion dose for the n' region as a parameter of implantation energy in Fig. 6. The σV_{th} increases monotonically as the Si^+ dose is increased. It is necessary for the concentration of the n' region to be low enough in order to get a small σV_{th}. In this experiment, we choose the Si^+ dose of 2×10^{12} cm^{-2}. For this dose, the implantation energy affects the σV_{th} only slightly in the range between 30 and 50 keV as shown in Fig. 6. In Fig. 7, the σV_{th}'s are also plotted against the Si^+ ion implantation energy for the n$^+$ regions. As seen from the figure, the σV_{th}'s stay almost constant at about 25 mV though the energy varies from 50 to 80 keV. In a limited range of the depth, we conclude that the depth of the n$^+$ region almost does not affect the V_{th} uniformity, since the n$^+$ region is clearly separated by the n' region of low concentration.

A. V_{th} Microscopic Uniformity

From the viewpoint of fabricating GaAs LSI's, especially fully functional SRAM's, the V_{th} uniformity should be evaluated in the scale of FET density in an LSI. It has been reported that the microscopic uniformity of V_{th} has been evaluated using a 60-μm-pitch FET array from the viewpoint of FET structure, which is the existence of the buried p-layer or the difference of gate length [1]. Using the same method, we examine the V_{th} microscopic uniformity of the BPLDD and compare the results with those for the BP. In Fig. 8 is shown the Mg^+ ion dose dependence of σV_{th} for both a BP and a BPLDD. With the same Mg^+ ion dose, the σV_{th} for the BPLDD is always smaller than that for the BP. The relationship between the σV_{th}

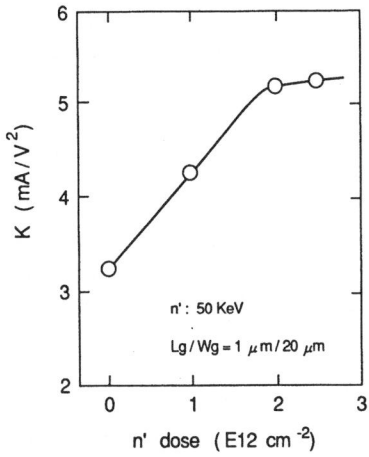

Fig. 5. Dependence of K value on Si^+ ion implantation dose for n'
regions.

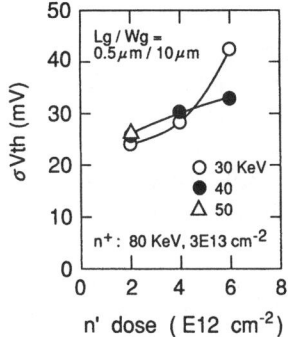

Fig. 6. Dependence of σV_{th} on Si^+ ion implantation dose for n' regions.

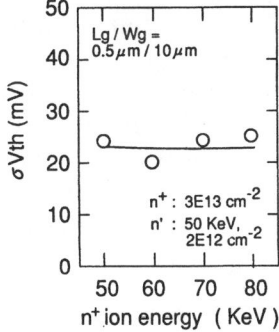

Fig. 7. Dependence of σV_{th} on Si^+ ion implantation energy for n^+ regions.

and the Mg^+ dose is very similar to that in Fig. 4(d), which is a macroscale relationship between them.

We find that the σV_{th}'s for the BPLDD with Mg^+ dose of 5×10^{11} cm^{-2} is smaller than that for the BP with 2×10^{12} cm^{-2}, even though the N_g for the BPLDD with Mg^+ dose of 5×10^{11} cm^{-2} is larger than that for the BP with the dose of 2×10^{12} cm^{-2}. The following reason may explain the result. In addition to the dose, the σV_{th} is also affected by the change in the effective gate length due to the lateral diffusion of the n^+ dopant, especially for BP's. Therefore, under the same Mg^+ ion dose, the σV_{th} of the BP is larger than that of the BPLDD. For the σV_{th}, we believe that the variation of the lateral diffusion has a

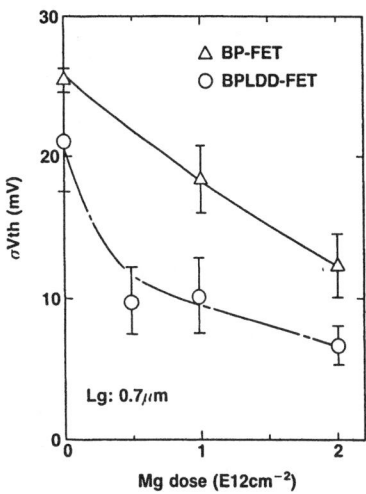

Fig. 8. Mg^+ ion implantation dose dependence of σV_{th} measured with 60-μm-pitch FET arrays.

different effect than the I_{sub}, which relates directly to the N_g. We therefore believe that it is possible that a difference in tendency between the N_g and σV_{th} exists.

V. ANALYSIS OF GATE CAPACITANCE

Since a BPLDD has a much smaller (by 90% in our experiment) carrier concentration near the gate edge than a BP, the gate-edge parasitic capacitance (C_{gpe}) of the BPLDD is smaller than that of the BP. Therefore, the gate–source capacitance (C_{gs}) of a BPLDD is expected to be smaller than that of a BP, especially for a short gate length. As shown in Fig. 9, C_{gpe} consists mainly of both the overlap capacitance ($C_{g\,ov}$) between the gate and the n' region and the fringing capacitance. Moreover, the latter is divided into outer fringing capacitance (C_{gfo}) through air at a side of the gate and the conventional inner fringing capacitance (C_{gfi}). In order to clarify the difference in C_{gpe} between a BPLDD and a BP and to find a forming condition for a faster FET operation, we examine C_{gs} hereinafter.

The gate-edge parasitic capacitance (C_{gpe}) is written as

$$C_{gpe} = C_{gfo} + C_{gfi} + C_{g\,ov}$$
$$\sim (2\epsilon_0/\pi) \ln(1 + H/d)W_g$$
$$+ \pi\epsilon_s W_g/2 + \epsilon_s \Delta L_g W_g/d \qquad (2)$$

where ϵ_0 is the permittivity of air, ϵ_s is that of GaAs, H is the gate thickness, d is the thickness of the depletion layer at the gate edge, and ΔL_g is the length of lateral diffusion of the n' region into the n-channel. Under ordinary conditions, C_{gfo} is much smaller than both C_{gfi} and $C_{g\,ov}$ because ϵ_0 is about 10 times smaller than ϵ_s. So, (2) is expressed as

$$C_{gpe} \sim (\pi/2 + \Delta L_g/d)\epsilon_s W_g. \qquad (3)$$

We measured C_{gs} with a YHP4194A Impedance/gain-phase analyzer. The measuring frequency was fixed at 1 MHz. Fig. 10 shows the gate voltage (V_{gs}) dependence of

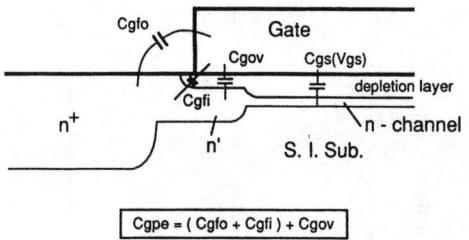

$$Cgpe = (Cgfo + Cgfi) + Cgov$$

Fig. 9. Schematic drawing of gate-edge parasitic capacitance (C_{gpe}).

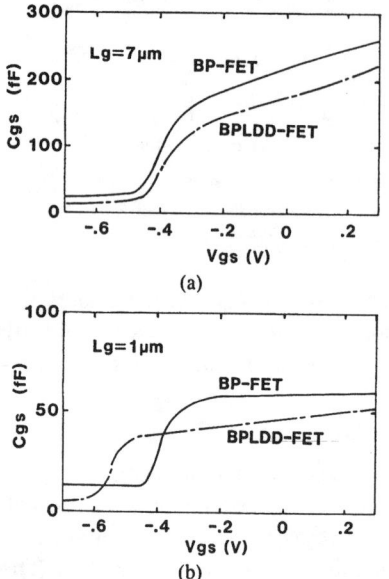

Fig. 10. Gate voltage dependence of C_{gs} in BP's and BPLDD's. (a) $L_g = 7 \ \mu$m. (b) $L_g = 1 \ \mu$m, $W_g = 10 \ \mu$m.

C_{gs} for both BP's and BPLDD's with L_g of 7 and 1 μm. Their W_g's are 10 μm. As seen from Fig. 10, for the FET's with 7-μm gates, the dependence curves of C_{gs} are almost fitted to the following equation [15] after the n-channel is on ($V_{gs} > V_{th}$):

$$C_{gs} = C_{gs}(V_{gs}) + C_{gp}$$
$$= A(V_{bi} - V_{gs})^{-1/2} + C_{gp} \qquad (4)$$

where $C_{gs}(V_{gs})$ is the intrinsic gate capacitance, C_{gp} is the parasitic gate capacitance, including C_{gpe}, A is a constant, and V_{bi} is the built-in voltage. In the original equation in [15], the parasitic gate capacitance (C_{gp}) is written as $\pi \epsilon_s W_g / 2$, which is C_{gfi} in (2). The other factors such as C_{gov} are not taken into account in the equation. For the FET's with 1-μm gates, on the other hand, the changes of C_{gs}'s for V_{gs} are much smaller than the value derived from (4). This indicates that C_{gp}, which depends little on V_{gs}, takes a large part of the total C_{gs} for the short gate length of 1 μm.

In Fig. 10, we also find that C_{gs}'s of BPLDD's are smaller than those of BP's for both long-gate (7-μm) (Fig. 10(a)) and short-gate (1-μm) (Fig. 10(b)) FET's. The capacitance difference between the BP and the BPLDD is related to C_{gp}, because the forming conditions (implantation and annealing) of an n-channel are the same for

both FET's. After the channel is on ($V_{gs} > V_{th}$) for 1-μm gate, the change of C_{gs} for V_{gs} in the BP is found to be smaller than that in the BPLDD. This is possibly explained as follows. After the channel is on, C_{gp} decreases as V_{gs} is increased because of the increase of gate leakage current near the gate edge, even though the intrinsic channel capacitance ($C_{gs}(V_{gs})$) increases as V_{gs} is increased. Therefore, for a short-gate length, the ratio of C_{gp} to C_{gs} becomes larger than that for a long-gate length. Consequently, the sum of $C_{gs}(V_{gs})$ and C_{gp} increases little as V_{gs} is increased.

It is suspicious that the parasitic capacitance related to a partially depleted p-layer (C_{gpp}) also takes a large part of the total C_{gp}. Therefore, we examine the increase in C_{gs} due to the p-layer with FATFET's ($L_g / W_g = 7 \ \mu$m/20 μm). For V_{th}'s of about -570 mV, the C_{gs} of the FET with a p-layer (4×10^{12} cm^{-2}, which is a partially depleted condition) is larger by at most 40% than that with no p-layer. The Si$^+$ ion dose for the n-channel of the FET with the p-layer, however, is larger by 130% than that with no p-layer. And the carrier density in the n-channel is calculated to be larger by about 90%, as the activation ratio of the n carrier decreases from about 65% to about 40% as the Si$^+$ dose increases. If the increase in C_{gs} due to the p-layer is effectively large, the increase in C_{gs} with both the p-layer and large Si$^+$ ion dose for the n-channel is quantitatively small. Therefore, we conclude that the increase in C_{gs} due to the p-layer is less effective than that due to the n-channel.

In the C_{gs} measurement, we set $V_{gs} = 0.3$ V because, in DCFL configurations, we assume that an average V_{gs} is an average value between V_H (0.5 V–0.6 V) and V_L (0–0.1 V). For simplicity, C_{gs} with V_{gs} of 0.3 V is denoted by C_{gon} hereafter.

We can estimate the gate-edge parasitic capacitance (C_{gpe}) by measuring the L_g dependence of C_{gon} because the gate-edge capacitance is independent of L_g. Fig. 11 shows such an L_g dependence at a fixed gate width (W_g) of 10 μm. From the figure, C_{gon} can be approximately described as

$$C_{gon} \sim C_{g1} + C_{g2} L_g \qquad (5)$$

where C_{g1} is the gate-edge parasitic capacitance (C_{gpe}), which is a constant, and C_{g2} is $\partial C_{gon} / \partial L_g |_{L_g}$. From (4) and (5), we believe $C_{g2} L_g = C_{gs}(V_{gs}) + C_{gpp}$. From Fig. 11, as L_g increases, C_{gon}'s for both the BP and BPLDD increase almost linearly and at nearly the same rate. Whereas the absolute value of C_{gon} in the BP is larger than that in the BPLDD. This indicates that the C_{g1} of the BP is larger than that of the BPLDD, and that the C_{g2} of the BP is nearly the same as that of the BPLDD. Therefore, the ratio of C_{gon} in the BPLDD to that in the BP becomes smaller as L_g is shortened. From the figure, we find that the ratio decreases from 89 to 74% as L_g is reduced from 1 to 0.5 μm. This means that BPLDD's are effective for shorter gate lengths in order to decrease their gate propagation delay times.

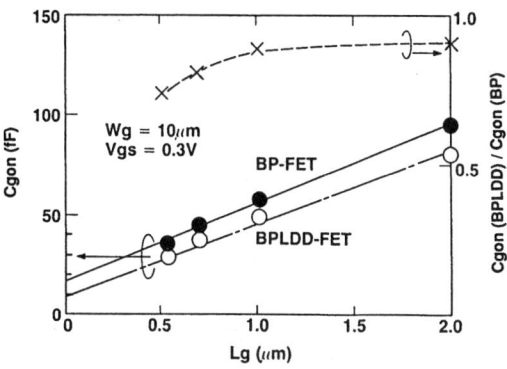

Fig. 11. Gate length dependence of C_{gon} in BP's and BPLDD's and of the C_{gon} ratio ($W_g = 10 \mu m$).

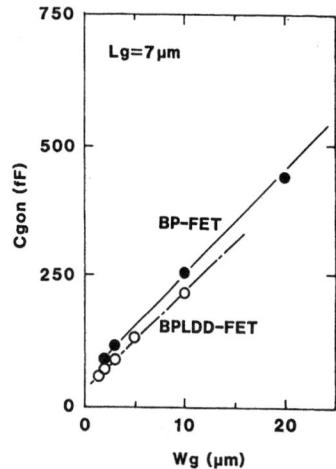

Fig. 12. Gate width dependence of C_{gon} in BP's and BPLDD's ($L_g = 7$ μm).

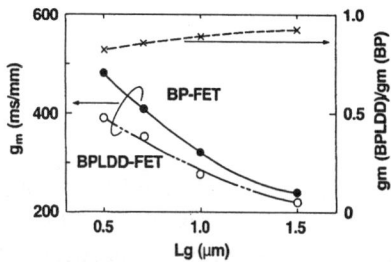

Fig. 13. Gate length dependence of current drivability (g_m) in BP's and BPLDD's and of the g_m ratio.

From the definition of (5), both C_{g1} and C_{g2} should be constant for different gate lengths. However, the differences between C_{g1}'s for L_g's of 0.7 μm, 1 μm, and longer lengths are not negligible. This is caused by the higher order dependence of C_{gon} on L_g. That is, $\partial C_{gon}/\partial L_g$ ($= C_{g2}$) is not constant but dependent on L_g. In our results, the higher order dependence is not taken into consideration. If we consider the effect of the dependence, we expect that for the L_g of less than 0.7-μm C_{g2} becomes larger than 3.5 fF/μm^2 and C_{gpe} ($= C_{g1}$) becomes less than 1.0 fF/μm as seen from Fig. 11.

Fig. 12 shows W_g dependence of C_{gon} for the FET's with L_g of 7 μm. By considering the C_{gon}–L_g curve in Fig. 11, C_{gon} can also be described approximately as

$$C_{gon} \sim W_g[\mu m](C_{gpe}^*[fF/\mu m]$$
$$+ C_{g2}^*[fF/\mu m^2]L_g[\mu m]) + C_{st} \quad (6)$$

where C_{st} is the stray capacitance, $C_{gpe}^* = (C_{gpe} - C_{st})/W_g$, and $C_{g2}^* = C_{g2}/W_g$. From Fig. 12, we find that C_{st}, which does not depend on both the W_g and L_g, exists. Therefore, in (5), it is found that $C_{g1} = C_{gpe} + C_{st}$, and we can evaluate C_{gpe} more precisely by (6). We measure each of C_{gpe}^* and C_{g2}^* by using (6) for L_g's of both 0.7 and 1.0 μm. The results are summarized in Table I. We find that the C_{gpe}^* of the BPLDD is smaller than that of the BP for both 0.7- and 1-μm gates. We also find that C_{gpe}^* of the BPLDD with a 0.7-μm gate is 0.5 fF/μm, which is a half of that of the BP. On the other hand, from (3), C_{gpe}^* is calculated to be about 0.6 fF/μm from our experimental results that d and ΔL_g are 500 and 1500 Å, respectively. Therefore, our measured C_{gpe}^* agrees well with the calculated value from (3).

VI. CUTOFF FREQUENCY f_T

To estimate the operating speed of a FET, it is necessary to measure the current drivability (g_m). Fig. 13 shows the gate length dependence of g_m's for both a BPLDD and a BP. We also plot the g_m ratios of BPLDD to BP. For an LDD structure, the R_s becomes larger than that for a non-LDD one because an n' intermediate region exists between the gate and the n^+ region. Then, the g_m for BPLDD is smaller than that for the BP. From Fig. 13, the

TABLE I
CALCULATION RESULTS OF C_{g2}^* AND C_{gpe}^*

FET Type		L_g (μm) = 0.7	1.0
BP	C_{g2}^* (fF/μm^2)	4.0	3.5
	C_{gpe}^* (fF/μm)	1.0	2.7
BPLDD	C_{g2}^* (fF/μm^2)	4.0	3.5
	C_{gpe}^* (fF/μm)	0.5	1.9

BPLDD to BP g_m ratio decreases as L_g is shortened. From Fig. 11, however, the ratio is found to be larger than that of C_{gon} for an L_g of less than 1 μm. Consequently, we expect that the BPLDD operates faster than the BP for the range.

We then measure the gate length dependence of f_T as shown in Fig. 14. From the figure, we find that the f_T of a BPLDD is higher than that of a BP at L_g of less than 0.6 μm. As mentioned in Section III, there is a difference in the effective gate length, that is, separation width between the source and drain n^+ regions, between a BPLDD and a BP for the same length of gate pattern. For a BP, we must estimate the effective gate length by subtracting the diffusion length (0.15 μm) of the n^+ region from the length of the gate pattern. We then plot the f_T against the effective gate length in Fig. 14. Even for the effective gate length for the BP, we find that the f_T of the BPLDD is larger than that of the BP for the range of less than 0.7

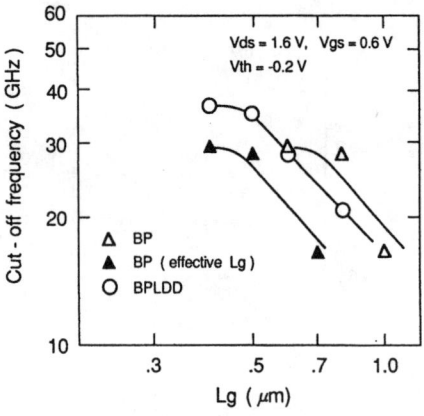

Fig. 14. Gate length dependence of cutoff frequency (f_T) in BP's and BPLDD's.

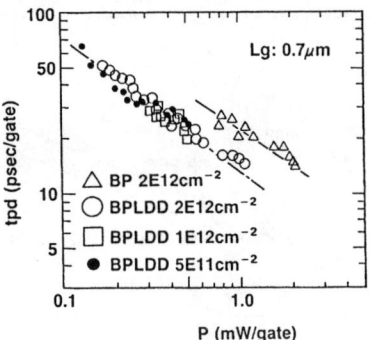

Fig. 15. Gate propagation delay time (t_{pd}) versus dissipation power per gate (P).

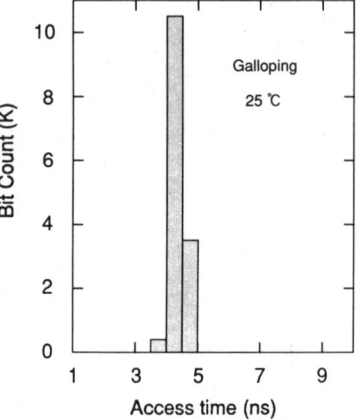

Fig. 16. Address access time (t_{acc}) distribution in a 16-kb SRAM.

μm. We believe that this results from the reduction of C_{gp} for the BPLDD at a short gate length as mentioned in Section V.

VII. GATE PROPAGATION DELAY TIME

We fabricated and measured 101-stage DCFL ring oscillators in order to compare the gate propagation delay times (t_{pd}'s) between BPLDD's and BP's. Notice that the output of each DCFL inverter is connected to the input of the next stage DCFL inverter in the ring oscillator. Therefore, the effect of Miller capacitance due to C_{gd} becomes larger in this case than that in the case of an f_T measurement. When the FET structure is changed from a BP to a BPLDD, C_{gd} decreases due to the reduction of C_{gpe}. Consequently, it is possible that t_{pd} decreases compared with the result of f_T measurement. Fig. 15 plots t_{pd} against dissipation power per gate (P). From the figure, we find that the t_{pd} of a BPLDD is smaller than that of a BP for the same L_g. At 1 mW/gate, the t_{pd} of the BPLDD for a 0.7-μm gate is 15 ps, which is about 65% of that for the BP. The reduction ratio is smaller than that of the f_T, which is about 73%. From Fig. 15, on the other hand, we find that t_{pd} depends little on Mg^+ ion dose. This fact indicates that additional parasitic capacitance due to a p-type neutral region, which is generated at a relatively high Mg^+ dose, is fairly small in this BPLDD structure. This result corresponds directly with the result for FATFET's dealing with added parasitic capacitance due to a p-layer in Section V.

VIII. APPLICATION TO SRAM'S

We used BPLDD FET's with 0.7-μm gates in 16-kb SRAM's [16], [17]. Their chip size and cell size are 5.05 $\times$ 6.00 mm^2 and 36 $\times$ 23 μm^2, respectively. The minimum size of line, space, and hole are 1.5 μm, 1.5 μm, and 1.6 $\times$ 1.6 μm^2, respectively. A triple-level interconnection is used [16], [17]. We obtain fully functional chips with a wafer yield of 10%. Fig. 16 shows the distribution of address access time (t_{acc}) in a 16-kb SRAM chip. Galloping is used as a test pattern. From Fig. 16, we confirm that the maximum t_{acc} is less than 5 ns at room tempera-

ture with a dissipation power of less than 2 W. We also confirm that the RAM's can operate at a temperature as high as 100°C. At 100°C, the maximum t_{acc} of 4.4 ns is obtained with power of 2 W. The realization of such a high speed is mainly due to 1) very small V_{th} scattering (σV_{th}) especially in a microscale range as small as 7 mV with a submicrometer gate length of 0.7 μm and 2) the reduction of gate-edge capacitance by the use of an LDD structure that has a small C_{gs} and C_{gd}.

IX. SUMMARY

We have improved and optimized a Buried p-layer Lightly Doped Drain (BPLDD) GaAs MESFET with such forming conditions of ion-implanted regions that improve its short-channel effects, V_{th} uniformity, and FET operating speed, simultaneously. We find that the concentrations of both buried p-layer and n' intermediate region must be optimized in order to improve the V_{th} uniformity. The optimum Mg^+ dose for the p-layer increases as the gate length is shortened. This agrees with the result of the scaling law. The σV_{th}'s of 7 mV in an area of 3 mm $\times$ 3 mm are realized for 0.7-μm gates. The reduction of gate-edge capacitance by the use of an LDD structure is effective so as to decrease the total gate capacitance and to make the FET operation faster. For a 0.7-μm gate, the gate-edge parasitic capacitance of the BPLDD is reduced to 0.5 fF/μm, which is half that of the BP. For the same gate length, the gate propagation delay time of the BPLDD can be reduced to 15 ps at a power dissipation of 1 mW/gate, which is about 65% of that of the BP. We have successfully obtained high-speed, fully functional, and stable 16-kb SRAM's using the BPLDD's with 0.7-μm gates. The maximum access times of less than 5 ns have been obtained with temperatures ranging from room temperature to 100°C.

ACKNOWLEDGMENT

The authors wish to thank T. Shimura for his advice on the condition of implantation for the BPLDD, Y. Nakajima for arranging the measurement of dc characteristics of FET's, K. Maemura for measuring the t_{pd} of FET's, T. Kato for arranging the measurement of f_T, H. Makino and S. Matsue for measuring the t_{acc} distributions of 16-kb SRAM's, and N. Tanino and Y. Kohno for helpful discussions. The authors also wish to express their appreciation for Dr. K. Fujikawa and Dr. S. Kayano for their continuous encouragement and useful advice.

REFERENCES

[1] M. Noda, K. Hosogi, K. Sumitani, H. Nakano, H. Makino, K. Nishitani, and M. Otsubo, "A high-yield 4Kb SRAM process technology using self-aligned gate MESFETs with a partially depleted p-layer," in *GaAs IC Symp. Tech. Dig.*, Nov. 1988, pp. 227–230.

[2] H. Makino, S. Matsue, M. Noda, N. Tanino, S. Takano, K. Nishitani, and S. Kayano, "A 7 ns/850 mW GaAs 4Kb SRAM fully operative at 75°C," in *GaAs IC Symp. Tech. Dig.*, Nov. 1988, pp. 71–74.

[3] F. Tokuyoshi, H. Takemura, T. Tashiro, S. Ohi, H. Shiraki, M. Nakamae, T. Kubota, and T. Nakamura, "A 2.3 ns access time 4K ECL RAM," in *ISSCC Tech. Dig.*, Feb. 1984, pp. 220–221.

[4] H. Ishiuchi, Y. Matsumoto, S. Sawada, and Ozawa, "Measurement of intrinsic capacitance of lightly doped drain (LDD) MOSFET's," *IEEE Trans. Electron Devices*, vol. ED-32, pp. 2238–2242, Nov. 1985.

[5] N. Matsunaga, M. Miyazaki, Y. Umemoto, J. Shigeta, H. Tanaka, and H. Yanazawa, "Gallium arsenide MESFET technologies with 0.7 μm gate-length for 4Kb 1ns static RAM," in *GaAs IC Symp. Tech. Dig.*, Oct. 1987, pp. 129–132.

[6] K. Ishida, T. Matsunaga, S. Miyano, A. Kameyama, and N. Toyoda, "A 5 Gb/s 4 bit shift register with 0.5 μm WN$_x$-gate GaAs MESFETs," in *20th Conf. on Solid State Devices and Materials* (Tokyo, Japan), 1988, pp. 129–132.

[7] S. Hayano, K. Nagashima, S. Asai, T. Maeda, and T. Furutsuka, "A GaAs 8 × 8 matrix switch LSI for high-speed digital communications," in *GaAs IC Symp. Tech. Dig.*, Oct. 1987, pp. 245–248.

[8] E. P. Finchem, W. A. Vetanen, B. Odekirk, and P. C. Canfield, "Reduction of the backgate effect in GaAs MESFETs by charge confinement at the backgate electrode," in *GaAs IC Symp. Tech. Dig.*, Nov. 1988, pp. 231–234.

[9] K. Inokuchi, M. Tsunotani, T. Ichioka, Y. Sano, and K. Kaminishi, "Suppression of sidegating effect for high performance GaAs ICs," in *GaAs IC Symp. Tech. Dig.*, Oct. 1987, pp. 117–120.

[10] Y. Umemoto, M. Masuda, J. Shigeta, and K. Mitsusada, "Improvement of alpha-particle-induced soft-error immunity in a GaAs SRAM by a buried p-layer," *IEEE Trans. Electron Devices*, vol. 35, pp. 268–274, Mar. 1988.

[11] S. Matsue, H. Makino, M. Noda, N. Tanino, S. Takano, K. Nishitani, and S. Kayano, "A soft error improved 7 ns/2.1 W GaAs 16Kb SRAM," in *GaAs IC Symp. Tech. Dig.*, Oct. 1989, pp. 41–44.

[12] C. T. M. Chang, T. Vrotsos, M. T. Frizzell, and R. Carroll, "A subthreshold current model for GaAs MESFET's," *IEEE Electron Device Lett.*, vol. EDL-8, pp. 69–72, Feb. 1987.

[13] Y. Awano, M. Kosugi, M. Kosemura, T. Mimura, and M. Abe, "Short-channel effects in subquarter-micrometer-gate HEMT's: Simulation and experiment," *IEEE Trans. Electron Devices*, vol. 36, pp. 2260–2265, Oct. 1989.

[14] P. M. Asbeck, C. P. Lee, and M. C. F. Chang, "Piezoelectric effects in GaAs FET's and their role in orientation-dependent device characteristics," *IEEE Trans. Electron Devices*, vol. ED-31, pp. 1377–1380, 1984.

[15] T. Takada, K. Yokoyama, M. Ida, and T. Sudo, "A MESFET variable-capacitance model for GaAs integrated circuit simulation," *IEEE Trans. Microwave Theory Tech.*, vol. MTT-30, no. 5, pp. 719–724, 1982.

[16] M. Noda, S. Matsue, M. Sakai, M. Sumitani, H. Nakano, T. Oku, H. Makino, K. Nishitani, and M. Otsubo, "A triple-level interconnection technology for high speed 16Kb GaAs SRAM," in *20th Conf. on Solid State Devices and Materials* (Sendai, Japan), 1990, pp. 71–74.

[17] H. Nakano, M. Noda, M. Sakai, S. Matsue, T. Oku, K. Sumitani, H. Makino, H. Takano, and K. Nishitani, "A high-speed GaAs 16Kb SRAM of 4.4 ns/2 W using triple-level metal interconnection," in *GaAs IC Symp. Tech. Dig.*, Oct. 1990, pp. 151–154.

GaAs DIGITAL VLSI DEVICE AND CIRCUIT TECHNOLOGY

James Mikkelson

Vitesse Semiconductor
741 Calle Plano   Camarillo, CA 93012

ABSTRACT

In recent years, GaAs digital IC technology has advanced from a high speed, high power, low integration complexity technology to a high speed, low power, very high level of integration circuit technology. This increase has been achieved by the use of simple MOS like circuit design topologies and silicon like process techniques. This paper provides an overview of ion implanted GaAs MESFET device fabrication, and the interconnect and process technologies developed for GaAs VLSI circuits. Circuit topologies and circuit results for GaAs VLSI are also presented.

DEVICE TECHNOLOGY

The predominant devices used to fabricate GaAs VLSI circuits are planar, ion implanted self-aligned refractory gate enhancement and depletion mode (E-D) Schottky gate MESFETs. Our current VLSI process, HGaAs III, produces high performance enhancement FETs with a gate length of 0.6μm and a transconductance of 250 mS/mm at a nominal threshold voltage of 250 mV. A subthreshold slope of 88 mV/decade and excellent short channel device characteristics are achieved by using a buried p layer, shallow, heavily doped channel and source and drain regions, and other scaling techniques analogous to those used for silicon MOS transistors. The I-V characteristics of a W=10μm L=0.6μm FET are shown in figure 1. The major performance characteristics of the transistors are described in table 1.

Both the enhancement and the depletion transistors are fabricated with the simple 4 mask process illustrated in figure 2. The wafer is first covered with a composite SiO_2/Si_3N_4 dielectric layer which protects the GaAs surface during processing and acts as an implant stop layer. The first mask level is used to pattern and etch active area regions in the dielectric layer. The second mask is used to define openings through a photoresist layer for the depletion transistor channel implant. After the photoresist is removed, both the depletion and enhancement transistor channels are implanted with the enhancement implant. This technique reduces the number of masking steps and provides for better E-D threshold voltage tracking. The channels are implanted with Si^{29}. After the channels are implanted, a tungsten based refractory metal is sputter deposited, masked, and etched to form the gate electrodes of the transistors and also a layer of local interconnections. After the gate electrodes are patterned, an LDD like implant is performed to increase the gate edge doping and to lower source resistance. Next, an oxide spacer is formed and the source and drain regions are implanted with Si^{28}. After the source-drain implant, the entire structure is capped with silicon nitride and annealed to activate the implants. A thin dielectric layer is deposited and patterned as a dielectric assisted lift-off mask for ohmic contact metal definition. After the ohmic metal has been deposited and patterned, all of the GaAs device specific processing is completed.

INTERCONNECT TECHNOLOGY

Preserving high speed device performance in VLSI circuit applications requires high density, high performance interconnect technology. To provide high performance and high yield, the well developed and characterized materials and methods of interconnect formation used for silicon VLSI circuit fabrication have been applied to GaAs VLSI. After the GaAs specific device fabrication has been completed, the remainder of the process consists of applying four full layers of aluminum metal interconnect for signal routing and power busing. The addition of more layers of metal greatly improves the circuit packing density and reduces the length of interconnections, thereby reducing the capacitive loading on the logic gates and increasing the overall circuit performance.

Aluminum metallization is used to provide low interconnect resistance while allowing the ability to pattern and etch fine geometries. Excellent electromigraton resistance is achieved by the addition of 1% copper, and by the use of refractory metal barriers and caps. The metallization is sputter deposited and dry etched. The interconnect capacitance is reduced by using thick layers of SiO_2 between the metal layers. Capacitance reduction is necessary to prevent RC delays in the interconnections from dominating the circuit performance. The inter-metal dielectric layers are deposited by PECVD and are planarized by the use of spin on glass.

In addition to the local signal routing capability provided by the gate metal, metal 1 is also often used for local interconnections and macro routing. Metal 1, metal 2, and metal 3 are used for signal routing. Metal 3 is also used for power and clock routing. A thick metal 4 layer is provided for reducing the IR drop in power distribution and for ease of routing gate arrays.

The total process for fabricating the devices and four layers of interconnections consists of only 13 mask levels. The process layout rules for the HGaAs III are given in table 2.

CIRCUIT TOPOLOGY

Historically, most GaAs integrated circuits have been designed with many different circuit topologies, but with only depletion mode transistors (1). These circuits are typically fast, but they consume large amounts of power and chip area. The self aligned refractory gate MESFET process provides the ability to repeatably fabricate high performance enhancement FETs. This allows the use of the direct coupled FET logic

Reprinted from *Tech. Dig. of IEDM*, pp. 231–242, 1991.

(DCFL) circuit topology for the production of GaAs VLSI circuits. The basic DCFL NOR gate schematic and operating I-V characteristics are shown in figure 3. The operation of this high speed, small size, low power circuit family is similar to that of E-D NMOS circuitry, except that the Schottky diode gate of the driven transistor acts as a diode clamp which limits the output high voltage level. The output voltage swing is 0.7 volts. This voltage swing allows high speed switching with low power dissipation and is equivalent to silicon bipolar ECL switching levels. DCFL is a ratioed logic family, so that transistor sizing as well as control of transistor threshold voltage and power supply distribution are critical factors for high performance and yield (2). Because the depletion FET current is steered through either the pulldown transistor or the driven gate, the supply current is constant and the switching noise on the power supply and ground buses is very small. Power consumption is low because the DCFL circuitry operates from a single 2 volt power supply. Typical unloaded NOR gate delays are 50ps and loaded delays are 100ps with a fan in and fan out of 2 at a power dissipation of only 0.25mW.

For large circuits with long interconnect lines, the metal interconnect capacitance can present significant loading to simple logic gates which can cause an unacceptable reduction of circuit performance. To minimize the capacitive loading effects, while preserving low power consumption, the dynamic driver circuit shown in figure 4 is used. This circuit provides extra current to charge or discharge capacitive loads during rising and falling signal transitions. Internal feedback in the buffer quickly returns the output current to its quiescent value, so that the static power dissipation is unchanged. The difference between buffered and unbuffered circuit performance when driving capacitive loads is shown in figure 5.

To convert between internal DCFL logic levels and the levels more commonly used in systems, I/O buffers are required. The input and output buffers for ECL I/O levels are shown in figure 6. These buffers consume significant area, power, and propagation time in GaAs integrated circuits, and are a major factor forcing the increased level of integration. The input receiver has a delay of 480 ps and consumes 4 mW of power in an area of 46,200 μm^2. The output driver has a delay of 750 ps and consumes 20 mW of power in an area of 77,000 μm^2. The I/O delays are 10 times larger than those of the internal logic gates, and the power and area consumption are 100 times larger. For small circuits, as much as 30% of the area and 50% of the power can be consumed by the I/O buffers. For VLSI circuits, the area penalty can be reduced to 10% and the power penalty can be reduced to 20%.

CIRCUIT IMPLEMENTATION

The process technology and circuit capability improvements described above have led to a dramatic increase in the level of integration of GaAs circuits. In the past five years, the complexity of DCFL logic circuitry has increased from 2000 gates to over 100,000 gates. Figure 7 illustrates the doubling of GaAs IC integration capability each year since 1986. The increasing specialization and customization of large scale circuits, and a more rapid design and implementation cycle have led to most GaAs VLSI circuits being customer specific gate array or standard cell ASICs. Mixed RAM and

logic are also available with the DCFL topology (3). Figure 8 is a photograph of a 45K gate standard cell GaAs integrated circuit. The high density of GaAs logic is illustrated by this circuit which has 256 I/Os and 396 bonding pads and is pad and I/O driver limited. To provide high logic density for gate arrays, a channelless sea of gates architecture has been developed (4). The cell area for a two input NOR gate is only 380 μm^2, allowing the incorporation of over 100,000 gates on the 8.5x7.1 mm die shown in figure 9.

CONCLUSION

In the past few years, the level of integration of GaAs circuits has been doubling on an annual basis. Significant improvements have been made in both GaAs circuit performance and power-delay product. The development and improvement of self-aligned refractory gate MESFET process technology, which produces high quality enhancement and depletion transistors, allows the use of simple, high performance DCFL circuit topology. This circuit family, coupled with advanced, high density, low capacitance interconnect technology provides the capability for very high performance, low power GaAs VLSI circuits. Silicon MOS processing techniques and equipment have been used for fabrication of the circuits, and silicon like circuit topologies, design tools, and design techniques have been used to design, verify, and test the circuits. Circuits of 100,000 gate complexity are currently available, and increased integration levels will be available within the next year.

ACKNOWLEDGMENTS

These results for the development and implementation of GaAs VLSI circuits represent the efforts of many people over several years. The efforts of the process development and manufacturing groups have been responsible for the rapid rate of integration increase and manufacturing advance. The circuit design and modelling groups have provided circuit design topologies and tools for the design of successful VLSI circuits. The test and product engineering groups have provided the necessary feedback for process and product characterization and improvement. Finally, the support of the entire Vitesse organization is acknowledged, because this work is really the result of the entire company.

REFERENCES

(1)   S. I. Long and S. E. Butler Gallium Arsenide Digital Integrated Circuit Design   McGraw-Hill 1990
(2)   T. Onodera, H. Onodera, S. Sugisaki, M. Okamoto, K. Suyama, I. Kuri, H. Nishi "GaAs MESFET LSI Design Using E/D-DCFL Circuits" 1990 GaAs IC Symposium pp. 219-222
3   R. Hinds, S. Canaga, G. Lee, A. Choudhury "A 20K GaAs Array With 10K of Embedded RAM"   IEEE Journal of Solid State Circuits   vol. SC 26 pp 245-256
4   G. Lee, B Donckles, A Grey, I Deyhimy   "A High Density Gate Array Architecture"   1991 Custom Integrated Circuits Conference pp. 14.7.1-14.7.4

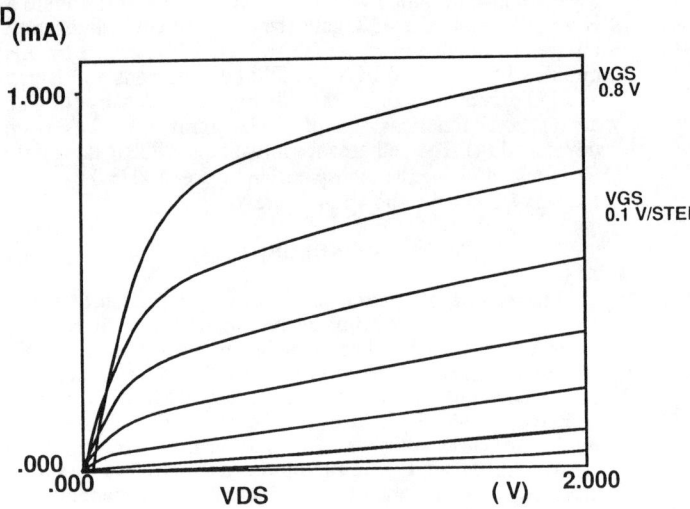

Figure 1 MESFET IV Characteristics

TABLE 1
H-GaAs III TRANSISTOR PARAMETERS

THRESHOLD VOLTAGE	0.25 V
TRANSCONDUCTANCE	250 mS/mm
K FACTOR	390 μA/μmV
OUTPUT CONDUCTANCE	9.5 mS/mm
SUBTHRESHOLD SLOPE	88 mV/decade
SOURCE RESISTANCE	0.8 Ω-mm

TABLE 2
H-GaAs III PROCESS LAYOUT RULES

FEATURE	LINE	SPACE
ACTIVE AREA	1.5	1.8
GATE METAL	0.6	1.5
VIA 1	1.2	1.5
METAL 1	1.2	1.8
VIA 2	1.8	2.0
METAL 2	1.8	1.8
VIA 3	2.6	3.0
METAL 3	3.0	4.0
VIA 4	4.0	4.0
METAL 4	-.-	-.-

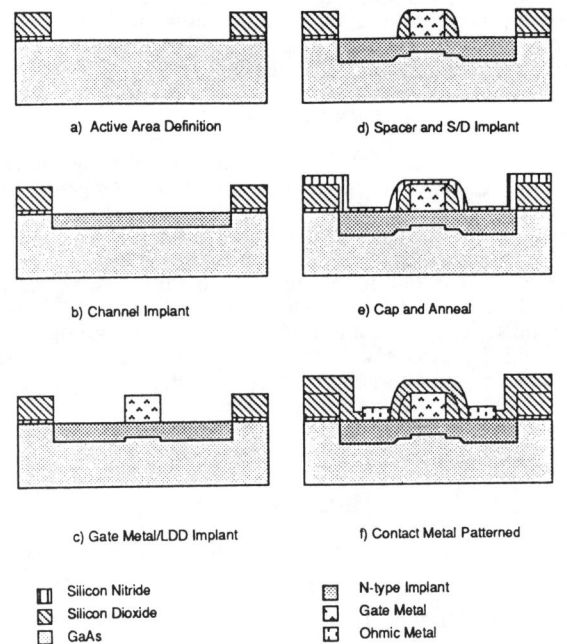

a) Active Area Definition

d) Spacer and S/D Implant

b) Channel Implant

e) Cap and Anneal

c) Gate Metal/LDD Implant

f) Contact Metal Patterned

Silicon Nitride
Silicon Dioxide
GaAs
N-type Implant
Gate Metal
Ohmic Metal

Figure 2   Transistor Fabrication Sequence

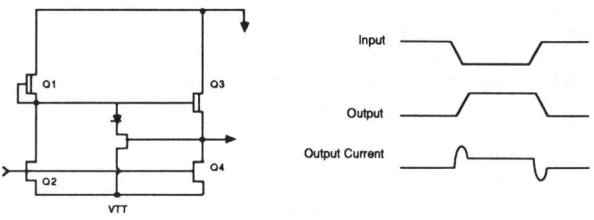

Figure 4   Dynamic Driver

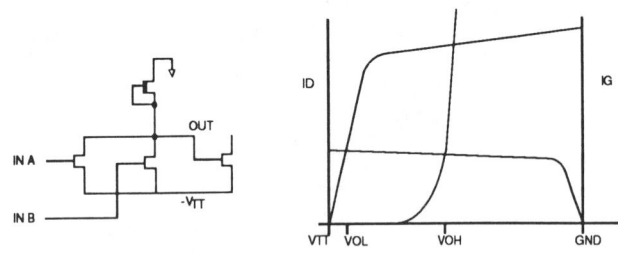

Figure 3   DCFL Topology
and Operating Characteristics

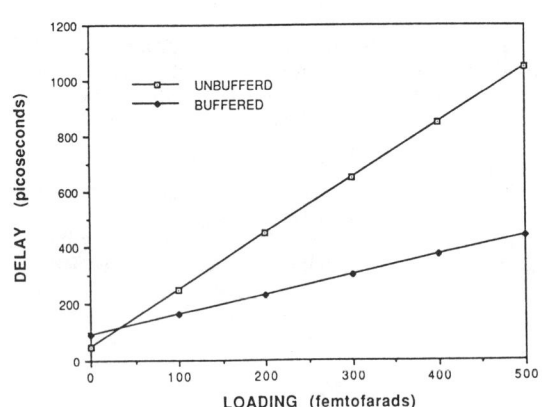

BUFFERED vs UNBUFFERED DELAYS

Figure 5   Line Drive Capability

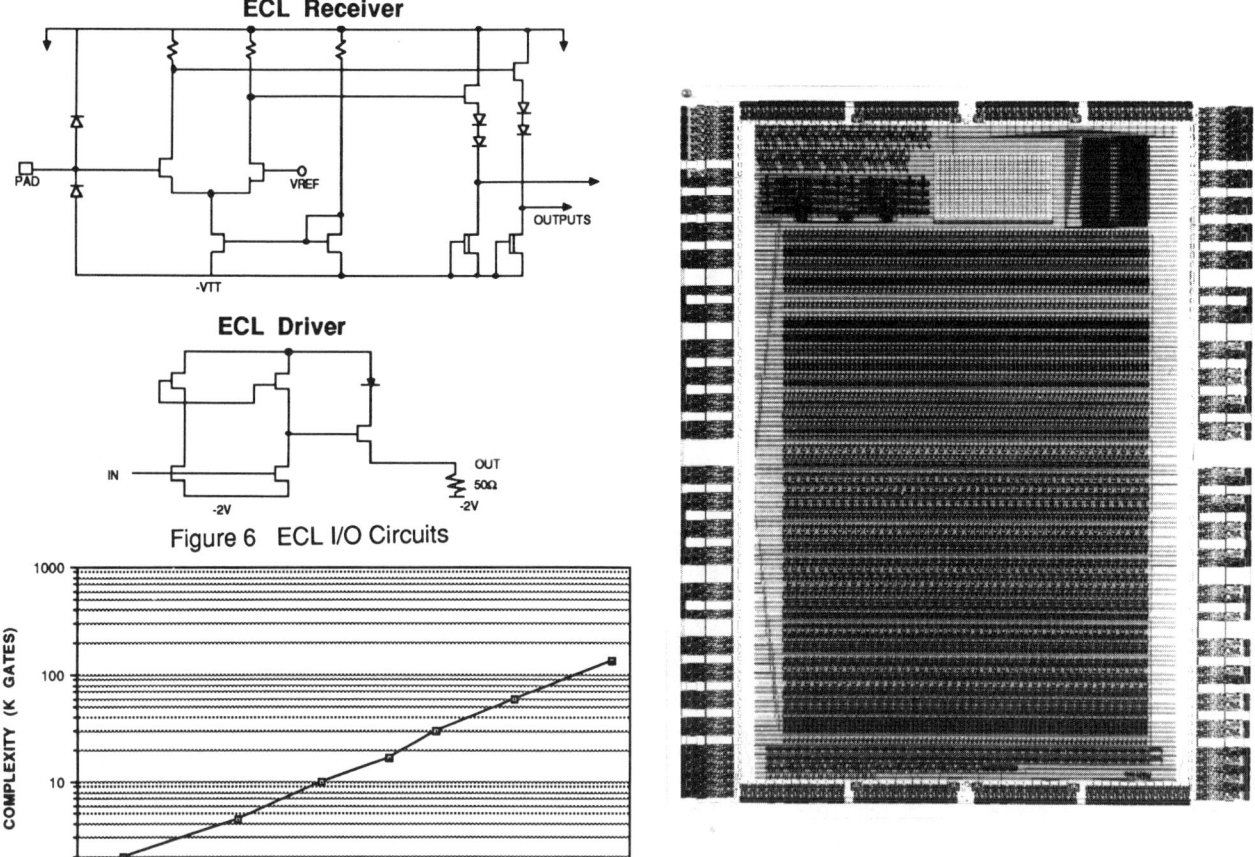

ECL Receiver
PAD
VREF
OUTPUTS
-VTT
ECL Driver
IN
OUT
50Ω
-2V
-2V

Figure 6 ECL I/O Circuits

COMPLEXITY (K GATES)
1000
100
10
1
1986
1987
1988
1989
1990
1991
1992

Figure 7 GaAs Integration Level vs Time

Figure 9 100K Sea of Gates Circuit

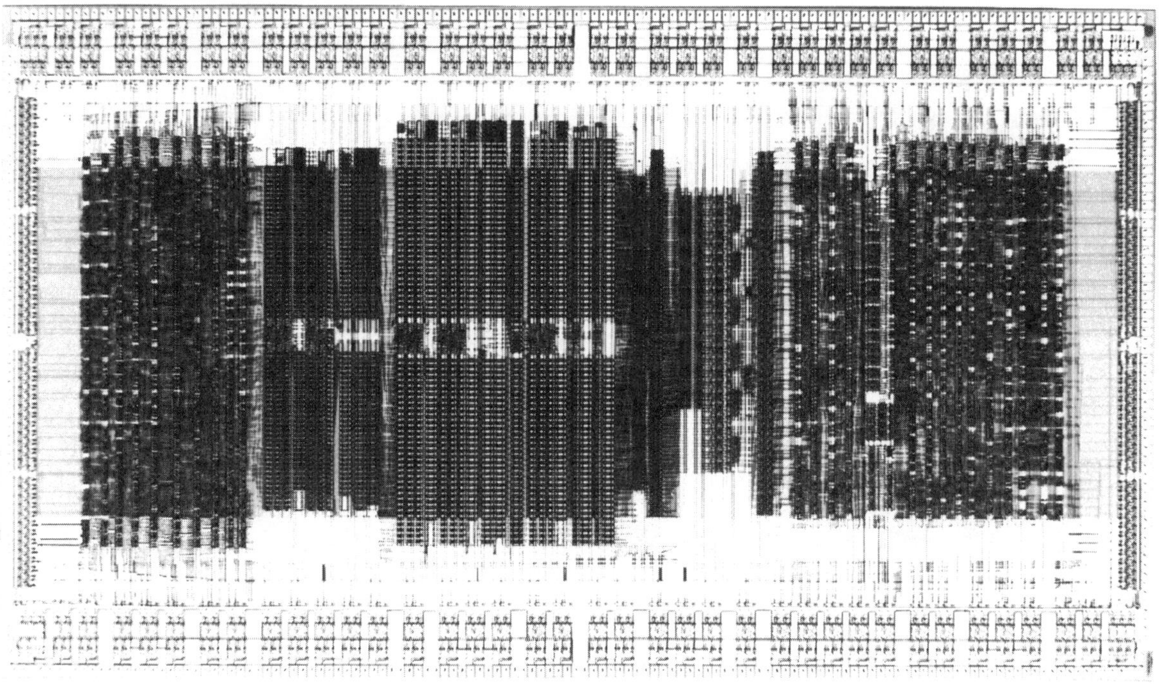

Figure 8 45K Gate Standard Cell Circuit

Part 3
HFETs

Electron energy levels in GaAs-Ga$_{1-x}$Al$_x$As heterojunctions

Frank Stern

IBM Thomas J. Watson Research Center, Yorktown Heights, New York 10598

Sankar Das Sarma

IBM Thomas J. Watson Research Center, Yorktown Heights, New York 10598
and Department of Physics and Astronomy, University of Maryland, College Park, Maryland 20742*

(Received 31 January 1984)

Calculated results for energy levels of electrons in GaAs-Ga$_{1-x}$Al$_x$As heterojunctions are presented and their sensitivity to various parameters—including acceptor doping level in the GaAs, heterojunction barrier height, effective-mass and dielectric-constant discontinuities, interface grading, and ambient temperature—is examined.

I. INTRODUCTION

Interest in the properties of electrons in GaAs-Ga$_{1-x}$Al$_x$As heterojunctions has led to a number of calculations[1-3] of their energy levels and other electronic properties. In this paper, calculated results of energy levels are presented for a range of temperatures and material parameters, including interface grading, which has not been treated in detail previously (however, see Price and Stern[4]). Values of the energy levels and of the Fermi level are needed to find the charge transfer from donors in the Ga$_{1-x}$Al$_x$As to the GaAs channel. Energy-level differences are more closely related to subband spectroscopy and to the onset of intersubband scattering. We find that the energy levels are affected to varying degrees by the values of the parameters, and that differences between them are affected less. The differences are most sensitive to the acceptor doping level in the GaAs. Many aspects of the calculations presented here have counterparts in calculations for silicon inversion layers and related two-dimensional electron systems, which have been reviewed by Ando *et al.*[5]

The next two sections give a description of the calculation, which follows familiar lines except for the treatment of the graded interface, and a discussion of the density-functional scheme that is used to take electron-electron interactions into account approximately. Section IV gives results of the calculations and Sec. V gives discussion and conclusions. A preliminary account of this work, but without electron-electron interaction effects, was presented last year (see Ref. 6).

II. OUTLINE OF THE CALCULATION

The conduction-band edge in a heterojunction, illustrated in Fig. 1(a), has a spatial dependence both because of electrostatic potentials and because of the energy-band discontinuity associated with the change in material across the heterojunction. In this calculation the heterojunction effect is modeled using a graded interface in which the barrier height as well as the effective mass and dielectric constant are assumed to change smoothly in a

transition layer whose thickness is specified. A physical interface between two materials may be crystallographically abrupt, but the bonding environment of the atoms adjoining this interface will change on at least an atomic scale. In place of a microscopic model of this transition, we use an approximate treatment in which the parameters vary smoothly but rapidly. A similar calculation has been used to treat the energy levels of electrons on the surface of liquid helium.[7]

An electron moves in an effective potential given by

$$V(z) = -e\phi(z) + V_h(z) + V_{xc}(z) + V_{im}(z) , \qquad (1)$$

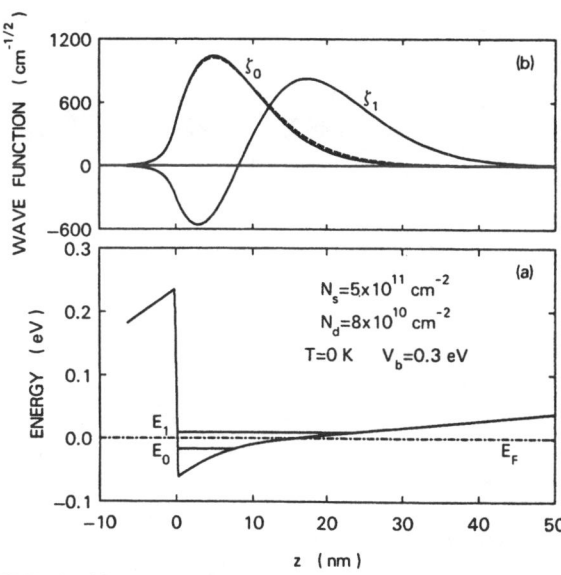

FIG. 1. (a) Conduction-band edge vs distance from a GaAs-Ga$_{0.7}$Al$_{0.3}$As heterojunction with 5×10^{11} cm^{-2} electrons in the GaAs channel, an acceptor doping of 3×10^{14} cm^{-3} in the GaAs, and a barrier height of 0.3 eV. E_0 and E_1 label the bottoms of the lowest and of the first excited subbands, respectively. All energies in this figure are relative to the Fermi energy E_F. (b) Normalized envelope wave functions (solid curves) $\zeta_0(z)$ and $\zeta_1(z)$ for electrons in the two lowest subbands calculated with inclusion of exchange and correlation effects. The dashed curve gives $\zeta_0(z)$ calculated without including exchange and correlation effects.

where $\phi(z)$ is the electrostatic potential, $V_h(z)$ is the effective potential energy associated with the (graded) heterojunction discontinuity, $V_{xc}(z)$ is the local exchange-correlation potential energy described in the next section, and $V_{im}(z)$ is the image potential energy. The normalized envelope function $\zeta_i(z)$ for an electron in subband i is assumed to be given by a Schrödinger equation of the BenDaniel-Duke[8] form:

$$-\frac{\hbar^2}{2}\frac{d}{dz}\frac{1}{m(z)}\frac{d\zeta_i(z)}{dz}+V(z)\zeta_i(z)=E_i\zeta_i(z) , \qquad (2)$$

where $m(z)$ is the position-dependent effective mass and E_i is the energy of the bottom of the ith subband. Some or all of the energy levels correspond to states that can penetrate the barrier into lower-energy regions in the Ga$_{1-x}$Al$_x$As and thus can be affected by the potential profile there, a situation treated by Vinter.[2] This does not have significant consequences for the heterojunctions in equilibrium treated here but is important for states whose energies lie near the top of the heterojunction barrier or when a current is flowing across the heterojunction.

Poisson's equation for the electrostatic potential (in SI units) takes the form

$$\frac{d}{dz}\epsilon_0\kappa(z)\frac{d\phi(z)}{dz}=e\sum N_i\zeta_i^2(z)-\rho_I(z) , \qquad (3)$$

$$N_i=\frac{m_ck_BT}{\pi\hbar^2}\ln\left[1+\exp\left[\frac{E_F-E_i}{k_BT}\right]\right] , \qquad (4)$$

where $\kappa(z)$ is the position-dependent dielectric constant,

N_i is the number of electrons per unit area in subband i, E_F is the Fermi energy, and m_c is the effective mass in the GaAs.[9] The potential is taken to be 0 at the interface and its derivative is taken to be 0 in the GaAs bulk. The impurity charge density in the GaAs ($z>0$) is taken to be $\rho_I(z)=-eN_{Ac}$, where N_{Ac} is the net acceptor density in the GaAs, and is here taken to vanish in the Ga$_{1-x}$Al$_x$As ($z<0$). Subscript c generally denotes the GaAs (channel) side of the heterojunction, and subscript b denotes the Ga$_{1-x}$Al$_x$As (barrier) side.

Three parameters in the theory change across the heterojunction: the effective heterojunction potential energy $V_h(z)$, the effective mass m, and the dielectric constant κ. The barrier V_b appears on the Ga$_{1-x}$Al$_x$As side of the heterojunction and vanishes on the GaAs side. Since the length scale of the system is given by atomic sizes, a mathematically abrupt transition is unphysical and, as noted below, can lead to mathematical difficulties. To smooth this transition we take a grading function that basically interpolates linearly between the values in the materials bounding the heterojunction, but has rounded corners. This functional form is the same as the one that was used to calculate the energy levels of electrons on the surface of liquid helium,[10] whose density is known to vary from its bulk value to the (very small) value in the vapor in a distance of order 1 nm. Other grading functions would presumably give comparable results. The grading of the interface barrier is taken to be

$$V_h(z)=[1-G(z)]V_b , \qquad (5)$$

$$G(z)=\begin{cases}(z+z_h)/z_t, & 0<z<z_h-a_t , \\[2mm] \dfrac{\{z+3z_h-a_t+(2a_t/\pi)\cos[\pi(z-z_h)/2a_t]\}}{2z_t}, & z_h-a_t<z<z_h+a_t , \\[2mm] 1, & z>z_h+a_t , \\[2mm] 1-G(-z), & z<0 , \end{cases} \qquad (6)$$

where V_b is the barrier height, z_t and a_t are parameters that characterize the transition layer ($z_t\geq 2a_t>0$), and $z_h=z_t/2$. The nominal thickness of the transition layer is z_t+a_t. The two remaining graded quantities are found similarly, using $G(z)$ to interpolate between the value in the Ga$_{1-x}$Al$_x$As and the value in the GaAs.

The image potential energy V_{im} is obtained from the graded dielectric constant using the method described in Ref. 7, and, as was intended, avoids the singularity of the conventional result

$$V_{im}(z)=\begin{cases}[(\kappa_c-\kappa_b)e^2]/[16\pi\epsilon_0\kappa_c(\kappa_c+\kappa_b)z], & z>0 \\[2mm] [(\kappa_c-\kappa_b)e^2]/[16\pi\epsilon_0\kappa_b(\kappa_c+\kappa_b)z], & z<0 \end{cases}$$

$$\qquad (7)$$

at the interface, $z=0$. The result in Eq. (7) cannot be used directly in the effective-mass Schrödinger equation, Eq. (2), when there is a finite barrier height because a finite value of the envelope wave function at the interface leads to a nonintegrable singularity in the energy. Most authors

avoid this problem by assuming an infinite barrier, as for the Si-SiO$_2$ interface,[5] or by ignoring the dielectric-constant change. One of the goals of the present calculation is to assess the magnitude of the error made in ignoring the dielectric-constant step.

Apart from the grading effects just noted, the calculation is conventional and will not be described further. In the next section the considerations that enter in a choice of the effective exchange-correlation potential are described.

III. EXCHANGE-CORRELATION EFFECTS

In this section we first describe the local-density-functional approximation for exchange-correlation effects—the effects of electron-electron interaction beyond the Hartree approximation—and then discuss some of the ways in which this approximation might be improved in connection with heterojunction calculations. A simple way of including many-body exchange-correlation effects in the calculation of electronic structure is the density-functional method[11] due to Hohenberg, Kohn, and Sham. This method has been used with great success in calculating the electronic structure of atoms, molecules, and solids.[12] The density-functional technique has also been applied quite successfully by Ando[13] and by Das Sarma and Vinter[14] in calculating the electronic subband structure of silicon inversion layers in Si-SiO$_2$ metal-oxide-semiconductor–field-effect-transistor systems.

Within the density-functional formalism, the electronic energy levels (including exchange-correlation effects) are obtained from the one-electron Kohn-Sham equation[11] which is formally the same as the Schrödinger equation for the Hartree problem except that an additional term, the so-called exchange-correlation potential energy V_{xc} first introduced by Kohn and Sham,[11] appears in the potential energy. The one-electron wave functions $\zeta_i(z)$ and energy levels E_i are usually identified as the real wave functions and energy levels of the system even though the theory is rigorously valid only for the total ground-state energy and density of the system. For justification of this identification we refer the reader to Ando's work[13] on silicon space-charge layers and to von Barth's recent investigation[15] of a more general nature.

The exchange-correlation potential energy $V_{xc}(z)$ is, in general, an unknown functional[11] of the electron density $n(z)$, given by the sum on the right-hand side in Eq. (3).[16] In reality, however, the simplest approximation to the exchange-correlation potential, the so-called local-density-functional approximation, works surprisingly well.[12-15] In this approximation one takes $V_{xc}(z) \equiv V_{xc}\{n(z)\} \equiv \mu_{xc}[n_0 = n(z)]$, where μ_{xc} is the exchange-correlation contribution to the chemical potential of a homogeneous electron gas having a uniform electron density n_0 which is equal to the local electron density $n(z)$ of the inhomogeneous system. V_{xc}, calculated within the local-density-functional approximation, has been parametrized by a number of authors[17] and different forms of V_{xc} give similar quantitative results for the subband energy levels. We therefore use the simple analytic parametrization[18] suggested by Hedin and Lundqvist:

$$V_{xc}(z) = -[1 + 0.7734 x \ln(1 + x^{-1})](2/\pi \alpha r_s) \text{ Ry}^* , \quad (8)$$

where $\alpha = (4/9\pi)^{1/3}$, $x \equiv x(z) = r_s/21$,

$$r_s \equiv r_s(z) = [\tfrac{4}{3}\pi a^{*3} n(z)]^{-1/3} , \quad (9)$$

$$a^* = \frac{4\pi\epsilon_0 \kappa \hbar^2}{me^2} , \quad (10)$$

$\kappa \equiv \kappa(z)$ is the local dielectric constant, and $m \equiv m(z)$ is the local effective mass. The unit of energy in Eq. (8) is the effective Rydberg, $\text{Ry}^* = (e^2/8\pi\epsilon_0 \kappa a^*)$, which is approximately 5 meV for GaAs.

One of the problems in applying the local-density-functional formalism to the calculation of the electronic structure of space-charge layers at semiconductor interfaces is that one is working within the effective-mass formalism and hence the corresponding homogeneous electron-gas calculation to obtain V_{xc} must somehow reflect the dielectric discontinuity at the interface. This is essential[13,14] for the Si-SiO$_2$ system where the static

dielectric constants for Si and SiO$_2$ are 11.5 and 3.9, respectively, giving rise to a substantial image interaction effect. In typical GaAs-Ga$_{1-x}$Al$_x$As heterostructures the change in dielectric constant is of order 10% or less and consequently the image interaction is much smaller than the direct Coulomb interaction. Thus it is a good approximation to neglect the image effect in calculating V_{xc} for the GaAs-Ga$_{1-x}$Al$_x$As system.

The formal issue of calculating a correct V_{xc} (even within the local-density-functional approximation) in the presence of a dielectric discontinuity, as encountered in the GaAs-Ga$_{1-x}$Al$_x$As heterojunction, remains open. Neglecting the tailing of the electronic wave function into the Ga$_{1-x}$Al$_x$As layer, one can write the electron-electron interaction between two electrons on the GaAs side as

$$v(\vec{r}_1, z_1; \vec{r}_2, z_2) = \frac{e^2}{4\pi\epsilon_0 \kappa_1 [(\vec{r}_1 - \vec{r}_2)^2 + (z_1 - z_2)^2]^{1/2}}$$
$$+ \frac{(\kappa_1 - \kappa_2)}{4\pi\epsilon_0 \kappa_1 (\kappa_1 + \kappa_2)}$$
$$\times \frac{e^2}{[(\vec{r}_1 - \vec{r}_2)^2 + (z_1 + z_2)^2]^{1/2}} , \quad (11)$$

where $(\vec{r}_1, z_1)$ and $(\vec{r}_2, z_2)$ with $\vec{r} \equiv (x, y)$ are the positions of the two electrons and κ_1 and κ_2 are to be identified with the dielectric constants κ_c and κ_b of GaAs and Ga$_{1-x}$Al$_x$As, respectively. Equation (11) defines an interaction that is clearly not translationally invariant, by virtue of the second term (the image term). To circumvent this problem Ando wrote[13]

$$(z_1 + z_2)^2 = (z_1 - z_2)^2 + 4z_1 z_2$$
$$\approx (z_1 - z_2)^2 + 4z^2 , \quad (12)$$

and identified this z as the same z defining the local density $n(z) = n_0$ in the calculation of V_{xc}. Thus V_{xc} becomes $V_{xc}[n(z); z]$, the explicit z dependence arising from the image interaction. Once Eq. (12) is put into Eq. (11), the resulting interaction can be used to obtain[13] μ_{xc} for given values of z and $n(z) = n_0$. This μ_{xc} is taken to be the relevant V_{xc} for the problem.

This particular way of handling the dielectric discontinuity works very well[13,14] for the Si-SiO$_2$ space-charge layer system, where the potential barrier at the interface is so large that electrons are confined in the Si with almost no effective wave-function tailing into the SiO$_2$ layer. It is, however, not obvious that this technique can be successfully used in a heterostructure with a dielectric discontinuity where wave-function tailing into the insulating layer is not negligible. The practical problem is that the form of the electron-electron interaction $v(\vec{r}_1, z_1; \vec{r}_2, z_2)$ now depends specifically on where the electrons are found (i.e., the formula for v depends on whether both electrons have $z > 0$ or $z < 0$ or whether they are on opposite sides of the interface). For the GaAs-Ga$_{1-x}$Al$_x$As system these image effects can safely be ignored because the dielectric constants differ so little. In an actual microscopic calculation (i.e., without the effective-mass approximation), this problem would not arise because the microscopic Coulomb interaction is well defined and translationally invariant everywhere. However, a microscopic calculation for this system may be prohibitively difficult for computational reasons.

Having a graded interface (i.e., going from κ_1 to κ_2 in a smooth fashion) helps one in defining a local dielectric constant $\kappa \equiv \kappa(z)$ and effective mass $m \equiv m(z)$, so that Eqs. (8)–(10) can be evaluated at each z without any discontinuity. However, the formal problem (when κ_1 and κ_2 are very different) of the image effect and the consequent lack of translational invariance in the electron-electron interaction still remains and should be investigated theoretically. The rapid variation of the effective mass in going from the κ_1 side of the interface to the κ_2 side also raises fundamental questions about the validity of the density-functional scheme as applied within the effective-mass approximation. This variation in the effective mass is handled simply by using the local effective mass in Eqs. (8)–(10). The validity of this approach in defining V_{xc} is unclear, but the graded-interface approximation may lead to errors even in the Hartree approximation and these are likely to overshadow the effects of grading on the exchange-correlation potential.

The finite-temperature generalization of the local-density-functional approximation to the electronic structure of space-charge layers has been investigated in detail by Das Sarma and Vinter[14] in the context of silicon inversion layers. Their basic conclusions should be valid for the GaAs heterostructure systems as well. Das Sarma and Vinter found that one can neglect the explicit temperature dependence of V_{xc}, keeping only the implicit temperature dependence which arises because $n(z)$ depends on temperature through the occupations of the self-consistent energy levels of the system. Unlike Si inversion layers where different subband ladders have different masses and one must do a "valley-polarized" local-density-functional calculation (when subbands in both ladders are occupied by electrons), GaAs is a single-valley isotropic system where the simple unpolarized local-density-functional approximation outlined above is applicable. If one wants to investigate the possibility of a ferromagnetic (or any other spin-related) instability in the GaAs heterostructure, one can readily generalize the formalism given here by using a suitable spin-polarized exchange-correlation potential.[19]

In this work we use the local-density-functional approximation as described above in calculating the electronic subband structure of the GaAs-Ga$_{1-x}$Al$_x$As heterojunction system. We use the local dielectric constant $\kappa \equiv \kappa(z)$ and the local effective mass $m \equiv m(z)$ in obtaining V_{xc} given by Eqs. (8)–(10). We neglect image effects in V_{xc}, which should be a good approximation for the GaAs-Ga$_{1-x}$Al$_x$As system because $\kappa_1 \approx \kappa_2$.

Finally, it should be emphasized that even though the local-density-functional approximation has had great empirical success in the calculation of electronic structure not only for space-charge layers in semiconductors but also for a wide class of systems including bulk solids, surfaces, atoms, and molecules, the condition for its validity is seldom obeyed in physical systems of interest. In particular, the validity of the local-density-functional approximation requires that the electronic density variation be small over distances of the order of a Fermi wavelength.

In a GaAs-Ga$_{1-x}$Al$_x$As heterojunction with an electron density of 10^{12} cm^{-2}, the Fermi wavelength is about 25 nm, whereas the typical wave-function width is about 10 nm for the lowest subband. Thus the electron density does not vary slowly on the scale of the Fermi wavelength, invalidating the condition for the local-density-functional approximation. It will therefore be interesting to explore the corrections to the exchange-correlation effects calculated in this paper by going beyond this approximation. We plan to calculate such corrections by using the nonlocal-density-functional scheme[20] recently developed by Langreth and Mehl, which gives improved results for the electronic structure in both atoms[20] and solids[21] when compared with the local-density-functional approximation.

IV. RESULTS

We first give results for a graded GaAs-Ga$_{1-x}$Al$_x$As heterojunction at absolute zero for several values of the net acceptor doping level N_{Ac} in the GaAs. The heterojunction parameters we used are

$$x = 0.3, \quad V_b = 0.3 \text{ eV} ,$$

$$m_c, m_b = 0.07, \quad 0.088 m_0 ,$$

$$\kappa_c, \kappa_b = 13.0, \quad 12.1 , \tag{13}$$

$$z_t, a_t = 0.4, \quad 0.1 \text{ nm} ,$$

where the last line corresponds to a 0.5-nm interface grading. Calculated energy levels for $N_{Ac} = 0.01$, 0.1, 0.3, 1.0, 3.0, and 10×10^{-15} cm^{-3}, corresponding to $N_d = 0.146$, 0.46, 0.80, 1.47, 2.56, and 4.69×10^{11} depletion charges per cm^2, respectively, are shown in Fig. 2. A representative conduction-band profile and the wave function for the lowest suband are shown in Fig. 1 for channel electron density $N_s = 5 \times 10^{11}$ cm^{-2} and $N_{Ac} = 3 \times 10^{14}$ cm^{-3}. The dashed curve in Fig. 1(b) gives the corresponding wave function calculated without the exchange-correlation potential energy V_{xc} in the effective potential energy of Eq. (1). These results show that the energy levels and energy-level differences are quite sensitive to the acceptor doping density in the GaAs even at densities in the 10^{14} cm^{-3} range.

Exchange-correlation effects in the energy levels can be seen from the differences between the solid and dashed curves in Fig. 2, calculated with and without these effects, respectively, for a net GaAs acceptor doping density $N_{Ac} = 0.3 \times 10^{15}$ cm^{-3}. The effects are also clearly seen in Fig. 3, which shows the value of N_s at which the second subband is just occupied versus the square root of the density of depletion charges N_d, calculated with and without the exchange-correlation potential. Inclusion of exchange and correlation increases the carrier concentration at which the Fermi level crosses into the second subband by about 2×10^{11} cm^{-2} at all values of N_d, both for AlAs fractions $x = 0.3$ and 0.4 (the larger value of x is needed for large values of N_d to keep the electrons in the first excited subband from leaking into the barrier to such an extent that the energy-level calculation is significantly affected). Occupation of the second subband opens a new

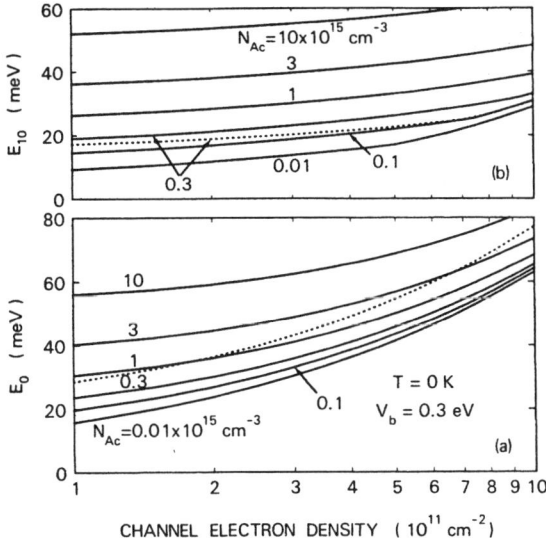

FIG. 2. Calculated values of (a) the energy E_0 of the bottom of the lowest subband and (b) the energy difference E_{10} between the first excited subband and the lowest subband at absolute zero vs channel electron density N_s for the six indicated values of net acceptor doping N_{Ac} in the GaAs. The corresponding values of N_d, the density of charges in the depletion layer, are 0.146, 0.46, 0.80, 1.47, 2.56, and 4.69×10^{11} cm^{-2}, respectively. The dashed curves give the values calculated without inclusion of exchange and correlation effects for $N_{Ac} = 0.3 \times 10^{15}$ cm^{-3}.

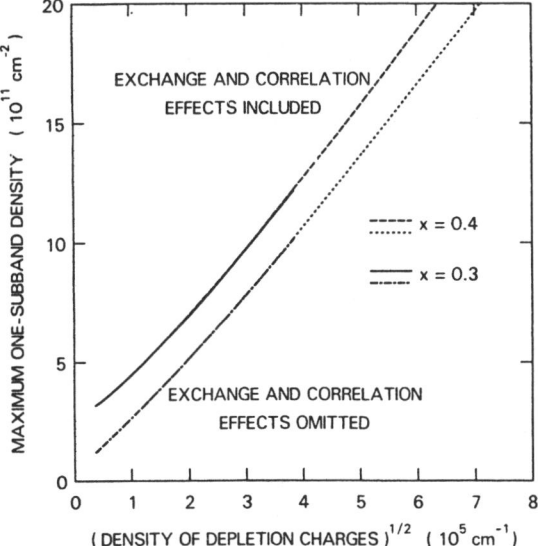

FIG. 3. Channel electron density at which the second subband is just occupied at $T = 0$ K vs the square root of the density of depletion charges N_d, calculated with (solid and dashed curve) and without (dot-dashed curve and dotted curve) inclusion of the exchange-correlation potential, for GaAs-Ga$_{1-x}$Al$_x$As heterojunctions with AlAs fractions x of 0.3 and 0.4. The results for values of N_d below about 10^{10} cm^{-2} approximate those expected for semi-insulating or n-type GaAs. Results for the largest values of N_d are calculated with $x = 0.4$ because the higher barrier reduces electron leakage through the barrier.

scattering channel and lowers the mobility.[22-24] Values of the channel density at the beginning of the occupation of the second subband have been calculated by Ando[1] with and without exchange for two values of N_d, and by Bastard,[25] using an extension of the Fang-Howard variational wave function in the Hartree approximation, for a range of values of N_d.

Figure 4 shows how the energy levels change with temperature for several values of N_s in a heterojunction with net acceptor doping $N_{Ac} = 3 \times 10^{14}$ cm^{-3} in the GaAs. As in the case of silicon inversion layers,[5,26] the subband energies increase with increasing temperature because increasing the temperature increases the occupation of higher subbands, thereby increasing the average electric field in the well.

Some representative values of the average distance

$$z_{av} = N_s^{-1} \int_{-\infty}^{\infty} z \sum N_i \zeta_i^2(z) dz \qquad (14)$$

of electrons from the interface are shown in Fig. 5. Fractional occupations of the subbands are shown in Fig. 6 and confirm Vinter's[3] finding that most of the channel electrons are in the lowest subband even at room temperature.

An important quantity in determining the transfer of electrons from donor impurities in the Ga$_{1-x}$Al$_x$As to the GaAs channel is the position of the Fermi level relative to the bottom of the GaAs conduction band at the interface. For this purpose we take the Fermi level as found in our calculation and refer it to the extrapolated conduction-band edge at the center of the graded interface layer, ignoring the image and exchange-correlation potentials in the determination of the band edge but not in the energy-level calculation. Values found in this way are given in Fig. 7 for $T = 0$, 77, and 300 K. Also shown is the value estimated for 0 K using the triangular-well approximation with the average electric field in the channel, given by $e(N_d + 0.5N_s)/\epsilon_0 \kappa_c$. This approximation, which ignores

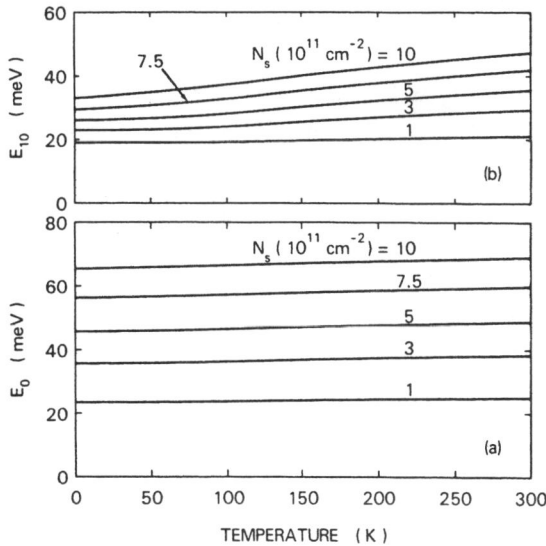

FIG. 4. Temperature dependence of (a) the energy E_0 of the bottom of the lowest subband and (b) the energy difference E_{10} between the two lowest subbands, for net acceptor doping $N_{Ac} = 3 \times 10^{14}$ cm^{-3} in the GaAs for several values of N_s, the electron density in the channel.

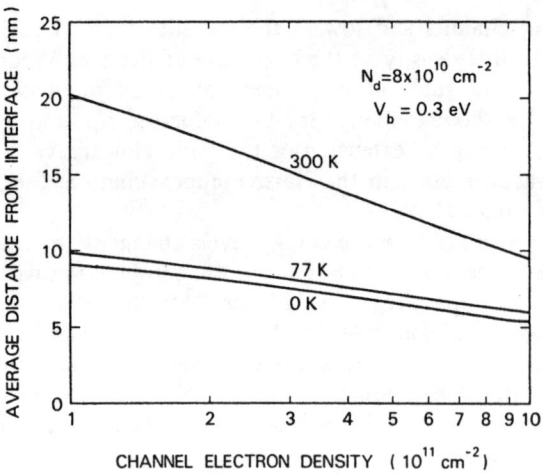

FIG. 5. Average electron distance from the interface, z_{av}, as a function of channel electron density N_s for 0, 77, and 300 K for the same acceptor doping as in Fig. 4.

image and exchange-correlation effects, is the one used for the next-to-last row of Table I.

The contribution of the image potential to the energy levels was tested by repeating the calculation with the dielectric constant of the Ga$_{1-x}$Al$_x$As taken to be equal to the dielectric constant of the GaAs. The change is so small that it would not be easily visible on a plot of the energies themselves. Therefore, we give in Table I the energy levels calculated with the standard parameters and also with the dielectric constants set equal to each other. As noted before, the image potential cannot be properly included in a calculation for a sharp interface because of the divergence that arises there. The image potential has therefore been omitted in most other heterojunction calculations. Table I shows that this leads to very small errors in the energies.

We can also test the sensitivity of the energy levels to a number of other parameters in the calculation. For these tests, the image potential is omitted. First, we look at the sensitivity of the results to the effective mass in the Ga$_{1-x}$Al$_x$As. Table I gives a few representative results for $m_b = 0.07 m_0$ and for the value $0.088 m_0$ used in most of the calculations. Note that the effect on energy-level

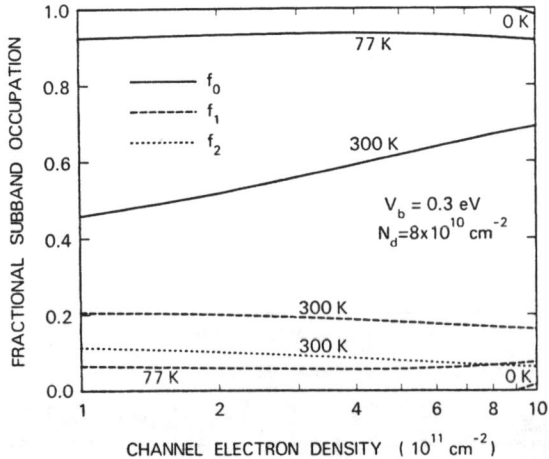

FIG. 6. Fractional occupation $f_i = N_i / N_s$ of the three lowest subbands vs N_s. Parameters are the same as for Fig. 5.

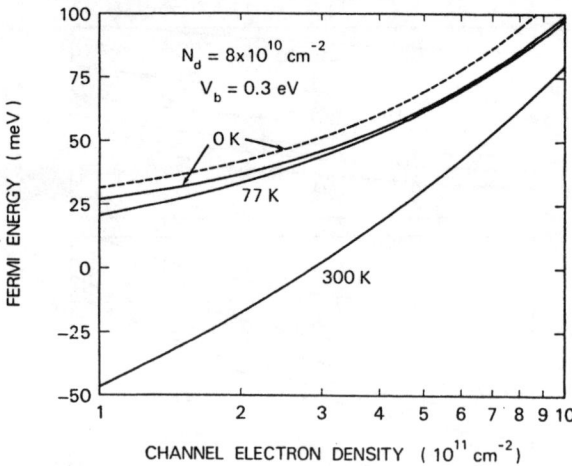

FIG. 7. Calculated Fermi energy relative to the extrapolated conduction-band edge at the interface vs N_s. Parameters are the same as for Fig. 5. The dashed curve gives the value found using the triangular-well approximation with the average field in the well, as in the next-to-last line of Table I.

differences is smaller than the effect on the individual levels, an example of a general result[26] that small changes in boundary conditions tend to shift levels together (note that here, in contrast to the case of silicon inversion layers, there is only one ladder of levels because there is no valley degeneracy at the bottom of the GaAs conduction band).

The dependence of the energy levels on the thickness $z_t + a_t$ of the interface transition layer is also given in Table I. The effects are small and again are even smaller for energy-level differences than for the individual energies.

Figure 8 shows the dependence of the energy levels on the fraction x of AlAs in the barrier. We take the barrier height in eV to be equal to x, and take the effective mass

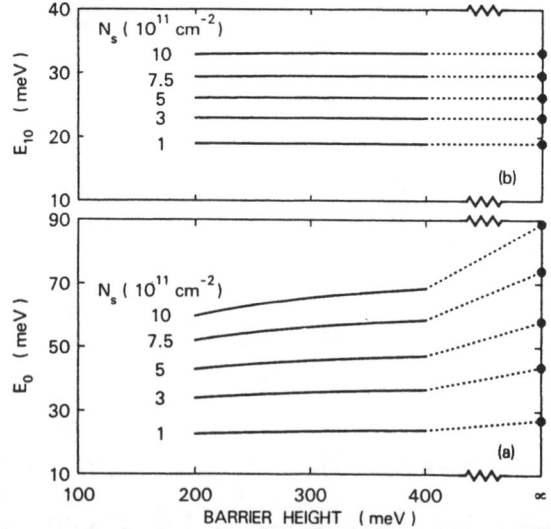

FIG. 8. Dependence of (a) energy levels and (b) energy-level differences on interface barrier height at $T = 0$ K for several values of channel electron density N_s. The barrier height (in eV) is taken equal to the AlAs fraction x. See the text for the effective-mass values in the Ga$_{1-x}$Al$_x$As. The points at the right edge of the figure are calculated for infinite barrier height. The GaAs acceptor doping is $N_{Ac} = 3 \times 10^{14}$ cm^{-3}.

TABLE I. Some calculated results to show their sensitivity to changes in parameters. The values shown are the first two energy levels, E_0 and E_1, and their difference E_{10}, all in meV, and the average distance z_0 (in nm) of electrons in the lowest subband from the GaAs-Ga$_{1-x}$Al$_x$As interface for electron densities N_s of 10^{11} and 5×10^{11} cm^{-2}. The "standard" parameter assumptions, used to calculate the values in the first row, are $T=0$ K, interface barrier height $V_b = 0.3$ eV, dielectric constants $\kappa_c = 13$ and $\kappa_b = 12.1$, effective masses $m_c = 0.07 m_0$ and $m_b = 0.088 m_0$, GaAs net acceptor doping $N_{Ac} = 3 \times 10^{14}$ cm^{-3} (for which there are $N_d = 0.8 \times 10^{11}$ charges per cm^2 in the depletion layer), and nominal interface transition layer thickness $z_t + a_t = 0.5$ nm. Exchange and correlation effects are included except where "no xc" is noted. The last three rows give approximate results, without image and exchange-correlation effects, calculated using the triangular-well approximation and the self-consistent variational approximation. More significant figures are given than are warranted by the precision of the calculation, to allow differences to be seen.

| | $N_s = 1 \times 10^{11}$ cm^{-2} | | | | $N_s = 5 \times 10^{11}$ cm^{-2} | | | |
Parameter changes	E_0	E_1	E_{10}	z_0	E_0	E_1	E_{10}	z_0
None	23.57	42.55	18.98	9.109	45.72	71.87	26.15	6.586
$\kappa_b = 13$ (no image)	23.37	42.40	19.03	9.071	45.40	71.62	26.22	6.565
$\kappa_b = 13$, $m_b = 0.07 m_0$	23.75	42.77	19.02	9.218	46.49	72.68	26.18	6.696
$\kappa_b = 13$, $z_t + a_t = 0$	23.34	42.37	19.03	9.062	45.33	71.56	26.22	6.557
$\kappa_b = 13$, $z_t + a_t = 0.15$	23.34	42.37	19.03	9.061	45.32	71.54	26.22	6.555
$\kappa_b = 13$, $z_t + a_t = 0.3$	23.35	42.38	19.03	9.064	45.35	71.57	26.22	6.558
$\kappa_b = 13$, $z_t + a_t = 0.5$	23.37	42.40	19.03	9.071	45.40	71.62	26.22	6.564
$\kappa_b = 13$, $z_t + a_t = 0.7$	23.40	42.43	19.03	9.083	45.49	71.71	26.22	6.575
$\kappa_b = 13$, $z_t + a_t = 0.9$	23.44	42.47	19.03	9.099	45.61	71.82	26.21	6.590
No xc	28.47	45.66	17.19	9.346	54.56	77.20	22.64	6.806
$\kappa_b = 13$, $V_b = \infty$, no xc	32.09	49.33	17.25	10.83	66.68	89.42	22.75	8.336
Triangular well[a]	35.0	61.5	26.5	9.30	76.2	134.1	57.9	6.29
Triangular well[b]	28.1	49.5	21.4	10.36	52.3	93.1	40.7	7.60
Variational[c]	33.1	50.3	17.3	11.56	67.9	93.2	25.3	8.88

[a]Calculated using Eq. (10b) of Ref. 26 with effective field $F_s = e(N_d + N_s)/\epsilon_0 \kappa_c$.
[b]Calculated as in the previous line, but with $F_s = e(N_d + 0.5 N_s)/\epsilon_0 \kappa_c$.
[c]Calculated using Eqs. (15)—(17) of Ref. 26.

in the barrier to be $m_b/m_0 = m_c/m_0 + 0.06x$. In the calculations for Fig. 8 we take the interface to be mathematically sharp and therefore set $\kappa_b = \kappa_c = 13$. The results are shown in Fig. 8 for values of x from 0.2 to 0.4. Results are also given for infinite barrier height, in which case the results do not depend on the choice of m_b.

In some experiments it has proved possible to change the channel charge density by applying a voltage to an electrode in the GaAs.[27,28] This backgate bias or substrate bias, widely used in silicon inversion layers, changes the electric field in the depletion layer. The effect has been considered theoretically by Vinter[2] and will not be considered further here because it depends in part on the doping in the Ga$_{1-x}$Al$_x$As.

The results presented here give a sample of calculated results for an ungated GaAs-Ga$_{1-x}$Al$_x$As heterojunction on p-type GaAs, ignoring the effect of doping in the Ga$_{1-x}$Al$_x$As on the levels, which was found by Bastard[29] to be small. A calculation for accumulation layers in n-type GaAs is in general more difficult, except at low temperatures if the donors in the GaAs can be frozen out. In that case the fixed charge is determined by the compensating acceptor ions, leading to a very low value of N_d. Such calculations have been carried out by Ando[1] and Bastard[25] for GaAs-Ga$_{1-x}$Al$_x$As heterojunctions and by Bastard[29] for InP-(In,Ga)As heterojunctions. A few results for

small values of N_d have been given in Fig. 3 above.

Self-consistent calculations for quantum wells along the same lines as those presented here have been carried out, and results, including a comparison with a tight-binding calculation by Schulman[30] for a graded interface, are planned to be prepared for publication.

V. DISCUSSION AND CONCLUSIONS

In this section we comment on some of the features exhibited by the results of this work, discuss some of the limitations of the calculations, and point out some directions for future work. Where comparison is possible, our results agree with those of Ando[1] who treated the interface somewhat differently and did not include image effects.

In many cases, calculations motivated primarily by device considerations have made rather severe approximations to estimate the energy levels. The most severe is to use the approximate energy for a triangular potential, in which the wave function is assumed to vanish at the interface and the field in the GaAs is taken to be a constant equal to its value at the interface. Alternatively, one can take the field to be the average of the interface field and the depletion field. Somewhat less severe is to use approximate results [given in Eqs. (16) and (17) of Ref. 26] that still assume the wave function to vanish at the interface, but treat the band bending in the GaAs variationally. Table I compares the results of these approximations with results of the present self-consistent calculations, and also shows the sensitivity of the calculated results to some of the parameters. The first row of Table I gives the "best" calculated results, as presented in Fig. 2, and the fourth row from the bottom gives numerical results for a case with no image or exchange-correlation effects and with an infinite barrier height, the case with which the analytical approximations should be compared. The simpler results of the last three lines agree only roughly with the more accurate results, and give some feeling for the magnitude of the errors. The variational approximation gives the best results overall, but the assumption of infinite barrier height leads to energies that are higher than those for realistic barrier heights. A modified variational approach that allows the trial wave function to enter the barrier has been described by Ando.[31]

Although the present results are thought to give a fairly good description of the energy-level structure of single heterojunctions, they nevertheless have limitations which could become relevant in some cases. The use of the effective-mass approximation to describe the system omits contributions of higher-lying energy bands and associated effects near the interface. Some errors are also introduced by neglect of conduction-band nonparabolicity and the change in density-of-states effective mass associated with wave-function matching at the interface. The limitations of our simple local-density-functional approximation for effects of electron-electron interactions have already been discussed in Sec. III. These approximations could introduce errors of order 10% or more in the energies, with smaller errors expected in the energy differences. In addi-

tion, approximations in the numerical integrations could also introduce errors, but these should not exceed a percent or so if there have not been any errors in carrying out the calculations. With our present knowledge of GaAs-Ga$_{1-x}$Al$_x$As heterojunctions, these errors may well be overshadowed by the uncertainty in our knowledge of material parameters such as impurity concentrations and interface barrier heights.

Perhaps the main conclusion of this work is that some of the calculated results—especially the energy-level differences—are relatively insensitive to the values of many of the parameters that characterize the interface. On the other hand, they are sensitive to the net density of acceptors in the GaAs, which may not be easy to measure at levels near or below 10^{15} cm^{-3}. If one can establish confidence in the calculated results, then it may be possible to use spectroscopic measurements of level splittings[28,32] or transport measurements of the threshold for occupying the second subband[23] to determine this doping density, which can affect other properties of experimental interest such as charge balance between the GaAs and the Ga$_{1-x}$Al$_x$As and mobility of electrons in the channel. Note, however, that optical-absorption measurements and some Raman scattering measurements do not give the energy-level differences directly, but require corrections whose magnitude must be calculated.[1,5]

ACKNOWLEDGMENTS

We are indebted to G. Bastard, A. Pinczuk, J. N. Schulman, and B. Vinter for sending us unpublished results that bear on the subject of this paper, and to E. E. Mendez and P. J. Price for comments on the manuscript.

*Present address.

[1]T. Ando, J. Phys. Soc. Jpn. **51**, 3893 (1982).

[2]B. Vinter, Surf. Sci. **142**, 452 (1984); Solid State Commun. **48**, 151 (1983).

[3]B. Vinter, Appl. Phys. Lett. **44**, 307 (1984).

[4]P. J. Price and F. Stern, Surf. Sci. **132**, 577 (1983).

[5]T. Ando, A. B. Fowler, and F. Stern, Rev. Mod. Phys. **54**, 437 (1982).

[6]F. Stern, Bull. Am. Phys. Soc. **28**, 447 (1983).

[7]F. Stern, Phys. Rev. B **17**, 5009 (1978).

[8]D. J. BenDaniel and C. B. Duke, Phys. Rev. **152**, 683 (1966).

[9]Nonparabolicity and density-of-states corrections associated with wave-function penetration into the barrier are ignored, except that we take an effective mass ($m_c = 0.07 m_0$) slightly larger than the value at the conduction-band minimum in bulk GaAs.

[10]The grading function $G(z)$ uses a different origin from the function $S(z)$ used in Ref. 7, so that the center of the graded region now lies at $z = 0$.

[11]P. Hohenberg and W. Kohn, Phys. Rev. **136**, B864 (1964); W. Kohn and L. J. Sham, *ibid.* **140**, A1133 (1965); L. J. Sham and W. Kohn, *ibid.* **145**, 561 (1966).

[12]A. R. Williams and U. von Barth, in *Theory of the Inhomogeneous Electron Gas,* edited by S. Lundqvist and N. H. March (Plenum, New York, 1983).

[13]T. Ando, Phys. Rev. B **13**, 3468 (1976).

[14]S. Das Sarma and B. Vinter, Phys. Rev. B **23**, 6832 (1981); **26**, 960 (1982); **28**, 3639 (1983).

[15]U. von Barth, in *The Electronic Structure of Complex Materials* (NATO Advanced Study Institute, Ghent, 1982).

[16]The electron density which enters here is only the density of the mobile carriers in the GaAs channel. In the effective-mass approximation, the effects of the bonding and core electrons are incorporated in the effective mass and dielectric contant.

[17]See Refs. 12 and 15 for extensive reference to work on the calculation of V_{xc}.

[18]L. Hedin and B. I. Lundqvist, J. Phys. C **4**, 2064 (1971).

[19]U. von Barth and L. Hedin, J. Phys. C **5**, 1629 (1972).

[20]D. C. Langreth and M. J. Mehl, Phys. Rev. Lett. **47**, 446 (1981).

[21]U. von Barth and R. Car (unpublished).

[22]S. Mori and T. Ando, Phys. Rev. B **19**, 6433 (1979); J. Phys. Soc. Jpn. **48**, 865 (1980).

[23]H. L. Störmer, A. C. Gossard, and W. Wiegmann, Solid State Commun. **41**, 707 (1981).

[24]N. T. Thang, G. Fishman, and B. Vinter, Surf. Sci. **142**, 266 (1984).

[25]G. Bastard, Surf. Sci. **142**, 284 (1984).

[26]F. Stern, Phys. Rev. B **5**, 4891 (1972).

[27]H. L. Störmer, A. C. Gossard, and W. Wiegmann, Appl. Phys. Lett. **39**, 493 (1981).

[28]Z. Schlesinger, J. C. M. Hwang, and S. J. Allen, Jr., Phys. Rev. Lett. **50**, 2098 (1983).

[29]G. Bastard, Appl. Phys. Lett. **43**, 591 (1983).

[30]J. N. Schulman (unpublished).

[31]T. Ando, J. Phys. Soc. Jpn. **51**, 3900 (1982).

[32]A. Pinczuk and J. M. Worlock, Physica **117B&118B**, 637 (1983).

Classical Versus Quantum Mechanical Calculation of the Electron Distribution at the n-AlGaAs/GaAs Heterointerface

JIRO YOSHIDA

Abstract—Four kinds of numerical models have been developed to investigate the quantum mechanical effect on charge control in HEMT's at room temperature. In spite of the two-dimensional nature of the channel electrons, a classical approach using Fermi statistics is shown to be able to predict the device performance with good accuracy, while the triangular-well approximation sometimes introduces serious errors.

I. INTRODUCTION

The accurate estimation of two-dimensional electron gas (2DEG) density at the modulation-doped AlGaAs–GaAs heterojunction interface is a fundamental problem in designing and analyzing the high electron mobility transistor (HEMT). The most accurate approach is to solve Schrödinger's equation and Poisson's equation self-consistently [1]–[3]. From a device modeling viewpoint, however, this approach highly complicates the numerical procedure, especially for modeling in two-dimensional space. In the numerical models for HEMT's reported hitherto, the subband structure of the 2DEG was incorporated only approximately [4] or was ignored [5]. The errors in the electron density introduced by approximate treatments of the 2DEG and the limitations of the models have not been clarified yet. Lee *et al.* showed the importance of the quantum effect on the 2DEG density [6]. However, their results were based on a triangular-well approximation. In the present paper, various approximate models for the 2DEG density are compared with a self-consistent quantum mechanical calculation to reveal how the quantum effect influences the HEMT performance.

II. MODELS

The following four kinds of one-dimensional numerical models were developed to investigate the 2DEG density dependence on the gate voltage:

1). Classical model in which the subband structure is neglected and Boltzmann statistics are used (3-D Boltzmann model). (The models using three-dimensional and two-dimensional density of states are referred to as the 3-D and 2-D models, respectively.)

2) Classical model in which the subband structure is neglected but Fermi statistics are used (3-D Fermi model).

3) Self-consistent quantum mechanical model including many-body exchange and correlation effects with the local density-functional approximation [1] (2-D exact model).

4) Approximate quantum mechanical model using the triangular-well approximation [7] (2-D triangular model).

In these models, the electrostatic potential V is numerically given by the solution of Poisson's equation

$$\frac{d}{dx}\left(\epsilon(x)\frac{dV(x)}{dx}\right) = q(N_D^+(x) - N_A^-(x) + p(x) - n(x)) \quad (1)$$

Manuscript received May 20, 1985; revised July 8, 1985.

The author is with Toshiba Research and Development Center, 1 Komukai Toshiba-cho, Saiwai-ku, Kawasaki, 210 Japan.

IEEE Log Number 8405701.

where ϵ is the position-dependent dielectric constant. Since the charge density on the right-hand side of (1) depends on the potential variation, an iterative calculation is required. The ionized donor density N_D^+ is expressed as

$$N_D^+ = N_D\left[1 - \frac{1}{1 + 1/2 \exp((E_D - E_f)/kT)}\right] \quad (2)$$

where E_D is the donor energy level and E_f is the Fermi level. The donor level in $Al_{0.3}Ga_{0.7}As$ is assumed to be 32.4 meV below the conduction band edge based on an experiment [8]. N_A^- represents the 100-percent ionized residual acceptors in the undoped GaAs layer. In the classical models (1 and 2), the electron concentrations are described as

$$n(x) = N_c(x) \exp((E_f - E_c(x))/kT) \quad (3)$$

and

$$n(x) = N_c(x) F_{1/2}((E_f - E_c(x))/kT) \quad (4)$$

respectively, where N_c is the three-dimensional density of states and $F_{1/2}$ is the Fermi–Dirac integral. In the present study, only the Γ valley is incorporated in the calculation and the nonparabolicity of the valley is neglected.

In model 3), Schrödinger's equation is numerically solved together with Poisson's equation to derive the electron distribution. Schrödinger's equation is

$$\left[-\frac{\hbar^2}{2m^*}\frac{d^2}{dx^2} + V_s(x)\right]\zeta_i(x) = E_i\zeta_i(x) \quad (5)$$

where m^* is the effective mass and V_s is the potential including the local exchange-correlation potential [1]. Strictly speaking, the envelope function ζ depends on the wave vector parallel to the heterointerface because of the effective mass difference between GaAs and AlGaAs [1]. This effect is neglected in the present study. The envelope function in AlGaAs and in GaAs are connected at the heterointerface to conserve the flux [1]. The electron concentration is given by

$$n(x) = \sum_i \int \frac{1}{2\pi^2} d\vec{k} \frac{1}{1 + \exp\left[\left(E_i + \frac{\hbar^2 k^2}{2m^*} - E_f\right)/kT\right]}\left|\zeta_i(x)\right|^2 \quad (6)$$

where E_i and ζ_i are the energy level and the envelope function of the ith subband, respectively. The integral is over the wave vector parallel to the heterointerface. From (6), $n(x)$ is given as

$$n(x) = \sum_i \frac{m^*kT}{\pi\hbar^2}\ln[1 + \exp((E_f - E_i)/kT)]|\zeta_i(x)|^2. \quad (7)$$

The effective mass of GaAs is used in calculating $n(x)$.

In model 4), the subband energy levels and the envelope functions are analytically given in terms of the interface electric field

Reprinted from *IEEE Trans. Electron Devices*, vol. 33, no. 1, pp. 154–156, Jan. 1986.

TABLE I
PHYSICAL PARAMETERS USED FOR CALCULATION

Parameter	GaAs	$Al_{0.3}Ga_{0.7}As$
Band gap (eV)	1.425	1.814
Electron effective mass	0.067	0.0919
Conduction band density of states at 300K (cm^{-3})	4.350×10^{17}	6.988×10^{17}
Valence band density of states at 300K (cm^{-3})	9.313×10^{18}	9.985×10^{18}
Relative dielectric constant	12.9	12.0

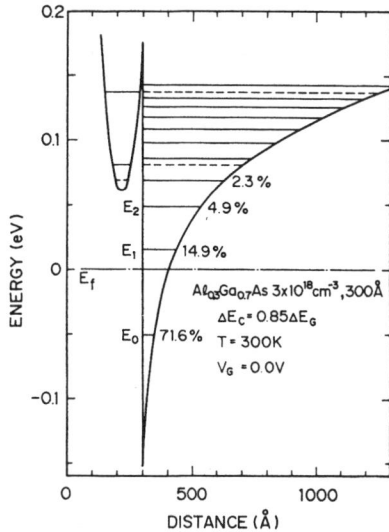

Fig. 1. Self-consistent conduction band-edge profile and subband energy levels at $V_G = 0.0$ V. The numerals in the figure indicate the fractional occupations of the subbands.

E, i.e.

$$E_i \sim \left(\frac{\hbar^2}{2m^*}\right)^{1/3} \left[\frac{3\pi q\mathrm{E}}{2}\left(i + \frac{3}{4}\right)\right]^{2/3} \quad (8)$$

$$\zeta_i(x) = Ai\left(\frac{2m^*q\mathrm{E}}{\hbar^2}\left(x - \frac{E_i}{q\mathrm{E}}\right)\right) \quad (9)$$

where Ai denotes the airy function. An empirical parameter f between zero and one is often used to replace E to $f\mathrm{E}$. In the present study, $f = 0.5$ and 1.0 are employed.

The gate bias dependence of the 2DEG density is calculated under the following boundary conditions. The conduction band edge at the gate electrode on the AlGaAs layer is fixed to the value corresponding to the Schottky-barrier height (1 eV) and the applied voltage. In the GaAs layer, the conduction band edge at a point sufficiently apart from the channel is fixed to its thermal equilibrium value. The channel is assumed to be grounded so that the Fermi level is fixed to be zero throughout the calculation. In model 3), the envelope function at both boundaries is fixed to be zero. In model 4), the electron concentration in the AlGaAs layer is obtained by solving Schrödinger's equation within this layer with the boundary condition that the envelope function at the heterointerface is zero.

Various physical parameters utilized in the present work are listed in Table I.

III. CALCULATED RESULTS

Calculations were carried out for a device with 3×10^{18} cm^{-3} doped and 300-Å-thick $Al_{0.3}Ga_{0.7}As$ gate layer operating at room temperature. No spacer layer was introduced between the gate layer and the undoped GaAs channel layer. The residual acceptor density in the GaAs layer was assumed to be 1×10^{14} cm^{-3}. The conduction band edge discontinuity at the heterointerface was assumed to be 85 percent of the bandgap discontinuity [9] in the present study.

Fig. 1 shows the conduction band edge profile and the subband energy levels calculated by using model 3). The numerals shown in the figure indicate the fractional occupation of the subbands. It is confirmed that more than 85 percent of the channel electrons are in the lowest and the first excited subbands, indicating the formation of the 2DEG system at the heterointerface. Subbands are also formed in the AlGaAs layer, as pointed out by Vinter [2]. The corresponding distribution profile of electrons in the conduction band is shown in Fig. 2. Classical results obtained by using model 2) are also shown in the figure (dashed line). It is noteworthy that the classically calculated profile coincides fairly well with the quantum mechanical results except in the region within 50 Å around the heterojunction.

Fig. 3 shows the channel electron concentration dependence on the applied gate voltage calculated by using various models. The differences among the calculated curves are enhanced with increase in the electron concentration. As expected, classical models (1) and

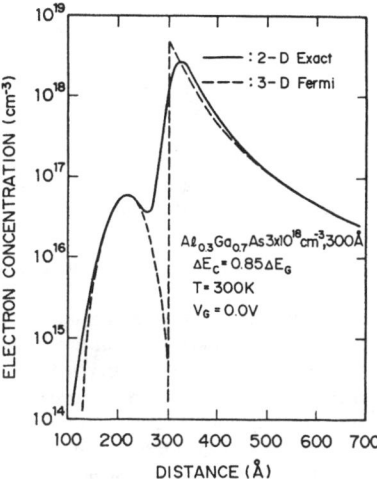

Fig. 2. Self-consistent quantum mechanical and classical electron profiles in the vicinity of the heterojunction.

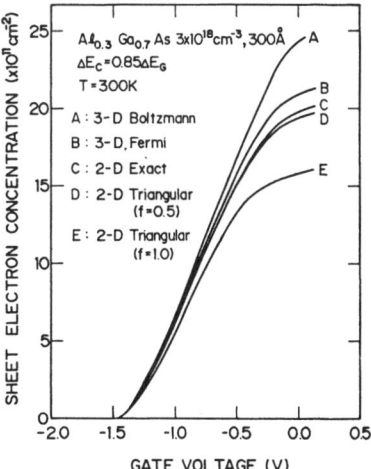

Fig. 3. Sheet electron concentration in the channel versus gate voltage calculated by using various models.

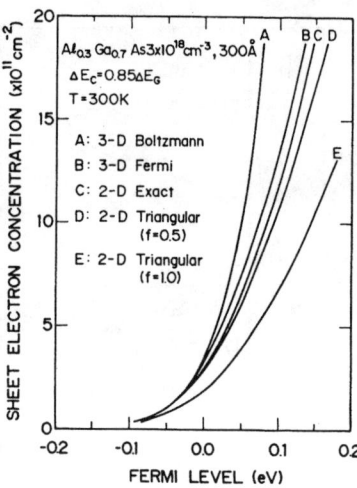

Fig. 4. Sheet electron concentration in the channel versus Fermi level relative to the conduction band edge at the interface calculated by using various models.

2)) overestimate the electron concentration. However, when the electron concentration does not exceed 1.5×10^{12} cm^{-2}, even a classical model using Boltzmann statistics can evaluate the electron concentration within 10-percent error. By adopting Fermi statistics, the accuracy of the classical model is remarkably improved. The difference between curves B and C is less than 5 percent even for the electron concentration of 2×10^{12} cm^{-2}. A similar calculation performed for the 77-K operation confirms that the difference is not enhanced at low temperature. On the other hand, the triangular-well model with $f = 1.0$ seriously underestimates the electron concentration. The accuracy of the triangular-well approximation varies widely depending on the choce of the f value. In the heterojunction system dealt with in this paper, $f = 0.5$ gives good agreement with the self-consistent quantum mechanical calculation.

Fig. 4 shows the sheet electron concentration versus Fermi energy relative to the conduction band edge at the interface. Corresponding to the results shown in Fig. 3, curves B, C, and D coincide fairly well with each other. This is in contrast with the results reported by Lee *et al.* [6] in which a classical calculation was compared with a triangular-well approximation. The reason for the discrepancy may be attributed to the choice of the f-value.

IV. Conclusion

Four kinds of numerical models have been developed to investigate the quantum mechanical effect on the channel electron concentration in HEMT's at room temperature. By comparing the calculated results, the classical approach using Fermi statistics was found to give a reasonably good description of the channel electron concentration dependence on the gate voltage. The classical approach using Boltzmann statistics has been shown not to introduce serious errors when the channel electron concentration is less than 1.5×10^{12} cm^{-2} at 300 K. The triangular-well approximation sometimes gives worse results depending on the choice of the empirical parameter. These results indicate that the classical description of the channel electrons can be used in device modeling for HEMT's to predict the macroscopic *I–V* and *C–V* characteristics.

References

[1] T. Ando, "Self-consistent results for a GaAs/Al$_x$Ga$_{1-x}$As heterojunction. I. Subband structure and light-scattering spectra,"*J. Phys. Soc. Japan.*, vol. 51, pp. 3893–3899, 1982.

[2] B. Vinter, "Subbands and charge control in a two-dimensional electron gas field-effect transistor," *Appl. Phys. Lett.*, vol. 44, pp. 307–309, 1984.

[3] F. Stern and S. D. Sarma, "Electron energy levels in GaAs-Ga$_{1-x}$Al$_x$As heterojunctions," *Phys. Rev.*, vol. B30, pp. 840–848, 1984.

[4] D. Widiger, K. Hess, and J. J. Coleman, "Two-dimensional numerical analysis of the high electron mobility transistor," *IEEE Electron Device Lett.*, vol. EDL-5, pp. 266–269, 1984.

[5] J. Yoshida and M. Kurata, "Analysis of high electron mobility transistors based on a two-dimensional numerical model," *IEEE Electron Device Lett.*, vol. EDL-5, pp. 508–510, 1984.

[6] K. Lee, M. S. Shur, T. J. Drummond, S. L. Su, W. G. Lyons, R. Fischer, and H. Morkoç, "Design and fabrication of high transconductance modulation-doped (Al, Ga)As/GaAs FET's," *J. Vac. Sci. Technol. B*, vol. 1, pp. 186–189, 1983.

[7] F. Stern, "Self-consistent results for n-type Si inversion layers," *Phys. Rev.*, vol. B5, pp. 4891–4899, 1972.

[8] M. O. Watanabe, K. Morizuka, M. Mashita, Y. Ashizawa and Y. Zohta, "Donor levels in Si-doped AlGaAs grown by MBE," *Japan. J. Appl. Phys.*, vol. 23, pp. L103–L105, 1984.

[9] R. Dingle, W. Wiegmann and C. H. Henry, "Quantum states of confined carriers in very thin Al$_x$Ga$_{1-x}$As-GaAs-Al$_x$Ga$_{1-x}$As heterostructures," *Phys. Rev. Lett.*, vol. 33, pp. 827–830, 1974.

Nonlinear Charge Control in AlGaAs/GaAs Modulation-Doped FET's

WILLIAM A. HUGHES AND CHRISTOPHER M. SNOWDEN, MEMBER, IEEE

Abstract—A new model for nonlinear charge control in normally on modulation-doped field-effect transistors (MODFET's) is proposed. It is shown that conventional charge control models are insufficient to describe MODFET's with large negative pinchoff voltages, and that the depletion approximation is inaccurate in circumstances where the layer dimensions become of the order of a Debye length. The new model is based upon a one-dimensional numerical solution of Poisson's equation and the drift-diffusion equation. It also takes into account the shift in the 2DEG position with gate bias, and parallel conduction in the doped AlGaAs layer. The effect of nonlinear charge control on MODFET transconductance is considered by combining the new model with a two-dimensional analytic representation of the MODFET.

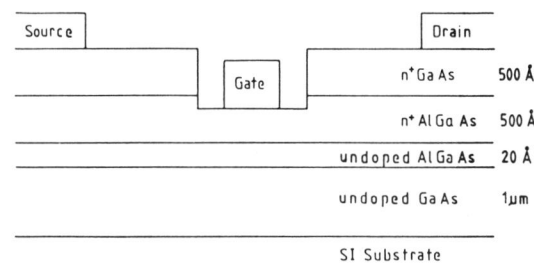

Fig. 1. Cross-sectional view of a typical AlGaAs/GaAs MODFET.

I. INTRODUCTION

IN RECENT YEARS considerable progress has been made in the fabrication of modulation-doped field-effect transistors (MODFET's). The first discreet devices were announced in 1980 [1], [2] and have subsequently demonstrated a performance superior to GaAs MESFET's in both analog [3] and digital [4] integrated circuits. They have been used to fabricate large-scale circuits such as a 4K SRAM that demonstrated an access time of 2 ns at 77 K [5]. MODFET's provide very low noise operation for small signal amplifiers; for example, Mishra *et al.* recently reported a 0.25-μm gate-length MODFET with 1.3-dB noise figure and 12-dB associated gain at 18 GHz [3].

The MODFET performance is due principally to the technique of modulation doping in which the doping is introduced only into the wide bandgap AlGaAs layer of the device. The channel electrons are physically separated from their parent donor ions forming a two-dimensional electron gas at the GaAs side of the AlGaAs/GaAs heterojunction. The material structure of a typical MODFET is shown in Fig. 1.

Significant improvements in device performance have increased interest in using these devices and stimulated an interest in the modeling of such structures [6]-[14]. The first kind of models for MODFET's were analytical and one-dimensional [6], [7]. Two-dimensional numerical models have been developed in order to describe more accurately a number of phenomena that could not be ana-

lyzed with simpler analytical models, especially when the device geometry is reduced [8]-[11]. More recently attention has been directed toward Monte Carlo simulation of these devices [12]-[14].

Analytic models are helpful in that they provide physical insight into the operation of the device and are generally faster to evaluate than numerical models. The requirement for good analytic models is that the simplifying assumptions necessary to enable solution of the problem are sufficiently accurate. The charge control theory, which relates carrier concentrations to electrostatic potentials, for the MODFET is central to the construction of realistic device models as it affects device transconductance, gate–source capacitance, and gate–drain capacitance, which are the critical equivalent circuit elements determining device performance. Most theories developed to date use a simple linear charge control model derived in [6] and developed further in [7] and [15]. We will show that MODFET's with sufficiently large negative pinchoff voltages cannot be described by this model. In addition, it is shown in this paper that the depletion approximation is inaccurate when applied to MODFET's where layer thicknesses can be of the order of a Debye length.

In this paper a new charge control model is proposed based on the solution of Poisson's equation and the drift-diffusion equation. These are solved numerically in one dimension to derive the two-dimensional electron gas (2DEG) concentration, and the parallel AlGaAs current component as a function of gate bias. The model is extended to take into account the position of the 2DEG as it moves away from the heterojunction into the undoped GaAs near pinchoff. This heterojunction-2DEG separation has conventionally been assumed constant [7]. Two-dimensional effects are considered by developing a three-region analytic model that takes into account the effect of the depletion region beyond the edge of the gate.

Manuscript received November 5, 1986.
W. A. Hughes is with GEC Research Limited, Hirst Research Centre, East Lane, Wembley, Middlesex HA9, 7PP, United Kingdom.
C. M. Snowden is with the Microwave Solid State Group, University of Leeds, Leeds LS2, 9JT, United Kingdom.
IEEE Log Number 8715170.

Reprinted from *IEEE Trans. Electron Devices*, vol. 34, no. 8, pp. 1617-1625, Aug. 1987.

II. EXISTING CHARGE CONTROL THEORIES

The conventional charge control equation, based on the Poisson equation

$$n_s = \frac{\epsilon_A}{qd_A}(V_{GS} - V_p) \quad (1)$$

was first derived in detail in [6]. Here n_s is the 2DEG sheet carrier concentration, ϵ_A and d_A are the AlGaAs permittivity and thickness, respectively, q is the electronic charge, and V_{GS} is the gate–source potential, which is normally negative for a depletion-mode device. The device pinchoff voltage V_p is given by [6]

$$V_p = V_{bi} - \Delta E_c - \frac{qN_d}{2\epsilon_A}(d_A - e)^2 + E_F \quad (2)$$

where V_{bi} is the Schottky-barrier built-in potential, ΔE_c is the conduction band discontinuity, N_d is the doping density in the AlGaAs layer, e is the undoped AlGaAs spacer layer thickness, and E_F is the Fermi-level position.

This theory was further developed in [7] and [15] to take into account the position of the 2DEG that does not reside exactly at the GaAs–AlGaAs interface. A correction factor Δd of approximately 80 Å was proposed resulting in the corrected equation for charge control

$$n_s = \frac{\epsilon_A}{q(d_A + \Delta d)}(V_{GS} - V_p). \quad (3)$$

The saturated current flowing from source to drain in the device is given approximately by

$$I_{ds} = qn_s W_g v_s \quad (4)$$

where W_g is the gate width and v_s is the average saturated transit velocity of an electron traveling from source to drain. It should be noted that diffusion is neglected in this equation. From (3) and (4), the intrinsic transconductance g_{m0} is given by

$$g_{m0} = \frac{\partial I}{\partial V_{GS}} = \frac{\epsilon_A W_g v_s}{(d_A + \Delta d)} \quad (5)$$

and the gate–source capacitance is given by

$$C_{gs} = q\frac{\partial n_s}{\partial V_{gs}} \times A = \frac{A\epsilon_A}{(d_A + \Delta d)} \quad (6)$$

where A is the total effective gate area.

Equations (5) and (6) suggest that g_{m0} and C_{gs} are constant with varying gate potential. Fig. 2(a) shows measured transconductance as a function of V_{GS} for a 0.5-μm gate-length MODFET. It is clear from this figure that g_m depends strongly on gate potential. Hence a new charge control theory must be developed if such behavior is to be understood and incorporated as part of a full device model.

The transconductance curve of Fig. 2(a) may be divided into two regions: region A where the g_m is increasing with increasing negative gate bias, and region B where the g_m decreases. The behavior in region B can be partly explained by using a two-dimensional version of (3) given

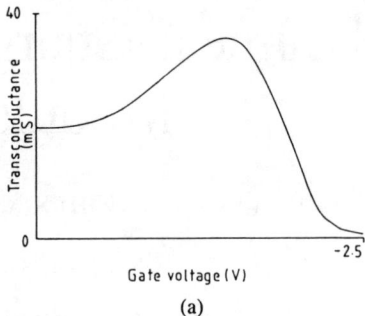

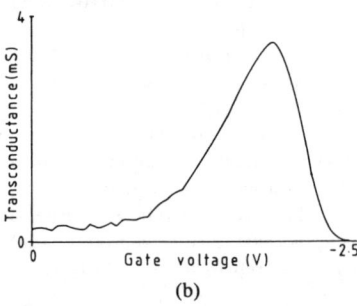

Fig. 2. Measured transconductance (g_m) characteristic for drain potentials of (a) 3 V and (b) 50 mV. In both cases a characteristic peak in the g_m is obtained. The curve may be divided into two regions: region A where the g_m increases with decreasing gate bias, and region B where the g_m decreases.

in [6]

$$n_s = \frac{\epsilon_A}{q(d_A + \Delta d)}(V_{GS} - V_p - V_c(x, V_{GS})) \quad (7)$$

where $V_c(x, V_{GS})$ is the channel potential as a function of x, the distance along the channel from the source end of the gate. This channel potential is a nonlinear function of the gate potential, and also of the position along the channel at which velocity saturation occurs. This will be considered in more detail in Section V.

The increase in g_m with V_{GS} in region A has been attributed ([16] and [17]) to parallel conduction in the doped AlGaAs. This is best explained with reference to Fig. 3 that represents the band diagram of the MODFET under operating conditions with a small negative gate bias. In this case, the depletion regions d_g and d_j, associated with the gate and heterojunction potentials, respectively, are insufficient to deplete the AlGaAs, resulting in a parallel conduction path of doped AlGaAs of thickness

$$d_p = d_A - e - d_g - d_j. \quad (8)$$

Equation (3) was derived assuming that the doped AlGaAs was fully depleted. Hence this equation for charge control can no longer be used in this case. However, we may presume that the gate is now ineffective in its control of the 2DEG, resulting in a lowering of the g_m at low V_{GS}.

We will now consider the effect of parallel conduction on g_m using the conventional depletion approximation for the two depletion regions d_g and d_j. Neglecting diffusion, the parallel current flowing in the AlGaAs region is then given by a modified form of (4) using v_{SA} for the saturated velocity of carriers in AlGaAs. To make further progress

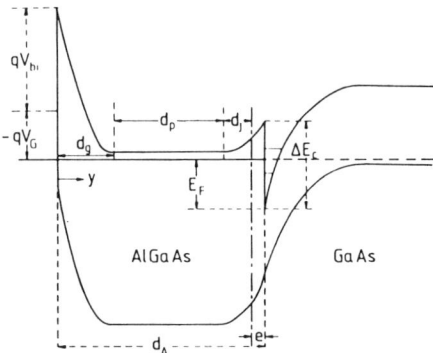

Fig. 3. Band diagram of a normally-on MODFET operating with a small negative gate bias. The gate and junction depletion regions do not overlap so that the gate is ineffective in its control of the 2DEG. A parallel conduction path d_p is formed in the AlGaAs.

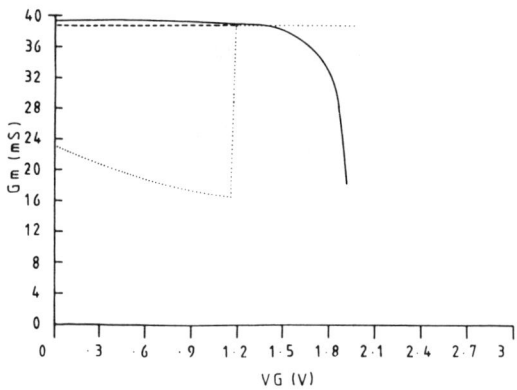

Fig. 4. Simplistic representations for the g_m characteristic. The dashed curve is the constant g_m approximation, the solid curve is the two-dimensional representation, and the dotted line takes into account parallel AlGaAs conduction using the depletion approximation.

using the depletion approximation, we must now assume that the sheet carrier concentration in the 2DEG remains constant with the gate potential until the depletion regions overlap. This follows because the band bending at the junction will not be affected by the gate potential until depletion region overlap occurs. The g_m for $d_p > 0$ is then given by

$$g_m = \frac{\partial I_p}{\partial V_G} = \epsilon_A W_g v_{SA} \sqrt{\frac{qN_d}{2\epsilon_A(V_{bi} - V_{GS})}}. \quad (9)$$

Once the two depletion regions overlap, we may use (5) to determine the g_m, with the parallel contribution from the doped AlGaAs reduced to zero as $d_p = 0$. From (9), therefore, we obtain a decrease in g_m with increasing negative gate bias until $d_p = 0$ when the g_m acquires the value given in (5).

In order to model region B correctly, the analysis must allow for the variation in the 2DEG position with gate bias. In this case, Δd becomes a function of the gate–source potential, increasing with increasingly negative V_{GS}. Equation (7), therefore, must be used in place of (3), which no longer leads to an analytic expression for g_m and C_{gs}. The effect of $\Delta d(V_{GS})$ on g_m and C_{gs} will be considered in Section IV.

The effect of the considerations given above is shown in Fig. 4. Clearly, the $g_m - V_g$ characteristic is not well represented by any of these curves. In the analysis given so far substrate current has not been taken into account. However, the $g_m - V_G$ characteristic of Fig. 2(b) was measured at 50-mV drain bias at which one would expect the substrate current contribution to be small. A characteristic peak in the curve is still obtained that suggests that the behavior is due to some other effect. The behavior of the g_m characteristic in region A can be understood better from consideration of the Debye lengths appropriate to the device. For the case shown in Fig. 2, at zero gate bias $d_p = 80$ Å. The debye lengths associated with d_g and d_j are approximately 40 Å. Hence, the effect of gate bias on the sheet carrier concentration may be significant even when the depletion regions, as calculated from the depletion approximation, do not overlap. Hence the de-

pletion approximation is inaccurate when applied to these devices.

III. NONLINEAR CHARGE CONTROL

The charge control theory presented here is based upon the solution of Poisson's equation, which, neglecting deep levels, is given by

$$-\frac{\partial^2 V(y)}{\partial y^2} = \frac{q}{\epsilon_A}(N_d - n(y)) \quad (10)$$

and the drift-diffusion equation

$$J = q\mu n(y) E(y) + qD \frac{\partial n(y)}{\partial y} \quad (11)$$

in one dimension. Energy transport effects that require a modified current density equation have been considered in recent publications [11]. However, the simplified drift-diffusion equation provides a useful approximation [11]. Fig. 3 shows the distance y, which is zero at the Schottky gate and is equal to d_A at the heterojunction. Here, $V(y)$ is the potential, $E(y)$ is the electric field, μ is electron mobility, D is the diffusion constant, and $n(y)$ is the free electron concentration.

In equilibrium, the current density J is zero and (11) becomes

$$n\frac{\partial V}{\partial y} = -\frac{kT}{q}\frac{\partial n}{\partial y} \quad (12)$$

where we have used the Einstein relationship

$$D = \frac{kT}{q}\mu \quad (13)$$

and

$$E = -\frac{\partial V}{\partial y}. \quad (14)$$

We now integrate (12) with respect to y to give the potential at y

$$V(y) = V(0) - \frac{kT}{q}\ln(n(y)/n(0)) \quad (15)$$

where $V(0)$ and $n(0)$ are the potential and free carrier concentration at the Schottky gate. We may now obtain an equation in $n(y)$

$$n(y) = n(0) \exp\left(\frac{-q}{kT}(V(y) - V(0))\right). \quad (16)$$

Substituting in (10) gives

$$-\frac{\partial^2 V(y)}{\partial y^2} + \frac{qn(0)}{\epsilon_A} \exp\left(\frac{-q}{kT}(V(y) - V(0))\right) = \frac{qN_d}{\epsilon_A}. \quad (17)$$

We must solve this equation for $V(y)$, subject to the boundary conditions discussed below, in order to obtain n_s as a function of V_G.

The free carrier concentration

$$n(0) = \frac{N_d}{\exp\left(\frac{q}{kT}V(0)\right)} \quad (18)$$

is obtained from (16) by substituting $n(\infty) = N_d$. The boundary condition for $V(0)$ is obtained simply from the Schottky-gate parameters

$$V(0) = V_{bi} - V_{GS} - V_c(x) \quad (19)$$

where we have taken into account parasitic source resistance effects by including $V_c(x)$, the channel potential. To solve the Schottky barrier, we use $V(\infty) = 0$. To solve the MODFET case, we must use

$$qV(d_A) = -E_F + \Delta E_c \quad (20)$$

where ΔE_c is the conduction band discontinuity (see Fig. 3) and E_F is the Fermi-level position, which is itself a function of the gate potential (see later). Note that we have taken the Fermi-level position as the zero of potential.

It has been shown [6] that by a simple solution of Poisson's equation at the GaAs side of the heterojunction we obtain

$$qn_s = \epsilon_G E_G = \epsilon_A E_A \quad (21)$$

where ϵ_G is the permittivity in GaAs and E_G and E_A are the electric fields at the GaAs and AlGaAs sides of the heterojunction, respectively. Hence we obtain

$$n_s = \frac{\epsilon_A}{q}\left(\frac{\partial V}{\partial y}\right)\Big| y = d_A \quad (22)$$

where $(\partial V/\partial y) | y = d_A$ is the slope of the solution of (17) at the AlGaAs side of the heterojunction.

Lastly we must relate n_s to E_F such that the boundary condition of (20) may be determined. It has been shown (see Section IV) that the sheet carrier concentration in the 2DEG may be approximated by

$$n_s = D_s \frac{kT}{q} \sum_n \log\left(1 + \exp\left(\frac{q}{kT}(E_F - E_n)\right)\right),$$
$$n = 0, 1, 2 \ldots \quad (23)$$

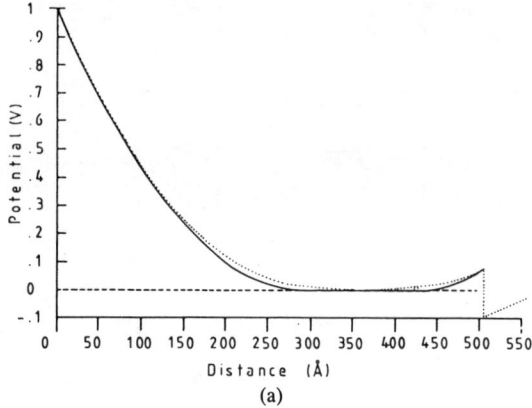

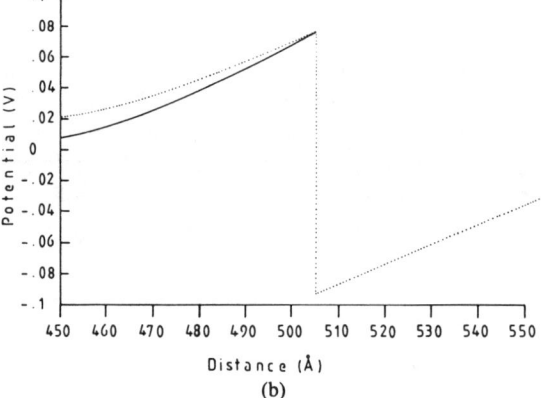

Fig. 5. (a) Potential distribution in a MODFET at zero gate bias. The solid curve is the depletion approximation and the dotted line is the result of nonlinear charge control. Interaction between the two depletion regions is obtained for the numerical solution indicating that the gate does control the 2DEG even when the depletion regions (as calculated from the depletion approximation) do not overlap. (b) Expanded view of the heterojunction showing the different potential slopes of the two solutions. The 2DEG sheet carrier concentration is determined by the slope of these curves.

where D_s is the two-dimensional density of states and E_n are the energies of the quantized energy states in the 2DEG.

We may now solve (17) subject to the boundary conditions of (19) and (20) by an iterative method. At the end of each iteration, n_s is calculated from (22) and used to obtain the source–drain current from (4), and the Fermi-level position from (23). The boundary condition values may then be updated via (19) and (20) and the process repeated. In reality, we also allow for the change in Δd (see Section IV) and for two-dimensional effects (see Section V).

Fig. 5 shows the result of the above calculations in terms of the solution $V(y)$ at zero gate bias compared with the depletion approximation solution (see Section II). Clearly, in the depletion approximation at low gate bias, the gate potential has no effect on the 2DEG that remains at its equilibrium value n_{s0} determined by the band bending shown and from (23). This is because the two depletion regions, calculated from the depletion approximation, do not overlap. However the nonlinear charge control solution demonstrates that the gate potential does, in fact,

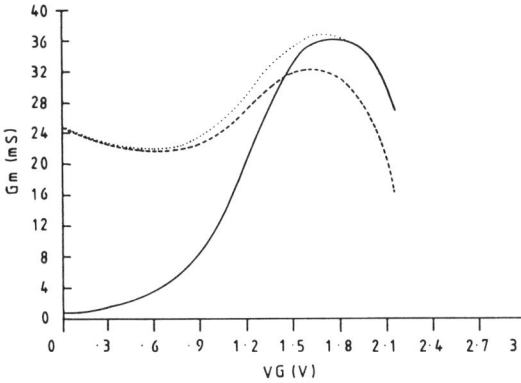

Fig. 6. Results of the model in terms of transconductance characteristics: the solid line is the g_m due to the 2DEG alone, the dotted line includes the effects of parallel conduction, and the dashed line includes the variation in the 2DEG offset. Two-dimensional effects are not included.

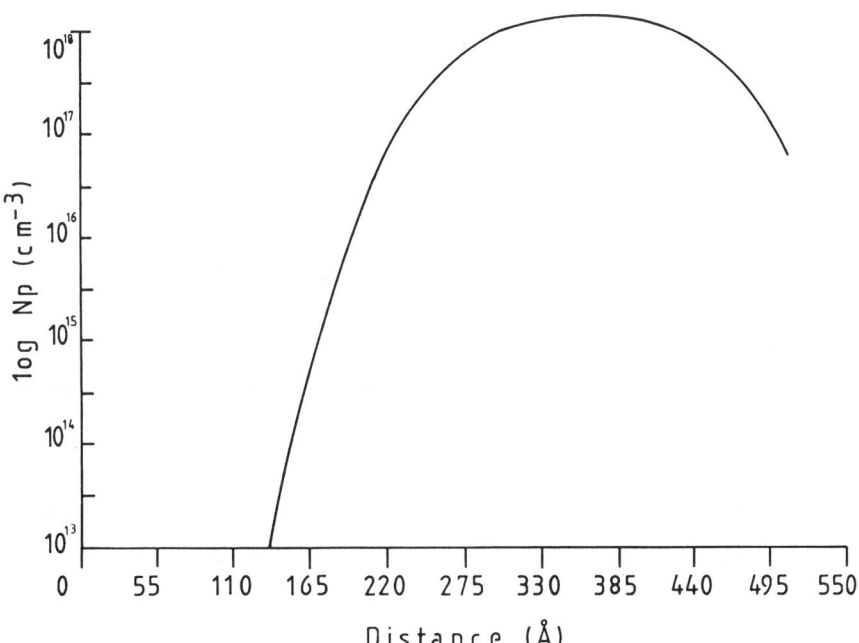

Fig. 7. Log plot of free carrier concentration in AlGaAs as a function of distance from the edge of the gate for a doping density of 1.5×10^{18} cm^{-3} at zero gate bias. The depleted regions due to the gate and heterojunction are clearly seen. This distribution may be integrated to give the sheet parallel charge that gives rise to parallel conduction.

modulate the 2DEG concentration, as the slope of the V-y curve has deviated from the equilibrium (depletion approximation) case. It should be noted here that it is the slope of the V-y curve that determines n_s via (22) and so the solution presented here deviates significantly from the depletion approximation case even though the absolute change in potential is small.

By solving for n_s as a function of V_G, the $I - V_G$ and $g_m - V_G$ characteristics at 3 V may be obtained for the theory described so far. Fig. 6 shows the $g_m - V_G$ characteristic due to this calculation alone (solid line). It is evident that this solution is similar in form to the measured curve of Fig. 2. In order to complete the model, parallel conduction, the shift in the 2DEG position, and two-dimensional effects must be taken into account.

The free carrier concentration in the AlGaAs may be determined from (16) using the solution $V(y)$. The free carrier concentration for the case shown in Fig. 5 is given in Fig. 7. This may be integrated to give the total parallel sheet carrier concentration n_{sp} which is then used to give the parallel current component via a modified form of (4). This parallel component may only be determined approximately in practice as the value of the AlGaAs saturation velocity V_{sa} has not been widely reported. A value of 5×10^6 cm/s was used in this work (see Section VI). The total transconductance is then given by the sum of the components due to the 2DEG current and the parallel AlGaAs current

$$g_m = \frac{\partial I}{\partial V_G} + \frac{\partial I_p}{\partial V_G}. \qquad (24)$$

The effect of this parallel component of the g_m is shown in Fig. 6 as a dotted curve.

163

IV. 2DEG Offset Calculation

The 2DEG offset has been traditionally taken as 80 Å after [7], independent of gate potential. However, as the gate potential becomes more negative, the slope of the conduction band on the GaAs of the heterojunction decreases. This results in an increase of the 2DEG offset that leads to a decrease in g_m according to (5).

For a given value of n_s, calculated according to Section III, the slope of the 2DEG quantum well (see Fig. 8) may be obtained from (21) and (22)

$$E_G = -\left(\frac{\partial V}{\partial y}\right) = \frac{qn_s}{\epsilon_G} \qquad (25)$$

and the Fermi-level position is calculated from (23). The positions of the quantized energy levels, shown in Fig. 8, may be calculated according to [18]

$$E_n = \left(\frac{\hbar^2}{2Ml}\right)^{1/3}\left(\frac{3}{2}\pi qE_G\right)^{2/3}\left(n + \frac{3}{4}\right)^{2/3},$$
$$n = 0, 1, 2, 3 \ldots \qquad (26)$$

where Ml is the effective mass in the 2DEG (assumed to be equal to the GaAs effective mass). Hence we may calculate the set of spacings d_n shown in Fig. 8 from

$$d_n = \frac{E_n}{qE_G}. \qquad (27)$$

The centroid of the sheet charge occupying the nth energy level is approximately $\frac{2}{3}d_n$ from the heterojunction [18], and the occupancy of each energy level n_{sn} is given by [6]

$$n_{sn} = \frac{D_s kT}{q}\log\left(1 + \exp\left(\frac{q}{kT}(E_F - E_n)\right)\right). \qquad (28)$$

We now calculate the 2DEG offset from the weighted average of all the energy levels

$$\Delta d = \frac{\frac{2}{3}\sum_n n_{sn}d_n}{n_s}, \quad n = 0, 1, 2, 3, \ldots . \qquad (29)$$

Clearly

$$n_s = \sum_n n_{sn}, \quad n = 0, 1, 2, 3, \ldots . \qquad (30)$$

In the above derivation we have used the approximation that the slope of the conduction band in the quantum well is constant (being equal to E_G). This is clearly not the case for higher energy levels where the curvature of the conduction band away from a straight line becomes significant. However, in practice the occupancy of energy levels above $n = 2$ is negligible such that their contribution to the summation in (29) is small.

Fig. 9 shows the 2DEG offset as a function of gate potential calculated using this model. The 80-Å value for Δd is indicated for reference. Near pinchoff, indicated as a dashed line, Δd becomes asymptotic as expected since the 2DEG quantum well disappears. Clearly the effect of

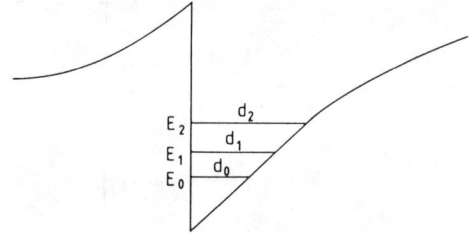

Fig. 8. Schematic representation of the heterojunction quantum well showing energy levels E_n and spacing d_n. The 2DEG offset is calculated from a weighted average of d_n.

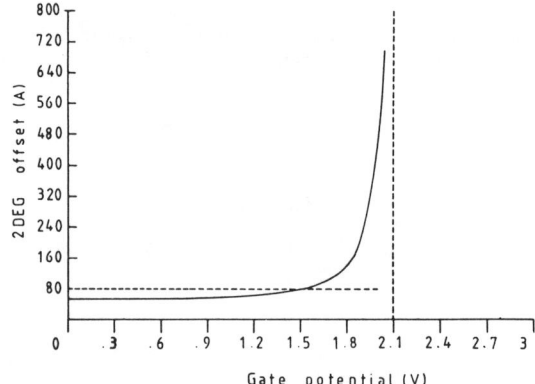

Fig. 9. Calculated values of the 2DEG offset as a function of gate potential. The solution becomes asymptotic near pinchoff (dashed line). The constant value of 80 Å is included for reference.

this variation in Δd on g_m is significant near pinchoff, as shown in the dashed curve of Fig. 6.

V. Two-Dimensional Considerations—An Analytic Device Model

To determine accurately the effect of nonlinear charge control on device characteristics, two-dimensional effects must be accounted for. In particular, the boundary condition of (19) must be modified to take into account the channel potential under operating conditions, which will, in reality, depend upon the drain potential and the gate length.

A three-region analytic model has been developed for this purpose, similar to a model developed for the inverted MODFET [19], but adapted to describe the device of interest here. Fig. 10 shows the three operating regions of the device: region 1 in which the velocity of carriers depends linearly upon the electric field, region 2 in which the carriers travel at their saturated drift velocity, and region 3 beyond the drain edge of the gate in which the electrons continue to travel at their saturated velocity. This third region of the device was first proposed by Hill et al. [19] and is important for accurate analytic models of the MODFET. Also shown in Fig. 10 are schematic electric field and potential distributions.

To achieve an analytic solution, we must employ an analytic expression for charge control. As described in Section III, it is impossible to represent nonlinear charge control by a simple analytic formula. We will, therefore, employ a corrected form of the conventional control equa-

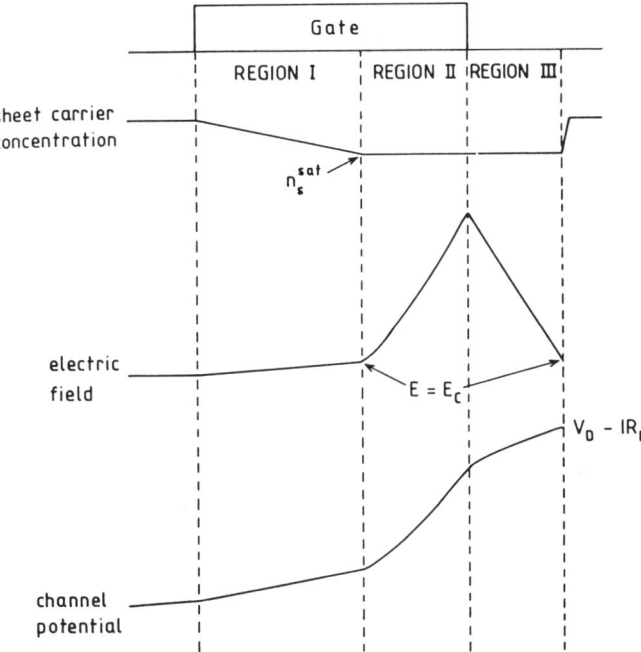

Fig. 10. Three regions used in the analytic model: region 1—the linear region, region 2—in saturation, and region 3—the extended depletion region. Schematic views of electric field and potential are included.

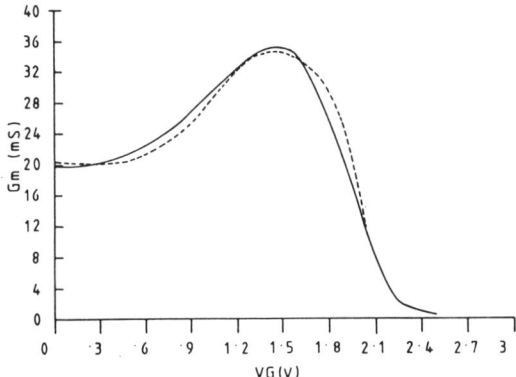

Fig. 11. Measured (solid curve) and calculated (dotted curve) transconductance characteristics for a 0.5-μm gate-length normally on MODFET. Deviations from the model are observed near pinchoff, possibly due to substrate current and gate leakage effects.

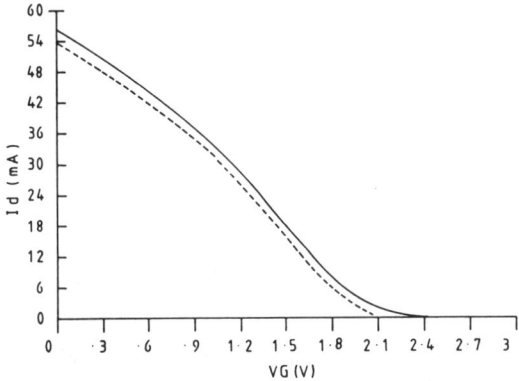

Fig. 12. Drain current against gate potential for a measured (solid curve) and modeled (dotted curve) MODFET. The current is slightly underestimated, possibly due to substrate current and gate leakage which also soften the pinchoff behavior.

tion given in (7) as follows:

$$n_s(x, V_{GS}) = \frac{\epsilon_A}{q(d_A + \Delta d(V_{GS}))} \left(V_{GS_{eff}} - V_p - V_c(x, V_{GS}) \right). \quad (31)$$

Here we have used the variable 2DEG offset, $\Delta d(V_{GS})$ as described in Section IV and an effective gate–source potential $V_{GS_{eff}}$ calculated from the nonlinear charge control model of Section III. This equation is solved iteratively in conjunction with the equations of Sections III and IV. First we solve the nonlinear charge control model to give us $n_s(x, V_{GS})$. We then use (31) to give $V_{GS_{eff}}$ where $V_c(x, V_{GS})$ is obtained from the model described later in this section. Equation (31) is then used, with the calculated value of $\Delta d(V_{GS})$ from Section IV and $V_{GS_{eff}}$, in the analytic model to determine the new value of $V_c(x, V_{GS})$. This process is repeated until convergence is obtained.

The analytic solution is obtained from models of each of the three regions of the device as described in the Appendix.

VI. RESULTS AND DISCUSSION

The results of the complete model are given in Figs. 11 and 12 (dotted curves) compared with measured characteristics (solid curves) of a 0.5-μm gate-length MODFET. This device had a gate width of 150 μm, a spacer layer thickness of 20 Å and an AlGaAs doping density of 1.5 $\times 10^{18}$ cm^{-3}. Values of 2×10^7 cm/s and 5×10^6 cm/s were assumed for the 2DEG and AlGaAs velocities respectively in line with published results [20]–[23]. Accurate values for these velocities have not been deter-

mined and as such they may be regarded as fitting parameters. The thickness of the doped AlGaAs was determined from the observed pinchoff voltage of the device that was calculated from an extrapolation of the $I_D - V_G$ curve of Fig. 12. A barrier height of 1 eV was assumed [23]. Measured values of the source and drain resistances of 0.9 $\Omega \cdot$mm were used. A good fit with the measured characteristics is obtained in both cases, although some discrepancy is observed near device pinchoff. This is believed to be due to gate leakage and substrate current which both become more significant in this region where large electric fields are present at the drain edge of the gate. This may also explain the slightly low current obtained from the model in Fig. 12.

The peak in the $g_m - V_G$ curve predicted by the model agrees well with the observed peak in the measured device characteristic. The position of this peak is important in determining the gain and noise properties of the device as a function of gate bias and, in particular, affects the minimum noise figure bias conditions.

VII. CONCLUSION

We have developed a nonlinear charge control model that includes the effects of parallel conduction in the AlGaAs layer, the shift in the position of the 2DEG away from the heterojunction near pinchoff and two-dimensional effects including the extended depletion region beyond the edge of the Schottky gate. Numerical methods are used to determine the sheet carrier concentration in the 2DEG as a function of gate bias and channel potential. It has been shown that parallel AlGaAs conduction is important for normally on MODFET's at large gate voltages, and that the depletion approximation is insufficient for describing these devices accurately. Good agreement between the model and measured device characteristics was obtained.

APPENDIX

The analytic model is based on models of each of the three main device regions:

Region 1

In this region the velocity of the channel electrons is given by

$$v = \mu E(x) \tag{32}$$

and (4) may be modified to give

$$I = \mu \frac{W_g \epsilon_A}{(d_A + \Delta d(V_{GS}))} \cdot (V_{GS_{eff}} - V_p - V_c(x, V_{GS})) \frac{\partial V_c}{\partial x} \tag{33}$$

where we have used (31) and replaced $E(x)$ with $\partial V_c / \partial x$. Equation (33) can be integrated with respect to x [6] to give an expression for $V_c(x, V_{GS})$

$$V_c(x, V_{GS}) = V_{GS_{eff}} - V_p$$
$$- \sqrt{(V_{GS_{eff}} - V_p - V_c(0))^2 - \frac{2(d_A + \Delta d(V_{GS}))Ix}{\mu W_g \epsilon_A}}. \tag{34}$$

The electric field is obtained by differentiating (34)

$$E(x) = \frac{(d_A + \Delta d(V_{GS}))I}{\mu W_g \epsilon_A} \left((V_{GS_{eff}} - V_p - V_c(0))^2 - \frac{2(d_A + \Delta d(V_{GS}))Ix}{\mu W_g \epsilon_A} \right)^{-1/2}. \tag{35}$$

The boundary of region 1, $x = X_s$, occurs when the electric field becomes equal to the critical electric field E_c. Hence, from (35)

$$X_s = \frac{\mu W_g \epsilon_A}{2(d_A + \Delta d(V_{GS}))I} \left[(V_{GS_{eff}} - V_p - V_c(0))^2 - \left(\frac{(d_A + \Delta d(V_{GS}))I}{\mu W_g \epsilon_A E_c} \right)^2 \right]. \tag{36}$$

This value of X_s can then be substituted into (34) to give V_1 the potential dropped in region 1.

Region 2

In region 2, since $E(x) > E_c$, the electron velocity is constant at its saturated value v_s. Applying the current continuity condition, then (4) implies that $n_s(x)$ must be constant in this region at a value given by

$$n_s^{sat} = \frac{\epsilon_A}{q(d_A + \Delta d(V_{GS}))} (V_{GS_{eff}} - V_p - V_1) \tag{37}$$

from (31).

Poisson's equation in this region becomes

$$\frac{\partial Ex}{\partial x} + \frac{\partial Ey}{\partial y} = \frac{q}{d_A \epsilon_A} (N_d(d_A - e) - n_s^{sat}). \tag{38}$$

It can be shown that by assuming the sheet positive charge density $(N_d(d_A - e) - n_s^{sat})$ contributes to the x and y directed field equation (38) reduces to

$$\frac{\partial Ex}{\partial x} = \frac{q}{d_A \epsilon_A} (N_d(d_A - e) - n_s^{sat}) - 2 \frac{(V_c(x) - V_G)}{(d_A - e)^2} \tag{39}$$

where $V_c(x)$ is the channel potential in region 2. Equation (39) may be reduced to a second order differential equation in E_x that may be readily solved to give the electric field

$$E(x) = E_c \cosh \frac{\sqrt{2}}{(d_A - e)^2} (x - X_s) \tag{40}$$

which, when integrated gives the potential drop in region 2

$$V_2 = E_c \frac{(d_A - e)}{\sqrt{2}} \sinh \frac{\sqrt{2}}{(d_A - e)} (L - X_s) \tag{41}$$

where L is the gate length.

Region 3

Since $E(x) > E_c$ in region 3, both n_s and the velocity remain constant at n_s^{sat} and v_s, respectively. Because most of the electric field in this region is x directed, we make the simplifying assumption [19]

$$\frac{\partial Ey}{\partial v} = 0 \tag{42}$$

and (38) becomes

$$\frac{\partial Ex}{\partial x} = \frac{q}{d_A \epsilon_A} (N_d(d_A - e) - n_s^{sat} + n_{ss}) \tag{43}$$

where n_{ss} is the surface state density (in this case, the value of n_{ss} has been taken as zero).

This is readily solved for the electric field

$$E(x) = E(L) - \frac{q}{d_A \epsilon_A} (N_d(d_A - e) - n_s^{sat} + n_{ss}) (x - L) \tag{44}$$

where $E(L)$ is the electric field at the drain edge of the gate, given by (40). The boundary of region 3, X_d, occurs when $E(x) = E_c$. Hence, from (44) we obtain

$$X_d = \frac{d_A \epsilon_A (E(L) - E_c)}{q(N_d(d_A - e) - n_s^{sat} + n_{ss})} + L. \quad (45)$$

The potential dropped in region 3 is given by integrating equation (44) from $x = L$ to $x = X_d$

$$V_3 = E(L)(X_d - L) - \frac{q}{2d_A \epsilon_A} (N_d(d_A - e)$$

$$- n_s^{sat} + n_{ss})(X_d - L)^2. \quad (46)$$

The solution of the above equations is obtained by comparing the total potential dropped across the three regions $(V_1 + V_2 + V_3)$ with the required potential at $x = X_d$ of $(V_D - IR_D)$. The current is adjusted until we obtain the condition

$$V_1 + V_2 + V_3 = V_D - IR_D. \quad (47)$$

The g_m value at a given V_{GS} and V_{DS} is determined by calculating the current $I(V_{GS}, V_{DS})$ at $(V_{GS} - \Delta V)$ and $(V_{GS} + \Delta V)$ and applying the equation

$$g_m = \frac{I(V_{GS} + \Delta V, V_{DS}) - I(V_{GS} - \Delta V, V_{DS})}{2\Delta V}. \quad (48)$$

ACKNOWLEDGMENT

The authors would like to thank J. A. Barnard and P. H. Ladbrooke for invaluable technical discussions, and C. Lau, M. Jain, and B. Alves for help with device measurements.

REFERENCES

[1] T. Mimura, S. Hiyamizu, T. Fujii, and K. Nanbu, "A new field-effect transistor with selectivity doped GaAs/n-Al$_x$Ga$_{1-x}$As heterojunctions," *Japan. J. Appl. Phys.*, vol. 19, pp. L225–L227, May 1980.

[2] D. Delagebeaudeuf, P. Delescluse, P. Etienne, M. Laviron, J. Chaplart, and N. T. Linh, "Two-dimensional electron gas MESFET structure," *Electron. Lett.*, vol. 16, pp. 667–668, 1980.

[3] U. K. Mishra, S. C. Palmateer, P. C. Chao, P. M. Smith, and J. C. M. Hwang, "Microwave performance of 0.25 μm gate length high electron mobility transistors," *IEEE Electron Device Lett.*, vol. EDL-6, no. 3, pp. 142–145, Mar. 1985.

[4] S. Pei, N. Shah, R. Hendel, C. Tu, and R. Dingle, "Ultra high speed integrated circuits with selectively doped heterostructure transistors," in *Tech. Dig. IEEE GaAs Integrated Circuit Symp.* (Boston, MA), Oct. 1984.

[5] S. Kuroda, T. Mimura, M. Suzuki, N. Kobayashi, K. Nishiuchi, A. Shibatomi, and M. Abe, "New device structure for 4Kb HEMT SRAM," in *Tech. Dig. IEEE Gallium Arsenide Integrated Circuit Symp.* (Boston, MA), Oct. 1984.

[6] D. Delagebeaudeuf and N. T. Linh, "Metal-(n) AlGaAs–GaAs two-dimensional electron gas FET," *IEEE Trans. Electron Devices*, vol. ED-29, pp. 955–960, June 1982.

[7] T. J. Drummond, H. Morkoç, K. Lee, and M. Shur, "Model for modulation doped field effect transistor," *IEEE Electron Device Lett.*, vol. EDL-3, pp. 338–341, Nov. 1982.

[8] J. Yoshida and M. Kurata, "Analysis of high electron mobility transistors based on a two-dimensional numerical model," *IEEE Electron Device Lett.*, vol. EDL-5, pp. 508–510, Dec. 1984.

[9] D. Widiger, I. C. Kizilyalli, K. Hess, and J. J. Coleman, "Two-dimensional transient simulation of an idealized high electron mobility transistor," *IEEE Trans. Electron Devices*, vol. ED-32, pp. 1092–1102, June 1985.

[10] J. Y. F. Tang, "Two-dimensional simulation of MODFET and GaAs gate heterojunction FET's," *IEEE Trans. Electron Devices*, vol. ED-32, pp. 1817–1823, Sept. 1985.

[11] D. Loret, R. Baets, C. M. Snowden, and W. A. Hughes, "Two-dimensional numerical models for the high electron mobility transistor," in *Proc. Int. Conf. Simulation Semiconductor Devices Processes* (Swansea, U.K.), July 1986.

[12] M. Al Mudares, "Computer simulation studies of microwave MESFET's," Ph.D. dissertation, Dept. Physics, Univ. Surrey, England.

[13] U. Radaioly and D. K. Ferry, "MODFET ensemble Monte Carlo model including the quasi-two-dimensional electron gas," *IEEE Trans. Electron Devices*, vol. ED-33, no. 5, pp. 677–681, May 1986.

[14] M. Al Mudares, D. Loret, and C. M. Snowden, "HEMT characteristics based on a 2D quantum Monte Carlo model," to be published.

[15] K. Lee, M. S. Shur, T. J. Drummond, and H. Morkoç, "Current-voltage and capacitance-voltage characteristics of modulation-doped field-effect transistors," *IEEE Trans. Electron Devices*, vol. ED-30, no. 3, pp. 207–212, Mar. 1983.

[16] J. P. Harrang, R. J. Higgins, R. K. Goodall, R. H. Wallis, P. R. Jay, and P. Delescluse, "Charge control and geometric magnetotransconductance of a gated AlGaAs/GaAs heterojunction transistor," *J. Appl. Phys.*, vol. 58, no. 11, pp. 4431–4437, Dec. 1985.

[17] K. Lee, M. S. Shur, T. J. Drummond, and H. Morkoç, "Parasitic MESFET in (Al,Ga)As/GaAs modulation doped FET's and MODFET characterization," *IEEE Trans. Electron Devices*, vol. ED-31, no. 1, pp. 29–35, Jan 1984.

[18] T. Ando, A. B. Fowler, and F. Stern, "Electronic properties of two-dimensional systems," *Rev. Mod. Phys.*, vol. 54, no. 2, pp. 437–672, Apr. 1982.

[19] A. J. Hill, P. H. Ladbrooke, S. Ransome, and D. Westwood, "Large signal characteristics of inverted high electron mobility transistor," presented at the 12th Int. Symp. Gallium Arsenide Related Compounds, Karuizawa, Japan, Sept. 1985.

[20] P. J. Tasker, L. H. Camnitz, L. M. Lunardi, H. Lee, P. A. Maki, P. Enquist, and L. F. Eastman, "Progress toward performance limits for MODFET and HBT devices," in *Proc. GaAs Related Compounds* (Biarritz, France), 1984.

[21] T. J. Drummond, W. Kopp, H. Morkoç, and M. Keever, "Transport in modulation-doped structures (Al$_x$Ga$_{1-x}$As/GaAs) and correlations with Monte Carlo calculations (GaAs)," *Appl. Phys. Lett.*, vol. 41, no. 3. pp. 277–279, Aug. 1982.

[22] K. Inoue, S. Hiyamizu, M. Inayama, and Y. Inuishi, "Analysis of 2D electron transport at a GaAs/AlGaAs interface," in *Proc. 14th Solid State Dev.* (Tokyo), 1982; see also *Japan. J. Appl. Phys.*, suppl. 22-1, pp. 357–363, 1983.

[23] M. Harano, Y. Takanashi, and T. Sugeta, "Current-voltage characteristics of an AlGaAs/GaAs heterostructure FET for high gate voltages," *IEEE Electron Device Lett.*, vol. EDL-5, no. 11, pp. 496–499, Nov. 1984.

Pulse-Doped AlGaAs/InGaAs Pseudomorphic MODFET's

NICK MOLL, MEMBER, IEEE, MARK R. HUESCHEN, MEMBER, IEEE, AND ALICE FISCHER-COLBRIE

Abstract—We have made MODFET's based on heterojunctions consisting of AlGaAs and pseudomorphic InGaAs, grown on GaAs substrates. The large conduction-band discontinuity (about 0.46 eV for 25-percent In and Al concentrations) leads to a 2D electron density as high as 2.3×10^{12} cm^{-2}, with electron mobilities of 7000 and 16000 cm^2/V · s at 300 and 77 K. Such a high electron density in combination with reasonable transport properties leads to MODFET's with exceptional characteristics. Devices with 0.15–0.25-μm gate length have room-temperature drain currents as high as 600 mA/mm and room-temperature transconductance as high as 500 mS/mm. The f_T is as high as 98 GHz, as determined by 20-dB/decade extrapolation of microwave data taken to 25 GHz. A comparison of the effect of bias on the total delay through standard and pseudomorphic MODFET's suggests that the excellent microwave performance exhibited by the pseudomorphic device arises from a reduction in parasitic and drain delays, and not from a higher electron velocity under the gate.

I. INTRODUCTION

FOR several years, material systems that are lattice-matched to InP and include InGaAs as the channel material have attracted attention because of their perceived advantages for modulation-doped field-effect transistors (MODFET's), particularly because of the higher expected electron velocity in InGaAs [1], [2]. The results on those devices have never quite measured up to expectations, and in fact have never even approached the best results on AlGaAs/GaAs until quite recently [3], [4]. Attention has also been focused on pseudomorphic InGaAs layers grown on GaAs substrates. The first MODFET's made on layers of this type use GaAs as the high bandgap material, and seem to have better properties than their InP-matched predecessors, but still fall short of AlGaAs/GaAs MODFET's in most respects [5]–[7]. Ketterson *et al.* and Henderson *et al.*, however, have shown that, by using AlGaAs as the high bandgap material in a pseudomorphic InGaAs MODFET [8]–[10], they can obtain results that rival the best AlGaAs/GaAs MODFET's in terms of gain or frequency response [11].

We have recently improved on these results [12]. In this paper, we report and analyze, in detail, our own results for AlGaAs/InGaAs pseudomorphic MODFET's. We believe that these devices, which have extrinsic transcon-

Manuscript received October 19, 1987; revised February 10, 1988.
The authors are with Hewlett-Packard Laboratories, Palo Alto, CA 94304.
IEEE Log Number 8820833.

ductance in excess of 500 mS/mm, maximum drain current density in excess of 600 mA/mm, and an f_T up to 98 GHz, are among the best made and represent a substantial advance over AlGaAs/GaAs MODFET's. We hope to show, through a comparison of our results on the two types of MODFET's, that this advance arises in part as a consequence of the large sheet density of electrons in these devices.

II. EPITAXIAL STRUCTURE AND GROWTH

Fig. 1 shows the basic epitaxial structure, which was grown by molecular-beam epitaxy in a Varian Gen II on an undoped (100) GaAs LEC substrate. In order of growth, we deposit a 1-μm undoped GaAs buffer layer, a 75–100-Å undoped In$_{0.25}$Ga$_{0.75}$As layer, and a 380-Å Al$_{0.25}$Ga$_{0.75}$As layer that contains a 100-Å 5×10^{18} cm^{-3} n-type pulse spaced 30 Å from the pseudomorphic channel. The final layer is 300 Å of undoped GaAs, to protect the AlGaAs from air exposure. The first part of the growth is done at a standard substrate temperature of 620°C; just before the growth of the InGaAs layer and for subsequent layers this temperature is reduced to 535°C. The growing surface is arsenic stabilized throughout the growth. Further details and discussion of the growth procedure will be published elsewhere [13].

This growth procedure typically results in films that have sheet electron densities of 2.3×10^{12} cm^{-2}, with Hall mobilities in the range of 6000–7000 cm^2/V · s at room temperature, and 13000–17000 cm^2/V · s at 77 K. Because of the particular pulse-doped structures that we chose to use, we expect that all the electrons are contained in the 2D electron gas and no electrons are present as AlGaAs conduction-band electrons. This idea is based on results of a numerical model based on a self-consistent quantum mechanical calculation of the occupation of the seven lowest sub-bands in the structure. Those results show that, for the structure described above, less than 1 percent of the total electron concentration resides in the AlGaAs at room temperature. They also show that, in structures that do contain electrons in the AlGaAs, more than 99 percent of those electrons freeze out at 77 K. Hall-effect measurements of the sheet electron density in our films at room temperature and 77 K vary at most 4 percent and typically much less, confirming the idea that AlGaAs electrons are essentially absent from our layers. As a consequence, we can vary growth and structural parameters

Reprinted from *IEEE Trans. Electron Devices*, vol. 35, no. 7, pp. 879–886, July 1988.

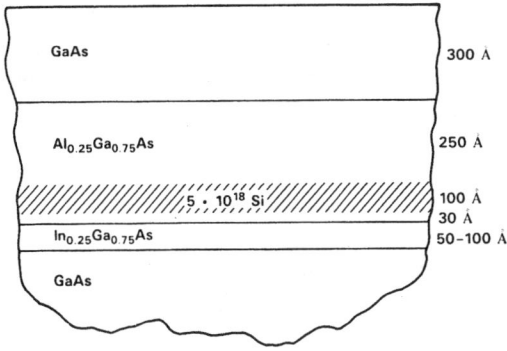

Fig. 1. Typical layer structure for pseudomorphic MODFET's. We studied a range of In concentrations, from 0 to 0.25 mole fraction.

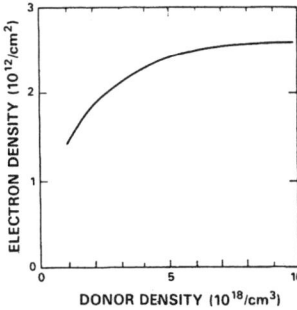

Fig. 2. Theoretical variation (based on a self-consistent 7-sub-band quantum mechanical calculation) of the 2D electron density as a function of donor density. The assumed conduction-band discontinuity is 0.5 eV. The doping pulse is spaced 30 Å from the heterojunction, and is varied in width to maintain the sheet dose at 5×10^{12} cm^{-2}.

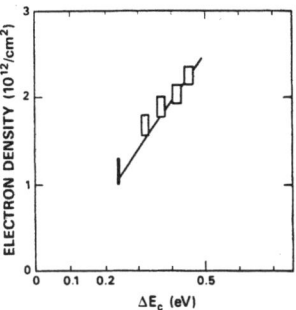

Fig. 3. Theoretical and experimental variation of the 2D electron density as a function of conduction-band discontinuity. The doping pulse is 5×10^{18} cm^{-3}, 100 Å wide, spaced 30 Å from the heterojunction. The In mole fraction for the experimental points varies from 0 to 0.25.

other than the silicon dose, e.g., substrate temperature, well width, or spacer width, and directly measure the effect on the room-temperature 2D electron gas without being confounded by changes in the behavior of AlGaAs electrons. This property was in fact used to determine the best growth conditions, well width, and In mole fraction, within the range explored, without going through full device fabrication.

The guiding principle that we followed in designing the epitaxial layers was to maximize the sheet electron concentration in the 2D gas without significantly degrading the mobility. There are several reasons for emphasizing this parameter in the development of microwave MODFET's. First, power is a key issue for the applications that interest us. While pulse doping leads to MODFET's with excellent breakdown voltage [11], equal or superior to that of comparable MESFET's, the maximum drain current generally falls short of that seen in MESFET's. Given the higher electron velocity that is characteristic of MODFET's, it should be possible to significantly reverse this situation; doing so requires the formation of a 2D electron gas with the maximum sheet concentration that can be successfully pinched off. Second, the higher sheet conductance associated with a high electron concentration results in lower source and channel resistances. This helps to minimize effects due to parasitic capacitances, which are particularly important in devices with short gates. The single heterojunction AlGaAs/GaAs MODFET is not particularly susceptible to improvements in electron concentration since we are already at the practical limits of doping density, donor depth, and conduction-band discontinuity in that material system. In the AlGaAs/InGaAs system, this situation is much changed and raises the prospect of superior power handling and improved frequency response in MODFET's.

There are only two controllable parameters that exert a strong influence on the electron density: the donor density and the conduction-band discontinuity. The theoretical effect of the donor density is shown in Fig. 2. That of the conduction-band discontinuity is compared to experiment in Fig. 3. While the theory that underlies that figure is based on the conduction-band discontinuity, the experimental points, in fact, represent different In mole frac-

tions. By reconciling the theory and experiment, we get the conduction-band discontinuity as a function of In mole fraction; for 25-percent In and Al this number is 0.46 eV ± 8 percent. The uncertainty depends on the accuracy of the In and Al mole fractions, the Si doping, and the donor energy, which is taken as 40 meV. In any case, the significance in both figures is clear; the most desirable structure is the one with the largest practical conduction-band discontinuity and donor density. In MODFET's that are not pulse doped, the maximum donor density is generally determined by considerations on the breakdown voltage. By using pulse doping, we remove this constraint; then the maximum donor density is purely a characteristic of the growth and material. Previous experience with AlGaAs/GaAs MODFET's has shown that this number is 5×10^{18} cm^{-3} in our laboratory; attempts at heavier doping result in films with reduced sheet conductivity. The maximum In mole fraction we have tried so far is 0.25, and that structure produced the highest carrier concentration.

III. DEVICE FABRICATION

Except for the gate step, device fabrication is conventional mesa-isolated lift-off technology [11]. The gate lithography step is based on an angled evaporation scheme and reactive-ion etching (RIE) and is done as follows. After the source and drain ohmic contacts are alloyed, the entire wafer is coated with 4000 Å of PECVD SiO$_2$. A

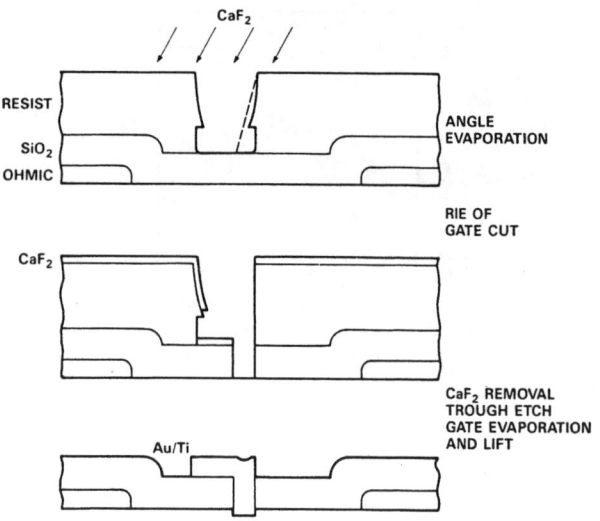

Fig. 4. Evolution of the device cross section during the gate fabrication step.

1-μm gate cut is then defined using a two-layer positive photo resist, with the gate cut typically set quite close to the source contact, as indicated schematically at the top of Fig. 4. Following definition of the gate cut, CaF$_2$ is evaporated at an angle of 10° toward the source. For our resist thickness, this results in a shadow of 0.25-μm nominal length. The CaF$_2$ acts as a mask for the subsequent RIE steps, which consist of an O$_2$ etch to straighten the wall on the drain side of the gate cut, followed by a CF$_4$ etch that opens a nominal 0.25-μm-wide trench through the SiO$_2$ down to the GaAs, as indicated in the middle of Fig. 4. Finally, the CaF$_2$ is removed, the gate trench is etched into the AlGaAs to the desired depth, and the gate metal, consisting of 350 Å of Ti followed by 3000 Å of Au, is evaporated and lifted. The resulting structure is shown in cross section at the bottom of Fig. 4. The active part of the gate is centered in the 2-μm gap between the source and drain and typically lies 300 Å above the heterojunction. Examination of several wafers by scanning electron microscope shows that this process consistently results in 0.15- to 0.25-μm-long gates that completely fill the oxide trench. The portion of the gate that is supported on top of the SiO$_2$ makes a negligible contribution to the gate capacitance, while contributing substantially to the gate conductance. The process that we have described thus results in a desirable device structure, and requires only standard processing equipment and techniques.

IV. Device Electrical Characteristics

With respect to effects on the dc electrical parameters, the use of an InGaAs channel is expected primarily to result in increased maximum drain current and decreased source resistance because of the higher sheet electron concentration. The lower source resistance should result in a somewhat higher extrinsic transconductance. There is the possibility of a higher electron velocity in the channel, arising either from properties intrinsic to the pseudomorphic channel, or from the higher barrier in the conduction band, which would allow more heating of the electrons and hence greater velocity before they scatter into the AlGaAs layer. This would somewhat increase the intrinsic transconductance.

Fig. 5 shows a comparison, for conventional and pseudomorphic MODFET's, of drain current normalized to the gate width, as a function of gate voltage. The structure of the conventional device is identical to that of the pseudomorphic device except for the absence of the InGaAs layer and the use of a 10-percent heavier doping pulse. Most of the pseudomorphic devices across the wafer show normalized drain currents in the range of 500 to 600 mA/mm—roughly twice as large as the typical value for conventional devices. Since we know, for instance from Hall results, that there are twice as many electrons in the pseudomorphic 2D gas as in the conventional well, this result suggests that there are not large differences in electron velocity. We shall show shortly that this is indeed the case. Fig. 6 shows representative common source characteristics for the pseudomorphic device of Fig. 5. The main feature of note here is that there are not strong short-channel effects; there is, however, some leakage at high drain voltage and low drain current. In addition there is the usual compression of transconductance at high and low currents, which is not associated with short-channel effects.

The peak transconductances of devices across the entire wafer correspond to normalized values in the range of 370–730 mS/mm. However, as is frequently the case with MODFET's, there is not perfect agreement between dc and RF measurements of this quantity; the dc transconductance is about 19 percent higher than that measured at 125 MHz. This difference is thought to be due to traps in the AlGaAs associated with the low growth temperature. The source resistance hardly varies, ranging from normalized values of 0.7 to 0.95 $\Omega \cdot$ mm with a mean value of 0.85 $\Omega \cdot$ mm.

The microwave characteristics of the devices were determined from S-parameter measurements made with an HP 8510 network analyzer and a Cascade Microtech probe card, over a frequency range of 0.125 to 25 GHz. The network analyzer was calibrated using on-wafer standards (with bonding pads identical to those of the transistors), which consist of 50 Ω loads, shorts, nominal opens, and a through line. The capacitance of the standard open was determined to be 5 fF by calculation; this value was then checked by measuring S_{11} of nonstandard opens of varying widths, and verifying that the capacitances of all the opens scaled properly with width.

The use of on-wafer standards completely eliminates bonding parasitics and obviates the need for, and controversies related to, de-embedding of the S-parameters [14], since each standard is at the same relative location, in the same microwave environment, as the actual device. The determination of f_T does depend somewhat on the accuracy of the open capacitance, which is why so much care was taken in its determination. We shall return to this point later.

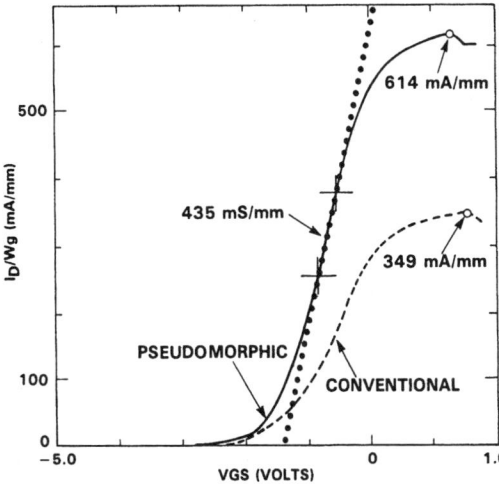

Fig. 5. Drain current density versus gate voltage at $V_{DS} = 4$ V, for 50-μm-wide pseudomorphic and conventional MODFET's. Neither device is light sensitive.

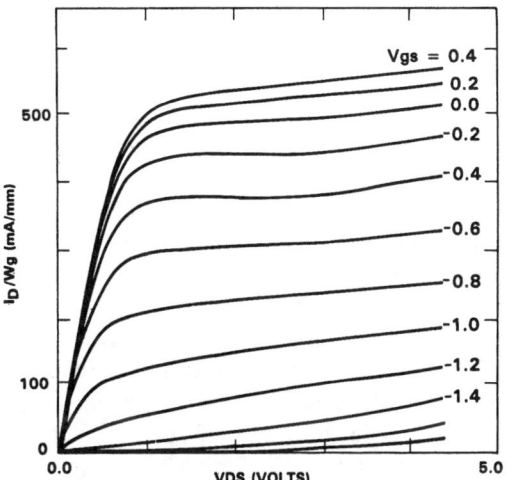

Fig. 6. Typical common source characteristics for the pseudomorphic transistor of Fig. 5.

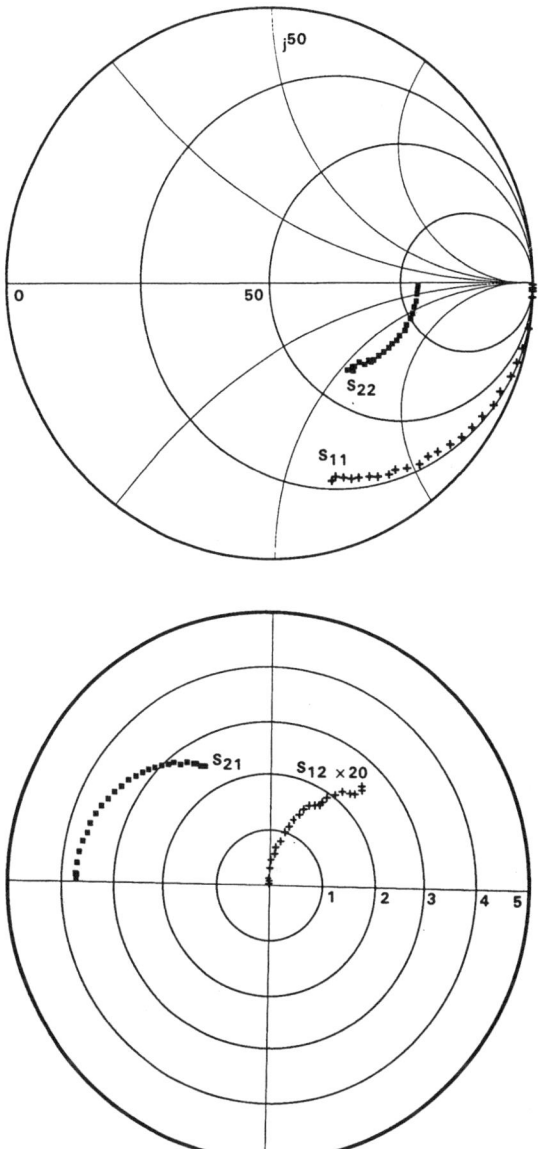

Fig. 7. Measured S-parameters up to 25 GHz for our best device. The bias point is at $V_{DS} = 1$ V, $I_D = 18.2$ mA, which is at the maximum in G_m, for the specified drain voltage. The gate width is 100 μm.

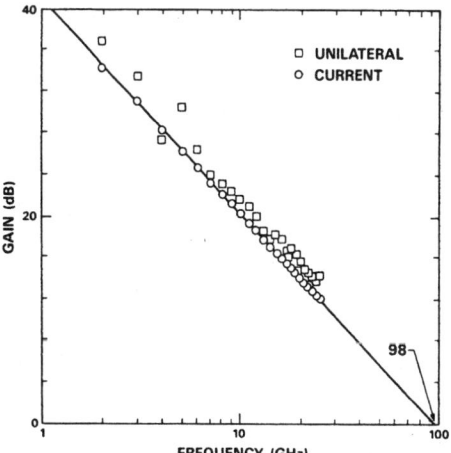

Fig. 8. Unilateral and current gain versus frequency for the same device, at the same bias, as in Fig. 7.

S-parameters resulting from the measurement of one particular device are shown in Fig. 7, and the unilateral gain and current gain calculated from them are shown in Fig. 8. Note that the current gain is quite well behaved: there is very little scatter from the fitted line, and the slope is 20 dB/decade of frequency. This nearly ideal behavior is an indication of the clean measurements that can be made with a Cascade probe and a good calibration. There are no inductive or transmission line length effects, which would produce an artificially shallow slope in the current gain at higher frequencies [14]. The repeatability is quite good as indicated by the small scatter. For the particular device whose gain is shown in the figure, the extrapolated intercept occurs at an f_T of 98 GHz; in that respect, this is one of our best devices. The transconductance for this device, as deduced from S_{21} and S_{22}, is 47 mS, and the total gate capacitance determined from G_m and f_T is 77 fF. The sensitivity of f_T to an error in the determination or calibration of the open capacitance is 1.3 percent/fF, with

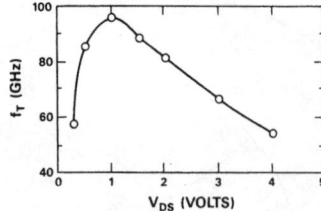

Fig. 9. f_T as a function of drain voltage at optimum drain current (maximum transconductance) for the device of Fig. 7.

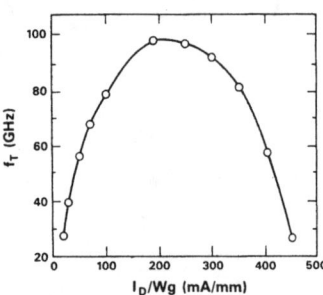

Fig. 10. f_T as a function of drain current at 1-V drain bias for the device of Fig. 7.

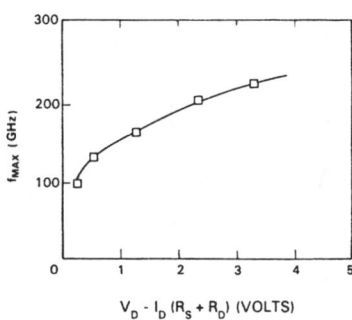

Fig. 11. f_{max} as a function of drain voltage, with I_D adjusted to optimize f_T.

an exaggerated value of the capacitance leading to an exaggerated value of f_T. Because of the care with which the open capacitance was determined, its small value, and the fact that we do not observe any variation in f_T among devices of different gate width (other than random device-to-device variation), we feel confident in the accuracy of this value. In general, the f_T of all devices across the wafer is more or less uniformly distributed from 80 to 98 GHz. We can only speculate that this variation comes about because of variations in gate length since we do not know the actual gate length of individual devices.

Both f_T and f_{max} of these devices (obtained by 20-dB/decade extrapolation of the unilateral gain) exhibit substantial bias dependence. Figs. 9 and 10 show the dependence of f_T on V_{DS} and I_D. In the first case, the drain current is adjusted to maximize f_T at each voltage; in the second the drain voltage is held constant at 1 V. The dependence of f_{max} on drain voltage is shown in Fig. 11. The increase with larger drain voltage is quite different from the behavior of f_T, and can be understood qualitatively in terms of the bias dependence of the drain-to-gate capaci-

tance and the output conductance. Both of these quantities, as deduced from S_{12} and S_{22}, decrease by a factor of 4 or more as the drain voltage is increased from 1 to 4 V. These changes more than outweigh the decrease in the f_T over the bias range. All this behavior strongly resembles that of conventional pulse-doped MODFET's, except that the f_T's of the pseudomorphic devices are larger at some bias points.

V. Discussion of Device Properties

The electrical results described above are substantially better than we obtain for standard pulse-doped MODFET's of similar geometry, particularly with respect to drain current and f_T. There are two obviously different features to the pseudomorphic MODFET: the 2D electron density is much higher, and the channel electrons move in $In_{0.25}Ga_{0.75}As$ rather than GaAs. The question that naturally arises is this: are the superior properties of the pseudomorphic device the consequence of a higher electron velocity (due to travel in the InGaAs) as some have suggested, or are they primarily a consequence of the high electron sheet density? Note that for the second hypothesis to be true, the sheet electron density would have to play a significant role in determining f_T, which runs counter to the behavior that is generally expected of FET's with gates short enough to be out of the Shockley regime of operation. There is a number of ways to test both of these hypotheses. Consider first the scaling behavior of both standard and pseudomorphic MODFET's. Depending on whether or not the gate fabrication is done using angle evaporation, we obtain gate lengths in the range of 0.15 to 0.25 μm, or 0.7 to 1.0 μm. For both standard and pseudomorphic MODFET's with longer gates, the f_T-gate length product is in the range of 20–25 GHz · μm. The devices on most wafers, including pseudomorphic ones, fall closer to the low than the high end of the range. For short-gate FET's, the pseudomorphic devices have an f_T-gate length product of 15–25 GHz · μm while standard MODFET's have a product of 10–15 GHz · μm. The similar gate length products for the longer gate devices suggest that the electron velocity is similar in pseudomorphic and standard MODFET's, while the more significant decline in that product for standard short-gate devices is consistent with the idea that there is greater parasitic delay in standard than in pseudomorphic devices. Both these ideas can be confirmed directly.

We can extract the electron velocity from the variation in device parameters across the wafer, as a consequence of our use of pulse doping. The intrinsic transconductance is given by

$$G_{mi} = \epsilon v_e W_g / X_c \quad (1)$$

where ϵ is the dielectric constant of AlGaAs, v_e is the peak electron velocity, which is reached near the drain edge of the gate, W_g is the gate width, and X_c is the gate-to-channel spacing. This transconductance describes the relationship between I_d and that voltage that appears between the gate and the drain end of the channel. Thus, the measured or extrinsic transconductance is given by

$$G_{mx} = G_{mi}/(1 + G_{mi}R_s) \qquad (2)$$

where R_s is the total dynamic resistance between the source and the drain edge of the gate. These equations can be recast as

$$1/G_{mx} = R_s + X_c/\epsilon v_e W_g. \qquad (3)$$

Since these structures are pulse doped, the threshold voltage V_T is a linear function of the gate-to-channel spacing;

$$V_T = qN_{dss}X_c/\epsilon + K \qquad (4)$$

where q is the electronic charge, N_{dss} is the donor sheet concentration, and the constant K embodies various geometric and physical factors such as pulse width and spacer-layer thickness. Both [3] and [4] are linear functions of X_c, and except for v_e the coefficients of X_c are known quite well. The gate trough etch leads to a natural large variation in X_c across the wafer; N_{dss} is extremely uniform by the nature of the growth. Consequently a scatter plot of V_T versus $1/G_{mx}$ for all the devices on a wafer exhibits a linear dependence, as shown in Fig. 12. The slope S of the line fitted to the data is given by

$$S = -qN_{dss}v_e W_g \qquad (5)$$

which allows us to calculate the electron velocity. For the pseudomorphic devices whose data are plotted, this quantity is 2.2×10^7 cm/s. As can be seen from the figure, this is quite close to the value we observe for our best standard pulse-doped wafers. There are thus three pieces of evidence (the scaling of maximum drain current with n_{ss}, the similarity of f_T for 1-μm gate devices, and the nearly identical channel velocities determined from V_T versus $1/G_m$ plots) that the electron velocity under the gate does not account for the difference between standard and pseudomorphic MODFET's. In trying to understand what other factors may play a role in the better performance of pseudomorphic MODFET's, it is useful to examine the total delay τ_d through the device, where

$$\tau_d = 1/2\pi f_T. \qquad (6)$$

We observed earlier that f_T exhibits a substantial bias dependence; if we replot the data as total delay versus different bias quantities, we see some rather striking behavior.

In Fig. 13 we show the total delay as a function of corrected drain to source voltage for a standard and a pseudomorphic MODFET, both with nominal 0.25-μm gates. At each voltage, the drain current has been adjusted for the best f_T. The resistive drop across the source and drain resistors has been accounted for, so the plotted voltage represents the potential drop across the channel and the drain depletion region. At high drain voltage there is a clear monotonic increase in the time required for electrons to drift across the drain depletion region. The increase presumably comes about both because of the higher field and wider depletion associated with higher voltages. At

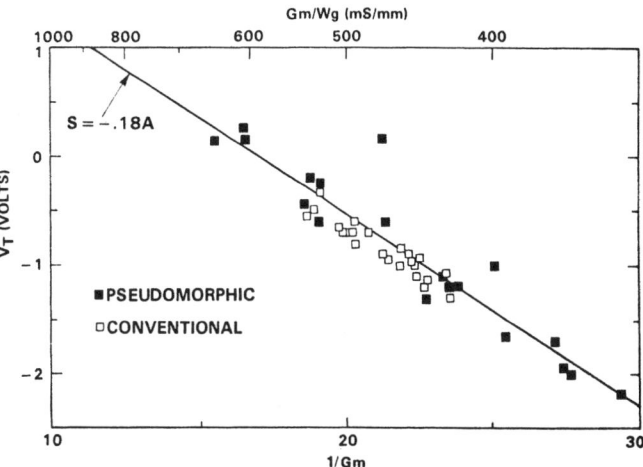

Fig. 12. V_T as a function of $1/G_{mx}$. In order to eliminate the confounding effects of traps, G_{mx} is calculated from S_{21} and S_{22} measured at 125 MHz.

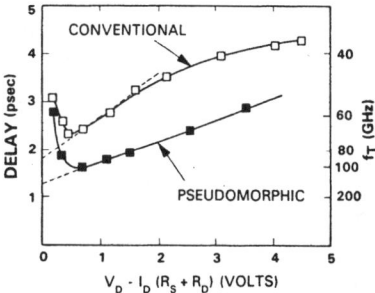

Fig. 13. Total delay as a function of voltage across the channel and drain depletion region, for conventional and pseudomorphic FET's. 1 V of applied bias corresponds to about 0.7-V corrected voltage.

the very lowest voltages, the delay increases dramatically, as the bias current is turned down. If we associate the linear part of the curve with drain delay, then the linearly extrapolated delay at zero bias is a combination of intrinsic and extrinsic delays that does not include drain delay. Taking this point of view, we then see that the difference between the extrapolated zero bias delay and the actual delay can be interpreted as delay associated with drift across the drain depletion region, which we call drain delay.

In Fig. 14, we show the total delay for the same devices plotted against the reciprocal of drain current density. The linearity of total delay at low, and even not so low, drain currents is quite striking. Since the drain current density is proportional to the electron density in the channel, assuming roughly constant electron velocity at the drain end of the gate, the behavior in Fig. 14 strongly suggests that we are observing the contribution of an RC time constant to the total delay, where R is the channel resistance. If we associate the linear part of this delay with this channel charging time, then the linearly extrapolated delay at zero channel resistance is a combination of delays, namely the intrinsic and drain delays, that does not include the channel charging time. The difference between the extrapolated and actual delay is exactly the channel charging delay.

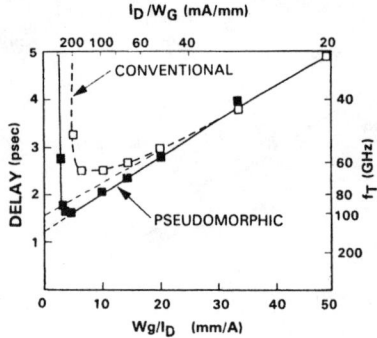

Fig. 14. Total delay as a function of the reciprocal drain current density at $V_{DS} = 1$ V, for conventional and pseudomorphic FET's.

TABLE I
SUMMARY OF THE INTRINSIC, DRAIN, CHANNEL CHARGING, AND TOTAL DELAYS τ_i, τ_d, τ_r, AND τ_T FOR PSEUDOMORPHIC AND CONVENTIONAL MODFET'S AT 1-V APPLIED DRAIN BIAS (0.7 V CORRECTED BIAS)
(For the conventional device, the drain and total delays decrease by 0.15 ps at 0.7-V applied drain bias (0.5-V corrected bias).)

	PSEUDOMORPHIC	CONVENTIONAL
τ_i	0.8 psec	0.9 psec
τ_d	0.4	0.7
τ_r	0.4	0.9
τ_T	1.6	2.5

At low drain voltage or high drain current, the delays plotted in both figures show a spectacular increase. The dramatic slowing at high drain current is most likely associated with the onset of source and drain resistor saturation, as well as the appearance of electrons in the AlGaAs conduction band, both of which we expect as the channel electron density approaches the maximum possible 2D electron density. The long delays at low current in Fig. 14 explain the increase in delay at low drain voltage observed in Fig. 13. In that plot, the drain current is constant for voltages greater than or equal to the voltage for minimum delay but is progressively reduced for lower drain voltages.

We can extend the analysis of these results to separately determine each source of delay in the devices. The difference between the extrapolated delay and the minimum delay in Fig. 13 gives us the minimum drain delay; the same difference in Fig. 14 gives us the minimum channel charging delay. Since the extrapolated intercepts are the sum of the intrinsic and channel charging delay (in Fig. 13) or the sum of the intrinsic and drain delay (in Fig. 14) we can determine the intrinsic delay from each intercept and thereby check for self-consistency. The results of such an analysis for 1-V applied drain bias are summarized in Table I. The intrinsic delay—which contains only the effect of the gate parallel plate and sidewall capacitance—is nearly the same for both devices. The largest difference is the channel charging delay, which is roughly half as long for the pseudomorphic device as the conventional one. This comes about because of the difference in electron densities, and accounts for much of the improvement in the f_T of pseudomorphic devices. The difference in the drain delay of the two devices also plays some role in that improvement; this difference must be related either to a difference in the space-charge configuration or to a difference in the electron velocity in the space-charge region. We do not understand at this point if this is an inherent difference between the two structures or a fortuitous occurrence. Note, finally, that the similarity in intrinsic delays supports the idea that the electron velocity in the channel is not very different between standard and pseudormorphic devices. That delay, of the order of 1 ps, would correspond to a channel velocity of about 2×10^7 cm/s for a 0.2-μm effective gate length.

VI. CONCLUSIONS

We have grown pseudomorphic AlGaAs/InGaAs/GaAs MODFET's with aluminum and indium mole fractions of 0.25. These layers exhibit 2D electron densities of 2.3×10^{12} cm^{-2}, and when fabricated into 0.15–0.25-μm gate MODFET's exhibit f_T's near 100 GHz. This excellent gain bandwidth product can be explained as resulting in large part from the relatively large electron density, and consequential reduction in parasitic time constants that exist in the channel region of these devices. It does not appear to be associated with a substantial difference between the channel electron velocity observed in $In_{0.25}Ga_{0.75}As$ and GaAs.

REFERENCES

[1] C. Y. Chen *et al.*, "Depletion mode modulation doped $Al_{0.48}In_{0.52}As$-$Ga_{47}In_{0.43}As$ heterojunction field effect transistors," *IEEE Electron Device Lett.*, vol. EDL-3, pp. 152–155, 1982.

[2] T. P. Pearsall *et al.*, "Selectively-doped $Al_{0.48}In_{0.52}As/Ga_{0.47}In_{0.53}As$ heterostructure field effect transistor," *IEEE Electron Device Lett.*, vol. EDL-4, pp. 5–8, 1983.

[3] C. K. Peng *et al.*, "Microwave performance of InAlAs/InGaAs/InP MODFET's," *IEEE Electron Device Lett.*, vol. EDL-8, pp. 24–26, 1987.

[4] U. K. Mishra *et al.*, "High performance sub-micron AlInAs-GaInAs HEMTs," presented at the Device Res. Conf., paper II A-6, 1987.

[5] T. E. Zipperian, L. R. Dawson, G. C. Osborn, and I. J. Fritz, "An $In_{0.2}Ga_{0.8}As/GaAs$, modulation-doped, strained-layer superlattice field-effect transistor," in *IEDM Tech. Dig.*, pp. 690–698, 1983.

[6] S. J. Rosenberg *et al.*, "An $In_{0.15}Ga_{0.85}As/GaAs$ pseudomorphic single quantum well HEMT," *IEEE Electron Device Lett.*, vol. EDL-6, pp. 491–493, 1985.

[7] T. E. Zipperian and T. J. Drummond, "Strained-quantum-well, modulation-doped field-effect transistor," *Electron. Lett.*, vol. 21, pp. 823–824, 1985.

[8] A. Ketterson *et al.*, "High transconductance InGaAs/AlGaAs pseudomorphic modulation-doped field effect transistors," *IEEE Electron Device Lett.*, vol. EDL-6, pp. 628–630, 1985.

[9] T. Henderson *et al.*, "DC and microwave characteristics of a high current double interface GaAs/InGaAs/AlGaAs psudomorphic modulation-doped field effect transistor," *Appl. Phys. Lett.*, vol. 48, pp. 1080–1089, 1980.

[10] T. Henderson *et al.*, "Microwave performance of a quarter-micron gate low-noise pseudomorphic InGaAs/AlGaAs modulation-doped field effect transistor," *IEEE Electron Device Lett.*, vol. EDL-7, pp. 649–651, 1986.

[11] M. Hueschen, N. Moll, E. Gowen, and J. Miller, "Pulse-doped MODFET's," in *IEDM Tech. Dig.*, pp. 348–351, 1984.

[12] N. Moll, A. Fischer-Colbrie, and M. Hueschen, "Pulse-doped AlGaAs/InGaAs pseudomorphic MODFETs," presented at the Devices Res. Conf., paper II A-5, 1987.

[13] A. Fischer-Colbrie *et al.*, "Growth and characterization of AlGaAs/InGaAs/GaAs pseudomorphic structures," *J. Vac. Sci. Tech.*, vol. B6, p. 620 ff., 1988.

[14] L. D. Nguyen, P. J. Tasker, and W. J. Schaff, "Comments on a new low-noise AlGaAs/GaAs 2DEG FET with a surface undoped layer," *IEEE Trans. Electron Devices*, vol. ED-34, pp. 1187, 1987.

Ensemble Monte Carlo Simulation of an AlGaAs/ GaAs Heterostructure MIS-Like FET

KAZUTAKA TOMIZAWA, MEMBER, IEEE, AND NOBUO HASHIZUME

Abstract—An ensemble Monte Carlo simulation of a heterostructure MIS-like FET is presented in which the quasi-two-dimensionality of electron gas in the heterostructure is taken into account by the lowest three subbands. An AlGaAs GaAs heterostructure MIS-like FET with a submicrometer-length gate was investigated. The electron transport in the MIS-like FET is discussed in detail. Also discussed are the dependences of device performance on the gate length and on the thickness of the AlGaAs semi-insulating layers.

I. INTRODUCTION

HETEROSTRUCTURE modulation-doped FET's and heterostructure MIS-like FET's have been of considerable interest for high-frequency and high-speed applications, and therefore have been investigated quite extensively [1]–[9]. It has been believed that the high performance of a heterostructure FET is caused by the high mobility of the quasi-two-dimensional electron gas (2DEG) rather than the saturation velocity. However, with the advance of the recent lithographic technologies, it has become possible to realize a heterostructure FET whose gate length is in the submicrometer range. It is considered that the electron transport in such a submicrometer FET is not totally quasi-two-dimensional, since the electric field component parallel to the heterointerface becomes substantially high; accordingly, hot-electron transport may take place in the device. Therefore, there is a need to study theoretically the electron transport in submicrometer heterostructure FET's so that the operation and design principles of the FET's are made clear.

So far, theoretical investigations of heterostructure FET's have been mainly done with high electron mobility transistors (HEMT's). Delagebeaudeuf *et al*. derived a formula for the current–voltage characteristics of a HEMT in which the lowest two subbands evaluated by a triangle potential model are taken into account. The electron concentration in the subbands are determined by assuming a Fermi–Dirac distribution [10]. Since their analysis assumed a Fermi–Dirac distribution at a constant temperature and a velocity-field curve for electron transport, the application of their results to a submicrometer HEMT is considered to be inappropriate. A similar model was also

developed by Drummond *et al*. [11], [12]. In order to treat high-field electron transport in HEMT, Widiger *et al*. proposed a numerical simulation in which the current continuity equation and the power balance equation are employed [13], [14]. A triangle potential model is utilized to determine the energy levels of the ground-state and first-excited subbands. Electrons whose energies are higher than the energy level of the first excited subband are assumed to be bulk-electrons (3DEG). The electron temperature in the device is evaluated by solving the power balance equation and the current continuity equation simultaneously. The electron temperature thus evaluated is used to determine the electron transport properties and the concentration of 2DEG and 3DEG in the channel with the help of the Fermi–Dirac distribution. They discussed the nonstationary electron transport in a 1-μm-long-gate HEMT in detail. Monte Carlo simulation of HEMT's, in which the two-dimensionality of electrons is taken into account, was undertaken for the first time by Tomizawa *et al*. [15]. An assumption has been made for a thickness of the size-quantized layer by using a triangle potential model, while the scattering matrices are calculated by the wave functions in a square-well potential. Taking into account the lowest two subbands for the transport of 2DEG, they calculated the current–voltage characteristics and the transconductance of a 1.2-μm-long-gate HEMT. A more sophisticated Monte Carlo program for HEMT has been developed by Ravaioli and Ferry [16]. They assumed the lowest two subbands for the 2DEG. The subband structures are calculated by a triangle potential model, in which a few assumptions are introduced to make the determination of the coefficients of the wave functions easy. They obtained some results by which the transport of the 2DEG in a 0.3-μm-long-gate HEMT was discussed. However, the device performance has not been investigated yet in detail.

In this paper, we present a novel device model of a heterostructure FET in which the transport of the 2DEG is more self-consistently treated. The model employed a simple potential function by which the wave functions for the lowest three subbands are determined. The details of Mote Carlo modeling of a heterostructure MIS-like FET are reported. Also reported are some results obtained for the electron transport in an AlGaAs /GaAs heterostructure MIS-like FET with a submicrometer channel and the device performance.

Manuscript received December 15, 1987; revised January 22, 1988.

K. Tomizawa is with Meiji University, School of Engineering, 1-1-1 Higashimata Tamaku, Kawasaki, Japan.

N. Hashizume is with the Electrotechnical Laboratory, 1-1-4 Sakuramura Umezono, Ibaraki, Japan.

IEEE Log Number 8821007.

Reprinted from *IEEE Trans. Electron Devices*, vol. 35, no. 7, pp. 849–856, July 1988.

II. SIMULATION METHOD AND ASSUMPTION

Fig. 1 shows a model of an AlGaAs/GaAs heterostructure MIS-like FET. The model consists of an undoped GaAs substrate, a couple of heavily doped n-type GaAs regions for the source and drain reservoirs, and an undoped AlGaAs layer for the gate insulator of the MIS-like structure. The GaAs substrate can be doped to be either p-type or n-type as required. However, it is assumed to be undoped in the present simulation. In the actual MIS-like FET, there is a heavily doped n-type GaAs layer sandwiched between the gate metal and the AlGaAs layer [9], which is ignored in the present model. Instead, we assumed a band discontinuity between the gate electrode and the AlGaAs layer equal to that between the GaAs substrate and AlGaAs layer. The gate length L_G is chosen to be either 0.34 or 0.7 μm considering the submicrometer gate length fabricated recently by many research workers.[1] The thickness of the undoped AlGaAs layer D is chosen to be 175, 350, or 500 Å. The doping densities in the heavily doped n^+ and n^{++} regions shown in the figure are 5×10^{17} cm^{-3} and 1×10^{18} cm^{-3}, respectively.

The Poisson equation is solved numerically in a finite-difference scheme. The spatial mesh size is variable, which is typically several tens of angstroms square in the vicinity of the heterointerface. The Dirichlet boundary condition is applied to the source, gate, and drain electrodes, and the Neumann boundary condition (zero normal derivative of the potential) is applied to the remaining device surfaces. For the boundary condition of electron motion, a reflecting mirror boundary is assumed at the surface of the device model except for those electrons penetrating into the source, gate, and drain electrodes. The charge neutrality is always maintained on the source and drain electrodes.

Fig. 2 shows a profile of the bottom of the Γ conduction band perpendicular to the heterointerface under the gate. As is shown in the figure, the lowest three subbands are taken into account for the transport of 2DEG. Our treatment for 2DEG is, so to speak, *a gradual approximation* in the sense that the electronic states of the 2DEG are assumed to depend solely on the potential profile perpendicular to the heterointerface. For rapid evaluation of the electronic states and the scattering matrices associated with the 2DEG, we introduced a simple potential function that approximates the numerical potential profile perpendicular to the heterointerface as follows:

$$V(z) = -V_0 \exp\left(-\frac{z}{z_0}\right) \quad (1)$$

where V_0 is the depth of the potential well and V_0/z_0 is the normal component of the electric field at the heterointerface. We expect that the exponential function of (1)

[1]See a recent review paper of MODFET's, for example, T. J. Drummond, W. T. Masselink, and H. Morkoç, "Modulation-doped GaAs/(Al,Ga)As heterojunction field-effect transistors: MODFET's," *Proc. IEEE*, vol. 74, pp. 733–822, 1986.

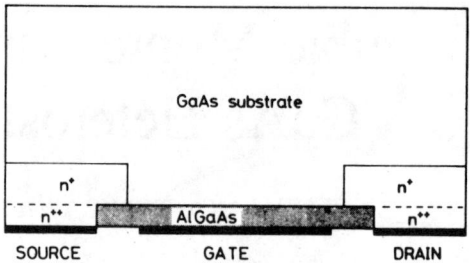

Fig. 1. Model of heterostructure MIS-like FET. $n^+ = 5 \times 10^{17}$ cm^{-3} and $n^{++} = 1 \times 10^{18}$ cm^{-3}. $L_G = 0.34$ or 0.7 μm. $D = 175, 350,$ or 500 Å. The thickness of the n^+ regions for the source and drain is 700 Å.

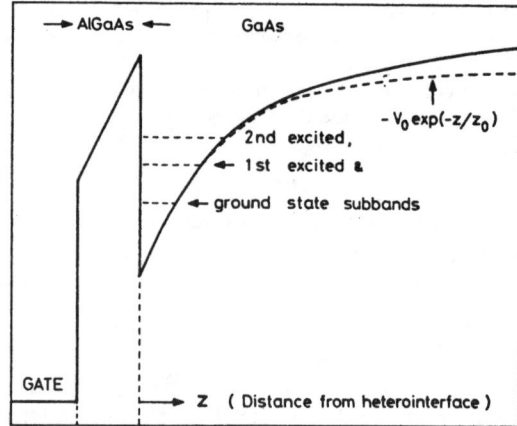

Fig. 2. The profile of the bottom of the Γ conduction band perpendicular to the heterointerface under the gate electrode. The barrier height of AlGaAs layer is assumed to be infinitely high for both the 2DEG and 3DEG in the present simulation.

represents the bending of the conduction band in the substrate, giving a suitable correction to a triangle potential model. These parameters V_0 and z_0 vary with the distance from the source.

Since we ignored the excited subbands equal to and higher than the third excited one, we introduced a rather artificial electron transfer between the 2DEG and 3DEG electronic states. Two-dimensional-electrons in the second-excited subband are assumed to transfer to the 3DEG states immediately after their energies become higher than eV_0, where e is the magnitude of electronic charge. On the other hand, 3D electrons within a certain distance from the hetero-interface are assumed to transfer to the second-excited 2DEG states if their energies become lower than eV_0, where the distance is chosen to be an expectation value of z obtained from the corresponding subband wave function given in (2). The energy conservation and the conservation of momentum components parallel to the heterointerface are taken into account in this artificial electron transfer. However, the momentum component perpendicular to the heterointerface is determined by a classical mechanical calculation using the conservations stated above, since the quantum mechanical determination is rather awkward.

The barrier height of the AlGaAs layer is assumed to be infinitely high for the 2DEG to make the calculation associated with the 2DEG easier, while it can be assumed to be either finite or infinite for the 3DEG depending on the real-space transfer of the 3DEG to be included or not in the model. In this paper, the barrier height for the 3DEG is also assumed to be infinite, since it is supposed in the case of an MIS-like FET that electrons in the AlGaAs layer (caused by the real-space transfer) may be swept away toward the gate electrode and may not contribute much to the parallel conduction (i.e., the drain current flowing in the AlGaAs layer). Accordingly, so far as the nature of the electron transport in the channel is concerned, we believe it may not be so sensitive to the real-space transfer.

The wave functions for the lowest three subbands are assumed as follows:

$$\Psi_0(z) = K_0 z \exp\left(-\frac{c_0 z}{2}\right)$$

$$\Psi_1(z) = K_1[z + a_1 z^2] \exp\left(-\frac{c_1 z}{2}\right)$$

$$\Psi_2(z) = K_2[z + a_2 z^2 + b_2 z^3] \exp\left(-\frac{c_2 z}{2}\right) \quad (2)$$

where the coefficients, a_n, b_n, c_n, and K_n are determined by a variational technique in advance for various V_0 and z_0. One might feel that it is rather awkward to evaluate these coefficients for various V_0 and z_0, but the parameters V_0 and z_0 can be reduced to only one parameter by suitable transformation of variables in the Schrödinger equation [17].

We took into account only the acoustic and polar optical phonons for the scattering processes of 2DEG since these are the dominant scattering processes for high-field transport. The phonons are assumed to be three-dimensional ones. The scattering rates due to the acoustic and polar optical phonons are evaluated, respectively, as [17]

$$W_{K,ac} = \sum_{n=0}^{n_{max}} \frac{\pi \Xi_0^2 k_B T}{\hbar c_L} \phi_{n,m}(\lambda) N(E_K) \quad (3)$$

$$W_{K,p0} = \frac{e^2 \omega_0}{8\pi^2 \epsilon_p}\left[n(\omega_0) + \frac{1}{2} \pm \frac{1}{2}\right] \frac{K}{E_K} \int_{q_{min}}^{q_{max}} \frac{F_{n,m}(q)}{\sin\theta(q)} dq \quad (4)$$

where k_B is the Boltzmann constant, T is the lattice temperature, and $\hbar$ is the Planck constant divided by 2π. Ξ_0 is the deformation potential of acoustic phonon scattering, c_L is the velocity of sound, ω_0 is the angular frequency of the polar optical phonon, and ϵ_p is defined as $1/\epsilon_p = (1/\epsilon_\infty - 1/\epsilon_0)$, respectively, for GaAs. K and E_K are the wave vector and energy of an electron. The functions $\phi_{n,m}(\lambda)$ and $F_{n,m}(q)$, which have the subscripts m and n, are only related to the coefficients of the wave functions of the mth and nth subbands. The impurity scattering for the 2DEG is not taken into account in the present model since the GaAs substrate and the AlGaAs layer are assumed to be undoped.

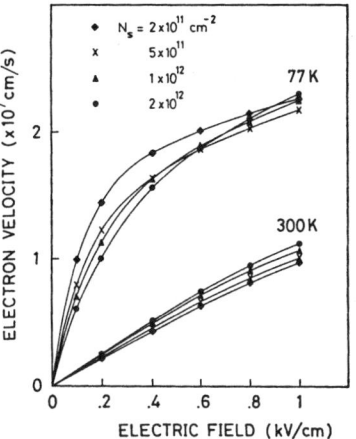

Fig. 3. The velocity-field curves of 2DEG at 77 and 300 K.

The degeneracy of the 2DEG and the 3DEG is ignored since the number of sample electrons (i.e., particles) distributed in a spatial mesh in the device model are not sufficient to represent the shape of the distribution function therein, even though the total number of particles employed reaches several tens of thousands. This is because the simulation is carried out in real space as well as in the momentum space. Also ignored is the size quantization of electrons in the L valleys. The X valleys are not taken into account in the model to save computing time.

In order to confirm the assumption made for the 2DEG and the 3DEG mentioned above, the velocity-field curves of the 2DEG in a single heterostructure are calculated at 77 and 300 K as shown in Fig. 3. The calculation is made such that the coefficients of the potential function (see (1)) and the wave functions (see (2)) are simultaneously determined by a variational method, where the density of the 2DEG in each subband is determined by an ensemble Monte Carlo calculation. The GaAs substrate is assumed to be undoped. Since the impurity scattering is not taken into account in the calculation, the results shown in Fig. 3 may not be applicable for the 2DEG in a modulation-doped structure. It is seen that the velocity-field curves at 77 and 300 K are slightly dependent on the electron concentration. Important features that we would like to point out are that the velocity-field curves obtained are not substantially different from those of pure bulk GaAs at 77 and 300 K, and those obtained by a more rigorous numerical calculation for 2DEG [18].

Fig. 4 shows the flow-chart of the simulation program. All sample electrons are assumed to be 3D electrons when their initial conditions are given. The iteration loop in the flow-chart shows the sequence of a self-consistent Monte Carlo calculation. Each iteration yields a time increment Δt of 5 fs. The space-charge density in the device is evaluated from the spatial distribution of sample electrons, where the distances of sample electrons for 2DEG from the heterointerface are determined by random numbers in conformity with the absolute square of the wave functions of (2). After the determination of the two-dimensional potential profile by solving the Poisson equation, the poten-

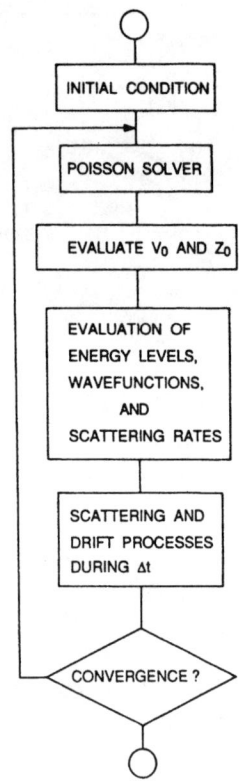

Fig. 4. Flowchart of the simulation program.

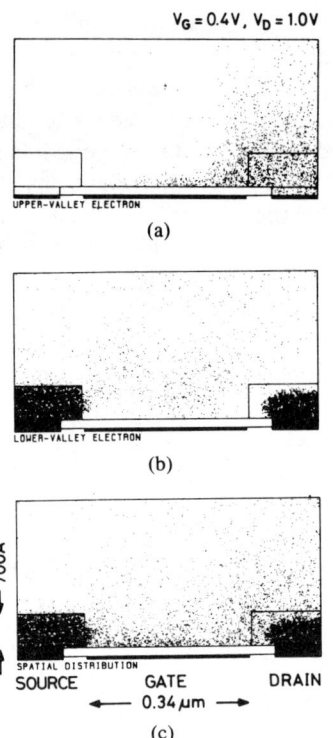

Fig. 5. The spatial distributions of electrons (a) in the Γ and L valleys, (b) in the Γ valley except the 2DEG, and (c) in the L valleys. $L_G = 0.34$ μm. $D = 175$ Å. $V_G = 0.4$ V and $V_D = 1.0$ V. $T = 77$ K.

tial profile perpendicular to the heterointerface is approximated by the potential function of (1); then the subband energies, wave functions, and scattering matrices are determined from V_0 and z_0 by the method previously reported [17]. Various quantities that are related to the scattering of the 2DEG are assumed to be constant during the time interval Δt and are employed for the calculation of the scattering and drift processes until these quantities are updated after solving the Poisson equation next time. The above calculation is repeated until the steady-state solution is established.

III. RESULT AND DISCUSSION

Fig. 5 shows the spatial distributions of electrons in an MIS-like FET. The gate length L_G is 0.34 μm, and the thickness of the undoped AlGaAs layer D is 175 Å. The lattice temperature T is assumed to be 77 K. Fig. 5(a), (b), and (c) shows, respectively, the distributions of the whole electrons, the electrons in the Γ valley except the 2DEG, and the electrons in the L valleys. It is seen in Fig. 5(a) that a slightly populated electron distribution is observed close to the heterointerface, which is not observed in Fig. 5(b). The distribution near the source n$^+$ region is due to mostly 2D electrons. It is seen in Fig. 5(b) and (c) that a number of electrons are widely distributed as 3D electrons in the GaAs substrate. These are partly caused by the diffusion of electrons from the n$^+$ regions into the substrate. However, these are also caused by hot electrons that are accelerated and scattered in the submicrometer channel. No electron is found in the AlGaAs layer, since the potential barrier of AlGaAs is

assumed to be infinitely high in the present model. It is also seen in Fig. 5(c) that quite a number of electrons in the L valleys are observed near the drain n$^+$ region.

Fig. 6(a), (b), and (c) shows the corresponding spatial distributions of 2DEG in the ground states, first excited, and second excited subbands, respectively. The surface-density profiles of the 2DEG are also shown in Fig. 7(a), (b), and (c). The spatial distribution of the 2DEG (perpendicular to the heterointerface) is determined by random numbers in conformity with the absolute square of the wave functions. It is seen in the figures that the spatial spreading of the 2DEG is about 500 Å from the heterointerface. It is seen in Figs. 6 and 7 that the number of the 2DEG decreases with the increase of the distance from the source. This is for the following two reasons. One is that the 2DEG becomes hot during the transit in the submicrometer channel; thus 2D electrons acquire enough energy to escape from the quantum well. The other is that the depth of the quantum well becomes smaller with the increase of the distance from the source; hence, it becomes easier for electrons to escape from the quantum well. It is also seen that the 2DEG disappears beyond a certain distance from the source. The complete disappearance of the 2DEG is due to the presence of "the pinchoff point" where the quantum well disappears. It is also seen in Fig. 7 that the surface density of the 2DEG is higher in the higher energy subbands, i.e., the population of the 2DEG is inverted. The inverted population of the 2DEG becomes less remarkable when the gate voltage is lower. However, the result shown in the figure re-

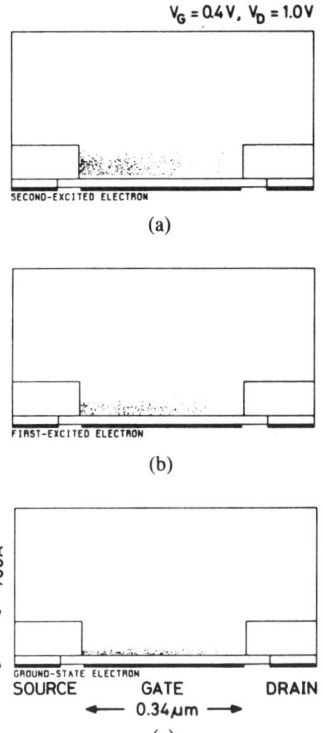

Fig. 6. The distributions of 2DEG (a) in the ground state, (b) in the first-excited subband, and (c) in the second-excited subband. $L_G = 0.34\ \mu m$. $D = 175$ Å. $V_G = 0.4$ V and $V_D = 1.0$ V. $T = 77$ K.

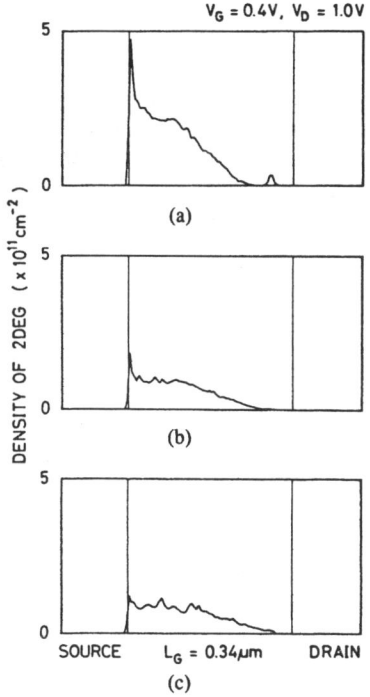

Fig. 7. The density profiles of 2DEG (a) in the ground state, (b) in the 1st-excited and (c) in the second-excited subband. $L_G = 0.34\ \mu m$. $D = 175$ Å. $V_G = 0.4$ V and $V_D = 1.0$ V.

veals that the energy distribution of the 2DEG is substantially different from the thermal equilibrium one in which the density of the 2DEG is much less in the higher excited subbands.

The surface-density profiles of the 2DEG shown in Fig. 7 are qualitatively the same as those reported for a submicrometer HEMT except that the population of the 2DEG is not inverted in case of HEMT [16]. This will be discussed later in connection with the doped and undoped AlGaAs layers for HEMT and MIS-like FET. The results shown in Figs. 6 and 7 suggest that the assumption of local quasi-equilibrium between the 2DEG and the 3DEG systems made by Widiger et al. [13], [14] may not be suitable in case of submicrometer heterostructure FET's.

Fig. 8 shows the equipotential lines in the FET. The bias condition is the same as those for the previous figures. It is seen that, near the drain n^+ region, there is an equipotential line that penetrates into the heterointerface and is almost perpendicular to the heterointerface in the vicinity of that. The position where the equipotential line lies perpendicular to the heterointerface is, so to speak, the pinchoff point, to the drain side of which the quantum well is not formed. It is also seen that the spacings of the equipotential lines are especially dense near the source and drain ends of the channel, showing that a high electric field component parallel to the heterointerface is present near the source and drain n^+ regions. Electrons are substantially accelerated by the high electric field near the source region as will be discussed later.

Fig. 9 shows the mean density profiles (calculated as bulk density) of all 2D electrons plus 3D electrons that are within a distance of 700 Å from the heterointerface, where a substantial number of electrons are accumulated. The drain voltage is fixed to be 1.0 V while the gate voltage is varied from 0.2 to 0.8 V. It is seen that the mean electron density changes substantially with the change of the gate voltage.

Fig. 10 shows the corresponding profiles of the mean electron velocity in the channel. It is seen that the mean electron velocity steeply increases in the vicinity of the source n^+ region and that it reaches as much as about 4 $\times 10^7$ cm/s, being much higher than those observed in a steady-state velocity-field curve for the 2DEG at 77 K. It is also seen that the profile of the mean electron velocity in the channel is modified by the change of the gate voltage, but the magnitude is less affected compared with the change of the electron density shown in Fig. 9. The results shown in Fig. 9 and 10 reveal that the control of the drain current of this MIS-like FET is mostly due to the modification of the electron density by the gate voltage. It is also seen that the mean electron velocity steeply decreases in the high-field region near the drain. This is due to the fact that electrons acquire enough energy for the intervalley transfer during the transit in the 0.34-μm-long channel. The results shown in Figs. 5, 6, 7, and 10 evidently reveal the presence of nonstationary and hot-electron transport in the submicrometer MIS-like FET.

However, the mean electron velocity observed in the present simulation does not reach as high as 8×10^7 cm/s as was reported in the simulation of a 1-μm-long gate HEMT [13], [14]. Such a high electron velocity is not reported even in the Monte Carlo simulation of a submi-

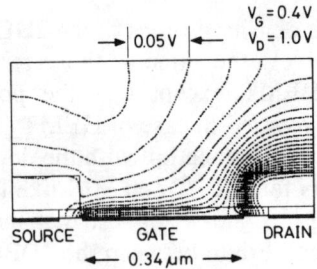

Fig. 8. The equipotential lines in the FET. $L_G = 0.34$ μm. $D = 175$ Å. $V_G = 0.4$ V and $V_D = 1.0$ V. $T = 77$ K.

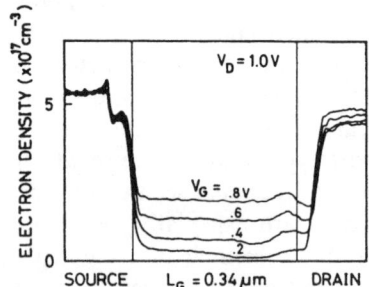

Fig. 9. The density profiles of all 2DEG plus 3D electrons that are within 700 Å from the heterointerface. $T = 77$ K. $L_G = 0.34$ μm. $D = 175$ Å. $V_D = 1.0$ V. V_G is varied from 0.2 to 0.8 V.

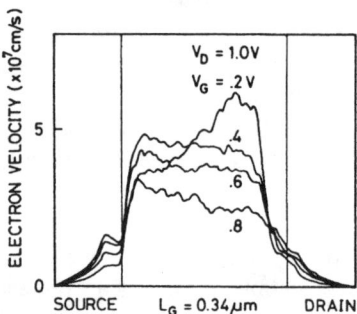

Fig. 10. The profiles of the mean velocity of all 2DEG plus 3D electrons that are within 700 Å from the hetero-interface in the channel. $T = 77$ K. $L_G = 0.34$ μm. $D = 175$ Å. $V_D = 1.0$ V. V_G is varied from 0.2 to 0.8 V.

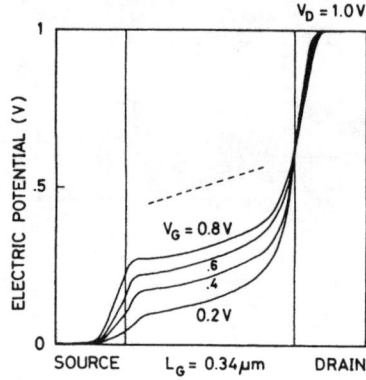

Fig. 11. The potential profiles along the channel. $T = 77$ K. $L_G = 0.34$ μm. $D = 175$ Å. $V_D = 1.0$ V. V_G is varied from 0.2 to 0.8 V.

crometer HEMT [16]. These results reveal that the electron transport equations whose transport properties (i.e., the electron mobility, etc.) are determined by Monte Carlo simulation "under constant bias electric field" may not be suited for the simulation of submicrometer devices.

The reason why the mean electron velocity shown in Fig. 10 increases steeply in the vicinity of the source and does not change appreciably in the remaining part of the channel can be understood when we see the potential profile in the channel shown in Fig. 11. Fig. 11 shows the potential profiles along the heterointerface. The bias condition is the same as for Figs. 9 and 10. It is seen in the figure that the slopes of the potential profiles are steep near the source and drain ends of the channel, showing that high electric field regions are present there. The high-field region near the source contributes to the steep in-

crease of the mean electron velocity near the source n^+ region. The electric field in the channel, except in the high-field regions mentioned above, does not change appreciably with the change of the gate voltage. This is due to the fact that the electric potential in the channel of the present MIS-like FET is substantially affected by the equipotential of the gate metal. The profiles clearly show why the mean electron velocity does not change appreciably in the channel. The slope of the broken line in the figure shows the magnitude of the electric field of 5 kV/cm. The profiles also show why the mean electron velocity in the channel (shown in Fig. 10) is slightly less when $V_G = 0.8$ V. This is due to the large potential drop near the source, which causes the transfer of some electrons into the L valleys.

It is also confirmed in the present simulation that the high-field region at the source end of the channel is less remarkable when the thickness of the AlGaAs layer is larger. However, it is still appreciable compared with the case of HEMT, where such high electric field seems to be scarcely observed [13], [14]. The potential profile along the channel of HEMT seems to be less affected by the gate potential compared with MIS-like FET. This remarkable difference in the potential profiles between HEMT and MIS-like FET is supposed to be due to the difference in the AlGaAs layers under the gate; one is doped and the other is updoped. Thus, the gate potential is supposed to be screened partly by the carriers in the doped AlGaAs layer in the case of the HEMT. The inverted population of 2DEG shown in Fig. 7 is believed to be caused by the nonuniform potential profile, i.e., the high electric field near the source.

Fig. 12 shows the current–voltage characteristics obtained by the present Monte Carlo simulation for a 0.34-μm-long-gate MIS-like FET. The thickness of the AlGaAs undoped layer is 175 Å. The transconductance g_m evaluated at $V_G = 0.2$–0.4 V and $V_D = 1.0$ V is about 1300 mS/mm. The enormously high g_m obtained is supposed to be due to the very thin AlGaAs undoped layer. This may not be expected in the case of a modulation-doped structure, since it requires a certain thickness for the doped AlGaAs layer to sufficiently provide the con-

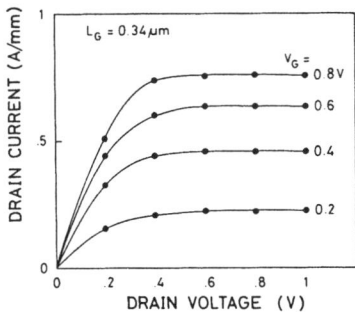

Fig. 12. The current–voltage characteristics of a MIS-like FET. $L_G = 0.34$ μm. $D = 175$ Å. $T = 77$ K.

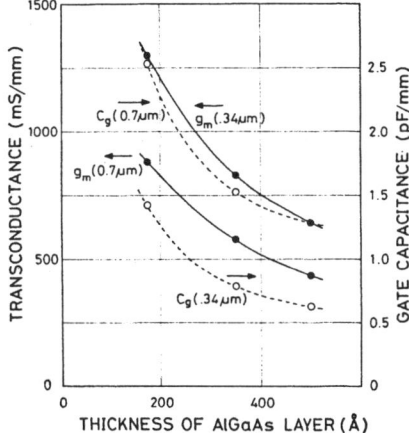

Fig. 13. Transconductance g_m and gate capacitance C_g of MIS-like FET's. $T = 77$ K.

centration of the 2DEG. The gate capacitance C_g evaluated at the same bias condition is 1.4 pF/mm, thus resulting in a unity-current-gain cut off frequency f_T, evaluated as $g_m/2\pi C_g$, of about 150 GHz.

Simulations for various FET's with different gate lengthes and thicknesses of the AlGaAs layer were also carried out. Fig. 13 shows the values of the transconductance g_m and the gate capacitance C_g. It is seen that the values of g_m for 0.7-μm-long-gate FET's are smaller by almost a factor of 2/3 than those for 0.34-μm-long-gate FET's, while the values of C_g for 0.7-μm-long-gate FET's are twice as large as those for 0.34-μm-long-gate FET's. This results in the values of f_T for the 0.7-μm-long gate FET's being about 1/3 of those for the 0.34-μm-long gate FET's. Thus, f_T's evaluated for 0.7-μm-long-gate FET's become about 50 GHz. The value of f_T does not change appreciably with the change of the thickness of AlGaAs layer within the range calculated.

These high g_m and f_T obtained might be slightly overestimated, since we ignored the real-space transfer. The real-space transfer in the case of the MIS-like FET may contribute to the decrease of the drain current because of the gate current taking place. It may also contribute to the increase of the gate capacitance because of the space-charge induced in the AlGaAs layer.

IV. Conclusion

The method of Monte Carlo simulation a heterostructure MIS-like FET is reported. Also reported are the results of simulations for submicrometer gate AlGaAs/GaAs heterostructure MIS-like FET's with various gate lengths and thicknesses of the AlGaAs undoped layer. The following features are observed in the results.

The electric potential in the channel of the present MIS-like FET is substantially affected by the equipotential of the gate metal, showing a nonuniform high electric field present in the submicrometer channel. The electron transport in the FET reflects this nonuniform high electric field, resulting in nonstationary and hot-electron transport in the submicrometer channel.

The control of the drain current is mostly due to the modification of the elctron density in the channel by the gate voltage. f_T's obtained for 0.34-μm-long-gate FET's reach as high as 150 GHz. f_T's obtained for 0.7-μm-long-gate FET's are about 50 GHz.

Acknowledgment

The authors acknowledge many useful conversations with Dr. T. Nakagawa and Dr. K. Matsumoto. We are indebted to Dr. K. Shimizu.

K. Tomizawa is indepted to the staff of the Computing Laboratory of Meiji University.

References

[1] T. Mimura, S. Hiyamizu, T. Fujii, and K. Nanbu, "A new field effect transistor with selectively doped GaAs/n-Al$_x$Ga$_{1-x}$As hetero-junctions," *Japan. J. Appl. Phys. Lett.*, vol. 19, pp. L225–L227, 1980.
[2] D. Delagebeaudeuf, P. Delescluse, P. Etienne, M. Leviron, J. Chaplart, and N. T. Linh, "Two-dimensional electron gas M.E.S.-F.E.T. structure," *Electron. Lett.*, vol. 16, pp. 667–668, 1980.
[3] P. Delescluse, M. Laviron, J. Chaplart, D. Delagebeaudeuf, and N. T. Linh, "Transport properties in GaAs-AlGaAs heterostructures and MESFET application," *Electron. Lett.*, vol. 17, pp. 342–344, 1981.
[4] D. Delagebeaudeuf and N. T. Linh, "Charge control of the heterojunction two-dimensional electron gas for MESFET applications," *IEEE Trans. Electron Devices.*, vol. ED-28, pp. 790–795, 1981.
[5] C. Chen, A. Y. Cho, K. Y. Cheng, T. P. Pearsall, P. O'Connor, and P. A. Garbinski, "Depletion mode modulation doped Al$_{0.48}$In$_{0.53}$As/Ga$_{0.47}$In$_{0.52}$As heterojunction field effect transistors," *IEEE Electron Device Lett.*, vol. EDL-3, pp. 152–155, 1982.
[6] K. Lee, M. S. Shur, T. T. Drummond, S. L. Su, W. G. Lyons, R. Fischer, and H. Morkoç, "Design and fabrication of high transconductance modulation-doped (Al,Ga)As/GaAs FETs," *J. Vac. Sci. Technol.*, vol. B1, pp. 186–189, 1983.
[7] T. J. Drummond, R. Fischer, S. L. Su, W. G. Lyons, H. Morkoç, K. Lee, and M. S. Shur, "Characteristics of modulation doped AlGaAs/GaAs field effect transistors: Effect of donor electron separation," *Appl. Phys. Lett.*, vol. 42, pp. 262–264, 1983.
[8] T. Mimura, K. Nishiuchi, M. Abe, A. Shibatomi, and M. Kobayashi, "High electron mobility transistors LSI circuits," *IEDM Tech Dig.*, pp. 99–102, 1983.
[9] K. Matsumoto et al., "Heterojunction GaAs MIS-like FET," in *Proc. 16th Int. Conf. Solid State Devices Materials* (Kobe, Japan), pp. 363–366, 1984.
[10] D. Delagebeaudeuf and N. T. Linh, "Metal-(n)AlGaAs-GaAs two-dimensional electron gas FET," *IEEE Trans. Electron Devices*, vol. ED-29, pp. 955–960, 1982.
[11] T. J. Drummond, H. Morkoç, K.Lee, and M. S. Shur, "Model for modulation doped field effect transistors," *IEEE Electron Device Lett.*, vol. EDL-3, pp. 338–341, 1982.
[12] K. Lee, M. S. Shur, T. J. Drummond, and H. Morkoç, "Current-voltage capacitance voltage characteristics of modulation doped field

effect transistors,'' *IEEE Trans. Electron Devices*, vol. ED-30, pp. 207–212, 1983.

[13] D. Widiger, K. Hess, and J. J. Coleman, ''Two-dimensional numerical analysis of the high electron mobility transistor,'' *IEEE Electron Device Lett.*, vol. EDL-5, pp. 266–269, 1984.

[14] ——, ''Two-dimensional transient simulation of an idealized high electron mobility transistor,'' *IEEE Trans. Electron Devices*, vol. ED-32, pp. 1092–1102, 1985.

[15] M. Tomizawa, A. Yoshii, and K. Yokoyama, ''Modeling for an AlGaAs/GaAs hetero-structure device using Monte Carlo simula-

tion,'' *IEEE Electron Device Lett.*, vol. EDL-6, pp. 332–334, 1985.

[16] U. Ravaioli and D. K. Ferry, ''MODFET ensemble Monte Carlo model including the quasi-two-dimensional electron gas,'' *IEEE Trans. Electron Devices.*, vol. ED-33, pp. 677–681, 1986.

[17] J.-P. Polonovski and K. Tomizawa, ''Phonon scattering of quasi-two-dimensional electron gas in a single heterostructure,'' *Japan. J. Appl. Phys.*, vol. 24, pp. 1611–1618, 1985.

[18] K. Yokoyama and K. Hess, ''Monte Carlo study of electronic transport in $Al_{1-x}Ga_xAs/GaAs$ single-well heterostructure,'' *Phys. Rev.*, vol. 33, pp. 5595–5606, 1986.

Reverse Modeling of E/D Logic Submicrometer MODFET's and Prediction of Maximum Extrinsic MODFET Current Gain Cutoff Frequency

HANS ROHDIN, MEMBER, IEEE

Abstract—A method is presented to estimate the source resistance, fringe capacitance, gate length, and effective saturation velocity from the microwave Y-parameters of MODFET's with known vertical structure. The scheme is applied to a variety of MODFET's fabricated on MBE material using a submicrometer enhancement/depletion (E/D) mode IC process. More than 100 MODFET's were measured and analyzed. Both the values and variances of the extracted parameters are very physical. In particular, it is found that the extracted saturation velocity 1) is independent of the gate length in the regime studied (0.25–0.91 μm); 2) is rather independent of process and threshold voltage variations; 3) is marginally higher when the Al mole fraction is increased from 20% to 28%; 4) is not significantly higher in pseudomorphic InGaAs than in GaAs; and 5) is quite a bit higher than is often assumed or extracted, with a value close to the stationary peak velocity in undoped GaAs. There is little sign of overshoot above this limit. Using the extracted peak velocity and a simple analytical MODFET model, the extrinsic current gain cutoff frequency (f_{Tx}) is predicted well in the gate-length regime studied. It is demonstrated that a denser and better confined 2DEG improves f_{Tx} as a result of reduced access resistances and output conductance.

I. INTRODUCTION

THIS PAPER deals with reverse modeling of MOD-FET's, i.e., with extraction of physical parameters that are used in MODFET (forward) modeling. We have adopted the term "reverse modeling" from Ladbrooke [1], but our approach is quite different. It assumes that the vertical device structure and the resulting charge control is known. This is a valid approach for an IC process where the excellent control of MBE is preserved by the use of selective RIE [2], [3] and careful choice of gate metallization [4]. Given the excellent process control, and the sophisticated Schrodinger–Poisson solvers developed (e.g., [5]), the charge control can indeed be calculated accurately. In this work, we use a program developed by Moll and Yeager [6]. By combining calculated or measured C–V data with measured FATFET I–V data, the low-field transport described by the mobility can be extracted. As the gate is reduced it becomes increasingly important

to also determine 1) the high-field transport properties as encompassed by the (effective) saturation velocity; 2) the electrical gate length; 3) the delay associated with the drain depletion region [7]; and 4) the parasitic elements (access resistances and fringe capacitances). In this work we focus on 1), 2), and 4). 3) becomes important when the MODFET drain bias is larger than at the point of maximum f_{Tx}.

Section II discusses the equivalent circuit, Y-parameters, and current gain cutoff frequency for the MODFET in the linear and saturated regions. This section sets up the framework for the reverse and forward modeling, and proposes a physically based equivalent circuit which contains elements representing the drain saturation region which are absent in other FET models. Section III describes the reverse modeling method. Section IV applies the method to more than 100 MODFET's with varying structures and processing. Section V outlines a forward modeling approach for MODFET's. Using this approach, the consistency of the results in Section IV allows us to predict, in Section VI, the maximum extrinsic MODFET current gain cutoff frequency for a wide variation in gate length, and for different epitaxial structures. Section VII summarizes the results.

II. MODFET EQUIVALENT CIRCUIT, Y-PARAMETERS, AND CURRENT GAIN CUTOFF FREQUENCY

Fig. 1 shows the proposed equivalent circuit for an intrinsic MODFET. $g_{dd'}$, $C_{dd'}$, and the broken current path are related to velocity saturation and will be discussed in Section II-B. Section II-A deals with the linear regime, where $g_{dd'} /\!/ C_{dd'}$ is a short ($g_{dd'} = \infty$), and the broken current path is indistinguishable from the G–D' path. Section II-C deals with the extrinsic MODFET (Fig. 2) where access resistances and fringe capacitances are present.

A. Equivalent Circuit Elements for the Intrinsic MODFET Biased in the Linear Region

In the linear part of the drain characteristic, the equivalent circuit elements for the intrinsic MODFET can be calculated using the Y-parameter expressions derived by

Manuscript received August 9, 1989; revised November 17, 1989. The review of this paper was arranged by Associate Editor S. Tiwari.

The author is with Hewlett-Packard Laboratories, Palo Alto, CA 94303-0867.

IEEE Log Number 8933858.

Reprinted from *IEEE Trans. Electron Devices*, vol. 37, no. 4, pp. 920–934, April 1990.

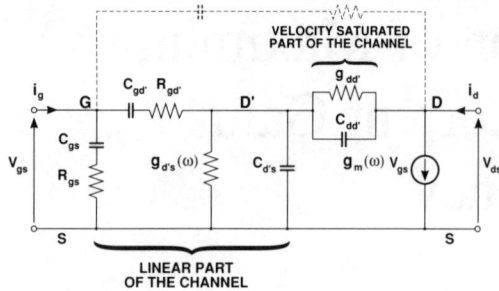

Fig. 1. Equivalent circuit of the intrinsic MODFET.

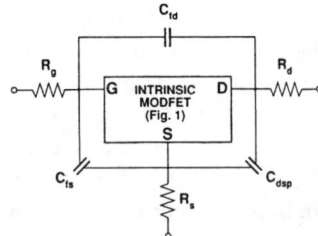

Fig. 2. Equivalent circuit of the extrinsic MODFET.

Roblin *et al.* [8]. Their model is valid for gate voltages near the maximum transconductance. This region includes the point of maximum intrinsic current gain cutoff frequency (f_{Ti}) which is of particular interest to us.

The two frequency ($f = \omega/2\pi$) dependent elements in Fig. 1 are the transconductance

$$g_m(\omega) = g_{m0}\left[1 - j\omega\tau_1 - \tfrac{1}{2}(\omega\tau_2)^2\right] \quad (1)$$

and the output conductance

$$g_{d's}(\omega) = g_{d's0}\left[1 - (\omega\tau_3)^2\right]. \quad (2)$$

The bias dependence of the parameters in (1) and (2), and those of the frequency-independent circuit elements in Fig. 1 can be described using the single parameter

$$k = \frac{V_{DS}}{V_{GS} - V_T} \quad (3)$$

where V_{DS} and V_{GS} are the intrinsic drain and gate biases, respectively, and V_T is the threshold voltage, k is a measure of the degree of pinch-down at the drain-end of the channel. $k = 0$ corresponds to the channel being fully open while $k = 1$ corresponds to classical (Shockley) saturation of the drain current by pinch-off. In reality, pinch-off will be preceeded by velocity saturation and $k = 1$ will never be reached. Of the ten parameters that determine the equivalent circuit, g_{m0} and $g_{d's0}$ have very simple k dependence

$$g_{m0} = \frac{k}{R_{ch}} \quad (4)$$

$$g_{d's0} = \frac{1-k}{R_{ch}} \quad (5)$$

where R_{ch} is the open channel resistance

$$R_{ch} = \frac{L}{\mu W c_g (V_{GS} - V_T)}. \quad (6)$$

μ is the mobility. L and W are the length and width of the gate, respectively. c_g is the maximum 2DEG capacitance per unit area

$$c_g = \frac{\epsilon}{d + \Delta d}. \quad (7)$$

ϵ is the dielectric constant of the wide-bandgap material, and d is its thickness. Δd is the effective distance from the heterojunction to the 2DEG and depends on the 2DEG concentration at the gate voltage where it is optimally modulated. The charge control program [6] yields c_g directly, and Δd can be calculated from (7). A larger and better confined 2DEG has a smaller Δd. In analytical charge control models Δd is often assumed to be 80 Å [9], but it varies from 50 to 170 Å for the epitaxial structures that we have modeled.

The remaining eight parameters in the equivalent circuit involve the dimensionless functions in [8] and are presented graphically in a normalized form in Fig. 3. The resistances are normalized to R_{ch}. The capacitances are normalized to C_g, the zero drain bias gate capacitance

$$C_g = c_g WL. \quad (8)$$

The time constants are normalized to $R_{ch}C_g = (2\pi f_{Ti0})^{-1}$, where f_{Ti0} is the intrinsic current gain cutoff frequency according to Shockley theory. As illustrated by Fig. 3, τ_1 and τ_2 track rather closely and do not vary much as k varies. Thus we have

$$g_m(\omega) \approx g_{m0} e^{-j\omega\tau} \quad (9)$$

where

$$\tau \approx \tau_1 \approx \tau_2 \approx \tfrac{1}{4} R_{ch} C_g. \quad (10)$$

Equation (9) is the typical approximation of the transconductance in the equivalent circuit of a FET. Note that the controlling voltage for the transconductive part of the drain current is the full intrinsic gate-to-source voltage. Fig. 3 also shows that R_{gs} is quite independent of k (and thus g_{m0})

$$R_{gs} \approx \frac{R_{ch}}{5} \quad (11)$$

contrary to [10]. $R_{gd'}$ varies more rapidly and goes to infinity as pinch-off is approached. C_{gs} and $C_{gd'}$ behave as in a MOSFET. The negative $C_{d's}$ can be thought of as a rather small correction necessary to fully account for the *distributed* gate-to-channel capacitance by *lumped* elements, or as a factor in a delay term in the output conductance [11].

184

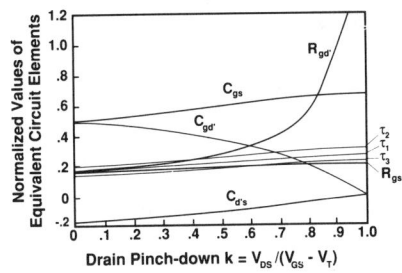

Fig. 3. The bias dependence of the normalized (see text) circuit elements in Fig. 1 and the time constants in (1) and (2).

B. Equivalent Circuit Elements for the Velocity Saturated Part of the Intrinsic MODFET

Before $k = 1$ is reached velocity saturation will set in. Beyond saturation, g_{m0} changes little. Comparing g_{m0} predicted by the saturation model in [12] with that given by (4) provides the value of k corresponding to onset of velocity saturation:

$$k_{\text{sat}} = \left[1 + \left(\frac{V_{GS} - V_T}{E_c L} \right)^2 \right]^{-1/2}. \quad (12)$$

E_c is the "critical" field at which the linear velocity–field curve is assumed to abruptly saturate at $v_{p0} = \mu E_c$. In our simple treatment of a MODFET in saturation (Section VI) we will use the equivalent circuit in Fig. 1 and evaluate the circuit elements discussed in Section II-A at $k = k_{\text{sat}}$. In saturation, at the drain end of the channel, a high-field region develops which will absorb most of a drain signal voltage. At the same time, most of the channel is still gradual, and the picture developed in Section II-A is still valid there. In the equivalent circuit in Fig. 1 the high-field region develops between nodes D and D'. Since the electron velocity is essentially saturated, the high-field region has a very low (compared to $g_{d's0}$) conductance $g_{dd'}$ which will limit the resulting output conductance. In addition, to account for charge modulation in the high-field region, there will be a capacitance $C_{dd'}$ in parallel with $g_{dd'}$. A delay associated with the saturated region will be included in τ. Due to the increased reverse bias of the gate–drain junction in saturation, and the behavior of $R_{gd'}$ and $C_{gd'}$ as $k \rightarrow 1$, we neglect the broken current path in Fig. 1. Applying the model described in [13] for various submicrometer gate lengths, one finds that the extent of the fully developed high-field region is rather constant. There are therefore reasons to expect both $g_{dd'}$ and $C_{dd'}$ to be only weakly dependent on the gate length in fully saturated MODFET's. Note that we model the high-field region as being in series with the rest of the channel, rather than essentially parallel [14]. Our representation is similar to that used in [15] for a domain at the drain side of a MESFET.

C. MODFET Y-Parameters and Current Gain Cutoff Frequency

The Y-parameters of the extrinsic MODFET in Fig. 2, expanded to $(j\omega)^1$, are given by

$$Y_{11} = j\omega C_{11} \quad (13)$$

$$C_{11} = \frac{1}{F_1} \left[C_{gs} + C_{gd'} + C_{fs} + C_{fd} + (R_s + R_d)\Delta_1 \right] \quad (14)$$

$$Y_{21} = G_{21} - j\omega C_{21} \quad (15)$$

$$G_{21} = \frac{g_{m0}}{F_1} \quad (16)$$

$$C_{21} = \frac{1}{F_1} \left[g_{m0}(\tau + \tau_p) + pC_{gd'} + C_{fd} + R_s\Delta_1 \right] \quad (17)$$

$$Y_{12} = -j\omega C_{12} \quad (18)$$

$$C_{12} = \frac{1}{F_1} \left[pC_{gd'} + C_{fd} + R_s\Delta_1 \right] \quad (19)$$

$$Y_{22} = G_{22} + j\omega C_{22} \quad (20)$$

$$G_{22} = \frac{g_{ds0}}{F_1} \quad (21)$$

$$C_{22} = \frac{1}{F_1} \left[C_{ds0} + C_{fd} + C_{dsp} + (R_g + R_s)\Delta_1 - g_{ds0}\tau_p \right] \quad (22)$$

where

$$g_{ds0} = \frac{g_{dd'} g_{d's0}}{g_{dd'} + g_{d's0}} \quad (23)$$

$$C_{ds0} = \frac{g_{dd'}^2 (C_{gd'} + C_{d's}) + g_{d's0}^2 C_{dd'}}{(g_{dd'} + g_{d's0})^2}. \quad (24)$$

$$p = \frac{g_{dd'}}{g_{dd'} + g_{d's0}} \quad (25)$$

$$F_1 = 1 + R_s g_{m0} + (R_s + R_d) g_{ds0} \quad (26)$$

$$\Delta_1 = g_{m0}(pC_{gd'} + C_{fd}) + g_{ds0}(C_{gs} + C_{gd'} + C_{fs} + C_{fd}) \quad (27)$$

$$\tau_p = \frac{1}{F_1} \bigg[R_s \big[C_{gs} + (1 - 2p)C_{gd'} + C_{fs} + C_{ds0} \\ - g_{m0}\tau + C_{dsp} \big] + R_d \big[C_{ds0} + C_{fd} + C_{dsp} \big] \\ + R_g \big[C_{gs} + C_{gd'} + C_{fs} + C_{fd} \big] \\ + \big[R_s R_d + R_s R_g + R_d R_g \big] \Delta_1 \bigg]. \quad (28)$$

g_{ds0} is the total intrinic dc output conductance and is the result of the series combination of the conductances of the gradual ($g_{d's0}$) and saturated ($g_{dd'}$) part of the channel. The dimensionless parameter p is less than 1 and approaches zero as $L \rightarrow 0$. This is because $g_{d's0}$ approaches ∞ as $L \rightarrow 0$ while $g_{dd'}$ remains finite. For the same reason the low-frequency intrinsic output capacitance C_{ds0} will approach $C_{dd'}$. Note that the full capacitance $C_{gd'}$ is only seen from the input (14), while from the output (19) it is damped by the factor p. For zero output conductance the factor F_1 becomes $1 + R_s g_{m0}$ which is the familiar denominator degrading the extrinsic dc transconductance. τ and the parasitic time constant τ_p will not affect f_{Tx} despite appearing in Y_{21}. This is because we insist on using a -20 dB/dec extrapolation of the current gain versus frequency to determine f_{Tx} [16], which becomes

$$f_{Tx} = \frac{G_{21}}{2\pi C_{11}} \tag{29a}$$

$$= \frac{g_{m0}}{2\pi \left[C_{gs} + C_{gd'} + C_{fs} + C_{fd} + (R_s + R_d)\Delta_1 \right]}. \tag{29b}$$

Note that the factor $1/F_1$ appears in both numerator (16) and denominator (14) of (29a), and thus cancels. It is clear from (29b) that in addition to the well-known degrading effect of fringe capacitances there is an additional effective parasitic capacitance C_r which is proportional to $R_s + R_d$

$$C_r = (R_s + R_d)\Delta_1 = (R_s + R_d) \left[g_{m0}(p C_{gd'} + C_{fd}) \right.$$
$$\left. + g_{ds0}(C_{gs} + C_{gd'} + C_{fs} + C_{fd}) \right]. \tag{30}$$

This term becomes important as the gate is shrunk, and its presence prevents the extraction of $C_{fs} + C_{fd}$ by extrapolating $1/f_{Tx}$ versus L to zero L. The gate resistance R_g does not degrade f_{Tx} and, more interestingly, neither does R_{gs}. An expression very similar to (29b) has recently been derived by Tasker and Hughes [17] using a conventional FET equivalent circuit.

The intrinsic ($R_s = R_d = 0$, $C_{fs} = C_{fd} = 0$) current gain cutoff frequency f_{Ti} is given by

$$f_{Ti} = \frac{g_{m0}}{2\pi (C_{gs} + C_{gd'})}. \tag{31}$$

In saturation, the intrinsic transconductance is given by

$$g_{m0} = W c_g v_p \tag{32}$$

where v_p is the effective peak velocity. This may deviate from the peak velocity v_{p0} of the steady-state velocity-field curve of GaAs because of velocity overshoot (for short gate lengths), gradual channel effects (for longer gate lengths), or Gunn domain formation. The intrinsic gate capacitance is given by

$$C_{gs} + C_{gd'} = W c_g (L - \delta L). \tag{33}$$

For long gates, δL approaches $L/3$ because of the voltage drop across the gradual part of the channel. For shorter gates, δL approaches 0, except when the FET is biased deep in saturation (large V_{DS}) and there is an increase in electrical gate length (negative δL). Typically, the maximum f_{Tx} occurs at rather low V_{DS}, and this effect is small. Combining (31)–(33), and introducing the effective transit

velocity v_t, we have

$$f_{Ti} \equiv \frac{v_t}{2\pi L} \tag{34a}$$

$$= \frac{v_p}{2\pi (L - \delta L)} \tag{34b}$$

and

$$v_t = \frac{L}{L - \delta L} v_p. \tag{35}$$

III. REVERSE MODELING METHOD

Table I shows the epitaxial structure for the GaAs based E/D MODFET process used for this study. The layers are grown by MBE. Except for the GaAs cap layer, the thickness and doping ranges correspond to design variations for the threshold voltages. The parameters d_{cap}, N_{cap}, and x have been defined in Table I for use in Section IV. The idealized device structure is shown in Fig. 4. The process is a low-temperature ($\leq 420°C$) self-aligned gate process that yields low source resistance without ion implantation and anneal. Standard lithography is used, and with the aid of oxide spacers submicrometer gates can be fabricated. The gates are deposited on the AlGaAs etch-stop layers after selective RIE [4].

A. Extraction of Source Resistance, Fringe Capacitance, and Intrinsic Current Gain Cutoff Frequency

Both source resistance R_s and fringe capacitance C_f are often estimated from independent measurements. The measurement of R_s [18] involves forward biasing the gate in the linear regime so that the gate current is appreciable. C_f can be estimated from Y_{11} data in pinch-off. Both measurements involve excursion outside the typical active biasing range for optimum microwave performance. The actual source resistance of a MODFET biased for optimum f_{Tx} could be larger due to a more restrictive electron access path. On the other hand, due to contributions from the gated part of the channel, the dc measurement could also overestimate the true source resistance. The total fringe capacitance for a MODFET biased in its active region is smaller than the capacitance in pinch-off since the 2DEG under the gate will terminate the field lines that in pinch-off would terminate on the ungated access regions and ohmic contacts, and deep in the buffer. Furthermore, the capacitance in pinch-off is often itself overestimated because of measurement limitations imposed by gate breakdown. In this subsection we propose a method that estimates R_s and C_f from microwave data with the MODFET in its active modulating bias range.

By reversing the analysis that led to (13)–(28) one can express the intrinsic Y-parameters in terms of the measured. Two results that we use are

$$C_{gs} + C_{gd'} + C_{fs} + C_{fd} = F_2 \left[C_{11} - (R_s + R_d)\Delta_2 \right] \tag{36}$$

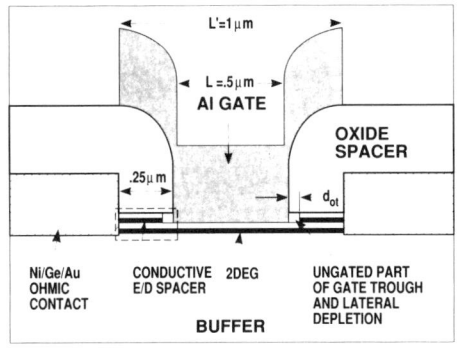

Fig. 4. Idealized device cross section.

TABLE I
GaAs-BASED E/D MODFET EPITAXIAL STRUCTURE

Layer	Al Mole Fraction	Thickness (Å)	Si Donor Conc. (cm⁻³)
GaAs cap	0	190-300 ($\equiv d_{cap}$)	$0\text{-}1 \times 10^{18} \equiv N_{cap}$
D-gate etchstop	28	40	$0\text{-}7 \times 10^{17} = .72 \times N_{cap}$
GaAs E/D spacer	0	25-150	$0\text{-}1 \times 10^{18} \equiv N_{cap}$
E-gate etchstop	28	70	0
Doping pulse	20-28 ($\equiv$ x)	85-150	$(3\text{-}4) \times 10^{18}$
Spacer	20-28 (=x)	20	0
Buffer	0	10,000	0

and

$$pC_{gd'} + C_{fd} = F_2[C_{12} - R_s\Delta_2] \qquad (37)$$

where

$$F_2 = \frac{1}{1 - R_sG_{21} - (R_s + R_d)G_{22}} \qquad (38)$$

and

$$\Delta_2 = C_{11}G_{22} + C_{12}G_{21}. \qquad (39)$$

Given the short gate lengths of interest, we assume that for a MODFET biased for maximum f_{Tx} the intrinsic gate capacitance equals that at zero drain bias, i.e., there is no decrease due to gradual channel effects. We also assume that the measured gate–drain capacitance is dominated by parasitic fringe capacitance, i.e., that the intrinsic gate–drain capacitance is negligible. In other words, we assume that $C_{gs} + C_{gd'} = C_g$ (i.e., $\delta L = 0$), and that $pC_{gd'} = 0$. With these assumptions, (36) and (37) (with the definitions (38) and (39)) have four unknowns: R_s, R_d, C_{fs}, and C_{fd}. If any two of these, say R_s and C_{fs}, are known from separate measurements, the other two can be easily calculated from the system of two linear equations. Considering the symmetric device structure of the process used (Fig. 4), and the introductory discussion in this subsection, we instead make the assumptions $R_d = R_s$ and $C_{fd} = C_{fs} = C_f$. In saturation, C_{fd} can be smaller than C_{fs} if a high-field drain domain extends beyond the metallurgical drain-side gate edge [15]. However, we look at the point of maximum f_{Tx} which typically occurs before such a domain becomes significant. With the assumptions made we

obtain for the source resistance and fringe capacitance

$$R_s = \frac{C_g - C_{11} + 2C_{12}}{C_g} \frac{1}{G_{21} + 2G_{22}} \qquad (40)$$

$$C_f = \frac{\Delta_2 - G_{22}C_g}{G_{21} + 2G_{22}}. \qquad (41)$$

The zero drain bias gate capacitance C_g is determined either by separate Y_{11} measurements on a FATFET, or by charge control modeling. Both require a knowledge of the electrical gate length of the FET under test. Extraction of this will be discussed in the following subsection.

Using R_s in (40) we can calculate the intrinsic transconductance and output conductance

$$g_{m0} = \frac{G_{21}}{1 - R_s(G_{21} + 2G_{22})} \qquad (42)$$

$$g_{ds0} = \frac{G_{22}}{1 - R_s(G_{21} + 2G_{22})}. \qquad (43)$$

These two expressions are equivalent to those derived in the dc case in [19]. The effective peak velocity can be calculated using (32) and (42).

The assumption $pC_{gd'} = 0$ should be valid as $L \to 0$, and one can show, without making the assumption $\delta L = 0$, that the intrinsic current gain cutoff frequency is given by

$$f_{Ti} = \frac{G_{21}}{2\pi(C_{11} - 2C_{12})}. \qquad (44)$$

f_{Ti} can thus be easily estimated directly from the measured Y-parameters.

B. Electrical Determination of the Gate Length

In this subsection we describe and compare two methods of electrically and nondestructively determining the gate length. Fig. 5 shows the measured C_{110}–V_{GS} characteristic of a representative (0.6 × 100) μm^2 D-MODFET. The subscript "0" indicates that $V_{DS} = 0$. The C–V curve exhibits several inflection points. The first two are marked and are of particular interest. Inflection point 1 occurs in the gate voltage range where the 2DEG is turning on, i.e., close to the threshold voltage. This is the point of most rapid variation of capacitance, and here C_{110} is entirely due to the 2DEG and fringe fields. At inflection point 2 the 2DEG has fully developed and is modulated close to maximum efficiency, meaning that the 2DEG part of the total (measured) capacitance is near its peak value (C_g). There is a component in C_{110} from "parasitic" charge (Q_{SL} in [20]) in the high-bandgap layer that prevents the total capacitance from saturating, but it does correspond to the point of minimum change in the total capacitance. While inflection point 2 overestimates the capacitance of interest ($C_g + 2C_f$), and inflection point 1 underestimates it, the capacitance C_{110x} corresponding to the intersection x of the to tangents in Fig. 5 is quite

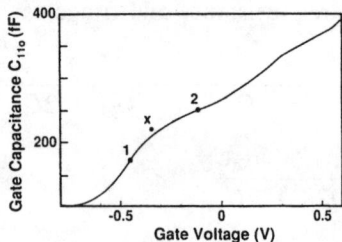

Fig. 5. Measured total gate capacitance of a representative 0.6 μm $\times$ 100 μm depletion-type MODFET with zero drain bias, as a function of gate voltage.

close to it

$$C_{110x} = \alpha C_g + 2C_f. \qquad (45)$$

The dimensionless parameter α is typically close to 0.9, but can be brought closer to unity if the MBE structure is improved for better charge confinement (e.g., pseudomorphic InGaAs channels). α is determined from the result of the numerical charge-control calculation. In the second method of the two to be discussed in this subsection the actual value of α will be important. However, in the first method the value is irrelevant and we will use the cross-point x only as a convenient reference for comparing capacitances of MODFET's with different gate lengths. An alternative choice of capacitance reference is inflection point 2 since the capacitance is the most constant there. However, C_{110x} is less sensitive than both inflection point capacitances to errors in the graphical determination of these.

Method 1 is based on a comparison of the short gate-length MODFET of interest with a reference MODFET (superscript r) which has a gate length that is larger by a known amount, but is otherwise identical. The gate length of the latter should not be so large that distributed effects and gate leakage distort the result. There are fringe capacitances for both FET's, and they have to be accounted for. The capacitances measured in pinch-off for the two FET's are denoted C_p and $C_p^{(r)}$, respectively. $C_p^{(r)}$ is larger than C_p and neither equals the total gate-length-independent fringe capacitance $2C_f$ that is of interest to us. A first-order gate-length term provides significant correction when determining the true gate-length-independent fringe capacitance $2C_f$ from the measured pinch-off capacitance C_p

$$C_p = 2(C_f + sL). \qquad (46)$$

The constant s is determined by pinch-off measurements of the two devices

$$s = \frac{1}{2} \frac{C_p^{(r)} - C_p}{L_0^{(r)} - L_0}. \qquad (47)$$

L_0 and $L_0^{(r)}$ are the *intended* gate lengths which need not equal the *actual*. However, the difference between the actual gate lengths is the same as the difference between the intended gate lengths

$$L^{(r)} - L = L_0^{(r)} - L_0. \qquad (48)$$

L is determined from the scaling criterion

$$\frac{C_{110x}^{(r)} - (C_p - 2sL)}{L^{(r)}} = \frac{C_{110x} - (C_p - 2sL)}{L} \qquad (49)$$

and (48). The term $C_p - 2sL$ appearing on both sides is the total true gate-length-independent fringe capacitance $2C_f$ in (46), with s given by (47). The numerators in (49) are therefore the *intrinsic* cross-point capacitances for the reference FET and the FET under study, and the denominators are the corresponding gate lengths. The gate length of the FET under study is given by

$$L = \frac{(L_0^{(r)} - L_0)(C_{110x} - C_p)}{(C_{110x}^{(r)} - C_p^{(r)}) - (C_{110x} - C_p)}. \qquad (50)$$

With L determined, C_f is calculated with (46).

Method 2 uses the numerical charge-control calculation to determine the factor α introduced above, and the maximum slope K of the 2DEG concentration with respect to the gate-to-channel voltage. $K = (dn_s/dV_{GS})_{max}$ gives the maximum 2DEG capacitance C_g

$$C_g = qKWL \qquad (51)$$

where q is the electron charge. With unknown gate length, (41) does not give the fringe capacitance explicitly. But together with (45) we can solve for C_f

$$C_f = \frac{\Delta_2 - G_{22}C_{110x}/\alpha}{G_{21} - 2G_{22}\left(\frac{1}{\alpha} - 1\right)}. \qquad (52)$$

Equations (45) and (51) gives us an expression for the gate length in terms of known quantities

$$L = \frac{C_{110x} - 2C_f}{\alpha qKW}. \qquad (53)$$

To compare the two methods described, we picked four MODFET's with dc characteristics typical for the particular wafer (C195). The C–V data in Fig. 5 were taken on one of the four devices. The nominal gate length is 0.6 μm, and the threshold voltage is ≈ -0.5 V. The results for the gate length and fringe capacitance are given in Tables II and III, respectively. With the exception of the fringe capacitance of device 33113 the two methods give very similar result with differences less than 10%. The more laborious Method 1 is only used in this subsection to quality Method 2 since the latter relies on charge-control modeling.

C. Summary of the Reverse Modeling Method

The comparisons in Tables II and III shows that the extraction technique proposed can be quite reliable. This is made possible by the high degree of control that can be achieved with today's GaAs IC processing techniques. MBE controls thicknesses to within a few percent [4]. The doping control is also good and can be checked by monitoring the threshold voltage. A couple of monolayers of $Al_xGa_{1-x}As (x \geq 0.2)$ is sufficient to stop CCl_2F_2/He

TABLE II
GATE LENGTH L (μm) EXTRACTED ON WAFER C195

DEVICE ID	METHOD 1	METHOD 2	% DIFFERENCE
33113	.60	.64	6.8
44113	.61	.61	0.6
55113	.66	.60	9.0
66113	.63	.60	5.1

TABLE III
FRINGE CAPACITANCE C_f (fF/mm) EXTRACTED ON WAFER C195

DEVICE ID	METHOD 1	METHOD 2	% DIFFERENCE
33113	397	328	19.0
44113	344	329	4.4
55113	330	313	5.5
66113	337	323	4.2

GaAs RIE [3]. A careful low-frequency (≤ 10 MHz) C–V study of aluminum-gated FATFET's ($100 \times 80 \ \mu$m^2) was made on a sister-wafer of C195 before and after anneal (1 min at 350°C + 2 hr at 325°C). The eight C–V patterns measured indicated a small (5 Å) increase in the gate-to-channel spacing. The standard deviation was of similar small magnitude (8 Å), so there is no evidence suggesting metal/semiconductor interdiffusion. Thus aluminum gates preserve the vertical device structure defined by the MBE growth and selective RIE. The numerical charge control becomes very applicable, and the parameters used in the analysis can be reliably extracted.

We summarize the steps involved in the reverse modeling: 1) determine α and K by charge-control modeling (one may want to compare the result with C–V measurement on a FATFET); 2) measure the Y-parameters, and use the ones at the bias point of maximum f_{Tx}, and at a frequency low enough (typically 2–8 GHz) so that higher order $j\omega$ terms are negligible and the current gain roll-off is -20 dB/dec; 3) measure $Y_{11} = j\omega C_{110}$ versus V_{GS} ($V_{DS} = 0$), at the same frequency and graphically determine C_{110x} (Fig. 5); 4) use (52) and (39) to extract the fringe capacitance C_f; 5) use (53) to extract the electrical gate length L; 6) use (40) and (51) to extract the source resistance R_s; 7) use (42) and (32) to calculate the effective peak velocity v_p; 8) use (44) and (34a) to calculate the effective transit velocity v_t; and 9) use (43) to calculate the intrinsic output conductance g_{ds0}. If several devices with different gate lengths are measured and the resulting $f_{Ti} - 1/L$ points fall along a straight line with good correlation, a better estimate of the saturation velocity can be calculated from the slope of the regression line

$$v_t = 2\pi \frac{df_{Ti}}{d(1/L)}. \tag{34a'}$$

Unlike the cordal estimates in steps 7) and 8), v_t in (34a') is not distorted by gradual channel effects, and should be a better estimate of v_{p0}. If, on the other hand, the $f_{Ti} - 1/L$ points indicate significant curvature for large $1/L$, this would indicate either significant velocity overshoot or δL. The validity of the assumptions made will be discussed in the next section, in light of the reverse modeling result presented there.

IV. REVERSE MODELING OF MODFET's WITH VARYING GATE LENGTH, STRUCTURE, AND PROCESSING

In this section we apply the reverse-modeling technique to a variety of MODFET's. The data were taken with an HP8510B Network Analyzer by on-wafer measurements using Cascade Microtech™ probes. Calibration was made on a Cascade Microtech™ standard substrate and special on-wafer calibration patterns. The current gain was extrapolated at -20 dB/dec to yield the extrinsic current gain cutoff frequency f_{Tx}. Fig. 6 illustrates the typical f_{Tx} bias dependence of an enhancement-mode MODFET on wafer C195. As the drain voltage increases from zero, f_{Tx} saturates rapidly, and is rather constant beyond 1 V (Fig. 6(b)). There is a slow decrease further into saturation which as been discussed [7] in terms of drain delays. Before the data were analyzed the pad capacitances were subtracted out. These were either measured on special patterns, or calculated with a three-dimensional Laplace solver [21].

A. Varying Gate Length

In order to test the proposed extraction techniques, a quarter of a wafer (C1904D) was allowed to have a variation in gate length. While this causes a variation in performance (as measured by current gain cutoff frequency), it should cause little variation in the extracted source resistance and fringe capacitance. The gate-length variation was accomplished by a misalignment of the gate mask.

Figs. 7 and 8 show the extracted source resistance and fringe capacitance, respectively, versus the extracted gate length for thirteen MODFET's ($V_T \approx 0.2$ V). The variations in R_s and C_f are not correlated with the gate length. This is illustrated by the two types of regression lines. The one with the smallest slope is the standard type of regression done on the x ($=L$) axis. It has the slope $r\sigma_y/\sigma_x$, where r is the correlation coefficient and $\sigma_x(\sigma_y)$ is the standard deviation of the x(y) parameter. The other line has the larger slope σ_y/σ_x. When the correlation is perfect ($r = 1$) the two lines are degenerate. The point of intersection determines the averages of the two parameters. The average source resistance and fringe capacitance are 0.79 $\Omega \cdot$ mm and 240 fF/mm, respectively. Note that, overall, R_s is very close to $R_{s1} = (\partial V_G/\partial I_D)_{IG = \text{const}}$ which is one common dc method of estimating the source resistance. The two calculated broken lines in Fig. 8 illustrate the small variation in fringe capacitance expected for the gate length range and device structure (Fig. 4). They were generated using a two-dimensional Laplace solver [21] assuming that all conductors, including the electron channels in the semiconductor, are perfect. Thus the undercut (d_{ot} in Fig. 4) should include a depletion of the electrons. We used $d_{ot} = 500$ Å for the calculation. Beyond 200 Å there is only a weak dependence of C_f on d_{ot}. The two broken lines are degenerate for $L > 0.5 \ \mu$m. The lower corresponds to the case where we, for $L < 0.5 \ \mu$m, allow L' in Fig. 4 to be shrunk

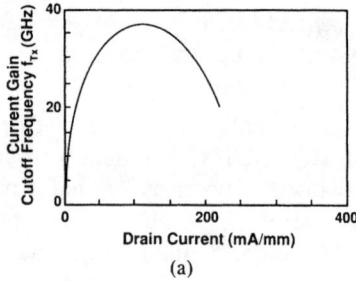

(a)

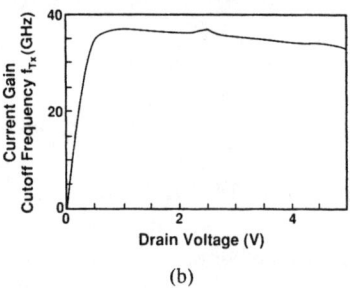

(b)

Fig. 6. Bias dependence of the extrinsic current gain cutoff frequency for a 0.6-μm enhancement-mode ($V_T = 0.2$ V) MODFET (wafer C195). (a) $V_D = 1.5$ V. (b) $V_G = 0.55$ V.

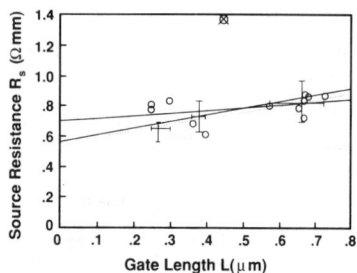

Fig. 7. Extracted source resistance versus extracted gate length for wafer B1904-D. The solid lines are the two types of regression (see text), excluding the crossed-out data point. The horizontal bars show the spread in extracted gate length in three distinct areas on the wafer. The vertical bars shows the spread in $R_{s1} = (\partial V_G / \partial I_D)_{I_G = \text{const}}$ measured on all dc test FET's in these areas. The bars intersect at the medians of L and R_{s1}, and the broken line is the regression of the median of R_{s1} on the median of L.

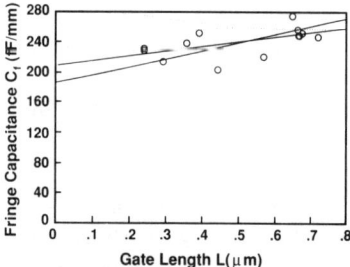

Fig. 8. Extracted fringe capacitance versus extracted gate length for wafer B1904-D. The solid lines are the regression lines. The two broken lines are calculated numerically (see text).

in order to equal the distance between the ohmic contacts, rather than letting the gate metal extend over the ohmic contacts. The calculation, although not rigorous, shows that the extraction gives very realistic values for the fringe capacitance. As illustrated by the scattergram in Fig. 9

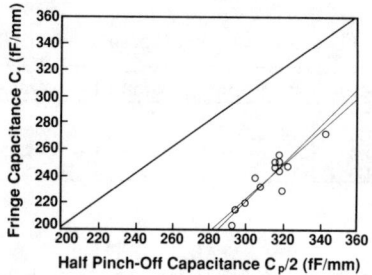

Fig. 9. Scattergram of the extracted fringe capacitance and half of the capacitance measured at zero drain bias and as negative gate bias as allowed by breakdown considerations.

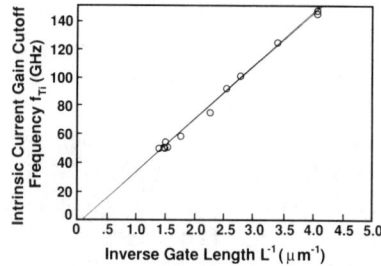

Fig. 10. Measured intrinsic current gain cutoff frequency (44) versus the inverse of the extracted gate length for wafer B1904-D.

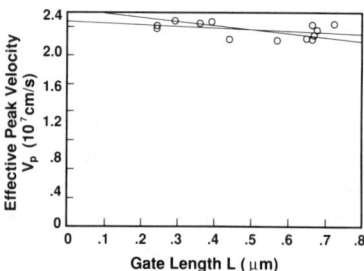

Fig. 11. Extracted effective peak velocity versus extracted gate length on wafer B1904-D.

the pinch-off measurements $C_p/2$ are significantly (≈ 70 fF/mm) larger than the extracted C_f.

In contrast to the parasitic elements, the intrinsic current gain cutoff frequency correlates with $1/L$ almost perfectly as illustrated in Fig. 10. From the slope of this the effective transit velocity is calculated: $v_t = 2\pi d f_{Ti}/d$ $(1/L) = 2.32 \times 10^7$ cm/s. There is no tendency of f_{Ti} to curve upwards. Thus there is no sign of velocity overshoot, or δL being significantly larger than zero. The essentially degenerate regression lines ($r \approx 1$) do not go through the origin. This is consistent with the fact that in the limit of long gate lengths f_{Ti} is proportional to $1/L^2$ rather than $1/L$. Fig. 11 shows the effective peak velocity versus gate length. There is little correlation and the average value ($v_p = 2.20 \times 10^7$ cm/s) is just 5% smaller than v_t. Also this indicates that δL is negligible for these gate lengths when the device is biased for maximum f_{Tx}.

Fig. 12 shows the intrinsic output conductance versus gate length. We can use the shorter gate length data points ($L \leq 0.6$ μm) to extract $g_{dd'}(L = 0) = 119$ mS/mm for the particular epitaxial structure used. Fig. 13 shows that $C_{ds0} + C_{dsp}$ has no significant gate length dependence in

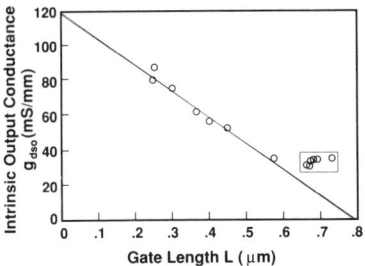

Fig. 12. Extracted intrinsic output conductance versus extracted gate length on wafer B1904-D. The framed data points were excluded when generating the two (essentially overlapping) regression lines.

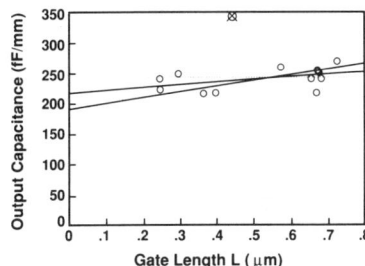

Fig. 13. Extracted output capacitance $C_{ds0} + C_{dsp}$ versus extracted gate length on wafer B1904-D. The crossed-out data point was excluded when generating the two regression lines. The broken line is calculated [21] assuming an 500-Å effective gate etch undercut and a 800-Å pinched-off drain-side region under the gate.

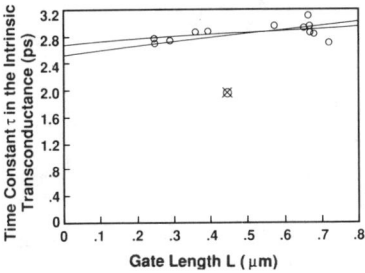

Fig. 14. Extracted time constant in the intrinsic transconductance phase delay versus extracted gate length for wafer B1904-D. The crossed-out data point was excluded when generating the two regression lines.

the regime studied, and compares very well with the numerically [21] calculated capacitance (broken line in Fig. 13). Also τ is essentially independent of L as Fig. 14 illustrates ($\tau \approx 2.9$ ps).

The gate length independence of the extracted parasitic and intrinsic parameters makes the task of modeling submicrometer MODFET's less daunting as will be discussed in Section V.

B. Varying Threshold Voltage and Gate-Trough RIE Time

Two recent articles [7], [20] argue that an increase in 2DEG carrier concentration improves the current gain cutoff frequency, even without an improvement in carrier transport properties. For logic circuits using enhancement-mode drivers, the performance could be severely degraded since ungated regions at the edges of the trough can act as current limiters. The problem becomes more

TABLE IV
PARAMETERS EXTRACTED FROM REVERSE MODELING ON WAFER C195

	4 EFETs		4 DFETs		4 EFETs + 4 DFETs	
	$\bar{x}$	σ	$\bar{x}$	σ	$\bar{x}$	σ
V_T (V)	.145	.004	-.497	.014	-	-
$L(\mu m)$	.686	.026	.612	.022	.649	.045
$C_f(fF/mm)$	290	10	323	8	306	20
$R_s(\Omega mm)$	.807	.054	.569	.063	.688	.138
$v_p(10^7 cm/s)$	2.092	.076	2.012	.039	2.052	.070
$\tau(ps)$	2.550	.151	2.713	.127	2.631	.156

serious the more positive the threshold voltage is. An additional effect is the gate leakage which contributes to limiting the drain current. In the context of this paper, a concern is that we may underestimate the electron saturation velocity. We investigated this possibility by reverse modeling MODFET's with varying threshold voltages, both depletion-mode (DFET's) and enhancement-mode (EFET's). In Table IV we summarize the result for four FET's of each kind on wafer C195. It shows the average ($\bar{x}$) and standard deviation (σ) of the extracted parameters for EFET's and DFET's separately, as well as joined. The most interesting averages have been framed. If, for a parameter, the standard deviation for the FET's as a joined group is not significantly larger than those for the two separate FET groups, we frame the joint average. Otherwise, as in the obvious case of the threshold voltage, we frame the separate averages. In spite of gate length, fringe capacitance, and source resistance being different for EFET's and DFET's, the effective peak velocity for the two is the same. v_p is within 10% of the result in the last subsection which dealt with a wafer with a different doping, yielding FET's with $V_T \approx -0.2$ V. The difference in extracted gate length between EFET's and DFET's is due to the different trough etch, and the fact that, for EFET's, the gate also has to modulate ungated regions at the edges of the trough. The edge modulation is less efficient than the gated-channel modulation which will make the effect apparent also as a higher source resistance. The 10% difference in fringe capacitance is related to a difference in trough etch.

We also applied the reverse modeling to a wafer (C197) with EFET threshold voltage so positive ($V_T \approx 0.6$ V) that there is an obvious degradation in dc performance compared to theoretical predictions. Judging from the dc transconductance one would expect $v_p \approx 1 \times 10^7$ cm/s which is half of what we have become accustomed to. However, the reverse modeling of the measured microwave Y-parameters yields $v_p = 2.27 \times 10^7$ cm/s, which closely matches the earlier results in this section. The discrepancy is directly related to the difference in dc and microwave extrinsic transconductance evident in Fig. 15 for an EFET on this wafer. The maximum extrinsic transconductance g_{mx} is a factor of two larger at microwave frequencies than at dc, and it occurs at larger gate voltages. For positive gate biases the ungated edge depletion notch (Fig. 4) will limit the current, and increase the small-signal source resistance. The latter results in a reduction in g_{mx}. At higher frequencies, the notch appears to be ca-

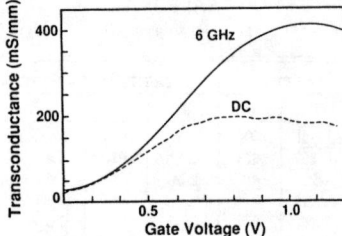

Fig. 15. The extrinsic saturated transconductance versus gate voltage for an enhancement-mode MODFET on wafer C197, at dc and 6 GHz.

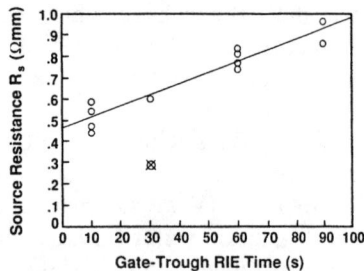

Fig. 16. Extracted source resistance of enhancement-mode MODFET's on wafer C99 versus gate trough RIE time. The crossed-out data point was excluded when generating the regression line.

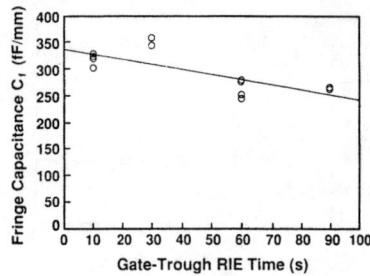

Fig. 17. Extracted fringe capacitance of enhancement-mode MODFET's on wafer C99 versus gate trough RIE time.

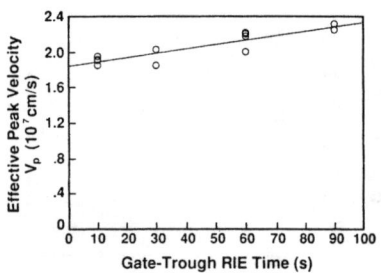

Fig. 18. Extracted peak velocity of enhancement-mode MODFET's on wafer C99 versus gate trough RIE time.

pacitively bypassed, thus effectively lowering the source resistance. Some contribution to the g_{mx} dispersion comes from dispersion in the donor occupancy, but the R_s dispersion appears to be the major cause.

To further demonstrate the effects of the ungated depleted notch Figs. 16–18 illustrate the result of excessive RIE on EFET's ($V_T \approx 0$ V). Wafer C99 was quartered, and four different etch times were used. Once the etchstop is reached the etch proceeds laterally, at a reduced rate. This worsens the trough-edge depletion notch, and results in a noticeable increase in the source resistance (Fig. 16) and a reduction in fringe capacitance (Fig. 17). The electrons communicate less with the gated supply layer, and the extraction of v_p (Fig. 18) may be less distorted by injected parasitic charge [22].

Also in this subsection, despite process variations that from a dc point of view affect the device performance, the extracted effective peak velocity is within 10% of 2.1 × 10^7 cm/s. This suggests that v_p is a property of the GaAs 2DEG. It is even more interesting that v_p essentially equals the equilibrium peak velocity (v_{p0}) in bulk GaAs. There is little sign of overshoot above this limit.

C. Varying Al Mole Fraction and Cap Layer Design

Table V summarizes the reverse modeling result for eight wafers processed identically, and differing only in the Al mole fraction in the doping pulse, and in the cap layer design. The parameter N in Table V is the number of EFET's ($V_T \approx 0$ V) that were measured at microwave frequencies, and analyzed. These MODFET's were picked to represent the much larger number of dc tested devices (126 per wafer). The numbers in parentheses are the standard deviations. v_p is quite independent of the cap layer design, but there is a marginally significant differ-

TABLE V
REVERSE MODELING ON WAFERS WITH DIFFERENT Al MOLE FRICTION AND CAP LAYER

wafer	C91	C94	C92	C93	C95	C97	C98	C99A
x (%)	28	28	28	28	20	20	20	20
d_{cap} (A)	300	300	300	190	300	300	300	190
$N_{cap}(cm^{-3})$	1×10^{18}	1×10^{17}	0	0	1×10^{18}	1×10^{17}	0	0
$V_T(V)$	-.007	.017	.062	-.009	-.003	.026	.050	-.006
	(.025)	(.044)	(.023)	(.030)	(.033)	(.035)	(.036)	(.022)
N	6	6	5	5	5	5	7	4
$f_{Tx}(GHz)$	42.1	41.5	36.3	39.5	41.4	33.9	32.3	31.4
	(1.6)	(0.5)	(0.8)	(1.0)	(1.8)	(0.6)	(2.1)	(0.7)
$L(\mu m)$	.48	.53	.63	.61	.49	.64	.65	.66
	(.04)	(.03)	(.01)	(.01)	(.06)	(.02)	(.03)	(.02)
$C_f(fF/mm)$	343	313	300	247	345	352	349	319
	(4)	(8)	(6)	(14)	(14)	(6)	(4)	(10)
$R_s(\Omega mm)$	.62	.67	.67	.77	.35	.43	.50	.51
	(.04)	(.03)	(.03)	(.02)	(.06)	(.03)	(.04)	(.07)
$v_p(10^7 cm/s)$	2.20	2.34	2.16	2.22	2.09	2.07	2.02	1.92
	(.07)	(.05)	(.03)	(.04)	(.03)	(.03)	(.09)	(.04)
$\tau(ps)$	2.65	2.83	2.89	2.93	3.53	3.61	3.32	3.87
	(.09)	(.06)	(.10)	(.03)	(.35)	(.42)	(.13)	(.89)

ence (in the statistical sense) between the two Al mole fractions. For 28% Al the average effective peak velocity is 2.24 × 10^7 cm/s, while for 20% Al it is 2.03 × 10^7 cm/s. The difference is 10%, but so is the three-sigma value within each group. Table V shows an increase in R_s as a result of a larger trough-edge depletion notch associated with less cap doping and thickness. For the same reason, the gate length increases, and the fringe capacitance tends to decrease, as the cap doping and thickness are reduced. It is interesting that the time constant τ in the transconductance is significantly larger for $x = 20\%$.

D. Pseudomorphic InGaAs Channel MODFET's

Fig. 19 shows the intrinsic current gain cutoff frequency versus inverse gate length for E- and DFET's on two wafers (C317 and C318) where the 2DEG is confined

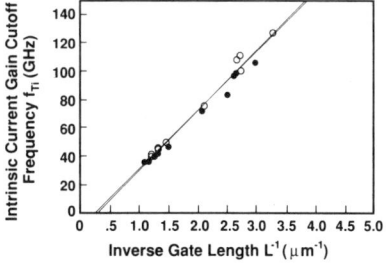

Fig. 19. Measured intrinsic current gain cutoff frequency (44) versus the inverse of the extracted gate length for wafers C317 and C318 (pseudomorphic InGaAs channel). Solid circles are EFET's, open cirles are DFET's.

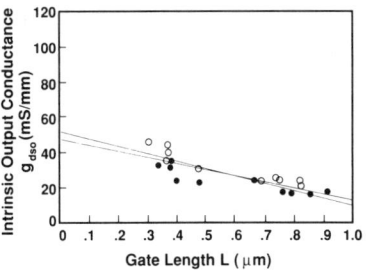

Fig. 20. Extracted intrinsic output conductance versus extracted gate length on wafers C317 and C318 (pseudomorphic InGaAs channel). Solid circles are EFET's, open cirles are DFET's.

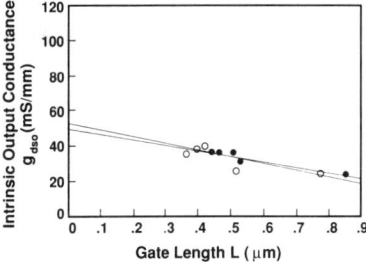

Fig. 21. Extracted intrinsic output conductance versus extracted gate length on wafer C308 (GaAs channel confined by two $Al_{0.28}Ga_{0.72}As$ layers). Solid circles are EFET's, open cirles are DFET's.

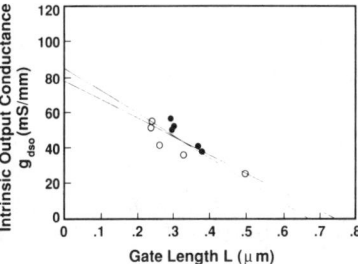

Fig. 22. Extracted intrinsic output conductance versus extracted gate length on wafer C315 (GaAs channel confined by two $Al_{0.20}Ga_{0.80}As$ layers). Solid circles are EFET's, open cirles are DFET's.

by a 100 Å pseudomorphic $In_{0.25}Ga_{0.75}As$ layer. The linear fit yields $v_t = 2.61 \times 10^7$ cm/s. This value is 12% higher than in the GaAs channel case, but considering 1) the larger spread in the data compared to Fig. 10; 2) the larger $1/L$ intercept compared to Fig. 10 despite a similar mobility; and 3) the $\pm 10\%$ expected uncertainty, it is not clear that this difference is significant. The modeling in Section VI indicates that it is not. The fact that the EFET's have slightly lower f_{Ti} than the DFET's is likely to be related to the larger drain voltage necessary to achieve maximum f_{Tx} (≈ 2.8 V versus ≈ 1.3 V for the DFET's). This results in a larger than typical electrical asymmetry of the device, and the assumptions $R_s = R_d$ and $C_{fs} = C_{fd}$ are more likely to be violated. Associated with the increased asymmetry is an increase in electrical gate length (≈ 300 Å/V) beyond that extracted at or below the drain saturation voltage. This shows up as an increase in the input capacitance, and is a manifestation of the drain delay discussed in [7]. The most significant difference between InGaAs pseudomorphic and GaAs channel is the much smaller output conductance of the former as illustrated by Fig. 20 and Fig. 12. The EFET's have lower output conductance than the DFET's due to the larger drain voltage.

E. Double-Heterojunction GaAs/AlGaAs MODFET's with Buried Superlattice and Additional Doping Pulse

Wafers C308 and C315 use a 200-Å GaAs channel confined below by a 10-period 20-Å GaAs/80-Å AlGaAs superlattice very similar to that described in [23]. The center of the last 80-Å AlGaAs layer has a additional 20-Å doping pulse to increase the channel electron concentration. The Al mole fraction is 0.2 in wafer C315, and 0.28 in wafer C308. The extracted transit velocities are very similar: $v_t = 2.11 \times 10^7$ cm/s for C315, and $v_t = 2.15 \times 10^7$ cm/s for C308. These numbers are only slightly lower than for the unconfined GaAs channel in Section IV-A. Electrons escaping into the superlattice are expected to have considerably smaller saturation velocity than in the GaAs channel, and the almost negligible degradation in saturation velocity would suggest that the electrons remain essentially confined to the channel under the gate. This is consistent with the reduction in output conductance (Figs. 21 and 22) as compared to the uncon-

fined GaAs channel (Fig. 12). For the two wafers C308 and C315, as for all other GaAs channel wafers, we did not have to go to high drain voltages to maximize f_{Tx} for the EFET's. These have as high f_{Ti} as the DFET's.

V. SIMPLE TRANSPORT PICTURE FOR EFFICIENT MODFET MODELING

The values extracted for the parasitic and intrinsic elements as well as the dependencies of these on controlled variations in device structure and processing are very physical. The assumptions made have been validated in the cases where the drain voltage is not excessive. The most interesting result is that the effective peak velocity v_p is essentially independent of gate length (for the regime studied) and process variations, and that it is close to the steady-state peak velocity v_{p0} of GaAs. de la Houssaye et al. [24] extracted from dc measurements similar values for even shorter gate lengths. Thus judging from the transconductance, there is little evidence for velocity overshoot (the roll-off in saturation velocity reported [24] for longer gate lengths is likely to be due to gradual channel effects rather than reduced saturation velocity). The large velocities extracted indicate that there is no formation of a longitudinal Gunn domain at the drain end of the channel. This is consistent with the modeling results of Awano [25], and with the moderate roll-off of f_{Tx} for large drain bias (Fig. 6(b)).[1] Thus there is also little evidence that electrons under the drain side of the gate slow down as the stationary velocity-field curve would suggest. Even if electron velocity are allowed to follow this curve, modeling [27] indicates that the performance is still largely governed by the peak velocity. For GaAs MESFET's this was anticipated early by Turner and Wilson [28], and it is likely to be even more appropriate for MODFET's due to the reduced tendency to form longitudinal domains. In spite of the lack of evidence for velocity overshoot, the velocities extracted for single-heterojunction MODFET's are larger than those measured [29] and predicted [30] for electrons confined to an ungated 2DEG. This may be related to the potential well opening up in the velocity saturated region [31], [32] which allows the electrons to move in bulk GaAs where they are faster. If so, real-space transfer of electrons from GaAs to the top AlGaAs is not the major saturation mechanism.

While the transconductance ($Wc_g v_p$) is quite independent of the gate length (Fig. 11), the output conductance (Fig. 12) is several times more sensitive to gate length variations than can be explained by a constant $g_{dd'}$. This is the case even when a barrier that reduces electron penetration into the buffer has been included (Figs. 20–22). One reason for the excess output conductance could be two-dimensional field effects at the drain-end. Another is that charge support by the electric field parallel to the transport (the longitudinal direction) become increasingly noticeable at shorter gate lengths. Modeling [33] indicates that these effects can only account for a fraction of the output conductance at shorter gate lengths. Thus there is a possibility that velocity overshoot does show up in the output conductance. The relative constancy of the output capacitance (Fig. 13) may lend credence to this.

For efficient modeling of MODFET's we would like to avoid having to account rigorously for nonstationary transport phenomena. It has been shown with Monte Carlo techniques [34] that the amount of overshoot for electrons accelerated in fields less than E_c is very small. So, barring velocity *under*shoot for very short gates, we can use the stationary velocity–field curve in the part of the channel where the longitudinal field is $< E_c$. Since E_c is quite small for the III–V compounds of interest, it is appropriate to use the gradual-channel approximation in this region. Beyond the gradual channel (towards the drain) it is difficult to assign an electron velocity, and it is inappropriate to tie it to the local (and very rapidly varying) electric field.

As mentioned, there are also other complications, related to the two-dimensionality of the electrostatics. The problem cannot be accurately treated with analytical means. A good approach is to model the electrostatics of the saturation region [35], and incorporating this self-consistently with the solution in the gradual channel [13]. The transport in the saturated region is not addressed by this approach, but it does correctly predict the onset of saturation, and thus the transconductance. An empirical output conductance $g_{dd'} = f(L)$ has to be included to get good agreement with experiment. This emulates the extremely complicated, but less critical, transport in the saturated region. As gates are shrunk, it may be tempting to increase the velocity used in the model [10], [36]. Experimental data, including ours, indicate that this is not necessary, and that v_{p0} is a good number for the maximum velocity in the gradual part of the channel, independent of the gate length.

The average electron energy corresponding to the electrons having been accelerated in electric field up to $E_c \approx 3.5$ kV/cm can be estimated from modeling work using nonstationary approaches [37], [38]. It varies between 50 and 90 meV at room temperature, and is considerable smaller than the Γ–L separation (0.36 eV) because of tailing of the electron energy distribution function and the large density of states in the L-valley. Combining this with the argument of [20], one can expect the simple modeling approach outlined to be useful at least down to ≈ 0.2 μm at room temperature.

VI. PREDICTION OF MAXIMUM EXTRINSIC MODFET CURRENT GAIN CUTOFF FREQUENCY

In this section, simple *forward* modeling, based on a constant saturation velocity, is compared with measured results. The extrinsic current gain cutoff frequency f_{Tx} predicted by (29b) is compared with the value obtained from a -20 dB/dec extrapolation of the measured current gain at optimum bias. At this bias we assume $\delta L = 0$, as suggested by the reverse modeling. We use [8] to calculate g_{m0}, $g_{d's0}$, and $C_{gs} + C_{gd'}$ at $k = k_{sat}$ given by (12) with $V_{GS} - V_T = n_{s0}/K$. The charge control (K, n_{s0}) is calculated numerically [6]. The low-field mobility used is that extracted from FATFET I-V measurements and the slope K. We use the high-field transport properties (v_t and $g_{dd'}(L = 0)$) extracted on wafers B1904-D, C317, and C318. Using $g_{dd'}$ extrapolated to zero gate length will overestimate the resulting output conductance, and the predicted f_{Tx} will be somewhat underestimated at longer gate lengths.

Fig. 23 shows the experimental and the modeled cutoff frequency versus $1/L$ for wafer B1904-D. Curve 1 corresponds to an ideal MODFET with no parasitic elements. It tracks the experimental f_{Ti} data much like the straight line fit in Fig. 10 did. The only difference is the $1/L^2$ regime for small $1/L$. As mentioned in Section IV-A, this corresponds to the Shockley limit. Curves 2, 3, and 4 correspond to increasing presence of parasitics. The fringe capacitance extracted in Section IV-A has been in-

---

[1]It is interesting at this point to recall Engelmann and Liechti's [15] analysis of GaAs and InP MESFET's. In the InP case where there was little sign of Gunn domain formation, the gate–drain capacitance can be accounted for by just the fringe capacitance. This supports the assumption $pC_{gd'} = 0$ which is also consistent with the results in [26].

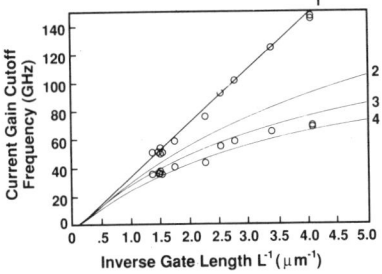

Fig. 23. Measured and modeled extrinsic and intrinsic current gain cutoff frequency versus the inverse of the gate length on wafer B1904-D. The curves are modeled with the following parameters: 1)—C_f, R_s, $g_{dd'} = 0$; 2)—$C_f = 240$ fF/mm, R_s, $g_{dd'} = 0$; 3)—$C_f = 240$ fF/mm, $R_s = 0.791$ $\Omega \cdot$ mm, $g_{dd'} = 0$; and 4)—$C_f = 240$ fF/mm, $R_s = 0.791$ $\Omega \cdot$ mm, $g_{dd'} = 119$ mS/mm. The common input parameters are: $n_{s0} = 1.26 \times 10^{12}$ cm^{-2}; $K = 2.00 \times 10^{12}$/V; $\mu = 7223$ cm^2/V$_s$; and $v_{p0} = 2.32 \times 10^7$ cm/s.

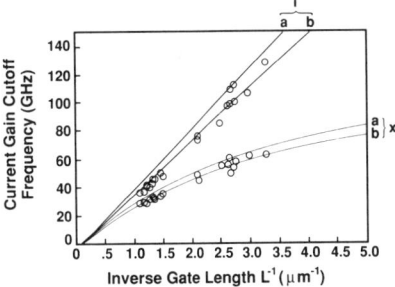

Fig. 24. Measured and modeled intrinsic (i: C_f, R_s, $g_{dd'} = 0$) and extrinsic (x: $C_f = 285$ fF/mm, $R_s = 0.74$ $\Omega \cdot$ mm, $g_{dd'} = 50$ mS/mm) current gain cutoff frequency versus the inverse of the gate length on wafers C317 and C318 (pseudomorphic InGaAs channel). Two sets of transport properties are used: a)—$\mu = 7600$ cm/V $\cdot$ s, $v_{p0} = 2.61 \times 10^7$ cm/s (measured/extracted on wafers C317 and C318); and b)—$\mu = 7223$ cm/V $\cdot$ s, $v_{p0} = 2.32 \times 10^7$ cm/s (measured/extracted on wafer B1904-D).

cluded to produce curve 2. Typically, this is the only degrading effect of parasitics considered. However, we see that we still overestimate the measured f_{Tx}, and we would have to underestimate the electron velocity to get good match. Curves 3 and 4 include the extracted source resistance, and we see that there is a significant reduction in the predicted f_{Tx}. The experimental data fall very close to these two curves. The difference between curve 3 and 4 is that in the latter the finite extracted conductance $g_{dd'}$ has been included. This is not an extrinsic element, but its increased presence at shorter gate lengths increases the degrading effect of the true parasitic elements as evident in (29b) and (30).

Fig. 24 shows the experimental and modeled intrinsic (i) and extrinsic (x) current gain cutoff frequency for the two wafers (C317 and C318) with pseudomorphic InGaAs channel. The values used for the parasitic elements (R_s, C_f) are the average extracted on the two wafers. For $g_{dd'}$ we used the value extrapolated to zero L in Fig. 20. The curves marked "a" use the transport parameters extracted on C317 and C318, while those marked "b" use the same mobility and saturated velocity as in Fig. 23 (GaAs). The latter model the data best, and the saturation velocity extracted in Section IV-D is probably a bit optimistic. As previously observed [7], the pseudomorphic InGaAs channel does not seem to have significantly better transport properties than GaAs.

Fig. 25 shows the modeled f_{Tx} for three different MODFET MBE structures resulting in different carrier concentration and confinement, and for gate lengths down to 0.1 μm. Curves marked S (for standard) correspond to the structure in Fig. 23 ($n_{s0} = 1.26 \times 10^{12}$ cm^{-2}, $K = 2.00 \times 10^{12}$ cm^{-2}/V). Curves marked R (for reduced bandgap discontinuity) correspond to a reduced Al mole fraction ($x = 0.15$) in the doping pulse and spacer. This results in less efficient 2DEG modulation ($n_{s0} = 9.00 \times 10^{11}$ cm^{-2}, $K = 1.64 \times 10^{12}$ cm^{-2}/V). Curves marked P (for pseudomorphic) correspond to the introduction of a thin strained InGaAs channel. This results in greatly improved 2DEG modulation ($n_{s0} = 2.32 \times 10^{12}$ cm^{-2}, $K = 2.32 \times 10^{12}$ cm^{-2}/V). We assume the same transport

properties in all three cases, those used in Fig. 23. However, for the InGaAs channel we do use the observed lower output conductance used in Fig. 24. Curves marked with subscript i (for intrinsic) show $f_{Tx} = f_{Ti}$ in the hypothetical case of no parasitics. Except for the Shockley limit (long channels) where carrier concentration determines g_{m0} there is essentially no difference in maximum f_{Ti} between the three cases. Curves marked with subscript x (for extrinsic) include the extracted parasitics for wafer B1904-D, and we see that this lifts the degeneracy of the curves: larger carrier concentration results in higher extrinsic cutoff frequency. A typical value for contact resistance is 0.1 $\Omega \cdot$ mm, and it is clear that the extracted source resistance is limited by semiconductor access resistance despite the rather tight geometry of Fig. 4. An increase in carrier concentration will therefore reduce the source resistance, and this effect has been taken into account in the two curves marked with subscript x' (0.1 $\Omega \cdot$ mm contact resistance assumed). This results in a more significant improvement in f_{Tx} as a result of the improved carrier concentration and confinement.

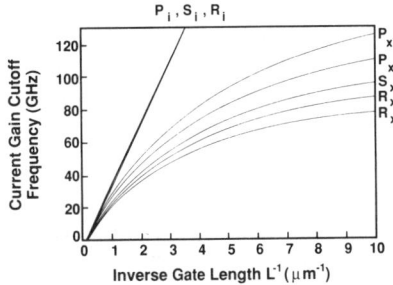

Fig. 25. Predicted current gain cutoff frequency for three MODFET epitaxial structures versus inverse gate length (see text for notation).

VII. SUMMARY

We have presented reverse and forward modeling of MODFET's with state-of-the-art performance [39]. Although the current gain cutoff frequencies set no records they are very respectable for an IC process. The reverse-modeling method is directly applicable to symmetric MODFET's with well-known vertical structure. Considering the complexity of nonstationary electron transport in a two-dimensional rapidly varying field, simplifying assumptions are necessary in order to extract a few meaningful numbers that describe the device. The assumptions made in the reverse modeling proposed here have been put to several tests by varying the gate length, epitaxial structure, and process conditions. They have been found valid for the process used in the study. For other processes, different assumptions may be better. A total of 115 MODFET's (both EFET's and DFET's) on 15 wafers with 7 distinctly different MBE structures were reverse modeled. If we restrict ourselves to those 95 MODFET's that have a GaAs channel (either single or double heterostructure) we find that the average extracted peak velocity v_p is 2.12×10^7 cm/s, with a standard deviation of 0.14×10^7 cm/s. The average value of the individually extracted v_p is within 10% of the typically more accurate v_t. It is interesting to note that the effective electron saturation velocity is rather tightly distributed close to the stationary peak velocity v_{p0} of GaAs. We have seen no evidence that would lead us to assume an increase in effective electron saturation as gate lengths are shrunk. However, in this limit the output conductance does increase faster than can be convincingly explained by simple models. While the extracted saturation velocity is essentially a constant, we have induced predictable variations in the extracted gate length, fringe capacitance, and source resistance with controlled changes in the fabrication process. The extracted fringe capacitance compares well with calculations, and the two methods for estimating the gate length yield results that are very similar. If anything, the reverse modeling may overestimate the source resistance, but we have demonstrated (Fig. 7) that conventional dc methods give similar values when g_{mx} dispersion is not significant. It is because of the careful determination of electrical gate length (rather than lithographical or metallurgical), fringe capacitance, and source resistance that we avoid underestimating the saturation velocity.

The restrictions of the reverse modeling do not apply to the forward modeling of the extrinsic cutoff frequency of MODFET's. Using a simple model with physical parameters and a transport picture where v_{p0} is of central importance, we have demonstrated very good agreement with experiment, down to quarter-micrometer gate length. This is helpful for process monitoring, device/circuit simulation, and realistic prediction of performance improvements. The degrading effects of the source and drain resistive and capacitive parasitics, which increase with increasing output conductance, are accounted for. The modeling offers an alternative explanation to the observed superior performance of MODFET's with larger conduction band offsets. The improvement does not rely on improved transport properties, but rather on the reduction in access resistances and output conductance resulting from a larger and better confined 2DEG.

ACKNOWLEDGMENT

The author wishes to thank A. Nagy for her creative development work, and her careful and efficient processing. He also greatfully acknowledges A. Fischer-Colbrie and M. Kanemura for well-controlled MBE growth; N. Moll for many enlightening discussions and for critical review of the manuscript; B. Hughes, P. Roblin, M. Hueschen, K. Seaward, and S. Camnitz for fruitful discussions; H. Yeager and L. Moresco for setting up the microwave measurement system and powerful analysis tools; R. Jaeger, B. Hughes, H. Yeager, S. Kofol, R. Kaneshiro, and J. Miller for reviewing the manuscript; and F. Mertz and D. Hirschnitz for performing essential process steps.

REFERENCES

[1] P. H. Ladbrooke, "Reverse modelling of GaAs MESFET's and HEMT's," *GEC J. Res. Inc. Marconi Rev.* (UK), vol. 6, p. 1, 1988.
[2] K. Hikosaka, T. Mimura, and K. Joshin, "Selective dry etching of AlGaAs-GaAs heterojunction," *Japan J. Appl. Phys.*, vol. 20, p. L847, 1981.
[3] K. L. Seaward, N. J. Moll, and W. F. Stickle, "The role of aluminum in selective reactive ion etching of GaAs on AlGaAs," *J. Vac. Sci. Technol.*, vol. B6, p. 1645, 1988.
[4] B. J. F. Lin, S. Kofol, C. Kocot, H. Luechinger, J. N. Miller, D. E. Mars, B. White, and E. Littau, "Threshold voltage control of MODFET IC," in *Tech. Dig. 1986 GaAs IC Symp.*, p. 51.
[5] B. Vinter, "Subbands and charge control in a two-dimensional electron gas field-effect transistor," *Appl. Phys. Lett.*, vol. 44, p. 307, 1984.
[6] N. Moll and H. Yeager, unpublished, program developed for internal HP use only.
[7] N. Moll, M. R. Hueschen, and A. Fischer-Colbrie, "Pulse-doped AlGaAs/InGaAs pseudomorphic MODFET's," *IEEE Trans. Electron Devices*, vol. 35, p. 879, 1988.
[8] P. Roblin, S. Kang, A. Ketterson, and H. Morkoç, "Analysis of MODFET microwave characteristics," *IEEE Trans. Electron Devices*, vol. ED-34, p. 1919, 1987.
[9] T. J. Drummond, H. Morkoç, K. Lee, and M. S. Shur, "Model for modulation doped field effect transistor," *IEEE Electron Device Lett.*, vol. EDL-3, p. 338, 1982.
[10] M. B. Das, "Millimeter-Wave performance of ultrasubmicrometer-gate field-effect transistors: A comparison of MODFET, MESFET, and PBT structures," *IEEE Trans. Electron Devices*, vol. ED-34, p. 1429, 1987.
[11] B. Hughes, private communication.
[12] K. Lee, M. S. Shur, T. J. Drummond, and H. Morkoç, "Current-voltage and capacitance-voltage characteristics of modulation-doped field-effect transistors," *IEEE Trans. Electron Devices*, vol. ED-30, p. 207, 1983.
[13] H. Rohdin and P. Roblin, "A MODFET dc model with improved pinchoff and saturation characteristics," *IEEE Trans. Electron Devices*, vol. ED-33, p. 664, 1986.
[14] R. J. Trew, "Equivalent circuits for high frequency transistors," in *Proc. 1987 IEEE/Cornell Conf. on Advanced Concepts High Speed Semiconductor Devices and Circuits*, p. 199.
[15] R. W. H. Engelmann and C. A. Liechti, "Bias dependence of GaAs and InP MESFET parameters," *IEEE Trans. Electron Devices*, vol. ED-24, p. 1288, 1977.
[16] L. D. Nguyen, P. J. Tasker, and W. J. Schaff, "Comments on 'A new low-noise AlGaAs/GaAs 2DEG FET with a surface undoped layer'," *IEEE Trans. Electron Devices*, vol. ED-34, p. 1187, 1987.
[17] P. J. Tasker and B. Hughes, "Importance of source and drain resistance to the maximum f_T of millimeter-wave MODFET's," *IEEE Electron Device Lett.*, vol. 10, p. 291, 1989.
[18] K. Lee, M. S. Shur, and A. J. Valois, G. Y. Robinson, X. C. Zhu, and A. van der Ziel, "A new technique for characterization of the 'end' resistance in modulation-doped FET's," *IEEE Trans. Electron Devices*, vol. ED-31, p. 1394, 1984.
[19] S. Y. Chou and D. A. Antoniadis, "Relationship between measured

and intrinsic transconductances of FET's,'' *IEEE Trans. Electron Devices*, vol. ED-34, p. 448, 1987.

[20] M. C. Foisy, P. J. Tasker, B. Hughes, and L. F. Eastman, ''The role of inefficient charge modulation in limiting the current-gain cutoff frequency of the MODFET,'' *IEEE Trans. Electron Devices*, vol. 35, p. 871, 1988.

[21] K. Lee, fcap2 and fcap3, programs developed for internal HP use only.

[22] J. Y.-F. Tang, ''Two-dimensional simulation of MODFET and GaAs gate heterojunction FET's,'' *IEEE Trans. Electron Devices*, vol. ED-32, p. 1817, 1985.

[23] M. Hueschen, N. Moll, and A. Fischer-Colbrie, ''High-current GaAs/AlGaAs MODFETs with f_T over 80 GHz,'' in *Tech. Dig. 1987 IEDM*, p. 596.

[24] P. R. de la Houssaye, D. R. Allee, Y.-C. Pao, D. G. Schlom, J. S. Harris, and R. F. W. Pease, ''Electron saturation velocity variation in InGaAs and GaAs channel MODFET's for gate lengths to 550 Å,'' *IEEE Electron Device Lett.*, vol. 9, p. 148, 1988.

[25] Y. Awano, ''New transverse-domain formation mechanism in a quarter-micrometre-gate HEMT,'' *Electron. Lett.*, vol. 24, p. 1315, 1988.

[26] D. Arnold, W. Kopp, R. Fischer, J. Klem, and H. Morkoç, ''Bias dependence of capacitance in modulation-doped FET's at 4GHz,'' *IEEE Electron Device Lett.*, vol. EDL-5, p. 123, 1984.

[27] M. F. Abusaid and J. R. Hauser, ''A comparative analysis of GaAs and Si ion-implanted MESFET's,'' *IEEE Trans. Electron Devices*, vol. ED-33, p. 908, 1986.

[28] J. A. Turner and B. L. H. Wilson, ''Implications of carrier velocity saturation in a gallium arsenide field-effect transistor,'' in *Proc. 2nd Int. Symp. GaAs* (Inst. Phys. Soc. Conf. Ser. No. 7, London, 1969), p. 195.

[29] W. T. Masselink, N. Braslau, W. I. Wang, and S. L. Wright, ''Electron velocity and negative differential mobility in AlGaAs/GaAs modulation-doped heterstructures,'' *Appl. Phys. Lett.*, vol. 51, p. 1533, 1987.

[30] K. Yokoyama and K. Hess, ''A Monte Carlo study of electronic transport in $Al_{1-x}Ga_xAs$/GaAs single-well heterostructures,'' *Phys. Rev. B*, vol. 33, p. 5595, 1986.

[31] J. Yoshida and M. Kurata, ''Analysis of high electron mobility transistors based on a two-dimensional numerical model,'' *IEEE Electron Device Lett.*, vol. EDL-5, p. 508, 1984.

[32] D. Loret, ''Two-dimensional numerical model for the high electron mobility transistor,'' *Solid-State Electron.*, vol. 30, p. 1197, 1987.

[33] P. Roblin, H. Rohdin, C. J. Hung, and S.-W. Chiu, ''Capacitance-voltage analysis and current modeling of pulse-doped MODFET's,'' *IEEE Trans. Electron Devices*, vol. 36, p. 2394, 1989.

[34] T. Wang, Ph.D. dissertation, University of Illinois at Urbana-Champaign, 1986.

[35] A. B. Grebene and S. K. Ghandhi, ''General theory for pinched operation of the junction-gate FET,'' *Solid-State Electron.*, vol. 12, p. 573, 1969.

[36] M. Weiss and D. Pavlidis, ''The influence of device physical parameters on HEMT large-signal characteristics,'' *IEEE Trans. Microwave Theory Tech.*, vol. 36, p. 239, 1988.

[37] I. C. Kizilyalli, K. Hess, J. L. Larson, and D. J. Widiger, ''Scaling properties of high electron mobility transistors,'' *IEEE Trans. Electron Devices*, vol. ED-33, p. 1427, 1986.

[38] M. Mouis, J.-F. Pone, P. Hesto, and R. Castagne, ''Aspect ratio phenomena in the high electron mobility transistor,'' in *Proc. 1985 IEEE/Cornell Conf. on Advanced Concepts High Speed Semiconductor Devices and Circuits*, p. 144.

[39] L. D. Nguyen, W. J. Schaff, P. J. Tasker, A. N. Lepore, L. F. Palmateer, M. C. Foisy, and L. F. Eastman, ''Charge control, dc, and RF performance of a 0.35-μm pseudomorphic AlGaAs/InGaAs modulation-doped field-effect transistor,'' *IEEE Trans. Electron Devices*, vol. 35, p. 139, 1988.

Deep-Level Analysis in (AlGa)As–GaAs 2-D Electron Gas Devices by Means of Low-Frequency Noise Measurements

L. LORECK, H. DÄMBKES, MEMBER, IEEE, K. HEIME, SENIOR MEMBER, IEEE, K. PLOOG, AND G. WEIMANN

Abstract—Low-frequency noise of (AlGa)As–GaAs heterostructures grown by molecular-beam epitaxy was investigated. The temperature of the samples was varied between 100 and 400 K. In the frequency range from 1 Hz to 25 kHz noise spectra can be described as superposition of several generation-recombination (GR) noise components. Four deep levels (E = 0.40, 0.42, 0.54, 0.60 eV) were detected, three of which are in agreement with those measured independently by deep-level transient spectroscopy (DLTS).

I. INTRODUCTION

AN IMPORTANT PROGRESS in the development of microwave and high-speed electron devices was the discovery of a quasi two-dimensional electron gas (TEG) located at the interface of certain semiconductor heterostructures [1]. The system (AlGa)As–GaAs is used for fabricating field-effect transistors (FET's) which are called two-dimensional electron gas FET's (TEGFET's) [2], high electron mobility transistors (HEMT's) [3] or modulation-doped FET's (MODFET's) [4]. Electrons are transferred from the n-doped (AlGa)As into the undoped GaAs and form the TEG near the interface. Therefore ionized donors and free electrons are spatially separated and the Coulomb scattering is reduced. Very high mobilities result especially at lower temperatures.

Manuscript received September 23, 1983. This work was supported by Stiftung Volkswagenwerk.

L. Loreck, H. Dämbkes, and K. Heime are with the Universitat Duisburg, D-4100 Duisburg 1, FRG.

K. Ploog is with the Max-Planck-Institut für Festkörperforschung, D-7000 Stuttgart, FRG.

G. Weimann is with Forschungsinstitut der Deutschen Bundespost, D-6100 Darmstadt, FRG.

The deep-level characteristics of the heterostructure are important, since electron traps reduce carrier concentration and mobility and increase noise. They can be characterized by low-frequency noise measurements. Generation-recombination (GR) noise is caused by fluctuation in the number of free electrons. Discrete energy levels in the forbidden gap are able to trap electrons or act as recombination centers. Theory explains that these fluctuations cause Lorentzian shaped GR noise contributions in the specturm of ac open-circuit voltage noise [5], [6]. Each trap has a characteristic spectrum defined by its low-frequency plateau value (amplitude) and the corner frequency, from the variation of which, with temperature, the activation energy of the deep level is deduced. The trap concentration is related to the amplitude. Only levels not more than a few kT away from the Fermi level contribute to noise. By changing the temperature the Fermi level is shifted across the forbidden gap.

II. EXPERIMENTAL PROCEDURE

Two types of heterostructures grown by MBE were investigated (Table I). Their main difference is the spacer layer of undoped (AlGa)As in sample D1 which enhances the mobility of the two-dimensional electron gas.

A test pattern allowed the measurement of electron Hall mobility and contact resistance from a transmission-line structure. For the noise measurements the samples consisted of two ohmic contacts with 60-μm separation and 100-μm contact width. Ohmic AuGe/Ni contacts with a specific contact resistance of about 10^{-4} $\Omega \cdot cm^2$ were used.

Samples were mounted in a cryogenic system allowing

Reprinted from *IEEE Electron Device Lett.*, vol. EDL-5, no. 1, pp. 9–11, Jan. 1984.

TABLE I
n_s SHEET CONCENTRATION OF THE TWO-DIMENSIONAL
ELECTRON GAS

sample type		S-3	D-1
top layer GaAs		22nm	20nm
		$p=10^{15}cm^{-3}$	$p\gtrsim 10^{14}cm^{-3}$
$Al_xGa_{1-x}As$		60nm	70nm
(Si-doped)		$n=1*10^{18}cm^{-3}$	$n=6*10^{17}cm^{-3}$
		$x=0.25$	$x=0.3$
spacer		–	6nm
undoped GaAs		1μm	1.1μm
		$p\gtrsim 10^{15}cm^{-3}$	$p\gtrsim 10^{14}cm^{-3}$
n_s/cm^2	300K	$1.2*10^{12}$	$5*10^{11}$
	77K	$7*10^{11}$	$6.5*10^{11}$
$\mu_{Hall}/cm^2/Vs$	300K	3600	7900
	77K	55 000	130 000

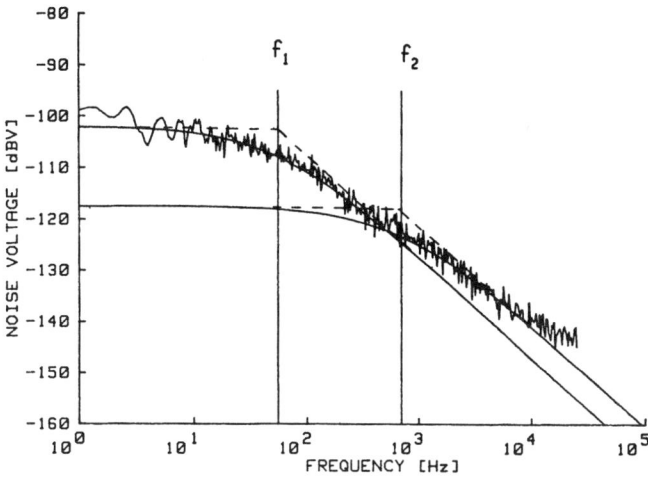

Fig. 1. Typical noise spectrum of sample $S3$ at 220 K and Lorentzian-shaped theoretical curves with corner frequencies f_1 = 55 Hz and f_2 = 700 Hz.

measurements from 20 to 400 K with 1-K accuracy. The equipment was situated in a shielded cabin. A desktop computer HP 9845 B was connected to the measurement apparatus inside the cabin via fiber-optic links.

Variable bias current was supplied by a BURSTER Digistant 6425 T current source with high impedance and low output capacitance, so that the sample noise could not be short circuited by the dc bias supply. A low-noise preamplifier ITHACO 1201 was connected in series with the HP 3582 A spectrum analyzer in order to achieve a higher input sensitivity. Smallest detectable noise voltages were about 10 nV (rms). The frequency range was 1 to 25 kHz. The computer program controlled frequency range and temperature automatically. Regular interferences such as line frequency and their multiples were suppressed by the computer evaluation.

III. MEASUREMENTS

Ohmic behavior of the samples was observed over the temperature range 200–400 K as long as the bias current was less than 5 mA. The spectrum analyzer bandwidth was 726 mHz over the whole frequency range, so that the measurement frequency was larger than the bandwidth.

No special calibration measurements were necessary, because the white noise of a metal film resistor with the same resistance as the sample was far less than that of the sample. Noise voltages were measured in decibels times volts, where 0 dB·V corresponds to 1-V (rms) single tone. Measurements were made at different temperatures with 10-K steps and stored on a magnetic tape.

IV. RESULTS AND DISCUSSION

A characteristic spectrum is shown in Fig. 1. The curves are evaluated by a superposition of several Lorentzian-type GR noise contributions. The amplitudes and characteristic corner frequencies varied with temperature. Corner frequencies are defined as $f_i = 1/2\pi\tau_i$. The activation energies ΔE_i are related to τ_i by

$$1/\tau_i \sim T^2 \exp\left(-\Delta E_i/kT\right)$$

and can be determined from an Arrhenius plot (Fig. 2(a) and (b)).

For comparison, results obtained by Hikosaka et al. [7], from deep-level transient spectroscopy (DLTS) with similar (AlGa)As-heterostructures are included in Fig. 2(a) and (b) as straight lines. Hikosaka et al., prove that the deep levels are located in the (AlGa)As layer. The good agreement between the results from noise and DLTS measurements leads us to the conclusion that the deep levels in our samples are located in the (AlGa)As layer, too.

No noise contribution from the TEG could be detected. This behavior is similar to the high-frequency noise of TEG-FET's which is considerably smaller than that of "normal" GaAs-MESFET's [9]. TEGFET's, therefore, are potentially very low noise devices.

Table II compares the levels detected by Hikosaka et al. [7], by Künzel et al. [10], and by the present authors [8]. Levels $E1$ and $E2$ could not be detected by the noise measurements, because in the available temperature range their corner frequencies are beyond the upper limit of the spectrum analyzer.

Level $E7$ was detected in sample $D1$ only (Fig. 2(a)). Since samples $S3$ and $D1$ are grown at different temperatures with different starting materials and in different MBE systems it is difficult to conclude whether $E7$ is an intrinsic or extrinsic level. The levels $E3–E5$ are common to all (AlGa)As samples. Therefore, they are not specific for the growth process or growth system, but could be either intrinsic defects or impurities common to present starting materials.

The deep-level concentrations cannot be deduced from the present noise spectra. In principle, the amplitude of the plateau value in homogeneous samples is definitely related to the trap concentration [5]. However, in the present samples a considerable transfer of electrons from the (AlGa)As layer into the TEG occurs. Transferred electrons do not contribute directly to the noise spectrum. Additional experiments on thicker (AlGa)As samples without TEG are necessary.

V. CONCLUSION

It was shown experimentally: 1) that low-frequency noise measurements are a useful method for deep-level analysis; 2)

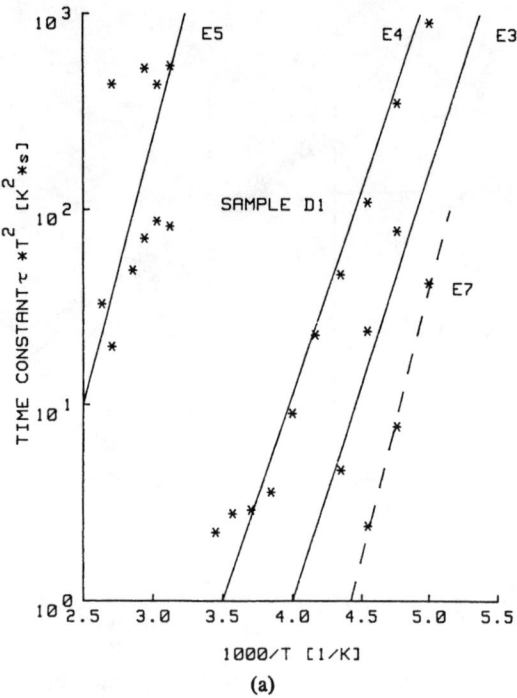

(a)

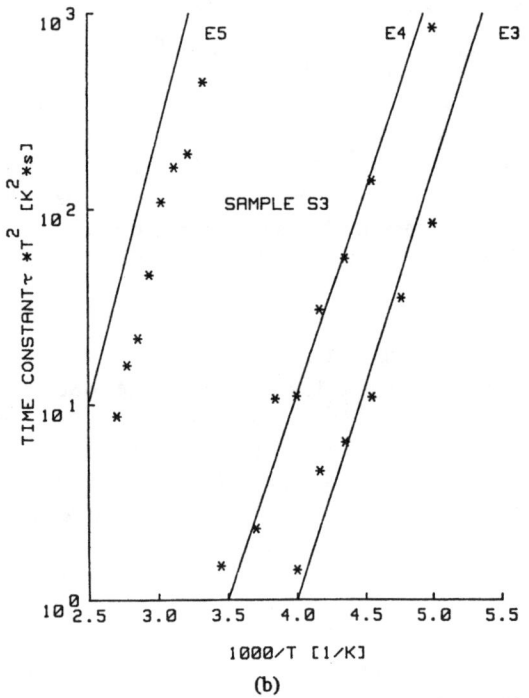

(b)

Fig. 2. Arrhenius plots of $\tau_i T^2$ versus $1/T$ for (a) sample $D1$ and (b) sample $S3$. * experimental values, this work; - - - best fit to experimental values of level $E7$; —— results from [7], [10] by DLTS.

TABLE II

Level	ΔE	Hikosaka et al./7/ (DLTS)	Künzel et al./10/ (DLTS)	this work (noise)
E 1	0.26	+	+	*
E 2	0.28	+	–	*
E 3	0.42	+	+	+
E 4	0.40	+	+	+
E 5	0.60	+	+	+
E 6	0.78	+	–	–
E 7	0.54	–	–	+

+ deep level present
– deep level not present
* deep level not detectable

that GR noise in (AlGa)As–GaAs heterostructures originates from the (AlGa)As only; and 3) that the levels correspond to those detected by DLTS. Deep-level concentrations and capture cross sections can be evaluated only if either an electron transfer into the TEG and a recombination back into the (AlGa)As is excluded experimentally or included into existing theories.

REFERENCES

[1] R. Dingle, H. L. Stormer, A. C. Gossard, and W. Wiegmann, "Electron mobilities in modulation doped semiconductor heterojunction superlattices," *Appl. Phys. Lett.*, vol. 33, pp. 665–667, Oct. 1978.

[2] D. Delagebeaudeuf and N. T. Linh, "Metal-(n) AlGaAs/GaAs two-dimensional electron gas FET," *IEEE Trans. Electron Devices*, vol. ED-29, pp. 955–960, June 1982.

[3] K. Joshin, T. Mimura, N. Niori, Y. Yamashita, K. Kosemara, and J. Saito, "Noise performance of microwave HEMT," in *IEEE Microwave Technology Theory Symp. Dig.* (Boston, MA), 1983, pp. 563–565.

[4] K. H. Duh, A. van der Ziel, and H. Morkoç, "1/f noise in modulation-doped field effect transistors," *IEEE Electron Device Lett.*, vol. EDL-4, pp. 12–13, Jan. 1983.

[5] G. Bosman and R. J. J. Zijlstra, "Generation-recombination noise in p-type silicon," *Solid State Electron.*, vol. 25, no. 4, pp. 273–280, 1982.

[6] A. van der Ziel, *Noise in Measurements.* New York: Wiley, 1976.

[7] K. Hikosaka, T. Mimura, and S. Hyamizu, "Deep electron traps in MBE-grown AlGaAs ternary alloy for heterojunction devices," *Inst. Phys. Conf. Ser.*, no. 63, pp. 233–238.

[8] L. Loreck, H. Dämbkes, K. Heime, K. Ploog, and G. Weimann, "Low frequency noise in AlGaAs-GaAs 2-D electron gas devices and its correlation to deep levels," presented at 7th Int. Conf. on Noise in Physical Systems and 3rd Int. Conf. on 1/f Noise, Montpellier, May 1983.

[9] N. T. Linh, "Microwave performance of two-dimensional electron gas FET," presented at 8th European Specialist Workshop of Active Microwave Semiconductor Devices, Maidenhead, 1983.

[10] H. Künzel, K. Ploog, K. Wünstel, and B. L. Zhou, "Influence of alloy composition, substrate temperature, and doping concentration on electrical properties of Si-doped n-Al$_x$Ga$_{1-x}$As grown by molecular beam epitaxy," to be published.

Excess Gate Current Due to Hot Electrons in GaAs-Gate FETs

D.J. Frank, P.M. Solomon, D.C. La Tulipe, Jr., H. Baratte, C.M. Knoedler, and S.L. Wright

IBM Thomas J. Watson Research Center, P.O. Box 218, Yorktown Heights, NY 10598, USA

We report on the first observations of hot electron effects in the gate current of GaAs-gate FET's. We have made and tested a variety of these FET's using MBE-grown material with 60nm and 35nm thick $Al_{.4}Ga_{.6}As$ gate insulator layers. In measurements at 300K and 77K these devices show drain-voltage-dependent gate current substantially exceeding that which would be expected on the basis of simple vertical transport measurements. We attribute this current to the real-space transfer of hot electrons from the channel of the device into the (Al,Ga)As/GaAs barrier, from which they are collected into the gate.

The GaAs-gate FET is a new variety of heterojunction transistor which has been described in several recent articles [1,2]. It consists of an n+GaAs gate, an (Al,Ga)As layer which acts as a gate insulator, and undoped GaAs for the channel. It offers several improvements over the conventional MOdulation-Doped FET (MODFET), including improved threshold control and the absence of DX-center related problems. We report new, detailed measurements of the characteristics of this device which show that it exhibits a large increase in gate current due to hot electron effects under certain conditions. A somewhat similar result has quite recently been reported for Heterostructure Insulated Gate FET's (HIGFET's) [3]. The details of these effects should prove quite valuable in the study of hot electron dynamics, which become increasingly more important to understand as dimensions continue to be scaled down.

We have investigated these effects in a variety of GaAs-gate FET's made on MBE material with either 60nm $Al_{.4}Ga_{.6}As$ or 35nm $Al_{.5}Ga_{.5}As$ insulator layers. These FET's were fabricated in the same manner described previously [1], using a self-aligned process in which the gate metallurgy serves as part of the mask for the ion-implanted source and drain. The resulting devices had gate lengths ranging from $0.7\mu m$ to $20\mu m$. Figure 1 shows a schematic cross section of the device.

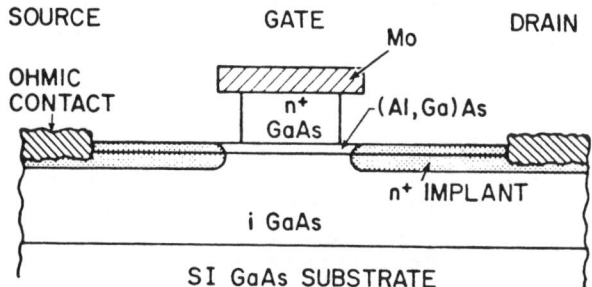

Figure 1. Schematic cross section of a GaAs-gate FET.

We have measured the drain and gate I-V characteristics of many of these GaAs-gate FET's at both room temperature and at 77K. Figure 2 shows the drain and gate I-V curves for a typical device at 77K. Qualitatively similar characteristics were obtained for nearly all of the working devices tested. We find several interesting features that have not previously been discussed in the literature in connection with these devices. There is a negative differential resistance (NDR) region in the drain current I_D vs drain voltage V_D characteristics at high gate voltage, as shown in Fig. 2(a). There is also slight NDR in the gate current I_G vs gate voltage V_G curves at moderate gate voltage and low

Reprinted with permission from *Springer Series in Electronics and Photonics*, pp. 140–143, Stockholm. Sweden, Aug. 7–9, 1986. ©1986 Springer-Verlag, Berlin-Heidelberg.

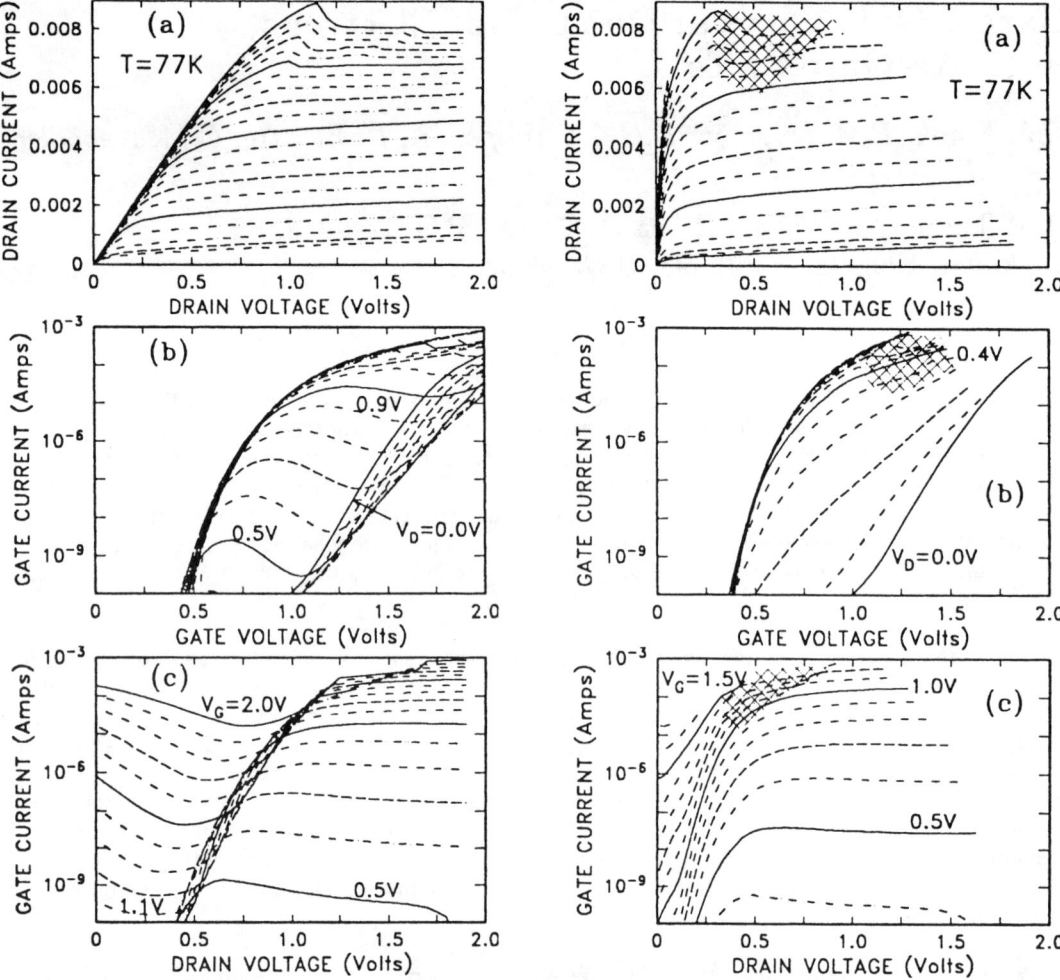

Fig. 2. I-V characteristics for a $1.5\mu m \times 25\mu m$ FET at 77K. (a) shows I_D vs V_D for $V_G = 0$ to 2.0V in steps of 0.1V. (b) shows I_G vs V_G for $V_D = 0$ to 1.9V in steps of 0.1V. (c) shows I_G vs V_D for $V_G = 0.5$ to 2.0V in steps of 0.1V (the lowest curve on the left is $V_G = 1.1V$ and on the right is $V_G = 0.5V$).

Fig. 3. I-V characteristics of the intrinsic FET deduced from Fig. 2, calculated using $R_s = 45.8\Omega$, $R_D = 48.8\Omega$, and a parasitic gate resistance of approximately 440Ω. (a) shows I_D vs V_D for $V_G = 0$ to 1.8V in steps of 0.1V. (b) shows I_G vs V_G for $V_D = 0$ to 1.6V in steps of 0.1V. (c) shows I_G vs V_D for $V_G = 0.4$ to 1.8V in steps of 0.1V.

drain voltage, as in Fig. 2(b). Finally, the gate current is generally higher than would be expected in the absence of hot electrons and depends strongly on V_D, as can be seen from Fig. 2(b) and (c). It decreases at low V_D primarily due to the voltage drop across the parasitic source resistance, which decreases the channel-to-gate voltage. It then increases sharply because of hot electron injection from the channel into the gate. This hot electron injection current finally saturates at larger values of drain bias, while still increasing with gate bias.

At room temperature the characteristics of these devices are similar. The drain I-V curves are qualitatively the same as Fig. 2(a), except that the magnitudes of the drain currents are reduced due to the lower transconductance. The gate leakage current continues to show the increase with V_D, but the thermal leakage is much higher and overwhelms the NDR effect seen in Fig. 2(b).

As a function of gate length, we find that at 77K the hot electron current is roughly constant for gate lengths of 0.7, 1.3 and $1.7\mu m$, decreases somewhat for $4\mu m$, and is still measurable in $20\mu m$

devices. At 300K, the current falls off more rapidly as a function of gate length. This is probably because more voltage drops along the channel of the device due to the lower room temperature mobility.

Unlike the HIGFET, where real-space transfer may directly cause the NDR [3], we find that in our GaAs-gate FET's the NDR features can be accounted for by the parasitic resistances at the three contacts. The NDR on the drain curves occurs when the rapidly rising I_G in this regime, passing through the gate contact resistance, causes the internal gate voltage to fall fast enough that the drain current actually decreases. This NDR causes circuit dependent RF oscillations which distort the average DC currents and produce the sharp slope discontinuities visible in the data. We have observed these oscillations to be in the 100kHz to 100MHz range, depending on circuit conditions. There should also be some decrease in I_D due to the transfer of current to the gate, but this effect seems to be masked by the mechanism above. The NDR in the gate I-V curves is due to the parasitic source and drain resistances. As the gate voltage increases, I_D increases, causing the internal source-to-drain voltage to decrease. If it decreases rapidly enough, the hot electron injection current can actually decrease, even though V_G is increasing. These negative resistances are not surprising when one notes that the hot electron injection effect is inherently regenerative, since the current flows in opposition to the drain bias. In general, when a regenerative effect is combined with passive external circuit elements it can lead to NDR characteristics.

To study the curves quantitatively and to uncover the hot electron injection characteristics of the intrinsic device, we used an HP4145A parameter analyzer to take detailed I-V curves (1599 points each for I_D and I_G) for several FET's at both 77K and 300K. These were then transferred to a mainframe computer which was used to subtract the parasitic resistance voltages. Interpolating among the resulting data yielded I-V curves that correspond to the intrinsic device.

The sum of the source and drain resistances, R_s+R_D, was determined from an extrapolation based on the initial slope of the drain I-V curves as V_G became large. $R_s - R_D$ was deduced by comparing the transconductance at moderate voltages with the transconductance at the same voltages when the source and drain were reversed, assuming the intrinsic FET to be symmetric. For the FET in Fig. 2 we obtained $R_s = 45.8\Omega$ and $R_D = 48.8\Omega$. When these two resistances are taken into account, the negative differential resistance in the gate I-V curves (Fig. 2(b)) is removed. The parasitic gate resistance I-V curve was estimated by assuming that the source transconductance, $g_s = \partial I_s/\partial V_G$, which typically peaks around $V_G = 0.6V$, ought to remain constant for higher values of V_G (due to velocity saturation), and that the extent to which it does not is solely caused by voltage developed on the parasitic gate resistance. This approximation neglects a slight decrease in intrinsic transconductance due to gate depletion with increasing V_G. The parasitic gate resistance estimated in this fashion for the FET in Fig. 2 is 440Ω for I_G below $500\mu A$ and decreases somewhat for higher currents. This is in reasonable agreement with estimates based on the resistivity of the n+GaAs gate and on measurements of the contact resistance of the gate metallurgy.

Figure 3 shows the intrinsic FET characteristics deduced for the device in Fig. 2 by removing the parasitic resistances. The only remaining NDR is in the cross hatched areas which indicate regions of distortion due to the RF oscillations which accompanied the original measurements. All other NDR effects are indeed removed by this process. Furthermore, it appears likely that if the distortions could be removed there would not be any NDR in the cross hatched areas either. This is not absolutely certain, however, since in other experiments we have observed NDR in the I_D of asymmetric FET's even in the absence of hot electron gate current. Having removed all of the parasitic effects, Figs. 3(b) and (c) now show very clearly the intrinsic excess gate current, which we attribute to hot electron injection.

As Fig. 3(c) shows, the hot electron injection current rises very rapidly for intrinsic V_D between 0.2 and 0.5V and tends to saturate for V_D above 0.5V. While the details of these characteristics are difficult to account for because of the complex nature of the two-dimensional FET geometry, the rapid rise in I_G can be explained in general terms as follows. As the source-drain field increases, it heats more and more of the electrons in the GaAs channel to energies exceeding the barrier height posed by the (Al,Ga)As insulator. Scattering then causes real-space transfer [4] of these electrons into the conduction band of the (Al,Ga)As. Since, unlike MODFET's, there is no barrier to prevent these electrons from flowing into the gate, they do so and are collected as leakage current. Since this mechanism requires a 'cold' electron at approximately source potential to be able to acquire enough energy to pass over the (Al,Ga)As barrier into the gate, it should be activated when both the source-drain voltage and the gate voltage exceed the barrier height (~ 0.3eV), as is indeed the case. We note that this effect is similar to those seen in NERFET's [5], which is to be expected, since the structures are similar. The reason the effects are smaller in GaAs-gate FET's is that they only collect those hot electrons that are directly under the gate, while in the NERFET, the collector electrode extends under the entire device and hot electrons are collected everywhere.

We believe the saturation of the electron injection at higher V_D values is due to the field distribution in the FET when drain voltages are high enough to create a stationary Gunn-domain. We estimate that domain formation should occur when V_D is about 0.5V (for ~1μm gate length), which roughly agrees with the voltage at which saturation occurs in our data. Furthermore, two dimensional device simulations using a drift/diffusion model [6] show that for $V_D > 0.5V$ the fields are such that only a limited number of hot electrons can both transfer into the (Al,Ga)As and be collected by the gate. As V_D increases, more electrons can real space transfer into the (Al,Ga)As, but the fields in the (Al,Ga)As direct most of them back into the channel, leaving I_G approximately constant. A different way of stating this is that most of the hot electron gate current comes from the leading edge of the domain, which is shielded from further increases in V_D by the domain itself. Other physical processes which may be involved in this behavior include the transfer of hot electrons to upper valleys and reduced wide angle scattering of hot electrons at high source-drain fields.

This hot electron gate current should not adversely effect the usefulness of the GaAs-gate FET in logic or memory applications. We have not seen semi-permanent threshold voltage shifts like those in MOSFET's, which leads us to conclude that electrons do not become trapped in the (Al,Ga)As. Furthermore, this excess gate current is only large when both V_G and V_D are large. In logic circuits the only time they are both large is during the switch-on transient of a device. Gate current during the transient can slow the switching slightly, but for reasonable logic levels even the maximum possible gate currents are insignificant. For example, if the logic level were 1.0V, the largest possible gate current for the device in Fig. 2 would only be about 10μA. Circuit simulations show that for supply voltages below 1.5V, the increase in delay is less than 1%.

In summary, we have displayed several interesting features in the I-V characteristics of GaAs-gate FET's. These include NDR in the drain and gate I-V curves and substantial excess gate current which is dependent on source-drain voltage. We have shown that the NDR results from the device currents coupled with the parasitic contact resistances of the device. We have carried out a computer-aided analysis of the data to remove the parasitic effects and to display the intrinsic device characteristics. These characteristics show an excess gate current which rises very rapidly for V_D and V_G of order the GaAs/(Al,Ga)As barrier height and saturates for somewhat higher drain voltages. We believe these effects to be caused by real-space transfer of hot electrons from the channel of the device into the (Al,Ga)As, from whence they are collected into the gate.

1.   P. M. Solomon, C. M. Knoedler and S. L. Wright: IEEE Elec. Dev. Lett., EDL-5, 379 (1984)
2.   K. Matsumoto, M. Ogura T. Wada, N. Hashizume, T. Yao and Y. Hayashi: Electron. Lett., 20, 462 (1984)
3.   M. S. Shur, D. K. Arch, R. R. Daniels and J. K. Abrokwah: IEEE Elec. Dev. Lett., EDL-7, 78 (1986)
4.   K. Hess, H. Morkoc, H. Shichijo and B. G. Streetman: Appl. Phys. Lett., 35, 469 (1979)
5.   A. Kastalsky, S. Luryi, A. C. Gossard and R. Hendel: IEEE Elec. Dev. Lett., EDL-5, 57 (1984)
6.   J. Y.-F. Tang: IEEE Trans. Elec. Dev., ED-32, 1817 (1985)

50-nm Self-Aligned-Gate Pseudomorphic AlInAs/ GaInAs High Electron Mobility Transistors

Loi D. Nguyen, *Member, IEEE*, April S. Brown, *Senior Member, IEEE*, Mark A. Thompson, and Linda M. Jelloian

Abstract—We report on the design and fabrication of a new class of 50-nm self-aligned-gate pseudomorphic AlInAs/ GaInAs High Electron Mobility Transistors (HEMT's) with potential for ultra-high-frequency and ultra-low-noise applications. These devices exhibit an extrinsic transconductance of 1740 mS/mm and an extrinsic current-gain cutoff frequency of 340 GHz at room temperature; the latter is the highest value yet reported for any three-thermal semiconductor device. We also report for the first time a detailed comparison of the small-signal characteristics of a pseudomorphic and a lattice-matched AlInAs/GaInAs HEMT with similar gate length (50 nm) and gate-to-channel separation (17.5 nm): the former demonstrates a 16% higher transconductance and a 15% higher current-gain cutoff frequency, but exhibits a 38% poorer output conductance. Finally, we offer an analysis on the high-field transport properties of ultra-short gate-length AlInAs/GaInAs HEMT's. This analysis shows that a reduction of gate length from 150 to 50 nm neither enhances nor reduces their average velocity. In contrast, the addition of indium from 53% to 80% improves this parameter by 19%, from approximately 2.6 to 3.1 × 10⁷ cm/s.

I. INTRODUCTION

EVER SINCE the advent of the field-effect transistor (FET), significant efforts have been devoted to improve its speed. Although these efforts often involve improvements in all aspects of the FET, the most straightforward approach has always been a reduction of gate length. A survey of the open literature clearly supports this trend. In a five-year period from 1984 to 1989, the gate length of state-of-the-art FET's (or more precisely HEMT's) shrank by more than a factor of 3, from 0.33 to 0.1 μm; as a result, their speed increased by a factor of 5, from 50 to 250 GHz [1]–[6].

The remaining factors that affect the speed of a FET (or HEMT) are material properties and device parasitics. In fact, this factor-of-5 increase in speed could not have been achieved without significant improvements in these two areas. For example, the material system of choice for ultra-short gate-length HEMT's has moved from AlGaAs/GaAs [1], [2] to pseudomorphic AlGaAs/GaInAs [3], [4] to lattice-matched or pseudomorphic AlInAs/ GaInAs [5]–[10]. The latter is most attractive in the

sub-0.1-μm regime because of its low Schottky-barrier height [11], high conduction-band discontinuity [12], high mobility [13], and high peak velocity [14]. Together these properties allow very close gate-to-channel separations, high transconductances, high intrinsic speeds, and low parasitic charging times. We have demonstrated some of these advantages in an earlier work, in which a transconductance (g_m) of 1150 mS/mm and a current gain cutoff frequency (f_T) of 250 GHz were achieved [15].

In this work, we have further advanced the art by demonstrating significant improvements in both g_m and f_T with a new class of 50-nm gate-length pseudomorphic Al-InAs/GaInAs HEMT's. Despite their extremely short gate length, these devices exhibit well-behaved characteristics and show no apparent sign of short-channel effects. In addition, we have successfully facilitated a detailed comparison of the small-signal characteristics of a lattice-matched (53% In) and a pseudomorphic (80% In) with similar gate length (50 nm) and gate-to-channel separation (17.5 nm). This comparison not only confirms, but also quantifies, the superior characteristics of short-gate-length pseudomorphic AlInAs/GaInAs HEMT's.

The main body of this paper is divided into five sections. Section II addresses the scaling of nanometer gate-length HEMT's. Section III describes the material growth. Section IV reports the device fabrication. Section V presents the device comparison. The last section, Section VI, discusses the significance of the work and offers suggestions for future work.

II. DEVICE SCALING

The scaling of FET's and HEMT's with gate-length begins with the intrinsic gate capacitance (C_g) and intrinsic transconductance (g_{m0}). The total delay (t_T) through a FET, which is given by $1/2\pi f_T$ by definition, can be approximately taken as the sum of the gate pad and gate fringe capacitance charging times (t_{pad}, t_{fringe}), the intrinsic delay ($T_i = C_g/g_{m0}$), and an additional electron transit time across the high-field drain depletion region (t_{drain}) [16]

$$t_T = t_{pad} + t_{fringe} + t_i + t_{drain}$$

$$= \frac{C_{pad}}{g_m W} + \frac{C_{fringe}}{g_{m0}} + \frac{C_g}{g_{m0}} + t_{drain} \qquad (1)$$

Manuscript received January 31, 1992; revised March 30, 1992. This work was supported in part by Air Force Rome Laboratory under Contract F19628-90-C-0171. The review of this paper was arranged by Associate Editor N. Moll.

The authors are with Hughes Research Laboratories, Malibu, CA 90265.

IEEE Log Number 9201829.

Reprinted from *IEEE Trans. Electron Devices*, vol. 39, no. 9, pp. 2007–2014, Sept. 1992.

205

where C_{pad} is the gate pad capacitance and W is the gate width; C_{fringe} and g_{m0} are the gate fringe capacitance and intrinsic transconductance per unit gate width. C_g is approximately given by

$$C_g = \frac{\epsilon L_g}{d + \Delta d} \qquad (2)$$

where ϵ is the dielectric constant, L_g is the physical gate length, d is the gate-to-heterojunction separation, and Δd is the displacement of the two-dimensional electron gas (2DEG) from the heterojunction interface.

Let v_{ave} be the time-average electron velocity throughout the entire length of the gate, namely

$$v_{\text{ave}} = \frac{L_g}{t_i} = \frac{L_g}{\displaystyle\int_0^{L_g} \frac{dx}{v(x)}} \qquad (3)$$

where $v(x)$ is the velocity at a coordinate x along the channel. The classic saturated-velocity model (SVM) gives

$$g_{m0} = \frac{\epsilon}{d + \Delta d} v_{\text{ave}} \qquad (4)$$

and

$$t_i = \frac{C_g}{g_{m0}} = \frac{L_g}{v_{\text{ave}}}. \qquad (5)$$

Fig. 1 shows an example of how this model might be used to design the HEMT epitaxial structure for a given gate length and material system, in which the extrinsic f_T's of a 50-nm gate-length pseudomorphic AlGaAs/GaInAs and a lattice-matched AlInAs/GaInAs HEMT are shown as functions of the gate-to-channel separation. We used an average velocity of 1.8 and 2.6×10^7 cm/s, respectively, and typical values for other parasitic elements (see Table I). Our calculation predicts an increase in f_T approximately from 140 to 260 GHz for the former, and from 200 to 340 GHz for the latter, as the gate-to-channel separation shrinks from 35 to 10 nm.

Three types of AlInAs/GaInAs HEMT structures were investigated during the course of this work. The first structure, as illustrated in Fig. 2, is a baseline design which consists of a relative-thick lattice-matched GaInAs channel (40 nm) and a uniformly doped donor layer (8 nm, 6×10^{18} cm^{-3}) [15], [17]. This design typically yields a 2DEG sheet density of 3.0×10^{12} cm^{-2} and a mobility of 10 500 cm^2/V · s at room temperature. The second structure employs a much narrower channel (20 nm) and a delta-doped donor layer ($\sim 5 \times 10^{12}$ cm^{-2}), but exhibits a similar 2DEG sheet density and mobility (2.9×10^{12} cm^{-2}, 10 050 cm^2/V · s). The third structure incorporates a thin pseudomorphic GaInAs layer (7.5 nm, 80% In) in the uppermost of its 20-nm-thick channel. This structure exhibits an approximately 20% increase in 2DEG sheet density, from 2.9 to 3.6×10^{12} cm^{-2}, and 26%

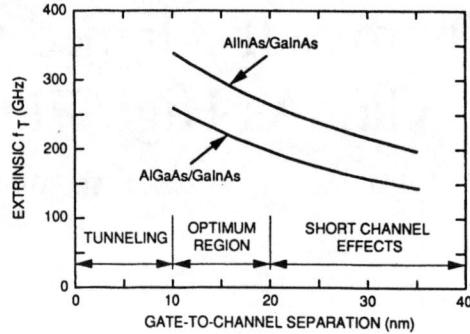

Fig. 1. Calculated extrinsic f_T for 50-nm gate-length pseudomorphic AlGaAs/GaInAs and lattice-matched AlInAs/GaInAs HEMT's.

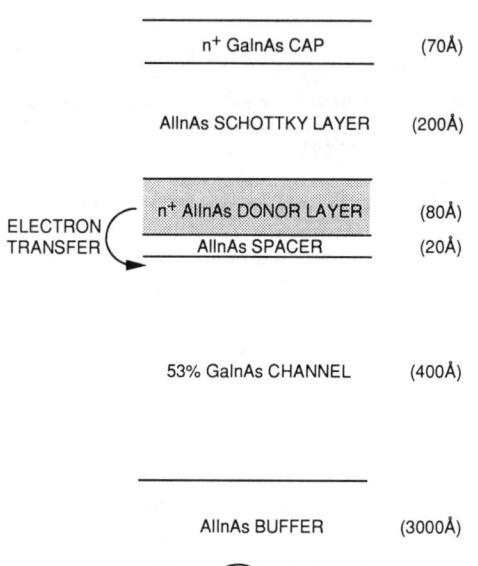

Fig. 2. A baseline lattice-matched AlInAs/GaInAs HEMT structure.

TABLE I
TYPICAL MATERIAL AND DEVICE PARAMETERS FOR 50-nm GATE-LENGTH PSEUDOMORPHIC AlGaAs/GaInAs AND LATTICE-MATCHED AlInAs/GaInAs HEMT'S

Parameters	AlGaAs/GaInAs	AlInAs/GaInAs	Unit
v_{ave}	1.8×10^7	2.6×10^7	cm/s
W	50	50	μm
R_s	0.3	0.2	$\Omega \cdot$ mm
t_{drain}	0.1	0.1	ps
C_{pad}	10	10	fF
C_{fringe}	300	300	fF/mm

increase in mobility, from 10 050 to 12 800 cm^2/V · s at 300 K.

III. MATERIAL GROWTH

The epitaxial layers used in this work were grown sequentially in a Riber 2300 molecular-beam epitaxy (MBE) system, typically at a substrate temperature of 500°C, a growth rate of 600 nm/h, and a V/III beam equivalent pressure ratio of 40. Further details of these standard growth conditions can be found in [18].

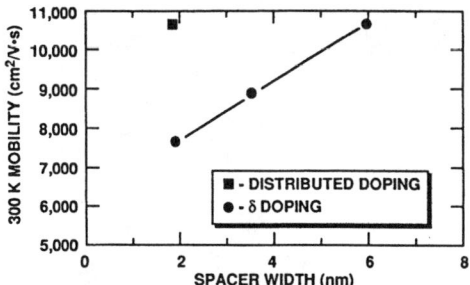

Fig. 3. Dependence of 300 K mobility on spacer thickness for uniformly and delta-doped AlInAs/GaInAs HEMT's.

The growth of the delta-doped structures deviates from the standard conditions in several ways. The first is a reduction of the substrate temperature during and after the growth of the delta-doped layer in order to minimize the surface segregation of silicon atoms in AlInAs. By controlling the growth temperature in this fashion, we are able to improve the abruptness of the silicon profile and achieve and increase of approximately 20 to 30% in electron transfer efficiency [19]. In this work, the growth was interrupted twice, approximately 1 min each, to accommodate the temperature changes. The first interruption occurred just before the deposition of the delta-doped layer, during which the substrate temperature was reduced to 320°C. The second interruption took place immediately after the growth of this delta-doped layer and the first 1.7 nm of the undoped AlInAs Schottky layer, during which the substrate temperature was reset to 500°C.

The second modification is the use of a wider spacer thickness. In order to maintain a reasonably high mobility, we increased the spacer thickness of the delta-doped structures from 2 to 6 nm. As shown in Fig. 3, their room-temperature mobility rapidly increases from approximately 7800 to 10 050 cm^2/V · s as a result of this modification; the latter value is comparable to that of a baseline uniformly doped structure with 2.0-nm spacer thickness.

Finally, the last, but most important, is a lower substrate temperature during the growth of the pseudomorphic channel. We grew this layer at a substrate temperature of 440°C, as opposed to the standard 500°C, to prevent the formation of three-dimensional nucleations and misfit dislocations [20]. The high quality of our pseudomorphic structures is clearly evident in their high mobilities, both at 300 and 77 K. As shown in Table II, the addition of indium, from 53 to 80%, improves the mobility by 27% at 300 K, from 10 050 to 12 800 cm^2/V · s, and as much as 45% at 77 K, from 33 000 to 48 000 cm^2/V · s. In comparing to previous work [13], our best 80% indium pseudomorphic structure exhibits a 100% higher 2DEG sheet density, 3.6 versus 1.8 × 10^{12} cm^{-2}, but suffers only a 19% lower mobility, 12 800 versus 15 300 cm^2/V · s.

We also used low-temperature photoluminescent (PL) as a means to evaluate the material quality. Fig. 4(a) and (b) shows the 10 K PL spectrum from the channel region

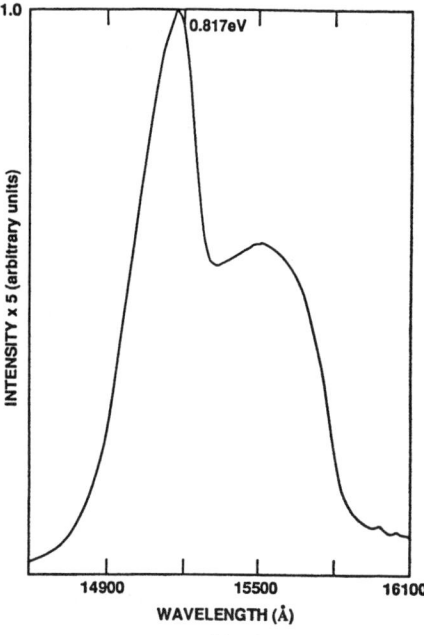

(a)

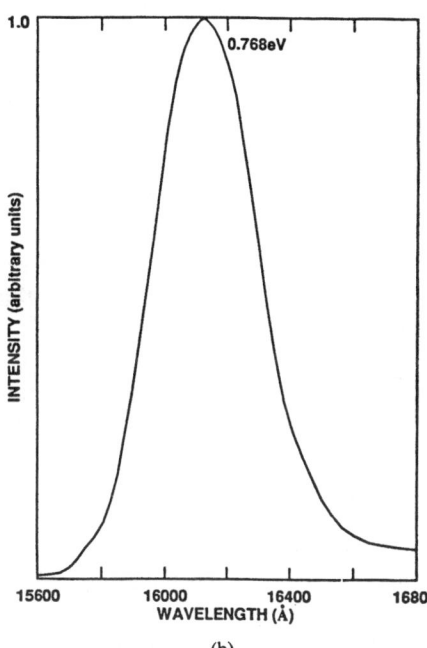

(b)

Fig. 4. 10 K photoluminescent spectrum of (a) lattice-matched and (b) pseudomorphic (80% In) AlInAs/GaInAs HEMT.

TABLE II
TYPICAL 2DEG SHEET DENSITIES AND MOBILITIES OF AlInAs/GaInAs
HEMT's

		300 K		77 K	
% In	W (nm)	n_s (cm^{-2})	μ_s (cm^2/V · s)	n_s (cm^{-2})	μ_s (cm^2/V · s)
53	40	3.3 × 10^{12}	10 657	3.3 × 10^{12}	39 000
53	20	2.9 × 10^{12}	10 050	3.1 × 10^{12}	33 000
65	20	3.2 × 10^{12}	11 900	3.4 × 10^{12}	39 000
80	7.5	3.6 × 10^{12}	12 700	3.8 × 10^{12}	48 000
80	7.5	3.1 × 10^{12}	12 800	3.6 × 10^{12}	43 000*

*Poor Hall sample contact.

of a lattice-matched and a pseudomorphic (80% In) Al-InAs/GaInAs HEMT. The optical quality of the pseudomorphic PL spectrum is in fact more superior. Its intensity is higher and its linewidth is narrower. Therefore, we believe that our pseudomorphic structures are of very high quality and essentially free of dislocations.

IV. DEVICE FABRICATION

We employed a self-aligned-gate process first proposed by Mishra *et al.* [6], which consists of five levels: 1) alignment marks, 2) device isolation, 3) T-shaped gate definition, recess, and metallization (Ti/Pt/Au), 4) ohmic definition, metallization (AuGe/Ni/Au) with the T-shaped gate serving as a shadow mask, and alloying, and finally, 5) overlay. We defined the sub-0.1-μm gates in a bi-layer of PMMA/P(MMA-MAA) electron-beam resist using a Philips EBPG-4 electron-beam lithography system operating at 50-kV acceleration voltage. Fig. 5(a) and (b) shows the cross section of a 50-nm T-shaped gate before and after the deposition of the self-aligned ohmic metal. The addition of the ohmic metal reduces its resistance from approximately 1500 to 1000 Ω/mm, which is the lowest resistance ever reported for a 50-nm-long gate.

We have successfully fabricated three lots of wafers using this process. The first lot consists of four baseline wafers with gate length between 100 and 80 nm [15]; the second lot includes a narrow-channel and a pseudomorphic wafer with gate length of 150 nm; and the third lot is a repeat of the second, but with gate length between 50 and 80 nm. We will, however, limit most of our discussions within the results obtained from the third wafer lot.

We obtained a total device yield of 67% for 1054 lattice-matched devices and 43% for 629 pseudomorphic devices, all with gate length between 50 and 80 nm. As shown in Fig. 6, the mean and standard deviation of the peak extrinsic g_m are 1270 and 88 mS/mm for the former, and 1580 and 325 mS/mm for the latter. The lower yield and wider g_m spread of the pseudomorphic devices are caused mostly by broken or partial gates, as opposed to variations in the material or gate recess depth.

Thirteen of the pseudomorphic devices exhibit a peak extrinsic g_m as high as 1700 to 1740 mS/mm, which is the highest value yet reported for a normal HEMT of any gate length. Fig. 7 shows the *I–V* characteristics of a typical pseudomorphic device with 50-μm gate width, which exemplify its good pinchoff characteristic, uniform transconductance, low knee voltage, and no apparent short-channel effects. The maximum drain–source operating voltage of this device is approximately 1.2 V, which is sufficient for most low-noise applications. The best pseudomorphic device exhibits an extrinsic f_T as high as 340 GHz, as shown in Fig. 8. To our knowledge, this is the highest value yet reported for any three-terminal semiconductor device at room temperature.

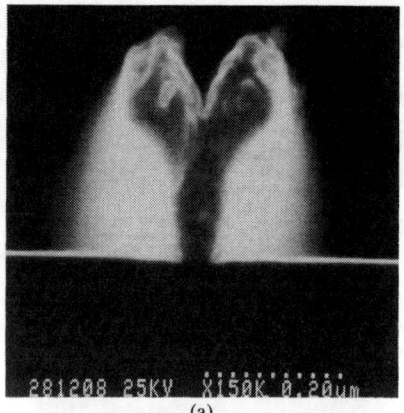
(a)

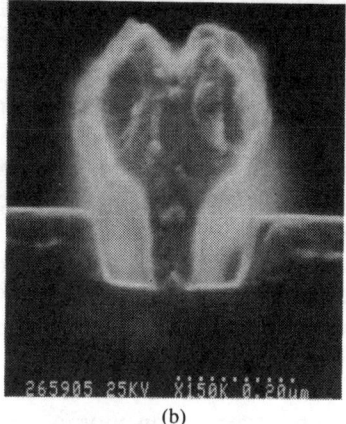
(b)

Fig. 5. Cross section of a 50-nm T-shaped gate (a) before and (b) after ohmic metal.

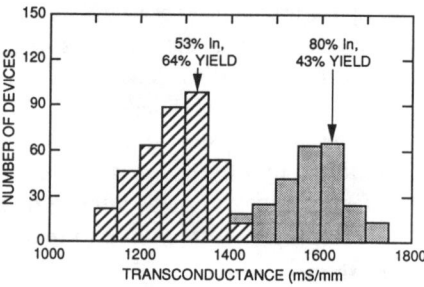

Fig. 6. Peak transconductance distributions for lattice-matched (53% In) and pseudomorphic (80% In) AlInAs/GaInAs HEMT's with 50- to 80-nm gate length.

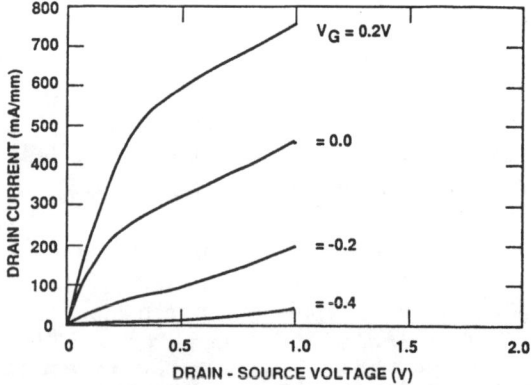

Fig. 7. *I–V* characteristics of a typical 50-nm gate-length pseudomorphic (80% In) AlInAs/GaInAs HEMT.

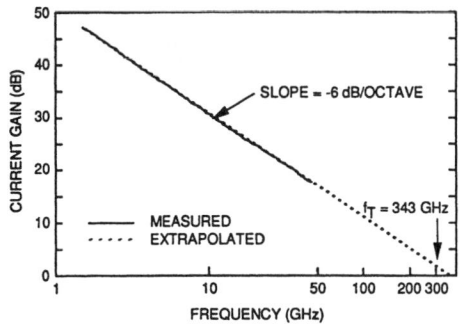

Fig. 8. Current gain versus frequency for a 50-nm by 50-μm pseudomorphic (80% In) AlInAs/GaInAs HEMT with f_T of 340 GHz. $V_d = 0.8$ V, $V_g = 0.0$ V, $I_d = 620$ mA/mm.

V. LATTICE-MATCHED VERSUS PSEUDOMORPHIC

Since its introduction in 1986 [21], the pseudomorphic AlInAs/GaInAs HEMT has consistently demonstrated higher mobility [13], [22], drift velocity [22], and current gain cutoff frequency [5], [10] than comparable lattice-matched devices. Recently, it has also demonstrated the best noise performance at 94 GHz and is currently the only three-terminal semiconductor device with useful gain above 100 GHz [9].

To date, however, no detailed analysis has been reported on the high-field transport properties of submicrometer gate-length pseudomorphic AlInAs/GaInAs HEMT's. Mishra et al. reported an increase of 11% and 17% in extrinsic g_m and f_T, respectively, for 0.1-μm gate-length devices with the addition of 9% indium (from 53% to 62%) [5]. Lai et al., on the other hand, reported an increase of as much as 77% in extrinsic f_T for 1.0-μm gate-length devices with the addition of 17% indium (from 53% to 70%) [10]. These apparent contradictory results (among others) clearly call for a more controlled experiment in which all major process variations can be either minimized or accounted for.

Our self-aligned-gate process provides an excellent environment for such an experiment. The HEMT's reported here employ extremely close gate-to-source and gate-to-drain spacings (~ 0.1 μm) and, as a result, their source resistance is hardly affected by an increase in indium composition. In fact, the source resistance of our devices is dominated by the contact resistance (~ 0.15 $\Omega \cdot$ mm), which can be easily measured with an independent test pattern.

In addition, we made every attempt to minimize all process variations, especially at the gate definition and recess steps. We also characterized the fabricated devices extensively to account for any process variations. We evaluated five extrinsic parameters: 1) RF transconductance (g_m), 2) gate-to-source capacitance (C_{gs}), 3) gate-to-drain capacitance (C_{gd}), 4) RF output conductance (g_{ds}), and 5) current-gain cutoff frequency (f_T). The first four parameters are deduced from the Y-parameters (obtained by first mea-

suring the S-parameters) as follows [23]:

$$g_m = |Y_{21} - Y_{12}|$$

$$g_{ds} = \text{Re}\{Y_{22} + Y_{12}\}$$

$$C_{gs} = -\cfrac{1}{\text{Im}\left\{\cfrac{1}{Y_{11} + Y_{12}}\right\}\omega}$$

$$C_{gd} = -\cfrac{1}{\text{Im}\left\{-\cfrac{1}{Y_{12}}\right\}\omega}$$

where $\omega = 2\pi f$ and f is the operating frequency. The last parameter, f_T, was obtained by extrapolating the current gain to unity at a standard 6-dB/octave slope.

The small-signal parameters of our devices were measured from 1 to 45 GHz using a Wiltron 360 network analyzer and Cascade Microtech on-wafer probes. Prior to each measurement session, the system was calibrated against a Cascade Microtech Impedance Standard Substrate using a Short-Load-Open-Through (SLOT) calibration procedure. Thus the results reported here include all parasitic elements, including all bonding pads. We randomly tested 20 devices from each wafer and then, based upon the test results, selected a matching pair of devices with similar gate length and gate-to-channel separation for subsequent analysis.

Fig. 9 shows the bias dependence of C_{gs} and C_{gd} of such a matching pair at a fixed drain voltage of 0.8 V. Except for a 64-mV shift in pinchoff voltage, which is approximately equal to the difference in their conduction band discontinuities, the capacitances of these two devices are virtually identical. The same is also true at 0-V drain bias at which the capacitance measurements are least affected by the presence of the series parasitic resistances and inductances. We estimated the gate length (L_g) by solving the following equation:

$$qn_s = \int_{V_{po}}^{V_k} \frac{C_{gs} + C_{gd} - C_{\text{fringe}} - C_{\text{pad}}}{WL_g} dV \qquad (6)$$

where q is the electronic charge, n_s is the maximum 2DEG sheet density as measured by the Hall effect technique, V_{po} is the pinchoff voltage, and V_k is the knee voltage at which the total gate capacitance—measured at 2 GHz and 0-V drain bias—sharply increases (see, for example, [24]). From these measurements, we estimated the gate length and gate-to-channel separation of both devices to be approximately 50 and 17.5 nm, respectively, both with an experimental error of approximately 10% (a more detailed modeling of the bias dependence of the transconductance and gate capacitance also agrees with this rough estimate).

Fig. 10 shows the g_m and g_{ds} characteristics of the above devices at 2 GHz. Since their gate lengths, gate-to-chan-

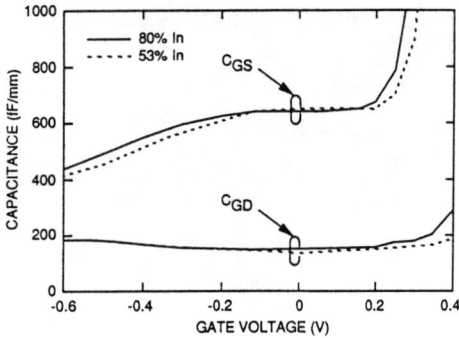

Fig. 9. C_{gs} and C_{gd} of a lattice-matched (53% In) and pseudomorphic (80% In) HEMT with identical gate length and gate-to-channel separation. $V_d = 0.8$ V, $f = 2.0$ GHz.

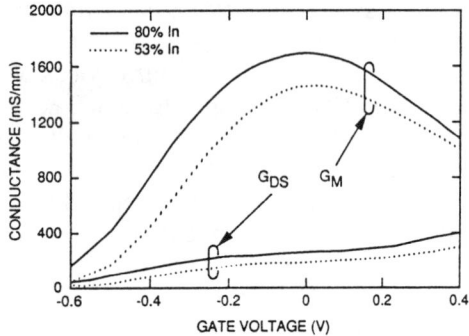

Fig. 10. g_m and g_{ds} of the devices in Fig. 10, $V_d = 0.8$ V, $f = 2.0$ GHz.

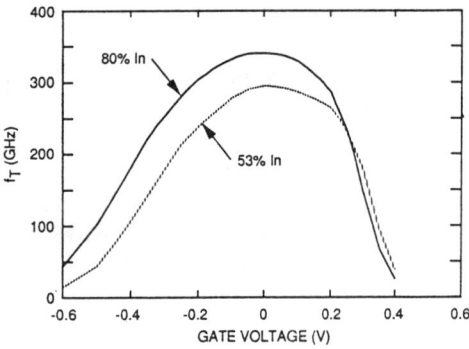

Fig. 11. f_T of the devices in Fig. 10, $V_d = 0.8$ V.

nel separations, and source resistances are similar, the 16% higher extrinsic g_m of the pseudomorphic device, 1690 versus 1460 mS/mm, must have come directly from its higher average velocity (v_{ave}). The g_m/g_{ds} ratio of our devices is approximately 6.5 for the pseudomorphic and 7.6 for the lattice-matched device, which is sufficient for most microwave and millimeter-wave applications. It is interesting, however, to note that the pseudomorphic device exhibits a lower g_m/g_{ds} ratio despite its higher conduction band discontinuity. The g_{ds} of the pseudomorphic is approximately 260 mS/mm, which is as much as 38% higher than that of the lattice-matched device. We are conducting a follow-on study to investigate this undesirable parasitic effect and will report the results at a later date.

The pseudomorphic device exhibits higher f_T over most of the operating range. Its f_T curve reaches a peak value approximately 15% higher, 340 versus 295 GHz, as shown in Fig. 11. By fitting their g_m and f_T curves, we obtained an average velocity of approximately 3.1×10^7 cm/s for the pseudomorphic and 2.6×10^7 cm/s for the lattice-matched device, both with an estimated uncertainty of $\pm 0.2 \times 10^7$ cm/s.

VI. DISCUSSION

The results achieved in this work clearly demonstrate the potential of 50-nm gate-length pseudomorphic Al-InAs/GaInAs HEMT's for near-future microwave and millimeter-wave applications. Presently, the RF performance of our devices is limited by a high gate resistance of 1000 Ω/mm. This parasitic resistance is about five times higher than that of our standard 0.1-μm process and limits the power gain cutoff frequency (f_{max}) to approximately 250 GHz. We believe that this limitation will be corrected in the near future with a wider gate cross section.

Further optimization of the layer structure should result in a lower output conductance for the pseudomorphic device and improve its f_{max}. In addition, a wider source-drain spacing, resulting from a wider gate cross section, should also help reduce the output conductance and improve the drain–source breakdown voltage. Presently, the intrinsic f_{max}/f_T ratio of our devices is approximately 1 (intrinsic $f_{max} = 500$ GHz, intrinsic $f_T = 550$ GHz) and the drain–source breakdown voltage is approximately 1.2 to 1.5 V.

Increasing the operating drain–source voltage, however, may result in a longer effective gate length and, consequently, a larger drain delay. At $V_{ds} = 0.8$ V, the effective gate length of our present devices is approximately 97 nm, which is already about twice their physical gate length. For most applications, however, this tradeoff is not a major concern as long as the f_{max} is improved.

The average velocities reported in this work are approximately equal to the peak electron velocities calculated by Thobel et al. for 53% and 80% indium composition [14]. This association has been made in the recent past for the AlGaAs/GaAs [3] and pseudomorphic Al-GaAs/GaInAs HEMT's [3], [4], [25], although the reported velocities differ from the calculated values by a multiplication factor called modulation efficiency [25]. Our present results provide an additional evidence to support this association.

Our work, however, contradicts those reported by de la Houssaye et al. and Han et al. [26], [27]. We observed no noticeable reduction [26] nor enhancement [27] in average velocity for devices with gate length between 150 and 50 nm. The estimated velocity of $2.6 \pm 0.2 \times 10^7$ cm/s for the 50-nm gate-length lattice-matched device is identical to that for devices with longer gate lengths (see, for example, [15], [16]). As shown in Fig. 12, we have successfully modeled the dependence of f_T on gate length

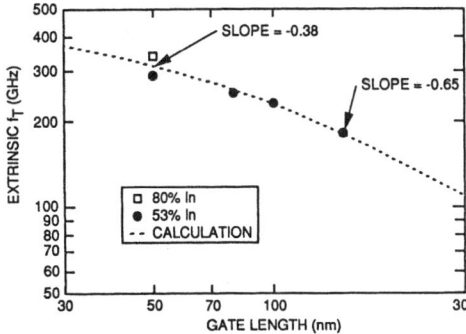

Fig. 12. Dependence of f_T on gate length for self-aligned-gate Al-InAs/GaInAs HEMT's: experimental versus modeled results.

with a single value for the average velocity using the analytical model described in Section II. The parameters used in the present calculation are: $C_{pad} = 10$ fF, $C_{fringe} = 300$ fF/mm, $W = 50$ μm, $d + \Delta d = 20$ mn, $\epsilon = 12.1$, $t_{drain} = 0.15$ ps, $R_s = 0.15$ $\Omega \cdot$ mm, and $v_{ave} = 2.6 \times 10^7$ cm/s, which are consistent with the physical parameters of the fabricated devices. Therefore, we believe that our analysis is valid and our model accurately describes the principles of operation of ultra-short gate-length HEMT's.

VII. SUMMARY

We have successfully designed and fabricated a new class of 50-nm self-aligned-gate pseudomorphic Al-InAs/GaInAs HEMT's, as well as accurately described their principles of operation. We believe that significant improvements in noise figure and gain at microwave and millimeter-wave frequencies will be achieved in the near future with this type of device.

ACKNOWLEDGMENT

The authors would like to acknowledge the valuable support of Dr. P. Greiling and Dr. L. Larson of the Hughes Research Laboratories, and R. Webster and Dr. P. Carr of the Air Force Rome Laboratory, Hanscom AFB, MA. Special thanks go to Dr. B. Hughes of Hewlett-Packard, Santa Rosa, CA, for many valuable comments and suggestions on the principles of operation of FET's and HEMT's.

REFERENCES

[1] L. H. Camnitz, P. J. Tasker, H. Lee, D. van der Merwe, and L. F. Eastman, "Microwave characterization of very high transconductance MODFET," in *IEDM Tech. Dig.*, Dec. 1984, pp. 360-363.

[2] A. N. Lepore, M. Levy, R. Tiberio, P. Tasker, H. Lee, E. Wolf, L. Eastman, and E. Kohn, "0.1 μm gatelength MODFETs with unity current gain cutoff frequency above 110 GHz," *Electron Lett.*, vol. 24, no. 6, pp. 364-366, Mar. 17, 1988.

[3] N. Moll, M. R. Hueschen, and A. Fischer-Colbrie, "Pulse-doped AlGaAs/InGaAs pseudomorphic MODFET's," *IEEE Trans. Electron Devices*, vol. 35, pp. 878-886, July 1988.

[4] L. D. Nguyen, P. J. Tasker, D. C. Radulescu, and L. F. Eastman, "Characterization of ultra-high-speed pseudomorphic AlGaAs/InGaAs (on GaAs) MODFET's," *IEEE Trans. Electron Devices*, vol. 36, no. 10, pp. 2243-2248, Oct. 1989.

[5] U. K. Mishra, A. S. Brown, and S. E. Rosenbaum, "Dc and rf performance of 0.1-μm gatelength $Al_{0.48}In_{0.52}As$-$Ga_{0.38}In_{0.62}As$ pseudomorphic HEMT's" in *IEDM Tech. Dig.*, Dec. 1988, pp. 180-183.

[6] U. K. Mishra, A. S. Brown, L. M. Jelloian, M. Thompson, L. D. Nguyen, and S. E. Rosenbaum, "Novel high performance self-aligned 0.1-μm long T-gate AlInAs-GaInAs HEMTs," in *IEDM Tech. Dig.*, Dec. 1989, pp. 101-104.

[7] C. K. Peng, M. I. Aksun, A. A. Ketterson, H. Morkoç, and K. R. Gleason, "Microwave performance of InAlAs/GaInAs/InP MODFETs," *IEEE Electron Device Lett.*, vol. EDL-8, no. 1, pp. 24-26, Jan. 1987.

[8] K. H. Duh, P. C. Chao, S. M. J. Liu, P. Ho, M. Y. Kao, and J. M. Ballingall, "A super low-noise 0.1 μm T-gate InAlAs-InGaAs-InP HEMT," *IEEE Microwave and Guided Wave Lett.*, vol. 1, no. 5, pp. 114-116, 1991.

[9] K. Tan, D. C. Streit, P. D. Chow, R. M. Dia, A. C. Han, P. H. Liu, D. Garske, and R. Lai, "140 GHz 0.1 μm gatelength pseudomorphic $In_{0.52}Al_{0.48}As/In_{0.60}Ga_{0.40}As/InP$ HEMT," in *IEDM Tech. Dig.*, pp. 239-242, Dec. 1991.

[10] R. Lai, P. K. Bhattacharya, S. A. Alterovitz, A. N. Downey, and C. Chorey, "Low-temperature microwave characteristics of pseudomorphic $In_xGa_{1-x}As/In_{0.52}Al_{0.48}As$ modulation-doped field-effect transistors," *IEEE Electron Device Lett.*, vol. 11, no. 12, pp. 564-566, Dec. 1990.

[11] L. P. Sadwick, C. W. Kim, K. L. Tan, and D. C. Streit, "Schottky barrier heights of n-type and p-type $Al_{0.48}In_{0.52}As$," *IEEE Electron Device Lett.*, vol. 12, no. 11, pp. 626-628, Nov. 1991.

[12] R. People, K. W. Wecht, K. Alavi, and A. Y. Cho, "Measurement of the conduction-band discontinuity of molecular-beam epitaxial growth $In_{0.52}Al_{0.48}As/In_{0.53}Ga_{0.47}As$ heterojunction by C-V profiling," *Appl. Phys. Lett.*, vol. 43, pp. 118-120, 1983.

[13] A. Chin and T. Y. Chang, "Achievement of exceptionally high mobilities in modulation-doped $Ga_{1-x}In_xAs$ on InP using a stress compensated structure," *J. Vac. Sci. Technol.*, vol. B8 no. 2, pp. 364-366, Mar./Apr. 1990.

[14] J. L. Thobel, L. Baudry, A. Cappy, P. Bourel, and R. Fauquembergue, "Electron transport properties of strained $In_xGa_{1-x}As$," *Appl. Phys. Lett.*, vol. 56, no. 4, pp. 346-348, Jan. 22, 1990.

[15] L. D. Nguyen, L. M. Jelloian, M. A. Thompson, and M. Lui, "Fabrication of an 80 nm self-aligned T-gate AlInAs/GaInAs HEMT," in *IEDM Tech. Dig.*, Dec. 1990, pp. 101-104.

[16] L. D. Nguyen and P. J. Tasker, "Scaling issues for ultra-high-speed HEMTs," *SPIE*, vol. 1288, *High Speed Electronics and Device Scaling*, pp. 251-257, 1990.

[17] L. D. Nguyen, A. S. Brown, M. J. Delaney, U. K. Mishra, L. E. Larson, L. M. Jelloian, M. E. Melendes, C. Hooper, and M. Thompson, "Vertical scaling of ultrahigh-speed AlInAs/GaInAs HEMTs," in *IEDM Tech. Dig.*, Dec. 1989, pp. 105-108.

[18] A. S. Brown, U. K. Mishra, J. A. Henige, and M. J. Delaney, "The impact of epitaxial layer design and quality on GaInAs/AlInAs high electron mobility transistor performance," *J. Vac. Sci. Technol.*, vol. B6, no. 2, Mar./Apr. 1988.

[19] A. S. Brown, L. D. Nguyen, R. A. Metzger, M. Matloubian, A. E. Schmitz, M. Lui, R. G. Wilson, and J. A. Henige, "Reduced Si movement in GaInAs/AlInAs HEMT structures with low temperature AlInAs spacers," to be published in *Proc. 1991 Int. Symp. on GaAs and Related Compounds* (Seattle, WA, Aug. 1991).

[20] P. R. Berger, K. Chang, P. Bhattacharya, and J. Singh, "Role of strain and growth conditions on the growth front profile of $In_xGa_{1-x}As$ on GaAs during the pseudomorphic growth regime," *Appl. Phys. Lett.*, vol. 53, no. 8, Aug. 22, 1988.

[21] J. M. Kuo, B. Lalevic, and T. Y. Chang, "New pseudomorphic MODFETs utilizing $Ga_{0.47-u}In_{0.53+u}As/Al_{0.48+u}In_{0.52-u}As$ heterostructures," in *IEDM Tech. Dig.*, Dec. 1986, pp. 460-464.

[22] W. P. Hong, G. I. Ng, P. Bhattacharya, D. Pavlidis, S. Willing, and B. Das, "Low- and high-field transport properties of pseudomorphic $In_xGa_{1-x}As/In_{0.52}Al_{0.48}As$ (0.5 < × < 0.65) modulation-doped heterostructures," *J. Appl. Phys.*, vol. 64, no. 4, Aug. 15, 1988.

[23] B. Hughes and P. J. Tasker, "Bias dependence of the intrinsic elements of MODFETs," *IEEE Trans. Electron Devices*, vol. 36, no. 10, pp. 2267-2273, Oct. 1989.

[24] L. H. Camnitz, P. J. Tasker, P. A. Maki, H. Lee, J. Huang, and L. F. Eastman, "The role of charge control on drift mobility in AlGaAs/GaAs MODFETs," in *Proc. IEEE/Cornell Conf. on Ad-

vanced Concepts in High Speed Semiconductor Devices and Circuits, 1985, pp. 199–208.

[25] M. C. Foisy, P. J. Tasker, B. Hughes, and L. F. Eastman, "The role of inefficiency charge modulation in limiting the current gain cutoff frequency of the MODFET," *IEEE Trans. Electron Devices*, vol. 35, no. 7, pp. 871–878, July 1988.

[26] P. R. de la Houssaye, D. R. Allee, Y. C. Pao, D. G. Schlom, J. S. Harris, Jr., and R. F. W. Pease, "Electron saturation velocity variation in InGaAs and GaAs channel MODFET's for gatelengths to 550 A," *IEEE Electron Device Lett.*, vol. 9, no. 3, pp. 148–150, Mar. 1988.

[27] J. Han, D. K. Ferry, and P. Newman, "Ultra-submicrometer-gate AlGaAs/GaAs HEMTs," *IEEE Electron Device Lett.*, vol. 11, no. 5, pp. 209–211, May 1990.

Recent Advances in Ultrahigh-Speed HEMT LSI Technology

MASAYUKI ABE, SENIOR MEMBER, IEEE, TAKASHI MIMURA, SENIOR MEMBER, IEEE,
NAOKI KOBAYASHI, MEMBER, IEEE, MASAHISA SUZUKI, MAKOTO KOSUGI,
MITSUO NAKAYAMA, KOUICHIRO ODANI, AND ISAMU HANYU

(Invited Paper)

Abstract—The current status of, and recent advances in, high electron mobility transistor (HEMT) technology for high-performance submicrometer VLSI are presented with a focus on material, self-aligned device fabrication, and HEMT LSI implementation. The HEMT is a very promising device for ultrahigh-speed LSI/VLSI applications because of the high-mobility GaAs/AlGaAs heterojunction structure. A 16-kbit static RAM and 4.1-kgate gate array have been developed, demonstrating high-speed operation. With submicrometer gates, as well as advanced material technologies, a HEMT 64-kbit static RAM with subnanosecond access operation and a 10-kgate logic LSI with sub-hundred-picosecond logic delays are just around the corner.

I. INTRODUCTION

HIGH electron mobility transistor (HEMT) technology has opened the door to new possibilities for ultrahigh-speed large-scale integration (LSI) and very large scale integration (VLSI) applications [1], required for future high-speed computers and signal-processing systems. The evolution of high-speed low-power GaAs-based HEMT integrated circuits is a result of continuous technological progress utilizing the superior electronic properties due to the supermobility in GaAs/AlGaAs heterojunction structure. Since the first HEMT was developed in 1980 [2], device, circuit design, processing, and material technologies have progressed and continued to grow rapidly. In 1981, a HEMT ring oscillator with a gate length of 1.7 μm demonstrated a 17.1-ps switching delay with 0.96-mW power dissipation per gate at 77 K, indicating that switching delays below 10 ps could be achieved with 1-μm gate devices [3]. A switching delay of 8.5 ps with 2.59-mW power dissipation per gate has been obtained with a 0.5-μm gate device at 77 K [4]. Delays of 5.8 ps with 1.76-mW power dissipation per gate at 77 K and 10.2 ps with 1.03-mW power dissipation per gate at 300 K have been achieved for a 0.35-μm gate device [5]. Even at room temperature, 9.2 ps with 4.2 mW per gate has been obtained with a 0.28-μm gate device [6].

For LSI level complexity, HEMT technology has made it possible to develop 4-kbit [7], [8] and 16-kbit static

Manuscript received April 10, 1989; revised June 9, 1989. This work is part of the National Research and Development Program on the "Scientific Computer System," conducted under a program of the Agency of Industrial Science and Technology, Ministry of International Trade and Industry of Japan.

The authors are with Fujitsu Laboratories, Ltd., Fujitsu Ltd., 10-1 Morinosato-Wakamiya, Atsugi 243-01, Japan.

IEEE Log Number 8930055.

RAM's [9] as memory circuits, and a 16 × 16-bit parallel multiplier [10], [11] as a logic circuit. The 4-kbit static RAM has an address access time of 500 ps at a power dissipation of 5.7 W per chip, with ECL-compatible levels [8]. The 16 × 16-bit parallel multiplier designed on a 4.1-kgate gate array has a multiply time of 4.1 ns with a power dissipation of 6.2 W. The HEMT has already jumped into the LSI/VLSI application field.

In this paper we review the high-speed performance of HEMT approaches, focusing on device structures in the submicrometer-dimensional range. Next, we describe a self-aligned HEMT technology for VLSI. Finally, we review the current status of HEMT LSI circuit implementation and project the future prospects of HEMT VLSI for ultrahigh-speed computer applications.

II. HEMT PERFORMANCE IN THE SUBMICROMETER DIMENSIONAL RANGE

The HEMT has a performance advantage over conventional devices because of the superior electron transport in HEMT channels and the unique electrical properties of the HEMT structure. During switching, the speed of the device is limited by both low-field mobility and saturated drift velocity. The low-field mobility routinely obtained is 8000 cm^2/V · s at 300 K and 40 000 cm^2/V · s at 77 K. Saturated drift velocity measured with HEMT structures at room temperature has been reported to be 1.7 to 2.0 × 10^7 cm/s [12], [13]. These superior transport properties in HEMT channels result in a high average current-gain cutoff frequency f_T value for a given gate length.

In Fig. 1, the current-gain cutoff frequency f_T versus gate length (L_G) summarizes the typical performance of experimental HEMT's and GaAs MESFET's reported so far [14]–[19]. At room temperature, the values of f_T were 38 GHz [14], 70 GHz [16], and 80 GHz [19] for HEMT's with gate lengths of 0.5, 0.27, and 0.25 μm, respectively. These are about twice those for GaAs MESFET's. No significant variation in threshold voltage with gate length was observed in the range from $L_G = 1.4$ μm to $L_G = 0.14$ μm [6], [20]. This lack of sensitivity indicates that reducing the geometry of HEMT's is an acceptable way to increase performance without short-channel effect problems.

For application of a FET device with submicrometer dimensions to integrated circuits, the short-channel effect

Reprinted from *IEEE Trans. Electron Devices*, vol. 36, no. 10, pp. 2021–2031, Oct. 1989.

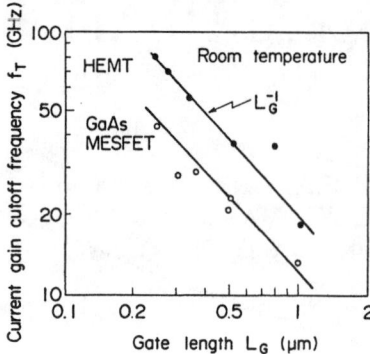

Fig. 1. Current-gain cutoff frequency versus gate length for experimental HEMT's and GaAs MESFET's.

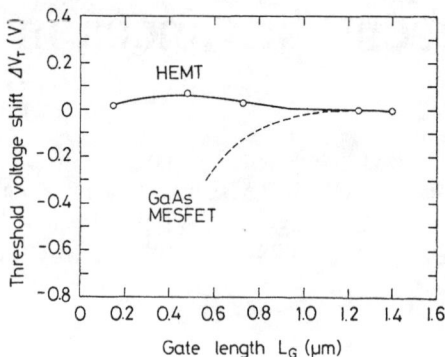

Fig. 2. Dependence of threshold voltage V_T on gate length L_G for HEMT and GaAs MESFET.

is one of the most serious problems. However, the HEMT structure has the advantage of reduced short-channel effect, i.e., the large aspect ratio (L_G/d) of the channel [20], where d is the spacing between the gate and the channel. This shields the gate from high drain fields, giving easily designable and stable current–voltage characteristics for gates in the submicrometer range. Fig. 2 shows how the threshold voltage varies with gate length for both HEMT's [20] and self-aligned gate GaAs MESFET's [21]. The threshold voltage variation of a submicrometer-gate HEMT is much smaller than that of a GaAs MESFET. For a HEMT the variation is almost negligible for gate lengths from 1.4 to 0.14 μm. Therefore, existing HEMT's potentially can allow a 0.15-μm gate LSI to be produced. To suppress the short-channel effect in a HEMT smaller than 0.15 μm, the thickness of the AlGaAs layer must be made thinner to raise the aspect ratio (L_G/d). Even in this case, since electrons in the channel are spatially isolated from the doped AlGaAs layer, although remote impurity scattering cannot be completely ignored, the electron mobility in the HEMT channel does not decrease significantly with increasing doping concentration in the thinner AlGaAs layer.

In LSI circuits with low power dissipation per gate, the logic voltage swing should be minimized. A high transconductance (g_m) value is required with the small logic voltage swing. The g_m in the gradual channel approximation is given by $g_m = K(V_{GS} - V_T)$, where the notations have their usual meanings. K is given by $K = (\epsilon \mu_n W_G / 2 d L_G)$, where ϵ is the dielectric constant, μ_n the electron mobility, and W_G the channel width. The dependence of the K factor and the g_m of E-HEMT's on gate length was measured at both 77 and 300 K. The K factor of a 0.5-μm gate HEMT at 77 K is 900 mA/V^2 per millimeter of gate width. This is about eight times higher than the K factor of conventional GaAs MESFET's. Below 1-μm gate length, at 300 K, the K factor and g_m deviate from the L_G^{-1} dependence as shown in Fig. 3. Velocity saturation and parasitic source resistances probably play a significant role in this behavior. The E-HEMT at 300 K exhibits a g_m of 360 mS/mm for a 0.5-μm gate length and 500 mS/mm for a 0.15-μm gate length [20]. The smaller level of logic voltage swing requires more

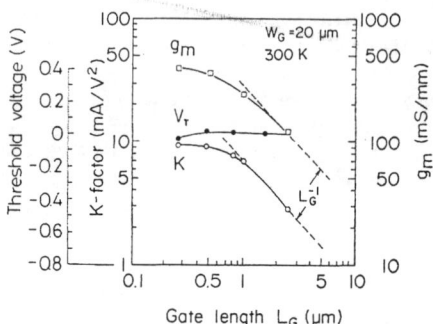

Fig. 3. Dependence of threshold voltage, K-factor, transconductance g_m on gate length L_G for HEMT.

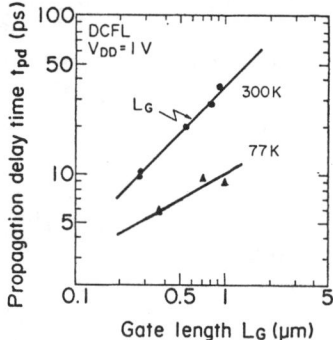

Fig. 4. Dependence of propagation delay time t_{pd} on gate length L_G for HEMT DCFL circuit.

precisely controlled threshold voltages with a smaller standard deviation. The state-of-the-art standard deviation of threshold voltages of enhancement mode (E) HEMT's is 11 mV over a full 3-in wafer, as described in the following section. This value corresponds to a controllability of better than 2 percent of the 0.8-V logic voltage swing.

It is very important to maintain large enough noise margins for LSI to operate stably. Fig. 4 shows the basic propagation delay time t_{pd} versus the gate length L_G for both high- and low-level noise margins larger than 200 mV at 300 K. The supply voltage V_{DD} is 1 V. It can be seen from Fig. 4 that t_{pd} increases proportionally with L_G. The average value of t_{pd} for $L_G = 0.28$ μm, over a 3-in wafer, was 13 ps with a standard deviation of 0.4 ps [16].

If the Si DRAM is the technology leader, we can expect a gate dimension of 0.5 μm at the production level in the near future. By then, HEMT 10-kgate gate arrays with gate delays in the subhundred-picosecond range and HEMT 64-kbit static RAM's with subnanosecond access operation will be feasible.

III. Self-Aligned HEMT Technology for VLSI

The development of any high-performance VLSI technology requries new technological breakthroughs. This section describes the advances that went into state-of-the-art self-aligned HEMT VLSI technology.

A. Heterostructure Material Technology

To grow high-quality material with molecular-beam epitaxy (MBE), we optimized the buffer layer between the semi-insulating GaAs substrate and the two-dimensional electron-gas channel layer. This layer is 0.6 μm thick. The electron mobility in this optimized heterostructure is 8×10^3 cm^2/V $\cdot$ s at 300 K and increases to 1.2×10^5 cm^2/V $\cdot$ s at 77 K due to reduced phonon scattering.

The surface defect problem of MBE is a serious one for the fabrication of circuits with LSI-level complexity [22]. Surface irregularities are called oval defects; they range in size from submicrometer size to several micrometers, which is comparable to the size of devices in LSI circuits. These oval defects seriously affect the current–voltage characteristics of HEMT's [23]. We have achieved a density of less than 10-cm^{-2} defects with a size of over 20 μm^2 by optimizing the growth conditions; this should allow enough complexity for 10-kgate logic and 64-kbit static RAM circuit level.

In HEMT structures, the AlGaAs layer, heavily doped with donors such as Si, contains DX centers [24] that behave as electron traps although their effect happens to be less dramatic at room temperature. Some anomalous behavior at low temperatures is believed to be related to these traps including distortion of drain I–V characteristics, unexpected threshold voltage shift, and highly sensitive persistent photoconduction. We have found that the distortion of drain I–V characteristics is related to the device structure. In a conventional partial gate HEMT structure that has exposed AlGaAs at both sides of the gate, electrons heated by the drain field have sufficient energy to transfer from the GaAs channel to the exposed AlGaAs surface on the drain side of the gate, where they are trapped at DX centers. As a result, space charge builds up at the drain side of the gate. This eventually increases the drain output resistance in the linear region of operation, leading to drain current collapse. To eliminate drain current collapse at low temperature, we adopt the self-aligned gate structure as shown in Fig. 5. The n-AlGaAs layer is completely covered by the n-GaAs top layer. There are no exposed surfaces at the drain side of the gate. In this structure, high-energy electrons can easily pass through the thin n-AlGaAs layer (30 nm) without being trapped and can reach the n-GaAs top layer, eliminating the anomalous drain I–V characteristics at low temperatures.

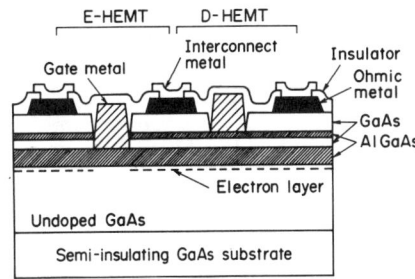

Fig. 5. Cross section of a typical self-aligned structure of E/D-HEMT's forming an inverter for DCFL circuit configuration.

Threshold voltage shifts of 0.2 to 0.3 V between 300 and 77 K have been reported in HEMT [3]. We have succeeded in eliminating the threshold voltage shift, the drain current collapse and the persistent photoconductive effect by optimizing device parameters with a reduced AlAs mole fraction of 0.2.

An important problem in fabricating HEMT LSI is implementation of highly uniform epitaxial wafer growth technology with high throughput and large wafer size. We optimized both the geometrical configuration between the sources and the substrates in the MBE system and the growth conditions to obtain highly uniform epitaxial layers on a 3-in-diameter semi-insulating GaAs substrate, with high throughput and high material quality. Selectively doped GaAs/n-AlGaAs heterostructures were grown on semi-insulating GaAs substrates mounted on a substrate holder with a diameter of 190 mm [25]. The substrate temperature during growth was held at 660°C. A uniformity of $\pm$1 percent for both the thickness and carrier concentration of the AlGaAs layer is achieved to control the threshold voltage of HEMT characteristics. It was also confirmed that the variations of the mobility and the sheet electron concentration in the selectively doped GaAs/n-AlGaAs heterostructure were less than $\pm$2 percent over the 190-mm diameter of the substrate holder.

Epitaxial material growth of HEMT LSI-quality AlGaAs/GaAs heterostructures has also been performed by atmospheric-pressure organic metal vapor-phase epitaxy (OMVPE) technology. Three-wafer growth is carried out by using an RF-heated rotating graphite susceptor with all wafers rotating simultaneously [26]. The susceptor rotates at 8 rpm and the wafers at 20 rpm. The Cr-doped GaAs substrates were oriented 2.5° off the $\langle 100 \rangle$ toward the $\langle 100 \rangle$. The source materials used in the hydrogen carrier were trimethylgallium (TMG), trimethylaluminum (TMA), AsH$_3$, and Si$_2$H$_6$.

The growth temperature was 660°C. The growth rates of GaAs and Al$_{0.28}$Ga$_{0.72}$As were 58 and 21 nm/min. To get abrupt interfaces, the compositions of the source gases were changed instantaneously by the so-called vent/run system [27]. The uniformity of the thickness was better than $\pm$2 percent, and that of the carrier concentration of n-AlGaAs films across a 3-in wafer was better than $\pm$1.5 percent. The sheet electron concentration and the electron mobility at 300 K of an AlGaAs/GaAs selectively doped heterostructure with a 2.5-nm spacer were 1.1×10^{12}

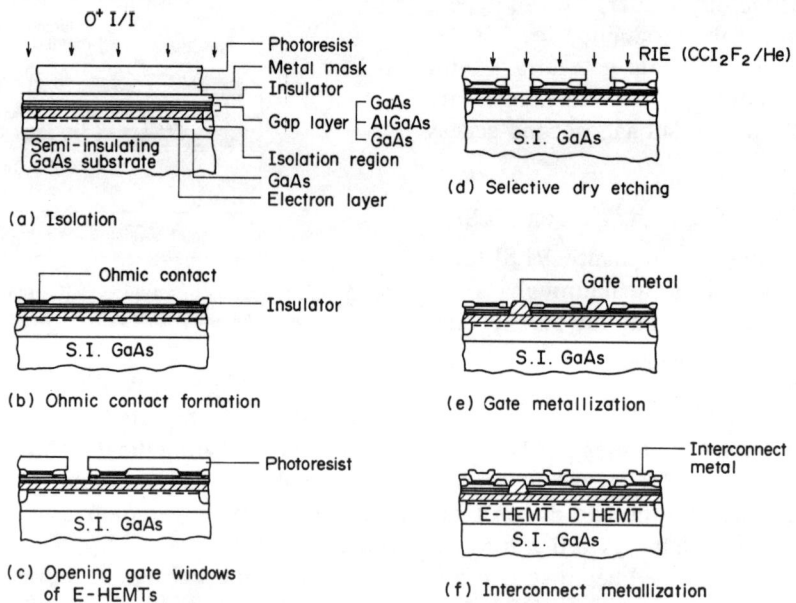

Fig. 6. Basic processing steps for HEMT LSI.

cm^{-2} and 6400 cm^2/V · s and those at 77 K were 9.3 × 10^{11} cm^{-2} and 46 000 cm^2/V · s. By increasing the spacer thickness to 7.5 nm, the electron mobility at 77 K was increased to 90 000 cm^2/V · s. We fabricated HEMT inverters with E-HEMT's and D-HEMT's. The standard deviations of the threshold voltage were 23 mV for E-HEMT's and 35 mV for D-HEMT's across a 3-in wafer. The transconductance and the current-gain cutoff frequency were 250 mS/mm and 23 GHz for an OMVPE-grown HEMT with a gate length of 0.8 μm. These values compare favorably with those for an MBE-grown HEMT.

B. LSI Fabrication Technology

Fig. 5 is a cross section of a typical self-aligned structure for E- and D-HEMT's forming an inverter for a DCFL circuit configuration [28]. The basic epilayer structure consists of a 600-nm undoped GaAs layer, a 30-nm Al$_{0.2}$Ga$_{0.8}$As layer doped to 2 × 10^{18} cm^{-3} with Si, without an undoped spacer layer, and a 70-nm GaAs top layer doped to 10^{18} cm^{-3}, grown on a semi-insulating GaAs substrate by MBE. The low-field electron mobility was found by Hall measurements to be 7200 cm^2/V · s at 300 K and 38 000 cm^2/V · s at 77 K. The concentration of the two-dimensional electron gas (2DEG) was 1.0 × 10^{12} cm^{-2} at 300 K and 8.4 × 10^{11} cm^{-2} at 77 K. The AlAs mole fraction was 0.2, although it can be expected that higher AlAs mode fractions will probably increase the maximum achievable concentration of the 2DEG, resulting in an increase in the transconductance of the HEMT's. The AlGaAs layer with a high AlAs mole fraction, however, exhibits inferior surface morphology and an increase in deep traps, making device fabrication difficult. A thin Al$_{0.3}$Ga$_{0.7}$As layer embeded in the top GaAs layer acts as a stopper against selective dry etching so E- and D-HEMT's can be fabricated in the same wafer. By adopting this new device structure, we can use the selec-

tive dry etching of GaAs to AlGaAs to precisely control the gate recessing for E- and D-HEMT's.

Fig. 6 shows the process sequence for the self-aligned gate process in the fabrication HEMT LSI's including E- and D-HEMT's forming an inverter for a DCFL circuit configuration. First, the active region is isolated by implanted oxygen at 130 keV with a dose of 10^{12} cm^{-2}. The source and drain for E- and D-HEMT's are metallized with AuGe eutectic alloy and a Au overlay, and they are alloyed to form ohmic contacts. The fine gate patterns are formed for E-HEMT's, and the top GaAs layer and thin Al$_{0.3}$Ga$_{0.7}$As stopper are etched off by nonselective chemical etching. The same resist is then patterned for the gates of the D-HEMT's. Selective dry etching is performed to remove the top GaAs layer for D-HEMT's and also to remove the GaAs layer under the thin Al$_{0.3}$Ga$_{0.7}$As stopper for E-HEMT's. Next, Schottky contacts for the E- and D-HEMT gates are formed by depositing Al, with the Schottky gate contacts and the GaAs top layer for the ohmic contact being self-aligned to achieve high-speed performance. Finally, Ti/Pt/Au electrical connections running from the interconnecting metal to the device terminals are provided through contact holes etched in a silicon oxynitride crossover insulator film deposited by plasma-enhanced CVD.

As described above, a unique epistructure in combination with a self-terminating selective dry recess etching makes it possible to fabricate extremely uniform E- and D-HEMT's simultaneously, reflecting the uniformity of MBE-grown expitaxial film. The key technique to achieve stable fabrication of self-aligned gate HEMT's is the selective dry etching of the GaAs/AlGaAs layer. The etching characteristics in CCl$_2$F$_2$ + He discharges achieved a selectivity ratio of more than 260, where the etching rate of Al$_{0.3}$Ga$_{0.7}$As is as low as 2 nm/min and that of GaAs is about 520 nm/min. A histogram of threshold uniform-

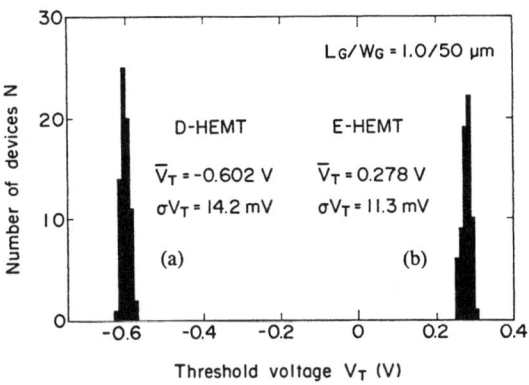

Fig. 7. Histograms of threshold voltages for (a) E-HEMT's and (b) D-HEMT's over a full 3-in wafer.

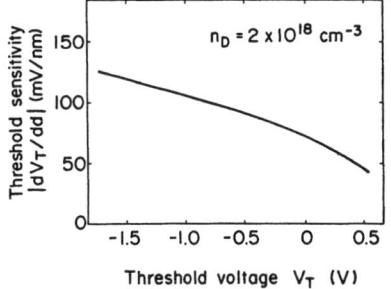

Fig. 8. Vertical threshold sensitivity as a function of threshold voltage.

ities for E- and D-HEMT's is shown in Fig. 7. The standard deviation in threshold voltage over a full 3-in wafer is 11 mV for E-HEMT's and 14 mV for D-HEMT's. The ratio of the standard deviation (11 mV) of theshold voltage to the logic voltage swing (0.8 V for DCFL) is 1.4 percent, indicating excellent controllability of MBE growth and the LSI fabrication process. This indicates the viability of these technologies for realizing IC's with LSI/VLSI-level complexities. The vertical threshold sensitivity is calculated as 60 mV/nm [28] at a V_T of 0.28 V and a doping concentration of 2×10^{18} cm^{-3}, shown in Fig. 8. The measured deviation in threshold voltage over the wafer for the E-HEMT's is 60 mV at a V_T of 0.28 V. This corresponds to an effective thickness deviation, including both actual thickness and doping variation, of only 1 nm over a full 3-in wafer, indicating excellent controllability of MBE growth and the device fabrication process.

IV. HEMT LSI CIRCUIT IMPLEMENTATION

The current implementation and recent advances of HEMT logic and memory LSI circuits are reviewed and discussed in this section.

A. Logic Circuits

A HEMT 4.1-kgate gate array with E/D type DCFL circuits has been designed and fabricated for logic circuits. This gate array consists of 156 I/O cells and 4096 basic cells. The basic cell includes one D-HEMT and three E-HEMT's with a gate length of 0.8 μm. It can be programmed as a three-input NOR gate. The cell size is 37.5 μm × 45 μm. The basic cell array has 32 columns of 128 cells each. Between the columns there are 15 interconnection tracks, each line being 2 μm wide with 2 μm spacing. The chip of this gate array has 100 pads, 72 for I/O signals and 28 for power supply. To obtain a sufficient noise margin, this chip is designed to minimize the V_{DD} voltage drop and ground (GND) voltage rise by careful arrangement of the power supply pads, so the chip has a relatively large number of power supply pads. The chip measures 6.3 mm × 4.8 mm, and contains 17 692 devices. The average value of the delay time of the inverter and the basic gate are 27 and 40 ps, respectively. The extra delay results from crossover capacitance between gate electrode and power supply lines. The power supply lines are on top of the basic gate to make the array more compact.

A 4.1-kgate gate array uses a 16 × 16-bit parallel multiplier as a test vehicle. A 16 × 16-bit multiplier constructed of 93 percent of this array consists of registers for a 16 × 16-bit multiplier product, 15 half-adders, 210 full-adders, and a carry look-ahead circuit. The multiplication time was 4.1 ns at 300 K, including a five-stage I/O buffer delay. The supply voltage V_{DD} was 1.1 V, and the total chip power dissipation was 6.2 W. This is the fastest multiplication time yet reported for a 16 × 16-bit parallel multiplier. From the simulation using the SPICE II program, we confirmed that the multiplication time was about 49 times the typical gate delay of 80 ps with a loading of 2.6 fan-outs and 363-μm interconnection line. This simulation gave an I/O buffer delay of about 8 percent of the multiplication time, so the intrinsic multiplication time is 3.8 ns. As a comparison between state-of-the-art GaAs MESFET [29], [30] and HEMT multipliers [10], [11], [31]–[34], the circuit gate delay/power performance is shown in Fig. 9 as a function of gate power dissipation.

The performance of the half-micrometer HEMT DCFL gates was measured by using different types of ring oscillators. The basic propagation delay time t_{pdo} was 22.5 ps with a power dissipation of 3.9 mW/gate and a supply voltage V_{DD} of 2 V. The standard deviation of t_{pdo} was 1.0 ps over a 30 mm × 30 mm area; the noise margin of the low level of the basic inverters was 220 mV, and that of the high level was 280 mV. The dependence of t_{pd} as a function of the supply voltage is shown in Fig. 10. Fig. 11 shows the propagation delay time as a function of loading conditions for a DCFL gate circuit. The loaded delay time (F/I = F/O = 3, 1 = 1 mm) is 84 ps.

A multibit data register circuit using HEMT's with 0.5-μm gate length has been developed [35]. The register synchronizes data signals for data transfer in a system with a high clock rate. Fig. 12 is a block diagram of a multibit data register circuit. The clock pulse is applied to each latch through the clock chopper, and then 4 × 9-bit latched input data are transferred to the output ports, synchronized with the clock signal. The input and output buffers were designed to provide signal levels compatible with the ECL interface. Fig. 13 is a microphotograph of the multibit data register. The chip contains 1137 gates, 99 signal pads, and 36 power supply pads, and measures

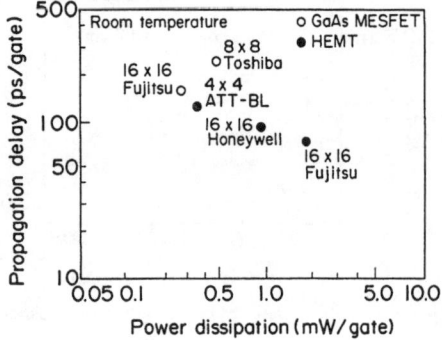

Fig. 9. Comparison of recent GaAs MESFET and HEMT multiplier gate propagation delays as a function of gate power dissipation.

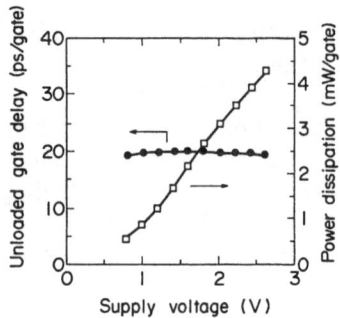

Fig. 10. Basic propagation delay time and power dissipation with unloaded condition as a function of supply voltage.

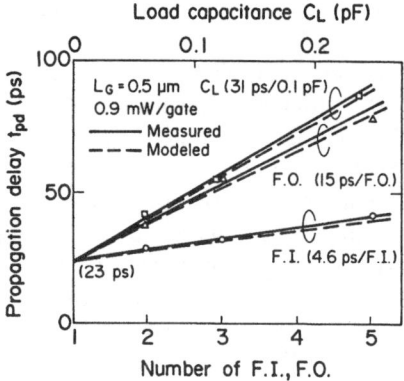

Fig. 11. Propagation delay time as functions of loading conditions for a DCFL gate circuit.

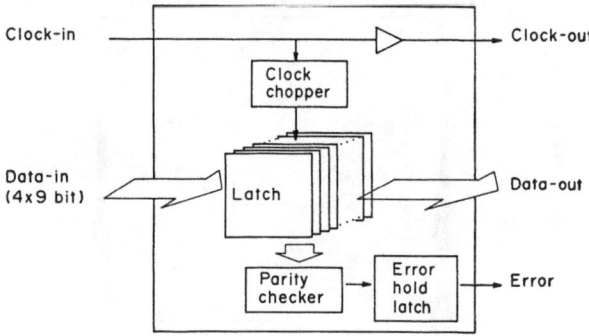

Fig. 12. Block diagram of multibit data register circuit.

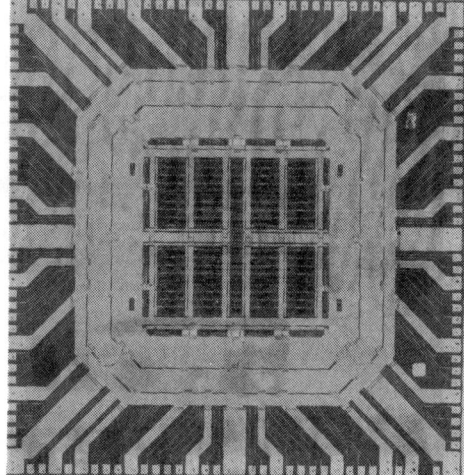

Fig. 13. Microphotograph of a multibit data register.

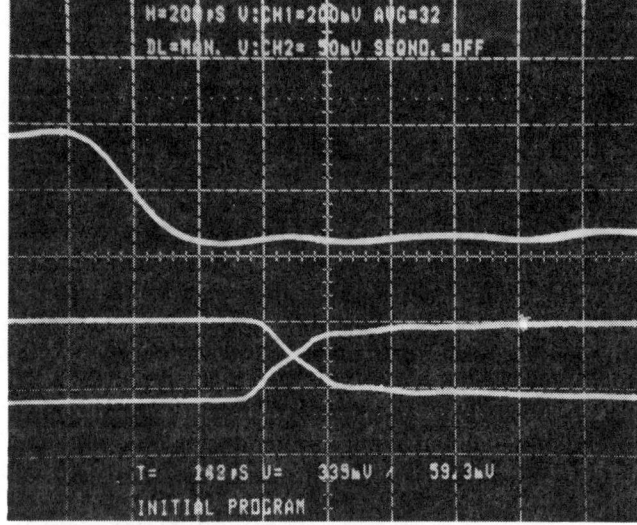

Fig. 14. Oscilloscope photograph measuring the delay time through the critical path from the clock input to the data output.

6.1 mm × 6.2 mm. This chip was designed to minimize the difference in the propagation delay from the clock input to each output. The width and spacing of the interconnecting lines were both 2 μm. The logic circuit is 2.4 mm × 2.4 mm, and contains 3335 HEMT's. The standard supply voltages are −2 and −3.6 V. The speed of the multibit data register was measured by the delay time from clock input to data output at room temperature, using a coaxial probe card. Fig. 14 is the oscilloscope photograph measuring the delay time through the critical path from the clock input to the data output. The delay time was 490 ps at room temperature, and the power dissipation was 4.12 W. The gate delay through the path from clock input to data output was estimated to be 43 ps/gate.

The state-of-the-art implementation of HEMT logic LSI circuits is summarized in Table I.

The performance of HEMT VLSI for high-speed logic in future computers can be projected and discussed based on the present results described above [1], [36], [37]. The chip delay time is the sum of the intrinsic gate delay, the

TABLE I
STATE-OF-THE-ART IMPLEMENTATION OF HEMT LOGIC LSI CIRCUITS

Company	Function	Complexity (gates)	Performance			Chip Size (mm)	Technology	Reference
			Delay (ns)	Power (W)	Temperature (K)			
Fujitsu	Gate Array (8 x 8 Multiplier)	1520	3.1	3.2	77	5.5 x 5.6	E/D DCFL L_G = 1.2 µm	33
	Gate Array (16 x 16 Multiplier)	4096	4.1	6.2	300	6.3 x 4.8	E/D DCFL L_G = 0.8 µm	10
	Data Register	1137	0.49	4.1	300	6.1 x 6.2	E/D DCFL, ECL Interface L_G = 0.5 µm	35
Oki	Gate Array (6 x 6 Multiplier)	1000	5.0	0.977	77	3.8 x 4.2	SBFL, Inverted Structure L_G = 0.8 µm	41
	8-bit D/A Converter	(919 devices)	(1.2 Gb/s)	2.4	300	3.0 x 3.0	SCFL, Inverted Structure L_G = 0.5 µm	42
AT & T Bell Lab.	4 x 4 Multiplier	141	1.6	0.053	300	1.1 x 1.2	DCFL L_G = 1.0 µm	31
Honeywell	5 x 5 Multiplier	350	1.08	0.75	77	1.86 x 1.54	DCFL L_G = 1.0 µm	32
	8 x 8 Multiplier	1350	3.2	1.9	300	2.7 x 3.9	DCFL L_G = 1.0 µm	34
	16 x 16 Multiplier	4500	(520 MHz)	4.0	300	7.8 x 8.5	DCFL L_G = 1.0 µm	11
Rockwell	A/D Converter D/A Converter	(1500 devices)	(140 Mega Sample/s)	–	300	–	L_G = 1.0 µm	43

logic layout delay on fan-out conditions, and the delay in the wiring on the chip. Chip delays are calculated based on experimental data for HEMT's with gate lengths of 0.5 and 0.25 µm at 300 K. Here, we assume that the fan-out is 3, the logic swing is 0.8 V, the wiring capacitance is 100 fF/mm, the average line length is 1 mm in the chip, and the heat flux is 20 W/cm^2, which requires liquid cooling. Fig. 15 shows the chip delay as a function of complexity at 300 K. At 10^4 gates, the chip delays are 70 and 40 ps under 0.5- and 0.25-µm design rules. This sub-100-ps performance is sufficient for future high-speed computer requirements.

B. Memory Circuits

A 1-kword × 4-bit static RAM using 0.5-µm-gate HEMT technology has been designed and fabricated [8]. Fig. 16 is a microphotograph of the RAM. A layout design rule of 1.5-µm line/2.0-µm space was used for interconnections. The smallest via hole is 2 µm × 2 µm.

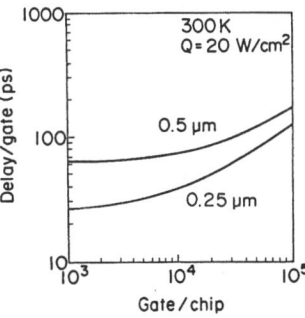

Fig. 15. Chip delay calculated as a function of LSI complexity, under 0.5-µm and 0.25-µm design rule HEMT technologies.

The memory cell is 24.5 µm × 23 µm, a very small size for GaAs LSI memory. The chip measures 2.8 mm × 3.0 mm.

A block diagram of a 1-kword × 4-bit static RAM is shown in Fig. 17. The memory cell array is divided into four 1-kibt memory planes, each with 32 rows and 32 col-

Fig. 16. Microphotograph of HEMT 1-kword × 4-bit static RAM, which measures 2.6 mm × 3.0 mm and contains 29 994 E/D-HEMT's.

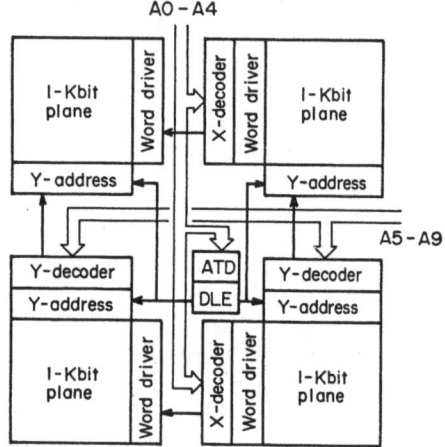

Fig. 17. Block diagram of HEMT 1-kword × 4-bit static RAM.

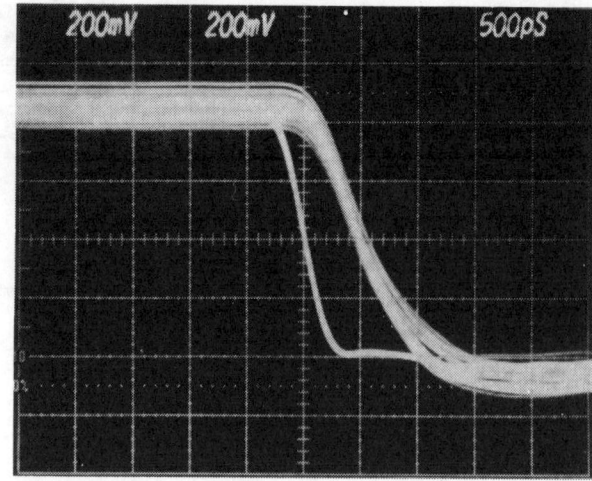

Fig. 18. Oscillograph for memory address access operations, showing 500-ps address access time and superimposed signals of address access in 1-kbit memory plane. The horizontal scale is 500 ps/div.

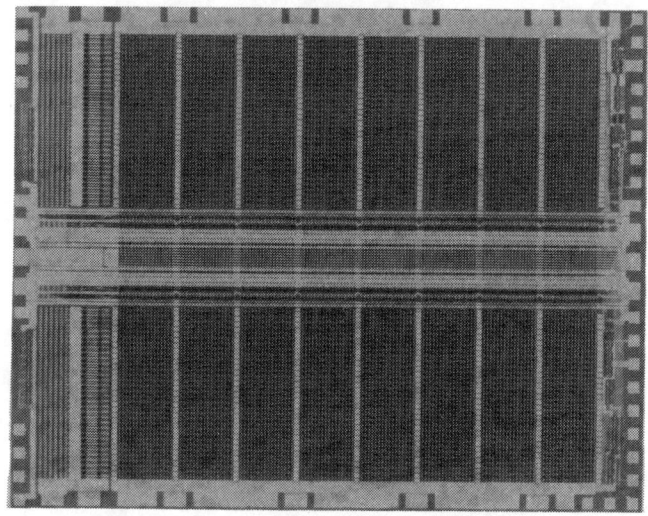

Fig. 19. Microphotograph of HEMT 16-kword × 1-bit static RAM, which measures 4.3 mm × 5.5 mm and contains 107 519 E/D-HEMT's.

umns. This configuration has reduced interconnections and a faster access time. The RAM has ECL-compatible I/O interface circuits, and the supply voltages are −2 and −3.6 V. An E/D type DCFL was used for the basic logic gate and the memory cell in the RAM circuit design, and source follower buffers were used for the driver circuits and level shifters. In this RAM circuit, data line equalization was adopted for reduction of the address access time. The address transition is detected by the address transition detector (ATD), and an ATD pulse is generated and used for the data line equalizer (DLE) as shown in Fig. 17. In the circuit simulation, an E-HEMT threshold voltage of 0.2 V and a D-HEMT threshold voltage of −0.6 V were used, and the propagation delay time of the basic DCFL gate (F/I = F/O = 1) was 22 ps. Using this DLE technique, the address access time was reduced from 0.68 to 0.54 ns. The design value of the chip power dissipation was 5 W. This power dissipation is rather large, but it can be reduced to less than 2 W by using a supply voltage of −1 V.

The address access time was 500 ps at room temperature with a chip power dissipation of 5.7 W. The chip select access time was 0.25 ns. Fig. 18 is an oscilloscope photograph of the superimposed 32 address signals and 32 outputs. The variation of the address access time in the 1-kbit memory plane was about 0.15 ns.

A HEMT 16-kword × 1-bit fully decoded static RAM has been successfully developed with the E/D-type DCFL circuitry [9]. Using D-HEMT's for load devices, E/D-type DCFL circuits were employed as the basic circuit. The memory cell is a six-transistor, cross-coupled flip-flop circuit with switching devices having gate lengths of 1.2 μm. As a result of the RAM layout design, the chip size is 4.3 mm × 5.5 mm, and the RAM cell size is 23 μm × 30 μm (690 μm^2). The RAM has a total device count of 107 519. Fig. 19 is a microphotograph of the 16-kword × 1-bit static RAM. Dynamic performance such as the address access time of the HEMT 16-kbit static RAM was evaluated both at room temperature and liquid-

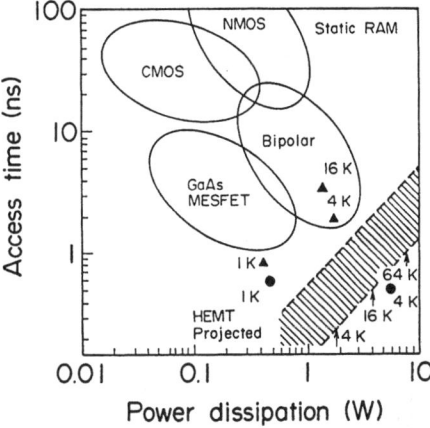

Fig. 20. Address access time and power dissipation of the static RAM, compared with Si MOS, bipolar, and GaAs MESFET static RAM's. The symbols ● and ▲ denote HEMT at 300 and 77 K. The shaded area shows projected performance with gate length between 0.5 and 0.25 μm at 300 K.

TABLE II
STATE-OF-THE-ART IMPLEMENTATION OF HEMT MEMORY LSI CIRCUITS

Company	Function	Complexity (gate)	Performance			Chip Size (mm)	Technology	Reference
			Access Time (ns)	Power (W)	Temperature (K)			
Fujitsu	SRAM	1 K	0.87	0.36	77	3.0 x 2.9	E/D DCFL L_G = 2.0 μm	39
	SRAM	4 K	2.1	1.6	77	4.76 x 4.35	E/D DCFL L_G = 2.0 μm	7
	SRAM	16 K	3.4	1.3	77	4.3 x 5.5	E/D DCFL L_G = 1.2 μm	9
	SRAM	4 K	0.5	5.7	300	2.8 x 3.0	E/D DCFL, ECL Interface L_G = 0.5 μm	8
Rockwell	SRAM	1 K	0.6	0.45	300	2.5 x 2.35	E/D DCFL L_G = 1.0 μm	40

nitrogen temperature. The minimum address access time was 3.4 ns, with a chip dissipation power of 1.34 W at 77 K. The GaAs MESFET 4-kword × 4-bit static RAM with 1-μm gate length devices has an address access time of 4.1 ns with a chip power dissipation of 2.52 W at 300 K [38]. The state-of-the-art implementation of HEMT memory LSI circuits is summarized in Table II. Fig. 20 compares the address access time and power dissipation of the static RAM's to Si MOS, bipolar, and GaAs MES-FET static RAM's. The plots show the performance of HEMT 1-, 4-, and 16-kbit static RAM's [7]–[9], [39], [40]. The shaded area shows the projected performance with a gate length between 0.5 and 0.25 μm. By using 0.25-μm-gate device technology, subnanosecond access operation can be projected for 64-kbit static RAM at room temperature.

V. SUMMARY

The current status and recent advances in HEMT technology for high-speed VLSI were reviewed with a focus on materials, self-aligned device fabrication, and HEMT LSI implementation.

HEMT's are very promising for VLSI because of their high speed and low power dissipation. The projected HEMT performance target suitable for VLSI is a fundamental switching delay below 10 ps. By evaluating the gate length dependence of the threshold voltage and the K-factor of short-channel HEMT's, short-channel effects were not found to be a problem in microstructures scaled down to the submicrometer dimensional range.

Since the first HEMT integrated circuit was developed in 1981, device, circuit design, processing and material technologies have progressed and continue to grow rapidly. The evolution of HEMT IC complexity is fourfold each year [28]. As the HEMT technology shifts from the phase of research and development toward commercialization, new developments in material technology or breakthroughs in growth technique will have to be focused to match not only the requirements of highly uniform and perfect epitaxial materials for LSI quality but also of wafer supply and production with high throughput and large wafer size.

A HEMT 4.1-kgate gate array with a 16×16-bit parallel multiplier has been developed to achieve a multiplication time of 4.1 ns. A HEMT 4-kbit static RAM with an address access time of 500 ps and a 16-kbit static RAM with an address access time of 3.4 ns demonstrate the feasibility of high-performance VLSI's. With the 0.25-μm device technology, a HEMT 64-kbit SRAM should achieve subnanosecond access operations. Using the experimental data on HEMT logic, we project an optimized chip delay of 40 ps at 10-kgate VLSI at room temperature. This performance will achieve speeds required for future main-frame computers.

ACKOWLEDGMENT

The authors thank Dr. T. Misugi and Dr. M. Kobayashi for their encouragement and support. The authors also thank their colleagues, whose many contributions have made possible the results described here.

REFERENCES

[1] M. Abe, T. Mimura, N. Yokoyama, and H. Ishikawa, "New technology towards GaAs LSI/VLSI for computer applications," *IEEE Trans. Electron Devices*, vol. ED-29, pp. 1088-1093, 1982.

[2] T. Mimura, S. Hiyamizu, T. Fujii, and K. Nanbu, "A new field-effect transistor with selectively doped GaAs/n-Al$_x$Ga$_{1-x}$As heterojunctions," *Japan. J. Appl. Phys.*, vol. 19, pp. L225-L227, 1980.

[3] T. Mimura, K. Joshin, S. Hiyamizu, K. Hikosaka, and M. Abe, "High electron mobility transistor logic," *Japan. J. Appl. Phys.*, vol. 20, pp. L598-600, 1981.

[4] N. C. Cirillo, Jr. and J. K. Abrokwah, "8.5-picosecond ring oscillator gate delay with self-aligned gate modulation-doped n$^+$-(Al,Ga)As/GaAs FET's," presented at the 43rd Annual Device Res. Conf., paper IIA-7, 1985.

[5] N. J. Shah, S.-S. Pei, C. W. Tu, and R. C. Tiberio, "Gate-length dependence of the speed of SSI circuits using submicrometer selectively doped heterostructure transistor technology," *IEEE Trans. Electron Devices*, vol. ED-33, pp. 543-547, 1986.

[6] Y. Awano, M. Kosugi, T. Mimura, and M. Abe, "Performance of a quarter-micrometer-gate ballistic electron HEMT," *IEEE Electron Device Lett.*, vol. EDL-8, pp. 451-453, 1987.

[7] S. Kuroda et al., "New device structure for 4Kb HEMT SRAM," in *Tech. Dig. IEEE GaAs IC Symp.*, 1984, pp. 125-128.

[8] S. Notomi et al., "A high-speed 1K $\times$ 4-bit static RAM using 0.5-μm gate HEMT," in *Tech. Dig. IEEE GaAs IC Symp.*, 1987, pp. 177-180.

[9] M. Abe et al., "Ultrahigh-speed high electron mobility transistor large scale integration technology," *J. Vac. Sci. Technol. A*, vol. 5, pp. 1387-1392, 1987.

[10] K. Kajii et al., "A 40ps high electron mobility transistor 4.1K gate array," in *CICC Dig. Tech. Papers*, 1987, pp. 199-201.

[11] T. Akinwande et al., "A 500 MHz 16 $\times$ 16 complex multiplier using self-aligned gate heterostructure FET technology," in *ISSCC Dig. Tech.*, 1989, pp. 232-233.

[12] T. Mimura, M. Abe, A. Shibatomi, and M. Kobayashi, "HEMT technology: Potential and advances," *Surface Sci.*, vol. 174, pp. 343-351, 1986.

[13] Y. Takanashi and N. Kobayashi, "AlGaAs/GaAs 2-DEG FET's fabricated from MO-CVD wafers," *IEEE Electron Device Lett.*, vol. EDL-6, pp. 154-156, 1985.

[14] K. Joshin et al., "Noise performance of microwave HEMT," in *IEEE MTT-S Dig.*, 1983, pp. 563-565.

[15] L. H. Camnitz, P. J. Tasker, H. Lee, D. V. D. Merwe, and L. F. Eastman, "Microwave characteristics of very high transconductance MODFET," in *IEDM Tech. Dig.*, 1984, pp. 360-363.

[16] H. Suehiro et al., "A new recessed-gate structure for HEMT LSIs," presented at WOCSEMMAD, Feb. 1989.

[17] M. Feng, H. Kanber, V. K. Eu, E. Watkins, and L. R. Hackett, "Ultrahigh frequency operation of Ion-implanted GaAs metal-semiconductor field-effect transistors," *Appl. Phys. Lett*, vol. 44, pp. 231-233, 1984.

[18] W. Chye and C. Huang, "Quartermicron low noise GaAs FET's," *IEEE Electron Device Lett.*, vol. EDL-3, pp. 401-403, 1982.

[19] P. C. Chao et al., "Millimeter-wave low-noise high electron mobility transistors," *IEEE Electron Device Lett.*, vol. EDL-6, pp. 531-533, 1985.

[20] Y. Awano, M. Kosugi, K. Kosemura, T. Mimura, and M. Abe, "Short-channel effects in subquarter-micrometer-gate HEMTs: Simulation and experiment," *IEEE Trans. Electron Devices*, this issue, pp. 2260-2266.

[21] K. Yamasaki, N. Kato, and M. Hirayama, "Buried P-layer SAINT for very high-speed GaAs LSI's with submicrometer gate lengths," *IEEE Trans. Electron Devices*, vol. ED-32, pp. 2420-2425, 1985.

[22] M. Abe, T. Mimura, K. Nishiuchi, A. Shibatomi, and M. Kobayashi, "Recent advances in ultrahigh-speed HEMT technology," *IEEE J. Quantum Electron.*, vol. QE-22, pp. 1870-1879, 1986.

[23] T. Mimura, M. Abe, and M. Kobayashi, "High electron mobility transistors," *Fujitsu Sci. Tech. J.*, vol. 21, pp. 370-379, 1985.

[24] D. V. Lang, R. A. Logan, and M. Jaros, "Trapping characteristics and a donor-complex (DX) model for the persistent-photoconductivity trapping center in Te-doped Al$_x$Ga$_{1-x}$As," *Phys. Rev. B*, vol. 19, pp. 1015-1030, 1979.

[25] J. Saito, T. Igarashi, T. Nakamura, K. Kondo, and A. Shibatomi, "Growth of highly uniform epitaxial layers over multiple substrates by molecular beam epitaxy," *J. Crystal Growth*, vol. 81, pp. 188-192, 1987.

[26] H. Tanaka et al, "Multi-wafer growth of HEMT LSI quality AlGaAs/GaAs heterostructure by MOCVD," *Japan. J. Appl. Phys.*, vol. 26, pp. L1456-L1458, 1987.

[27] E. J. Thrush et al., "Evidence for transient composition variations at GaAs/Ga$_{1-x}$Al$_x$As heterostructure interfaces prepared by metal-organic chemical vapor deposition," *J. Electron. Mater.*, vol. 13, pp. 969-988, 1984.

[28] M. Abe et al, "Ultrahigh-speed HEMT integrated circuits," in *Semiconductors and Semimetals*, R. K. Willardson and A. C. Beer, Eds., *Applications of Multiquantum Wells, Selective Doping, and Superlattices*, vol. 24, R. Dingle, Volume Ed. New York: Academic, 1987, pp. 249-278.

[29] Y. Nakayama et al., "A GaAs 16 $\times$ 16b parallel multiplier using self-aligned technology," in *ISSCC Dig. Tech.*, 1983, pp. 48-49.

[30] N. Toyoda et al., "A 42ps 2K-gate GaAs gate array," in *ISSCC Dig. Tech.*, 1985, pp. 206-207.

[31] A. R. Schlier, S. S. Pei, N. J. Shah, C. W. Tu, and G. E. Mahoney, "A high-speed 4 $\times$ 4-bit parallel multiplier using selectively doped heterostructures," in *Tech. Dig. IEEE GaAs IC Symp.*, 1985, pp. 91-93.

[32] D. K. Arch, B. K. Betz, P. J. Vold, J. K. Abrokwah, and N. C. Cirillo, Jr., "A self-aligned gate superlattice (Al,Ga)As/n$^+$-GaAs MODFET 5 $\times$ 5-bit parallel multiplier," *IEEE Electron Device Lett.*, vol. EDL-7, pp. 700-702, 1986.

[33] Y. Watanabe et al., "High electron mobility transistor 1.5K gate array," in *ISSCC Dig. Tech.*, vol. 29, 1986, pp. 80-81.

[34] N. C. Cirillo, Jr., et al, "8 $\times$ 8-bit pipelined parallel multiplier utilizing self-aligned gate n$^+$-(Al,Ga)As/GaAs MODFET IC technology," in *Tech. Dig. IEEE GaAs IC Symp.*, 1987, pp. 257-260.

[35] Y. Watanabe et al., "A HEMT LSI for a multibit data register," in *ISSCC Dig. Tech.*, 1988, pp. 86-87.

[36] M. Abe, T. Mimura, K. Nishiuchi, A. Shibatomi, and M. Kobayashi, "HEMT LSI technology for high-speed computers," in *Tech. Dig. IEEE GaAs IC Symp.*, 1983, pp. 158-161.

[37] M. Abe, T. Mimura, K. Nishiuchi, and N. Yokoyama, "GaAs VLSI technology for high-speed computers," in *VLSI Electronics; Microstructure Science*, vol. 11, *GaAs Microelectronics*, N. G. Einspruch and W. R. Wisseman, Eds. New York; Academic, 1985, pp. 333-365.

[38] Y. Ishii, M. Ino, M. Idda, M. Hirayama, and M. Ohmori, "Processing technologies for GaAs memory LSIs," in *Tech. Dig. IEEE GaAs IC Symp.*, 1984, pp. 121–124.

[39] K. Nishiuchi *et al.*, "A subnanosecond HEMT 1Kb SRAM," in *ISSCC Dig. Tech.*, 1984, pp. 48–49.

[40] N. H. Sheng, H. T. Wang, S. J. Lee, G. L. Sullivan, and D. L. Miller, "A high-speed 1 K-bit high electron mobility transistor static RAM," in *Tech. Dig. IEEE GaAs IC Symp.*, 1986, pp. 97–100.

[41] S. Nishi *et al.*, "Investigation of an inverted HEMT for LSIs," *Trans.* *IEICE (C)*, vol. J70-C, pp. 724–730, 1987.

[42] S. Seki, T. Saito, H. Fujishiro, S. Nishi, and Y. Sano, "An 8-bit 1GHz digital to analog converter using 0.5-μm gate inverted HEMTs," in *IEDM Tech. Dig.*, 1988, pp. 770–773.

[43] C. P. Lee, N. H. Sheng, H. F. Lewis, H. T. Wang, D. L. Miller, and J. Donovan, "GaAs/GaAlAs high electron mobility transistors for analog-to-digital converter applications," in *IEDM Tech. Dig.*, 1985, pp. 324–327.

Part 4

HBTs

Heterostructure Bipolar Transistors and Integrated Circuits

HERBERT KROEMER, FELLOW, IEEE

Invited Paper

Abstract—Two new epitaxial technologies have emerged in recent years (molecular beam epitaxy (MBE) and metal-organic chemical vapor deposition (MOCVD)), which offer the promise of making highly advanced heterostructures routinely available. While many kinds of devices will benefit, the principal and first beneficiary will be bipolar transistors. The underlying central principle is the use of energy gap variations beside electric fields to control the forces acting on electrons and holes, separately and independently of each other. The resulting greater design freedom permits a re-optimization of doping levels and geometries, leading to higher speed devices. Microwave transistors with maximum oscillation frequencies above 100 GHz and digital switching transistors with switching times below 10 ps should become available. An inverted transistor structure with a smaller collector on top and a larger emitter on the bottom becomes possible, with speed advantages over the common "emitter-up" design. Double-heterostructure (DH) transistors with both wide-gap emitters and collectors offer additional advantages. They exhibit better performance under saturated operation. Their emitters and collectors may be interchanged by simply changing biasing conditions, greatly simplifying the architecture of bipolar IC's. Examples of heterostructure implementations of I^2L and ECL are discussed. The present overwhelming dominance of the compound semiconductor device field by FET's is likely to come to an end, with bipolar devices assuming an at least equal role, and very likely a leading one.

"What is claimed is:

1) ...

2) A device as set forth in claim 1 in which one of the separated zones is of a semiconductive material having a wider energy gap than that of the material in the other zones."

Claim 2 of U.S. Patent 2 569 347 to W. Shockley,
Filed 26 June 1948,
Issued 25 September 1951,
Expired 24 September 1968.

I. INTRODUCTION

THIS IS A PAPER about an idea whose time has come: A bipolar transistor with a wide-gap emitter. As the introductory quote shows, the idea is as old as the transistor itself. The great potential advantages of such a design over the conventional homostructure design have long been recognized [1]–[3], but until the early 70's, no technology existed to build practically useful transistors of this kind, even though numerous attempts had been made [3], [4]. The situation started to change with the emergence of liquid-phase epitaxy (LPE) as a technology for III/V-compound semiconductor heterostructures, and in recent years reports on increasingly impressive true three-terminal heterostructure bipolar transistors (HBT's) have appeared at an increasing rate [5]–[14]. In addition, there is also a rapidly growing literature on *two*-terminal *photo*transistors with wide-gap emitters [15]. Many of the phototransistors employ InP emitters with a lattice-matched (Ga, In) (P,As) base.

Since the mid-70's, two additional very promising heterostructure technologies have appeared: molecular beam epitaxy (MBE) [16] and metal-organic chemical vapor deposition (MOCVD) [17]. Impressive results on MOCVD-grown (Al,Ga)As-GaAs phototransistors have already been published [18]; HBT's grown by MBE have also been achieved [19].

Because of the pre-eminence of silicon in current IC technology, there exists a strong incentive to incorporate wide-gap emitters into Si transistors, in a way compatible with existing Si technology. A possible approach—and the most successful one so far—has been the use of heavily doped "semi-insulating polycrystalline" silicon (SIPOS) as emitter [20], utilizing the wider energy gap of "polycrystalline" (really: amorphous) Si compared to crystalline Si. An alternate approach has been the use of gallium phosphide, which has a room-temperature lattice constant within 0.3 percent of that of Si, grown on Si either by CVD [21] or by MBE [22]. But the results reported for the GaP-Si combination have so far been disappointing.

Finally, the first reports have recently appeared, in which HBT's have been integrated on the same chip with other devices, such as double-heterostructure (DH) lasers [23] or LED's [24].

In view of these recent developments it appears that Shockley's vision is about to become a reality. In fact, one of the purposes of this paper is to show that the possibilities for HBT's go far beyond simply replacing a homojunction emitter by a heterojunction emitter.

To appreciate these possibilities, it is useful first to view the wide-gap emitter as a simple example of a more general *central design principle* of heterostructure devices; it is discussed in Section II of this paper. Discussions of future device possibilities must be based on technological premises; they are discussed in Section III. In Section IV and V the concept and the high-speed benefits of the wide-gap emitter are reviewed, including some recent conceptual developments that do not appear to have been widely appreciated. Section VI discusses the promising concept of an inverted transistor design, in which the collector is made smaller than the emitter and placed on the surface of the structure, similar to I^2L, but using a heterostructure design applicable to all transistors. In Section VII the idea of a single-heterostructure transistor with a wide-gap emitter is generalized to DH transistors with both wide-gap emitters and wide-gap collectors. Such a design appears to

Manuscript received June 30, 1981; revised August 31, 1981. This work was supported in part by the Army Research Office and by the Office of Naval Research.

The author is with the Department of Electrical and Computer Engineering, University of California, Santa Barbara, CA 93106.

Reprinted from *Proc. IEEE*, vol. 70, no. 1, pp. 13–25, Jan. 1982.

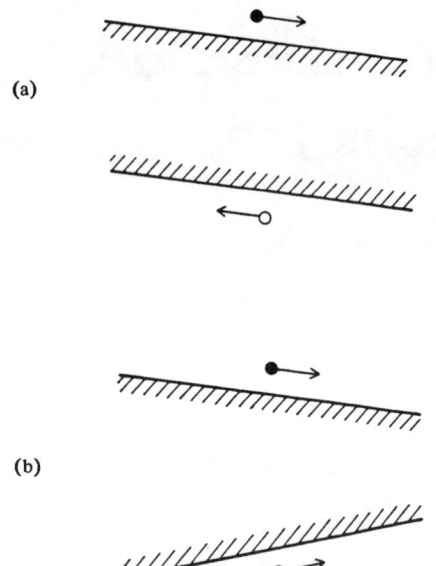

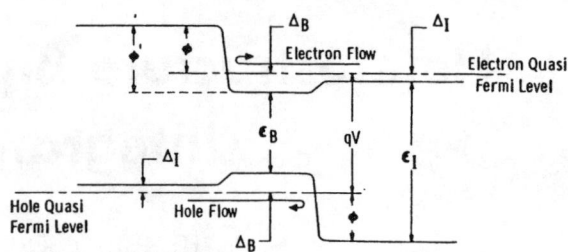

Fig. 2. Energy band diagram of a DH laser, showing the confinement forces driving both electrons and holes towards the active layer, on both sides of the latter. (From [25].)

Fig. 1. Forces on electrons and holes. In a uniform-gap semiconductor (top) the two forces are equal and opposite to each other, and equal to the electrostatic force $\pm q\vec{E}$. In a graded-gap structure, the forces in electrons and holes may be in the same direction.

offer surprisingly large advantages for both microwave and digital devices, and especially for digital IC's. As examples of potential IC advantages, heterostructure modifications of both I^2L and ECL architecture are discussed. Finally, Section VIII offers some speculations on the question of FET's-versus-bipolars, and related questions.

In line with the character of this Special Issue (integrated) digital HBT's are emphasized over (discrete) microwave devices, but not to the point of exclusion of the latter. It would be artificial to attempt a complete separation: Not only was much of the past development of HBT's oriented towards discrete microwave devices, but several of the newer concepts originating in a digital context would improve microwave transistors as well.

II. THE CENTRAL DESIGN PRINCIPLE OF HETEROSTRUCTURE DEVICES

If one looks for a general principle underlying most heterostructure devices, one is led to the following considerations. If one ignores magnetic effects, the forces acting on the electrons and holes in a semiconductor are equal (except for a sign in the case of electrons) to the slopes of the edge of the band in which the carriers reside (Fig. 1). In ideal homostructures the energy gap is constant; hence the slopes of the two band edges are equal, and the forces acting on electrons and holes are necessarily equal in magnitude and opposite in sign. In fact they are equal to the ordinary electrostatic force $\pm q\vec{E}$ on a charge of magnitude $\pm q$ in an electric field $\vec{E}$. In a heterostructure, the energy gap may vary; hence the two band edge slopes and with it the magnitudes of the two forces need not be the same, nor need they be in any simple way related to the electrostatic force exerted by a field $\vec{E}$. In fact, the two slopes may have opposite signs (Fig. 1), implying forces on electrons and holes that act in the same direction, despite their opposite charges.

In effect, heterostructures utilize energy gap variations in addition to electric fields as forces acting on electrons and holes, to control their distribution and flow. This is what I would

like to call the *Central Design Principle* of heterostructure devices. It is a very powerful principle, and one of the purposes of this paper is to give examples that show just how powerful it is.

Although by no means restricted to bipolar devices, the principle is especially powerful when, as in a bipolar transistor, the distribution and flow of *both* electrons and holes must be controlled. By a judicious combination of energy gap variations and electric fields it then becomes possible, within wide limits, to control the forces acting on electrons and holes, *separately and independently of each other*, a design freedom not achievable in homostructures.

The central design principle plays a role in almost all heterostructure devices, and it serves both to unify the ideas underlying different such devices, and as guidance in the development of new device concepts. No device demonstrates the central design principle better than the oldest and so far most important heterostructure device, the DH laser. This point is illustrated in Fig. 2, which shows the energy band structure of the device under lasing conditions, as anticipated (with only slight exaggeration) in the paper in which this device was first proposed [25], and from which Fig. 2 is taken. The drawing shows band edge slopes corresponding to forces that drive *both* electrons and holes towards the inside of the active layer, at *both* edges of the latter. This is the principal reason why the DH laser works, although it is not the only reason. The difference in refractive indices between the inner and outer semiconductors also plays an important role. Such a participation of additional concepts is not uncommon in other heterostructure devices either.

III. THE TECHNOLOGICAL PREMISE

Throughout its history, heterostructure device design has chronically suffered from a technology bottleneck. Even LPE, whatever its merits as a superb laboratory technology, has outside the laboratory been largely limited to devices, such as injection lasers for fiberoptics use, which could simply not be built without heterostructures, but which were needed sufficiently urgently to put up with the limitations of LPE technology. Already for the "ordinary" three-terminal transistor (i.e., excepting phototransistors), the necessary high-performance combination of LPE and lithography was never developed to the point that the resulting heterostructures would reach the speed capability of state-of-the-art Si bipolars, much less reach their own theoretical potential exceeding that of Si.

As a result of the emergence of two new epitaxial technologies in the last few years, the heterostructure technology bottleneck is rapidly disappearing, to the point that the incorporation of heterostructures into most compound semi-conductor devices will probably be one of the dominant themes of compound semiconductor technology during the remainder of the present decade.

The two new technologies are MBE [16] and MOCVD [17]. Although differing in many ways, for the purposes of this paper the commonalities of the two technologies are more important than their differences, and there is no need to enter here into the debate as to which of the two technologies will eventually be best for doing what.

Both technologies are capable of growing epitaxial layers with high crystalline perfection and purity, comparable to state-of-the-art results with LPE and halide-CVD. Highly controlled doping levels up to 10^{19} impurities per cm^3 and more can be achieved, and highly controlled changes in doping level are possible during growth without interrupting the latter, and with at most a minor adjustment in growth parameters. The doping may be changed either gradually or abruptly. Because of the comparatively low growth temperatures (especially for MBE), diffusion effects during growth are weak, and with certain dopants much more abrupt doping steps can be achieved than with any other technique, not only when doping is "turned on," but also when it is "turned off."

Most important in our context of heterostructures, it is possible in both technologies to change from one III/V semi-conductor to a different (lattice-matched) III/V semiconductor with greater ease than in any other technique. In both tech-niques, a change in semiconductor and hence in energy gap is not significantly harder to achieve than a change in doping level! In particular, the change can again be accomplished dur-ing growth without interruption, either gradually or abruptly and, if abruptly, over extremely short distances.

Finally, in both techniques the growth rates and hence the layer thicknesses can be very precisely controlled. Because the growth rates themselves are low (or can be made low), ex-tremely thin layers can be achieved, to the point that effects due to the finite quantum-mechanical wavelengths of the elec-trons can be readily generated. It is in the context of the study of such quantum effects that both techniques have demonstrated their so far highest capability level. With both MOCVD and MBE, GaAs-(Al,Ga)As structures with over 100 epitaxial layers have been built [26], [27], and essentially arbitrary numbers appear possible. With MOCVD, layer thick-nesses below 50 Å have been achieved, with MBE, below 10 Å. In either case, the capability far exceeds anything needed in the foreseeable future for transistor-like devices.

So far, these are laboratory results, mostly on GaAs-(Al,Ga)As structures. But it is the consensus of those working on the two technologies that much of this performance can be carried over into a production environment, with high yields and at an acceptable cost. Acceptable here means a cost low enough that it will not deter the use of the new technologies in most of those high-performance applications that need the performance potential of heterostructure devices.

An extension of both technologies to lattice-matched III/V-compound heterosystems beyond GaAs-(Al,Ga)As is an all but foregone conclusion, including GaAs-(Ga,In)P, InP-(Ga,In)(P,As), and InAs-(Al,Ga)Sb.

In view of these developments, the following scenario for the III/V-compound heterostructure technology of the 1990's is likely. Epitaxial technologies will be routinely available in which both the doping and the energy gap can be varied almost at will, over distances significantly below 100 Å, and covering

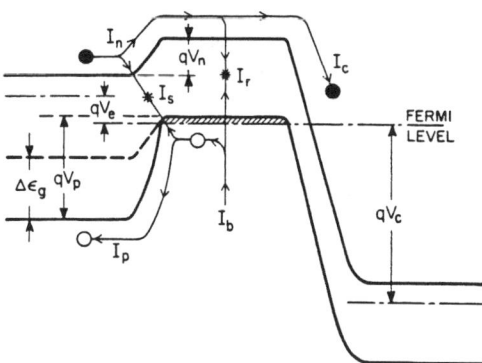

Fig. 3. Band diagram of an n-p-n transistor with a wide-gap emitter, showing the various current components, and the hole-repelling effect of the additional energy gap in the emitter.

a large fraction of their physically possible ranges, by what is essentially a software-controlled operation within a given growth run. The cost of the technology will be sufficiently low to encourage the development of high-performance devices that utilize this capability. The cost will be essentially a fixed cost per growth run, depending on the overall tolerance level but hardly at all on the number of layers and what they con-tain, similar to the cost of optical lithography, which has largely a fixed cost per masking step, almost independent of what is on the mask (at a given tolerance level). In particular, there will be only a negligible cost increment associated with using a heterojunction over using a homojunction (or no junc-tion at all), and hence there will be only a negligible economic incentive *not* to use a heterojunction.

What *will* be expensive, just as with masking, are multiple growth runs, in which the growth is interrupted and the wafer removed from the growth system for intermediate processing, with the growth to be resumed afterwards. Hence there will be a strong incentive to accomplish the desired device structure with the minimum number of growth runs, no matter how complicated the individual run might become.

The above scenario is the technological premise of the re-mainder of this paper. Although presented here in the context of bipolar transistors and IC's, this scenario, as well as the cen-tral design principle of Section II, obviously go far beyond these specific devices. Together, the two concepts might form the starting point for a fascinating speculation about the future of semiconductor devices beyond simple bipolar structures. However, such a discussion would go beyond the scope of this paper as well as of this Special Issue.

IV. THE WIDE-GAP EMITTER

A. Basic Theory

The basic theory behind a wide-gap emitter is simple [1]. Consider the energy band structure of an n-p-n transistor, as in Fig. 3. In drawing the band edges as smooth monotonic curves we are implicitly assuming that the emitter junction has been graded sufficiently to obliterate any band edge discontinuities or even any nonmonotonic variations of the conduction band edge. We will return to this point later. There are the following injection-related dc currents flowing in such a transistor:

a) A current I_n of electrons injected from the emitter into the base;

b) A current I_p of holes injected from the base into the emitter;

c) A current I_s due to electron–hole recombination within the forward biased emitter-base space charge layer.

   d) A small part of I_r of the electron injection current I_n is lost due to bulk recombination.

The current contribution I_n is the principal current on which the device operation depends; the contributions I_p, I_s, and I_r are strictly nuisance currents, as are the capacitive currents (not shown in Fig. 3) that accompany any voltage changes. We have neglected any currents created by electron–hole pair generation in the collector depletion layer or the collector body.

Expressed in terms of these physical current contributions, the net currents at the three terminals are:

Emitter current: $I_e = I_n + I_p + I_s$ (1a)

Collector current: $I_c = I_n - I_r$ (1b)

Base current: $I_b = I_p + I_r + I_s$.   (1c)

A figure of merit for such a transistor is the ratio

$$\beta = \frac{I_c}{I_b} = \frac{I_n - I_r}{I_p + I_r + I_s} < \frac{I_n}{I_p} \equiv \beta_{\max}. \qquad (2)$$

Here, $\beta_{\max}$ is the highest possible value of β, in the limit of negligible recombination currents. It is the improvement of $\beta_{\max}$ to which the wide-gap emitter idea addresses itself.

To estimate $\beta_{\max}$ we assume that emitter and base are uniformly doped with the doping levels N_e and P_b. We denote with qV_n and qV_p the (not necessarily equal) heights of the potential energy barriers for electrons and holes, between emitter and base. We may then write the electron and hole injection current densities in the form

$$J_n = N_e v_{nb} \exp(-qV_n/kT) \qquad (3a)$$

$$J_p = P_b v_{pe} \exp(-qV_p/kT). \qquad (3b)$$

Here v_{nb} and v_{pe} are the mean speeds, due to the combined effects of drift and diffusion, of electrons at the emitter-end of the base, and of holes at the base-end of the emitter.

In writing (3a, b) with simple Boltzmann factors, we have implicitly assumed that both emitter and base are nondegenerate. In a homojunction transistor the emitter might be degenerate; in a heterojunction transistor the base might be degenerate, as is in fact assumed in Fig. 3. This requires small corrections either in (3a) for the homojunction case, or (3b) for the heterojunction case, which we neglect here for simplicity. We have also neglected correction factors allowing for the differences in the effective densities of states of the semiconductors.

We are interested here only in the ratio of the two currents. If the energy gap of the emitter is larger than that of the base by $\Delta\epsilon_g$, we have

$$q(V_p - V_n) = \Delta\epsilon_g \qquad (4)$$

and we obtain

$$\frac{I_n}{I_p} = \beta_{\max} = \frac{N_e}{P_b} \frac{v_{nb}}{v_{pe}} \exp(\Delta\epsilon_g/kT). \qquad (5)$$

For a good transistor, a value $\beta_{\max} \gtrsim 100$ is desirable.

Of the three factors in (5), the ratio v_{nb}/v_{pe} is least subject to manipulation. As a rule

$$5 < v_{nb}/v_{pe} < 50. \qquad (6)$$

To obtain $\beta_{\max} \gtrsim 100$ it is therefore necessary that either

$$N_e \gg P_b \qquad (7)$$

or that $\Delta\epsilon_g$ is at least a few-times kT.

Energy gap differences that are many-times kT are readily obtainable. As a result, very high values of I_n/I_p can be achieved *almost regardless of the doping ratio*. This does not mean that arbitrarily high β's can be obtained. It simply means that the hole injection current I_p becomes a negligible part of the base current compared to the two recombination currents: $I_b \cong I_s + I_r$. To have a useful transistor, we must still have $I_r \ll I_n$. If we approximate I_e by I_n, we obtain

$$\beta = \frac{I_n}{I_r + I_s}. \qquad (8)$$

Based on the evidence from high-β HBT's that have been reported $(\beta \gtrsim 10^3)$,[1] the emitter–base hetero-interface can be made sufficiently defect-free to keep the interface recombination current I_s below $10^{-3}I_n$, at least at sufficiently high current levels I_n. At the same time, the base doping in a properly designed heterostructure transistor will be very high, and hence the minority carrier lifetime correspondingly low, to the point that the bulk recombination current I_r, rather than the interface recombination current I_s will dominate, in contrast to the situation in many homojunction transistors. We therefore neglect I_s beside I_r.

The bulk recombination current density may be written

$$J_r = \gamma n_e(0) w_b/\tau. \qquad (9)$$

Here $n_e(0)$ is the injected electron concentration at the emitter end of the base, w_b is the base width, and τ the average electron lifetime in the base. The factor γ is a factor between 0.5 and 1.0, indicating by how much the average electron concentration differs from the electron concentration at the emitter end. If we insert (3a) and (9) into (8), and neglect I_s, we obtain

$$\beta \cong \frac{1}{\gamma} \frac{v_{nb}\tau}{w_b}. \qquad (10)$$

This depends on the base doping only through the effect of the base doping on the lifetime. For heavy base doping levels the lifetimes may be short indeed.[2] Nevertheless, even for very short lifetimes, high β's should be achievable in transistors with a sufficiently thin base region, which is the case of dominant interest in any event. As an example, assume $w_b \cong 1000$ Å $= 10^{-5}$ cm. In such a transistor the electron velocity is likely to approach values close to bulk limited drift velocities $v_{nb} \cong 10^7$ cm $\cdot$ s^{-1}. Even for a lifetime as short as 10^{-9} s, this would lead to $\beta \cong 10^3$, a value that should satisfy even the most stringent demands. Evidently, no serious problems from reduced minority carrier lifetimes arise unless the latter drop to the vicinity of 10^{-10} s or lower, at least not for plausible base widths not exceeding 1000 Å.

Much of the remainder of this paper will deal with the tradeoffs made possible when high β-values can be obtained without a high emitter-to-base doping ratio. Before turning to these tradeoffs, it is instructive to return to (5) and to apply it to

---

[1] See, e.g., [7], [8], [9], [14], [18]. Even higher values have been found in some phototransistors. See [15] for further references.

[2] For GaAs, injection laser experience suggests lifetimes between 10^{-10} and 10^{-9} s for degenerate doping levels, slightly longer for nondegenerate doping.

energy gap variations in the conventional silicon transistor. The energy gap of Si, like that of the other semiconductors, is not strictly constant, but decreases slightly at the high doping levels that are desirable in the emitters of a homojunction transistor. As a result, a Si transistor is not strictly a uniform-gap transistor; it is itself a heterojunction transistor, but with a small yet highly undesirable *negative* value of $\Delta\epsilon_g$. The best available data, taken on actual transistor structures [28], indicate a gap shrinkage beginning at a doping level $N_d \sim 10^{17}$ cm^{-3}, and reducing the gap approximately logarithmically with doping level, reaching a gap shrinkage between 75 and 80 meV ($>3kT$) at $N_d \sim 10^{19}$ cm^{-3}. According to (5), an emitter gap shrinkage of $3kT$ reduces the ratio I_n/I_p by a factor $e^{-3} \sim 1/20$. The overall effect at this doping level is the same as if the emitter were doped only to 5×10^{17} cm^{-3}, without gap shrinkage. To obtain β-values larger than the ratio v_{nb}/v_{pe} (<50), the base region must be even less heavily doped than this value, which is far below what is metallurgically possible, and far below what would be desirable in the interest of almost all other performance characteristics, especially base resistance. Increasing the emitter doping beyond 10^{19} cm^{-3} improves β only very slowly, roughly proportionally to $N_e^{0.33}$. By pushing everything to the limit, state-of-the-art microwave transistors with P_b-values (averaged over the base region) of about 1×10^{18} cm^{-3} have been achieved [29]. But this is still far below what would be desirable.

Evidently, the conventional Si bipolar transistor behaves far less well than the naive uniform-gap textbook model would predict. In fact, the energy gap shrinkage and its consequences represent one of the dominant performance limitations of the device.

B. Graded Versus Abrupt Emitter Junctions

In Fig. 3, and in the discussion accompanying it, we had assumed that the emitter/base junction is compositionally graded, so as to yield smoothly and monotonically varying band edges. Such graded transistors are easily achieved, but unless the appropriate measures are taken to do so, the modern epitaxial technologies tend to produce abrupt transistors in which band edge discontinuities are present. As a rule, the conduction band on the wider gap side lies energetically above that on the narrower gap side. Applied to the wide-gap emitter in a transistor, this leads to the "spike-and-notch" energy band diagram shown in Fig. 4(a). Because the emitter-to-base doping ratio in an HBT tends to be low, most of the electrostatic potential drop will occur on the less heavily doped emitter side, and the potential spike will project above the conduction band in the neutral portion of the base, leading to a potential barrier of net height $\Delta\epsilon_B$. Such a barrier has both advantages and disadvantages, and a brief discussion is in order.

Consider first the potential notch accompanying the barrier on the base side. Such a notch will collect injected electrons, and therefore enhance recombination losses, a highly undesirable effect. Because of the low emitter-to-base doping ratio expected in an HBT, the notch will be quite shallow, with a depth given approximately by

$$\Delta\epsilon_N = (P_e/N_b) q V_n \qquad (11)$$

which will typically be of the order 5 meV $\ll kT$. Nevertheless, because of the danger of interface recombination defects, it would be desirable to eliminate the notch altogether, and perhaps even replace it by a slightly repulsive potential, as

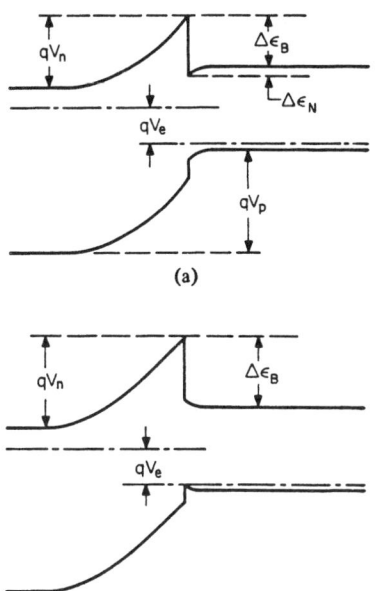

Fig. 4. Band structure of an abrupt wide-gap emitter, showing the spike barrier and the accompanying electron trapping notch (a) in the conduction band structure. The notch can be removed (b) by the incorporation of a planar acceptor doping sheet into the heterojunction.

shown in Fig. 4(b). This is easily accomplished by incorporating a very thin sheet with a very high acceptor concentration right at the interface. Typical required sheet doping concentrations will be of the order 10^{11} acceptors per *square* centimeter. The feasibility of such "planar doping" sheets has been demonstrated [30], at least with MBE, and there is little doubt that it can be accomplished by MOCVD as well.

As to the barrier itself, one minor drawback of its existence is the accompanying increase of the order $\Delta\epsilon_B/q$, in required emitter voltage to yield a given current density. More severe is the (related) drawback that the potential barrier $\Delta\epsilon_B$ drastically reduces the ratio J_n/J_p, from the value in (5), by a factor $\exp(-\Delta\epsilon_B/kT)$. Instead of (4a), we now have

$$q(V_p - V_n) = \Delta\epsilon_g - \Delta\epsilon_B \cong \Delta\epsilon_V. \qquad (4b)$$

The last equality results if the notch depth is small compared to kT, in which case $\Delta\epsilon_B = \Delta\epsilon_C$. Here $\Delta\epsilon_C$ and $\Delta\epsilon_V$ represent the conduction and valence band discontinuities. Instead of (5a), we obtain

$$\frac{I_n}{I_p} = \beta_{\max} = \frac{N_e}{P_b} \frac{v_{nb}}{v_{pe}} \exp(\Delta\epsilon_V/kT). \qquad (5b)$$

If the valence band discontinuity is sufficiently large, a major improvement remains. Unfortunately, in the system of largest current interest, the (Al,Ga)As/GaAs system, the valence band discontinuity is quite small, $\Delta\epsilon_V = 0.15\,\Delta\epsilon_g$ [31], and the reduction of the spike by grading is probably essential. A detailed discussion of the detrimental effects of the spike is found in a paper by Marty *et al.* [10].

The above drawbacks of the extra potential barrier accompanying an abrupt emitter/base junction are partially compensated by the fact that such a barrier would inject the electrons into the base region with a substantial kinetic energy, and hence with a very high velocity ($\sim 10^8$ cm/s). Because of the directional dependence of the polar optical phonon scattering that is the dominant scattering process in III/V-compounds, several collisions are required before the electrons have lost their high forward velocity. The result should be a highly efficient and very fast near-ballistic electron transport through the base. Such ballistic transport effects have been of great interest recently, and although their discussion has been largely in an FET context [32], [33], much of this discussion applies as well (or even more) to bipolar transistors with an emitter junction barrier that represents, in effect, a ballistic launching ramp.

Exactly what the balance between drawbacks and benefits will be for the abrupt emitter/base junction versus the graded one, remains to be seen. But it appears likely that ballistic effects will find their way into future transistors specifically designed around them.

An extreme case of high-energy electron injection into the base was discussed some time ago by Kroemer [34], in the form of a so-called *Auger transistor*. If the conduction band discontinuity $\Delta\epsilon_C$ becomes larger than the energy gap in the base, the electron injection may lead to Auger multiplication of electrons, and hence to a transistor with true current amplification in a grounded-base configuration $\alpha > 1$. Such a transistor might be of interest for power switching applications at very high microwave speeds. It remains to be seen what will come of this idea.

V. Speed Tradeoffs

A. The Emitter Capacitance Tradeoff

High beta-values above, say, 100 are of limited interest by themselves, except perhaps in phototransistors. The principal benefit of a wide-gap emitter is therefore not the ability to achieve high β-values, but the freedom to change doping levels in emitter and base without significant constraints by injection efficiency consideration, and thereby to re-optimize the transistor at a higher performance level.

We start our discussion with the choice of emitter doping. A wide-gap emitter permits a drop in emitter doping by several orders of magnitude without a deterioration of β, a prediction [1] that has been confirmed experimentally in almost all HBT's built. Now it is well known that the junction capacitance of a highly unsymmetrically doped p-n junction depends only on the doping level of the less heavily doped side. Suppose the base doping is initially kept fixed. If the emitter doping is now dropped below the base doping, the emitter capacitance of the transistor then depends principally on the emitter doping and drops with a decrease of the latter, roughly as

$$C_e \propto N_e^{1/2}. \tag{12}$$

Evidently, by dropping the emitter doping sufficiently far below the (initial) base doping a large reduction in emitter capacitance can be obtained [1], and this reduction remains if the base doping is subsequently increased. The result is an improvement in speed, but this effect is usually small, because the emitter capacitance is only one of several capacitances. The true significance of the reduction of the capacitance per unit area lies in two different facts. First, it permits an increase in the capacitive emitter area in the inverted transistor design discussed later, without increase in total emitter capacitance. Second, in HBT's for small-signal microwave amplification, a reduction in emitter capacitance will reduce the noise significantly [35].

Obviously, the doping in the emitter cannot be lowered arbitrarily far. Even if achievable crystal purities permitted it, the emitter series resistance would eventually become excessive, at least for a thick emitter body. However, under the technological scenario envisaged earlier, the weakly doped part of the emitter can always be kept very thin (say, a few-times 10^{-5} cm) to permit a drop in emitter capacitance per unit area by at least a factor 10 before emitter series resistance effects become serious.

A minor advantage of reduced emitter doping, mentioned by Milnes and Feucht [3], might be that the resulting emitters would have a significant reverse breakdown voltage. It is not clear how much of an advantage this would be.

B. The Base Resistance Tradeoff: Microwave Transistors

The most important single change made possible by a wide-gap emitter is a drastic increase in base doping, limited only by technological constraints and by the need to keep the minority lifetime in the base significantly above 10^{-10} s. The principal benefit is a major reduction in base resistance, which, in turn, increases the speed significantly [2]. A second benefit is a major improvement in overall transistor performance at high current densities [1], [3]–[5], including specifically an improvement in the speed-versus-power tradeoffs of microwave transistors.

Because we are principally interested in low-power speed aspects, we concentrate here on the effect of base resistance reduction. This effect is somewhat different in microwave transistors and in switching transistors.

For microwave transistors, Ladd and Feucht [2] have given a very detailed analysis, using the maximum oscillation frequency f_{max} as the figure of merit. It may be written in the form

$$f_{max} = \frac{1}{2}(f_t f_c)^{1/2} \tag{13}$$

where f_t has its familiar meaning as the frequency at which the current gain is reduced to unity, and f_c is the frequency equivalent of the RC time constant of the combination base resistance–collector capacitance,

$$f_c = 1/(2\pi R_b C_c). \tag{14}$$

Evidently, a reduction in R_b causes an increase in f_c and with it a smaller increase in f_{max}.

Ladd and Feucht's work was done in the late 60's and they give numerical values only for the "best" system known at the time, a GaAs emitter on a Ge base, of a construction previously demonstrated by Jadus and Feucht [36]. Because of severe limitations inherent in the then-available technology, the *external* base resistance (between the emitter edge and the base contact) could not be significantly decreased, and as a result, Ladd and Feucht concluded that only a negligible improvement in frequency could be achieved with the then-existing technology. If, however, the external base resistance problem could be solved, maximum oscillation frequencies f_{max} around 100 GHz would be achievable. Similarly high values can be predicted for other heterosystems such as (Al,Ga)As-on-GaAs or GaP-on-Si [37], [38]. There is little point in quoting more exact values, becaue the predictions depend noticeably on both technological and operating parameters whose choice would be applications-dependent. To pursue these matters in detail would lead us too far away from our principal interest in digital switching transistors.

C. The Base Resistance Tradeoff: Digital Switching Transistors

The quantity of interest in digital switching transistors is not the maximum frequency of oscillation but the (somewhat vaguely defined) switching time. Although one would expect that any structural measures that improve the maximum oscillation frequency will also improve the switching speed, there is no simple one-to-one relationship between the two. The modes of operation are just too different. For example, in microwave transistors a high output power is usually of interest, while in highly integrated digital switching transistors the opposite is the case.

A comparison is further complicated by the fact that switching time depends on the circuit, and no standard measure for switching time, comparable to the frequencies f_t and f_{max} for oscillatory operation, has been agreed upon. Probably the best measure of switching time applicable to HBT's is the estimate by Dumke, Woodall, and Rideout (DWR) [5], who estimate the switching time as

$$\tau_s = \frac{5}{2} R_b C_c + \frac{R_b}{R_L} \tau_b + (3C_c + C_L) R_L. \qquad (15)$$

Here R_b is the base resistance, C_c the collector capacitance, and τ_b the base transit time, while R_L and C_L are load resistance and capacitance of the circuit. The result (15) is based on Ashar's analysis [39] of a two-transistor circuit, modified by Dumke. Dumke's modification simply consists of the following [40]. The load resistance must be large enough to develop a potential change equal to the necessary emitter swing ΔV on the next stage. Therefore, $R_L = \Delta V/I = R_E$, where I is the current that is switched to. Making the appropriate substitutions in Ashar's expression yields (15). Dumke et al. apply (15) to estimate the switching time of a hypothetical (Al, Ga)As-on-GaAs transistor with the following parameters. Base width: 1200 Å; base doping: 3×10^{18} cm^{-3}; base and emitter stripe widths: 2.5 μm, separated by 0.5-μm gaps; collector doping: 3×10^{16} cm^{-3}; load resistance: 50 Ω; load capacitance: negligible compared to collector capacitance. These values lead to the following values for the three terms in (15): 8.3 ps, 1.4 ps, and 8.3 ps, combining into an overall switching time of ~18 ps. The authors state that this is "roughly a factor of 5 or 8 faster than that which might be realized from the current post alloy diffused Ge or double diffused Si technologies respectively." Today, nearly 10 years later, post-alloy diffused Ge technology is all but forgotten (it never made it into IC's), and much of the then-predicted advantage over Si remains.

Just as in the case of Ladd and Feucht's estimate of f_{max}, much of the improvement is due to the reduction in base resistance that is associated with the high base doping possible in an HBT. In fact, two of the three terms in (15) depend linearly on R_b rather than with the square root as does f_{max}. This means that as long as those terms dominate τ_s, a reduction of R_b is even more effective in a digital switching transistor than in a microwave transistor. Only after the base resistance reduction has been carried so far that the $R_L C_L$ term dominates, does a further reduction in R_b lead to no further benefit. The hypothetical device analyzed by DWR lies at the borderline between the two regimes.

The specific numerical values quoted above should be viewed as approximations. To obtain an expression as simple as (15), Ashar and Dumke had to make numerous simplifications, just as the expression (13) for f_{max} is based on gross simplifications. The importance of the Ashar–Dumke result (15) is that it indicates the relative significance of the most important transistor parameters. A more detailed analysis is certainly needed, in particular, one that investigates the extent to which the various approximations made in deriving (15) remain applicable in HBT's that have been drastically modified from conventional design.

The assumption of different structural transistor parameters would, of course, have led to different values of τ_b. But the values assumed by DWR were quite reasonable in 1972; they are easily within the range of today's technology, and hence conservative. Further reductions in τ_b to below 10 ps appear readily achievable.

One possibility for improvement is to strive for a lower load resistance than the ad hoc value of 50 Ω assumed by DWR. One sees readily from (15) that the switching time goes through a minimum for

$$R_L = [R_b \tau_b/(3C_c + C_L)]^{1/2} \qquad (16)$$

for which (15) reduces the

$$\tau_s = \frac{5}{2} R_b C_c + 2[(3C_c + C_L) R_b \tau_b]^{1/2}. \qquad (17)$$

For the structural values assumed in DWR one would need $R_L \cong 21$ Ω, which would yield $\tau_s \cong 15$ ps. The improvement is not large, and the low load resistance might not be easy to achieve [40]. A much larger improvement would result from a reduction of the collector capacitance, obtained by inverting the transistor. This possibility will be discussed later.

D. The External Base Resistance Problem

In their detailed analysis of the (microwave) performance potential of HBT's, Ladd and Feucht go to great lengths to discuss the special problem posed by the highly detrimental external portion of the base resistance. Because their considerations also apply to digital switching transistors, and because they appear not to have been fully appreciated by subsequent workers on heterostructure bipolar transistors [41], it appears proper to re-emphasize the problem raised by Ladd and Feucht here, and to offer a remedy.

In all real transistors only part of the base resistance lies underneath the emitter, part lies between the edge of the emitter and the base contact. Usually, the outer region of the base is appreciably thicker than the inner region, and the near-surface portion of the outer base is more heavily doped than the remainder (Fig. 5(a)). This design minimizes the outer base resistance. If one wishes to obtain the postulated advantages of a wide-gap emitter, it is essential that the outer base resistance is not permitted to dominate the overall base resistance. This is easier said than done. For example, suppose that technological changes associated with the change from a (diffused or implanted) homojunction emitter to a heterojunction emitter, forced a change in geometry from that in Fig. 5(a) to that in Fig. 5(b) with a thin outer base. This is in fact the geometry used in the HBT's reported in the literature, except for the transistors reported by Ankri et al. [11], [14] and by Katz et al. [23]. Even though the doping level in the

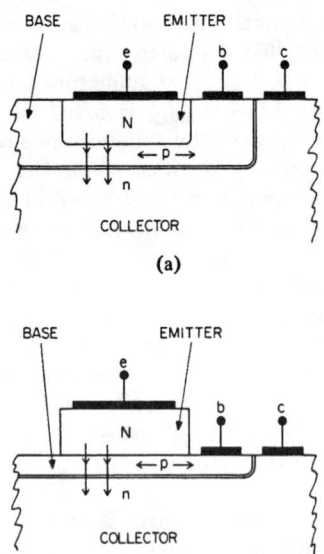

(a)

(b)

Fig. 5. In homojunction transistors of current technology (a), the base region is usually much thicker and more heavily doped outside the emitter than between the emitter and collector, reducing the external base resistance. This desirable feature would be lost in heterostructure transistors with the emitter island design shown in (b). To appreciate this point fully, recall that in actual structures the horizontal dimensions greatly exceed the vertical ones. In this drawing (and in Fig. 6) the vertical dimensions have been greatly exaggerated relative to the horizontal ones.

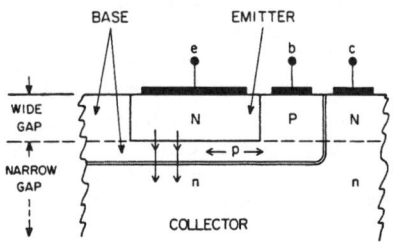

Fig. 6. Desirable emitter structure in which the p-n junction does not follow the planar hetero-interface, but is pulled up towards the surface.

outer base may have been increased, the beneficial effect of this change would be at least partially compensated by the reduction in thickness of the outer base. In unfavorable cases the outer base resistance might even have increased. Ladd and Feucht fully recognized the importance of this problem. They wrote "... *it is clear that the advantages of the low base resistance of the heterojunction devices will only be exploited if suitable geometries can be developed.*"

It is now important to realize that the wide-gap emitter configuration contains a built-in design possibility to keep the outer base resistance low [37], [38], [41], [42]. The design is shown in Fig. 6. Rather than constructing the wide-gap emitter as an island riding by itself on the top of a uniformly thin narrow-gap base layer, the wide-gap semiconductor may be extended beyond the emitter edge, forming part of the outer base region, with the emitter-base p-n junction pulled away from the heteroboundary and towards the surface. Such a configuration should be easily achievable by first growing the top wide-gap layer with the same relatively low n-type doping as the emitter, and then converting the region outside the emitter to heavy p-type doping by diffusion or ion implantation.

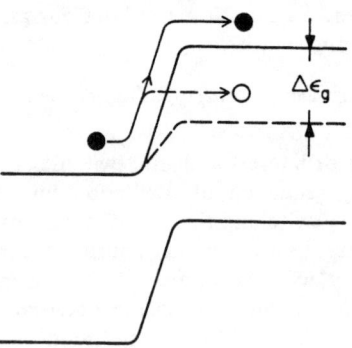

Fig. 7. Blocking of injection of electrons into the wide-gap portion of the base region in Fig. 6, due to the extra repulsive force generated by the wider energy gap.

In such a design the portion of the emitter that lies within the wide-gap region carries only a negligible current, compared to the wide–narrow portion. The reason for this is illustrated in Fig. 7. For injection into the wide-gap p-region, the electrons would have to climb a barrier that is higher by the energy gap difference $\Delta\epsilon_g$. But this reduces the injection current density by the same factor $\exp(-\Delta\epsilon_g/kT)$ that also reduces the hole injection into the wide-gap emitter.

This possibility does not appear to have been as widely recognized as it deserves; it *has* been used in the devices reported by Ankri *et al.* [11], [14], and by Katz *et al.* [23]. In both cases diffusion was used to convert the wide-gap portion of the base region to p-type.

VI. The "Inverted" Transistor

Since the first days of the alloy transistor, bipolar transistors have been built with a larger collector than emitter area, in the interest of efficient charge collection. In planar technology, the two junctions are necessarily of different area. The need for efficient charge collection then enforces the familiar configuration with the collector at the bottom and the emitter at the top. The exception to this rule is, of course, integrated injection logic (I^2L), where other considerations override this rule—at a price. I will say more about I^2L below. But apart from the I^2L exception, the "emitter-up rule" is so pervasive that it has become hard to imagine that a useful transistor could be built with the inverse order.

Now we have just seen that with a wide-gap emitter the emitter junction can be designed in such a way that part of the emitter-base junction does not inject carriers. Evidently, with such a design the need for efficient carrier collection can be met even with an emitter larger than the collector, IF those portions of the emitter-base junction that are not immediately opposite to a part of the collector-base junction are inactivated by pulling them onto the high-gap side of the hetero-interface. Once this is done, the transistor might just as well be "flipped," with the emitter on the substrate side and the collector on top, as shown in Fig. 8. The inverted configuration has several advantages, to the point that it might very well turn out the "canonical" configuration of future heterostructure bipolar transistor design [43].

The principal (but not the only) advantage of the inverted transistor is that it permits the use of a significantly smaller collector area, with an appropriately smaller collector capacitance. The consequences for the high-speed performance are obvious. Modern high speed transistors, both digital and (inter-

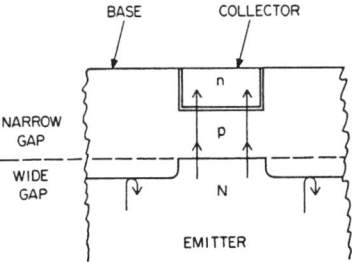

Fig. 8. Inverted "collector-up" transistor structure in which the emitter has a larger area than the collector, but the external portions of the emitter do not contribute to the injection, because there the p-n junction has been pulled into the wide-gap portion of the structure.

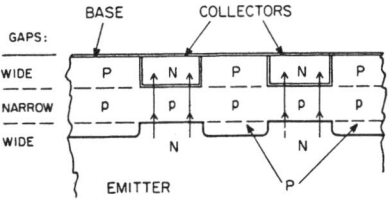

Fig. 9. A DH implementation of I^2L, combining wide-gap collectors with noninjecting emitter regions between the collectors.

digitated) microwave transistors, typically have a collector area close to three-times the active (emitter) area. Inverting the structure thus permits a reduction of the collector capacitance by close to a factor of 3. For example, in the hypothetical switching transistor analyzed by Dumke *et al.* [5], the emitter area was 3.4-times the emitter area. If, in that device, one reduces the collector area by a factor $\frac{1}{3}$ and leaves all other quantities unchanged, the two dominant terms in (15) are reduced by the same factor, and the switching time is reduced from ~18 ps to ~7 ps. Similar improvements would occur in microwave power transistors.

However, some care is in order: Because now the total emitter area is larger than the active area, the emitter junction capacitance will increase, at least compared to a heterostructure transistor of conventional emitter-up configuration. But, as we saw earlier, the emitter junction capacitance per unit area of a heterojunction transistor can in any event be made significantly less than for a homojunction transistor. Hence, compared to the latter, a net reduction in emitter capacitance may result even in the face of a larger (inactive) emitter area.

A second advantage of the inverted configuration is the possibility of a major reduction of the large lead inductance in series with the emitter that is present in the conventional emitter-up configuration. Again an improvement in high-frequency properties will result.

A third advantage of an inverted transistor configuration, for digital switching transistors, will emerge later.

Technologically, the inverted structure should be achievable in essentially the same way as the pulled-up emitter junction: By first growing the top layer lightly n-type doped throughout, and then converting the region outside the collector to heavy p-doping by diffusion or ion implantation. Obviously, the collector layer must be chosen thick enough to support the intended collector bias voltage. Converting part of the surface inside the collector region to n$^+$ might be desirable.

VII. DH TRANSISTORS

A. Introduction: The Wide-Gap Collector

A reading of Shockley's patent quoted at the beginning of this paper leaves no doubt that the "*one. . . zone . . . having a wider energy gap than. . . the other zones*" is the emitter of the transistor. The question was soon raised whether there might also be advantages to a wide-gap collector [1]; but only the trivial and insignificant advantage of a reduction in the reverse-biased collector saturation was recognized.

This assessment must be revised in the light of the anticipated technological scenario discussed in Section III of this paper,

and particularly in the light of the increased interest in highly integrated digital switching transistors. It appears that there are in fact several excellent reasons urging a wide-gap collector design, to the point that DH transistors with a wide-gap collector might very well be the rule rather than the exception for future bipolar transistor designs.

I give in this Section three examples that illustrate advantages to be gained by such a design. They fall into three groups:
a) Suppression of hole injection from base into collector in digital switching transistors under conditions of saturation;
b) Emitter/collector interchangeability in IC's;
c) Separate optimization of base and collector, especially in microwave power transistors.

The presentation does not attempt to give a complete and systematic critical evaluation of all aspects of DH transistor design. Its purpose is to initiate a discussion, not to end it.

B. Suppression of Hole Injection into the Collector under Saturated Conditions

In many digital logic families the collectors of the transistors are forward-biased during part of the logic cycle. If the base region is more heavily doped than the collector, as would normally be desirable, a copious injection of holes from the base into the collector takes place, which increases dissipation and slows down the switching speed. In a heterostructure technology, this highly deleterious phenomenon is easily suppressed the same way hole injection into the emitter is suppressed: By making the collector a wide-gap collector [40]. Such a design is an attractive alternative to the Schottky clamp in Schottky-TTL. Just as the wide-gap emitter, the wide-gap collector should be fairly lightly doped, in the interest of a low collector capacitance, and the base should remain heavily doped, in the interest of low base resistance. This choice of relative doping levels remains both possible and desirable in the inverted I^2L configuration, rather than calling for a weakly doped base to suppress collector injection, with its high base resistance penalty. In fact, in a recent paper [42], Kroemer has proposed a DH implementation of I^2L, which combines this idea with the idea of a selectively injecting emitter, discussed earlier. The structure is shown in Fig. 9. It avoids both the electron injection into those portions of the base where such injection is undesirable because of the absence of a collector opposite to the emitter, and the injection of holes into either collector or emitter. Even electrons spilling over at the edge of the active portion of the base region would not be able to penetrate into the upper part of the inactive portion of the base, because they would be repelled by the heterobarrier in the conduction band at the p–P interface. Because of the essentially complete suppression of parasitic charge storage, combined with greatly reduced RC-time constant effects due to the reduced base resistance, such an implementation of I^2L can be expected to have a much higher speed than the notoriously

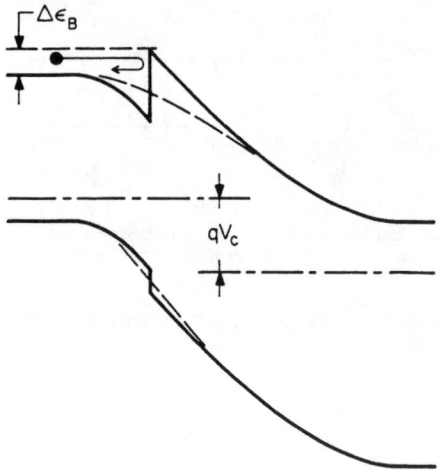

Fig. 10. Electron blocking action for low reverse bias at an abrupt p-n heterojunction collector. The blocking action can be prevented by grading the heterojunction, as indicated by the broken line.

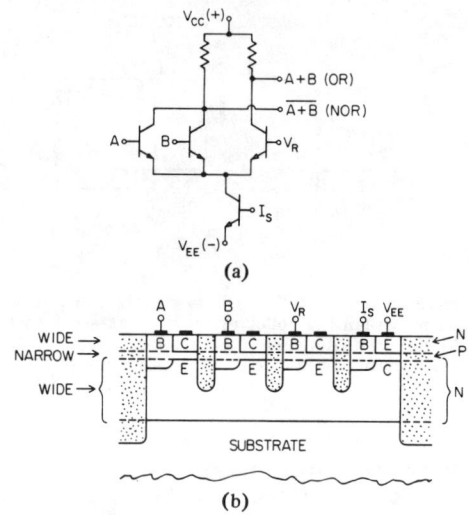

Fig. 11. Input stage of a DH implementation of ECL. The four transistors shown are implemented by three inverted and one noninverted transistor of identical structure, differing only in biasing. The dotted regions are isolation regions, prepared by proton bombardment or equivalent techniques.

slow homostructure implementations of I^2L, without increasing the highly desirable low dissipation levels of I^2L. Unfortunately, no quantitative estimates of the expected performance improvement have so far been published, but the possible improvements appear to be large.[3]

The referenced paper [43] also shows that the pnp horizontal transistor that serves as a current source in I^2L is easily incorporated into a DH design. It emerges as a rather peculiar structure that is basically a homostructure transistor with heterostructure sidewalls, which confine the current and improve the performance of the device.

There is one important restriction in the use of wide-gap collectors, which must not be overlooked. It is important that the free collection of electrons by the reverse-biased collector not be impeded by any heterobarrier due to a conduction band discontinuity (Fig. 10). Such barriers are easily eliminated by grading the heterostructure [44], [45].

C. Emitter/Collector Interchangeability

The advantages of a DH design for bipolar transistors are not restricted to the suppression of hole injection into the collector in saturating logic. A different advantage lies in the possibility of designing transistors in which the role of emitter and collector can be interchanged by simply changing the biasing conditions, while retaining the advantages of a wide-gap emitter regardless of which of the two terminal n-regions is used as the emitter. To achieve this freedom, the transistor need not be geometrically symmetrical: In the inverted structure shown earlier in Fig. 8, in which the active portion of the lower p-n junction covered the same area as the upper p-n junction; either the upper junction or the lower junction could be used as the emitter. While this might be no more than a mildly esoteric advantage in a discrete transistor, it offers a major new option in the architecture of digital IC's, be they of the saturating or nonsaturating variety: The DH design makes it possible, within a common three-layer n-p-n epitaxial layer structure, to integrate high-performance wide-gap emitter transistors having the conventional emitter-up configuration, with similar transistors

---

[3] I have been informed by an anonymous reviewer that K. T. Alavi, in an unpublished M.S. thesis (M.I.T., 1980) has estimated that "over a 10-fold improvement in speed–power product can be anticipated." I did not have access to this work.

having the I^2L-like inverted emitter-down configuration discussed previously.

The full power of this new option can probably not be appreciated without an example. The input stage of emitter-coupled logic (ECL), a nonsaturating logic family, serves admirably. Fig. 11(a) gives the basic circuit diagram of the parts of interest here. The top three transistors serve as a differential switch that compares the voltage levels of two logic signals A and B with a reference voltage V_R. The bottom transistor serves essentially as a constant-current source. (In some simpler versions of ECL it is replaced by a resistor.)

Evidently, the configuration calls for tying together the emitters of three transistors with the collector of a fourth. In a DH design, this integration is achieved easily, without sacrificing a high transistor performance, by implementing the top three transistors as inverted transistors, and the current supply transistor as a conventional emitter-up transistor, as shown in Fig. 11(b). The emitters of the three top transistors and the collector of the bottom transistor come together in a buried n-layer on top of the substrate. All four transistors are structurally identical; they differ merely in their biasing. Those readers who are familiar with ECL and its notorious integration difficulties will undoubtedly recognize the great integration advantages offered by what I would like to call HECL, for Heterostructure ECL.

A complete discussion of various other heterostructure modifications of ECL is intended for another place; the purpose of the present discussion was merely to demonstrate the central idea of the interchangeability of emitter and collector in a DH IC design.

D. Separate Optimization of Base and Collector

Except for the interrelated needs of a high mobility and a high saturated drift velocity for the electrons, the semiconductor properties desired for the base of a transistor are quite different from those for the collector and for the base/collector depletion layer. This is especially true in microwave power transistors. Evidently the different needs of base and collector regions can, at least in principle, be optimized best by selecting different materials in the two regions, that is, by a heterostructure collector. In practice, this tends to mean a semiconductor with a wider energy gap in the collector and in the base/collector layer, compared to the base region.

Again, an example is called for to illustrate this idea. Consider the question as to the semiconductor combination offering the highest speed in a room-temperature microwave power transistor. One can argue that the fastest possible such transistor would be a GaAs-Ge-GaAs transistor [46] —IF such a transistor could in fact be built, which is by no means certain.

The reason for the choice of Ge as the ideal semiconductor for the base region is its high hole mobility, unexcelled by any other group-IV or III/V-compound semiconductor. Also, Ge is easily doped very heavily p-type. Taken together, the two properties assure a much lower base resistance than any other known useable semiconductor.

Admittedly, Ge has a lower electron mobility than several III/V compounds one might consider. But in a microwave power transistor with its necessarily fairly thick collector depletion layer (in the interest of a high breakdown voltage and a low collector capacitance) the transit time through the base is only a minor speed limitation compared to that through the collector depletion layer. Hence the beneficial effects of the high hole mobility in a Ge base layer are much larger than the detrimental effects of the lower electron mobility compared to, say, GaAs. On the other hand, Ge is hardly a desirable semiconductor for the collector and the base/collector depletion layer: Apart from a somewhat low saturated electron drift velocity ($v_s \cong 5 \times 10^6$ cm/s) and a high dielectric constant ($\epsilon \cong 16$), its low energy gap would lead to a low breakdown field and high thermally generated currents. Here a wider gap semiconductor is needed. Lattice-matching considerations suggest GaAs, which would be near-ideal in any event. One might be inclined to argue that the narrow gap of Ge also rules Ge out as a base region material of acceptably low thermal current generation rate. However, this is not the case: In a practical GaAs-Ge-GaAs transistor the Ge base region would be so thin and so heavily p-type doped that the thermal generation of electrons in the base would not contribute an unacceptably high collector saturating current.

Unfortunately, it is not at all clear whether or not GaAs-Ge-GaAs transistors with an acceptably low density of interface defects can be grown. Our own work at UCSB with the MBE growth of GaAs on Ge, and GaP on Si, has shown that the defect-free growth of a polar semiconductor such as GaAs on a nonpolar substrate such as Ge faces a number of quite fundamental difficulties, which have so far not been surmounted, and which may, in fact, be insurmountable [47].

However, none of the experimental uncertainties affect the principal point of our discussion here: The desirability of different semiconductors for base and collector, implying a heterostructure collector, is likely to be the rule rather than the exception in the technology of the future.

VIII. SOME SPECULATIONS ABOUT THE FUTURE OF COMPOUND SEMICONDUCTOR DEVICES

A. Bipolar Transistors versus FET's

If one ignores injection lasers and other optoelectronic devices, today's compound semiconductor device world is a pure FET world with essentially no bipolar inhabitants. A paper that predicts what amounts to a bipolar revolution in this FET world cannot simply ignore FET's. This is true even more once one realizes that the same technologies that promise to revolutionize bipolar transistors will also improve FET's

[48]. In fact, very active and successful research into heterostructure FET's is already under way. However, on balance, heterostructures can be expected to benefit bipolar devices much more than they benefit FET's, and if so, this will naturally tend to shift the balance between the devices much more towards bipolars than past developments might suggest. There are several reasons for these expectations:

a) As was pointed out already in Section II, the Central Design Principle permits one to control the flow of electrons and holes separately and independently of each other. This makes heterostructures a very major advantage in bipolar devices (including lasers) in which there are in fact both kinds of carriers present. It does little for an FET, although a related benefit is obtained in FET's through the concept of modulation doping [49].

b) Every device has a dimension in the direction of current flow that controls the speed of the device. In FET's (other than VMOS) the current flow is parallel to the surface, and the critical control dimension is established by fine-line lithography. In a bipolar transistor, the speed-determining part of the current path is perpendicular to the surface (and to the epilayers), and to the first order, speed is governed by the layer thicknesses. Because vertical layer thicknesses can be easily made much smaller than horizontal lithography dimensions, there is, for given horizontal dimensions, an inherently higher speed potential in bipolar structures than in FET's. The two qualifiers "to the first order" and "for given horizontal dimensions" are important, though: Small horizontal dimensions are still needed to minimize speed-limiting *second-order* effects caused by horizontal resistive voltage drops in the thin base layers. These second-order effects are actually reduced in HBT's, due to the much higher base doping levels, and they are not as severe as the first-order limiting effects of the horizontal dimensions in FET's. But in any event, there is nothing in bipolar technologies that would require or even suggest the use of larger horizontal dimensions than in FET's. The same fine-line lithography technologies that are used for FET's, can and will be used for bipolar devices. The capability offered by the new epitaxial technologies is an *additional* capability, not an alternate.

c) Once sufficiently small dimensions have been achieved, "ballistic" effects become important [32], [33], and they are in fact extensively studied, so far predominantly in an FET context. On the whole, ballistic effects improve device performance by minimizing electron scattering. To obtain this benefit, two conditions must be satisfied. First, the electrons must be accelerated very quickly [32]. The most effective way to do this is by launching the electrons with a high kinetic energy from the conduction band discontinuity in a heterostructure, as discussed earlier. This is much more effective than acceleration by an ordinary nonuniform electric field, the rate of nonuniformity of which is limited by Debye-length considerations. Second, the path along which ballistic effects are to be utilized, must be short, at most a few thousand Angstrom units long. Evidently, both the abrupt launching and the short current paths call for a current flow perpendicular to the epitaxial layers rather than parallel to them, once again favoring the geometry of bipolar designs.

d) All digital switching transistors have a critical bias voltage (often called turn-on voltage), in the vicinity of which the switching action takes place. For high-performance digital IC's, especially VLSI circuits, it is important that this critical voltage be as reproducible as possible, not only across the chip in a single VLSI circuit, but also from wafer to wafer. This reproducibility is easier to achieve in bipolar transistors than in FET's. In bipolar transistors the turn-on voltage is almost fixed for a fixed energy gap of the semiconductor in the base region. It depends logarithmically on the base doping and, apart from temperature, on hardly anything else. Hence it is easy to keep stable. One might say with little exaggeration that it is close to being a natural constant. The turn-on voltage in an FET is, by contrast, purely "man-made," depending at least linearly on both the electron concentration in the channel and the channel thicknesses. To achieve reproducible turn-on voltages, at least two separate quantities must be controlled tightly. Considering that processing differences tend to be very important in IC technology, this particular difference between bipolars and FET's might well turn out to be as important as the more fundamental differences, strongly favoring bipolars [50].

The above arguments suggest strongly that bipolar devices will play a much larger role in the future that they have in the past, eventually assuming a leading role ahead of FET's. Exactly where the border between the two technologies will be, is something too hazardous to predict.

B. A Change in Technological Philosophy?

We have witnessed, since about 1964, a steady growth in III/V-compound semiconductor devices, principally GaAs devices. The driving force behind this development has been the high performance of such devices, not attainable with mainstream Si devices. If we ignore once again lasers and other optoelectronic devices, and restrict ourselves to purely electronic amplifying and switching devices, high performance has been largely synonymous with high speed, made possible by the high electron mobility of GaAs, and by the availability of semi-insulating GaAs as a substrate. However, not even the most ardent advocate of GaAs ever claimed that GaAs was used because it had an attractive technology. We used GaAs despite its technology, not because of it, and the threat was never far away that Si devices, with their much simpler and more highly developed technology, would catch up with GaAs performance, the fundamental advantages of GaAs notwithstanding.

It is exactly this imbalance between fundamental promise and technological weakness that is being removed by the new epitaxial technologies. If the technological scenario postulated in Section III of this paper is even remotely correct, it means nothing less but that the great future strength of III/V-compounds lies precisely in their new technology, which permits an unprecedented complexity and diversity in epitaxial structures, going far beyond anything available in Si technology! This new technological strength is thus emerging as more important than the older fundamental strengths of high mobilities and semi-insulating substrates. It is a remarkable reversal of priorities indeed.

None of this means even remotely that III/V compounds will replace Si. They will not do so any more than aluminum, magnesium, and titanium replaced steel. The analogy of Si to steel is due to M. Lepselter, who called Si technology "the new steel" [51], to bring out the similarity in the role of Si in the new industrial revolution of our own days, to the role of steel in the industrial revolution of the early-19th century. I would like to carry this excellent analogy a bit further. Just as the *structural* metallurgy of the 19th century found it necessary eventually to go beyond steel, to aluminum, magnesium, titanium, and others taking their place beside steel, so the *electronic* metallurgy of our own age is going beyond Si, to the III/V-compounds and probably further, to take their own place beside Si.

We continue to build locomotives, ships, and automobiles from steel, but if it is airplanes and spacecraft we want, we need the other metals besides. And, of course, it took us a while to go from locomotives to spacecraft. The analogy to semiconductors is too obvious to require elaboration; only the time scale will be compressed.

All along the way from steel to titanium there were those who argued that the next step, while perhaps possible, was one for which no foreseeable need existed: All foreseeable needs of man could presumably be met by improvements of the technologies already in hand. Well, this too has not changed.

ACKNOWLEDGMENT

The work of this paper has greatly benefitted from uncounted discussions, over several years, with numerous individuals. Foremost amongst them were R. C. Eden, D. G. Chen, and S. I. Long, all (at the time) at the Rockwell Electronics Research Center. Others at Rockwell to whom I am indebted for discussions are J. S. Harris, R. Zucca, D. L. Miller, and P. Asbeck. I am grateful to W. P. Dumke (IBM) for a copy of his unpublished work on the switching time of bipolar transistors, which clarified many questions I had about that difficult problem. The final version of the paper benefitted from intense discussions with Prof. H. Beneking (Aachen) and from comments made by two anonymous reviewers.

REFERENCES

[1] H. Kroemer, "Theory of a wide-gap emitter for transistors," *Proc. IRE*, vol. 45, no. 11, pp. 1535–1537, Nov. 1957.

[2] G. O. Ladd and D. L. Feucht, "Performance potential of high-frequency heterojunction transistors," *IEEE Trans. Electron Devices*, vol. 17, pp. 413–420, May 1970.

[3] For a review see A. G. Milnes and D. L. Feucht, *Heterojunctions and Metal-Semiconductor Junctions*. New York: Academic, 1972. (See especially ch. 3.)

[4] For additional references see B. L. Sharma and R. K. Purohit, *Semiconductor Heterojunctions*. Elmsford, NY: Pergamon, 1974. (See especially sect. 7.6.)

[5] W. P. Dumke, J. M. Woodall, and V. L. Rideout, "GaAs-GaAlAs heterojunction transistor for high frequency operation," *Solid-State Electron.*, vol. 15, no. 12, pp. 1339–1334, Dec. 1972.

[6] a) M. Konagai and K. Takahashi, "Formation of GaAs-(GaAl)As heterojunction transistors by liquid phase epitaxy," *Elect. Eng. Japan*, vol. 94, no. 4, 1974;
b) ——, "(GaAl)As-GaAs heterojunction transistors with high injection efficiency," *J. Appl. Phys.*, vol. 46, no. 5, pp. 2120–2124, May 1975.

[7] B. W. Clark, H.G.B. Hicks, I.G.A. Davies, and J. S. Heeks, "A (GaAl)As-GaAs heterojunction structure for studying the role of cathode contacts on transferred electron devices," *Gallium Arsenide and Related Compounds 1974* (Deauville), Inst. Phys. Conf. Ser., vol. 24, 1975, pp. 373–375.

[8] M. Konagai, K. Katsukawa, and K. Takahashi, "(GaAl)As/GaAs heterojunction phototransistors with high current gain," *J. Appl. Phys.*, vol. 48, no. 10, pp. 4389–4394, Oct. 1977.

[9] P. W. Ross, H.G.B. Hicks, J. Froom, L. G. Davies, F. J. Probert, and J. E. Carroll, "Heterojunction transistors with enhanced gain," *Electron Eng.*, vol. 49, no. 589, pp. 36–38, Mar. 1977.

[10] A. Marty, G. Rey, and J. P. Bailbe, "Electrical behavior of an n-p-n GaAlAs/GaAs heterojunction transistor," *Solid-State Electron.*, vol. 22, no. 6, pp. 549–557, June 1979.

[11] D. Ankri and A. Scavennec, "Design and evaluation of a planar GaAlAs-GaAs bipolar transistor," *Electron. Lett.*, vol. 16, no. 1, pp. 41–47, Jan. 1980.

[12] J-P. Bailbe, A. Marty, P. H. Hiep, and G. E. Rey, "Design and fabrication of high-speed GaAlAs/GaAs heterojunction transistors," *IEEE Trans. Electron Devices*, vol. ED-27, pp. 1160–1164, June 1980.

[13] H. Beneking and L. M. Su, "GaAlAs/GaAs heterojunction microwave bipolar transistor," *Electron. Lett.*, vol. 17, no. 8, pp. 301–302, Apr. 1981.

[14] D. Ankri, A. Scavennec, C. Besombes, C. Courbet, F. Heliot, and J. Riou, "High frequency low current GaAlAs-GaAs bipolar transistor," presented at Dev. Res. Conf., Santa Barbara, June 1981, unpublished.

[15] For an up-to-date account, containing essentially complete earlier references, see three of the most recent papers on the subject:
a) M. Tobe, Y. Amemiya, S. Sakai, and M. Umeno, "High-sensitivity InGaAsP/InP phototransistors," *Appl. Phys. Lett.*, vol. 37, no. 1, pp. 73–75, July 1980;
b) M. N. Svilans, N. Grote, and H. Benking, "Sensitive GaAlAs/ GaAs wide-gap emitter phototransistors for high current applications," *IEEE Electron Devices Lett.*, vol. ED-11, pp. 247–249, Dec. 1980;
c) J. C. Campbell, A. G. Dentai, C. A. Burrus, Jr., and J. F. Ferguson, "InP/InGaAs heterojunction phototransistors," *IEEE J. Quant. Electron.*, vol. QE-17, pp. 264–269, Feb. 1981.

[16] For two reviews see:
a) A. Y. Cho and J. R. Arthur, "Molecular beam epitaxy," *Prog. Solid State Chem.*, vol. 10, pt. 3, pp. 157–191, 1975;
b) K. Ploog, "Molecular beam epitaxy of III-V compounds," in *Crystals: Growth, Properties, and Applications*, H. C. Freyhardt, Ed. New York: Springer-Verlag, 1980, vol. 3, pp. 73–162.

[17] For a review with complete references to earlier work, see: R.D. Dupuis, L. A. Moudy, and P. D. Dapkus "Preparation and properties of Ga$_{1-x}$Al$_x$As-GaAs heterojunctions grown by metal-organic chemical vapor deposition," *Gallium Arsenide and Related Compounds 1978* (St. Louis), Inst. Phys. Conf. Ser., vol. 45, pp. 1–9, 1979.

[18] R. A. Milano, T. H. Windhorn, E. R. Anderson, G. E. Stillman, R. D. Dupuis, and P. D. Dapkus, "Al$_{0.5}$Ga$_{0.5}$As-GaAs heterojunction phototransistors grown by metalorganic chemical vapor deposition," *Appl. Phys. Lett.*, vol. 39, no. 9, pp. 562–564, May 1979.

[19] D. L. Miller, personal communication.

[20] a) T. Matsushita, N. Oh-uchi, H. Hayashi, and H. Yamoto, "A silicon heterojunction transistor," *Appl. Phys. Lett.*, vol. 35, no. 7, pp. 549–550, Oct. 1979;
b) N. Oh-uchi, H. Hayashi, H. Yamoto, and T. Matsushita, "A new silicon heterojunction transistor using the doped SIPOS," *IEDM Dig.*, pp. 522–524, Dec. 1979;
c) T. Matsushita, H. Hayashi, N. Oh-uchi, and H. Yamamoto, "A SIPOS-Si heterojunction transistor," *Japan. J. Appl. Phys.*, vol. 20, suppl. 20-1, pp. 75–81, Jan. 1981 (Proc. 12th Conf. Solid-State Devices, Tokyo, Aug. 1980).

[21] T. Katoda and M. Kishi, "Heteroepitaxial growth of gallium phosphide on silicon," *J. Electron Mat.*, vol. 9, no. 4, pp. 783–796, Apr. 1980.

[22] S. L. Wright and H. Kroemer, to be published.

[23] J. Katz, N. Bar-Chaim, P. C. Chen, S. Margalit, I. Ury, D. Wilt, M. Yust, and A. Yariv, "A monolithic integration of GaAs/GaAlAs bipolar transistor and heterostructure laser," *Appl. Phys. Lett.*, vol. 37, no. 2, pp. 211–213, July 1980.

[24] H. Beneking, N. Grote, and M. N. Svilans, "Monolithic GaAlAs/ GaAs infrared-to-visible wavelength converter with optical power amplification," *IEEE Trans. Electron Devices*, vol. ED-28, pp. 404–407, Apr. 1981.

[25] H. Kroemer, "A proposed class of heterojunction injection lasers," *Proc. IEEE*, vol. 51, pp. 1782–1783, Dec. 1963.

[26] See, e.g., J. J. Coleman, P. D. Dapkus, N. Holonyak, Jr., and W. D. Laidig, "Device-quality epitaxial AlAs by metalorganic-chemical vapor deposition," *Appl. Phys. Lett.*, vol. 38, no. 11, pp. 894–896, June 1981. This paper quotes only structures containing about 80 layers; much larger numbers have been achieved in unpublished work (personal communication).

[27] For two recent reviews see:
a) L. L. Chang and L. Esaki, "Semiconductor superlattices by MBE and their characterization," *Prog. Cryst. Growth Charact.*, vol. 2, no. 1, pp. 3–12, 1979;
b) A. C. Gossard, "Molecular beam epitaxy of superlattices in thin films," in *Thin Films: Preparation and Properties*, K. N. Tu and R. Rosenberg, Eds. New York: Academic, to be published.

[28] J. S. Slotboom and H. C. de Graaf, "Measurement of bandgap narrowing in Si bipolar transistors," *Solid-State Electron.*, vol. 19, no. 10, pp. 857–862, Oct. 1976.

[29] See, e.g.,
a) J. A. Archer, "Design and performance of small-signal microwave transistors," *Solid-State Electron.*, vol. 15, no. 3, pp. 249–258, Mar. 1972.
b) J. M. Gladstone, P. T. Chen, P. Wang, and S. Kakihana, "Computer aided design and fabrication of an X-band oscillator transistor," Int. Electron Devices Meeting (IEDM) 1973, *IEDM Dig.*, pp. 384–386, Dec. 1973.
c) T. W. Sigmon, "Characteristics of high performance microwave transistors fabricated by ion implantation," Int. Electron Devices Meeting (IEDM) 1973, *IEDM Dig.*, pp. 387–389, Dec. 1973.

[30] C.E.C. Wood, G. Metze, J. Berry, and L. F. Eastman, "Complex free-carrier profile synthesis by atomic-plane doping of MBE GaAs," *J. Appl. Phys.*, vol. 51, no. 1, pp. 383–387, Jan. 1980.

[31] R. Dingle, "Confined carrier quantum states in ultrathin semiconductor heterostructures," *Festkörperprobleme/Advances in Solid State Physics*, vol. 15, pp. 21–48, 1975.

[32] For a review, see H. Kroemer, "Hot electron relaxation effects in devices," *Solid-State Electron.*, vol. 21, no. 1, pp. 61–67, Jan. 1978.

[33] M. S. Shur and L. F. Eastman, "Ballistic and near ballistic transport in GaAs," *IEEE Electron Devices Lett.*, vol. EDL-1, pp. 147–148, Aug. 1980.

[34] H. Kroemer, "Heterojunction device concepts," U.S. Air Force Tech. Rep. AFAL-TR-65-243, Oct. 1965, unpublished. A published description is found in Milnes and Feucht [3], pp. 28–29.

[35] R. E. Yeats, personal communications.

[36] D. K. Jadus and D. L. Feucht, "The realization of a GaAs-Ge wide band gap emitter transistor," *IEEE Trans. Electron Devices*, vol. 16, pp. 102–107, Jan. 1969.

[37] H. Kroemer, Dev. Res. Conf. 1978, Santa Barbara; see *IEEE Trans. Electron Devices*, vol. ED-25, p. 1339, Nov. 1978.

[38] ——, *Bull. Amer. Phys. Soc.*, vol. 24, p. 230, Mar. 1979.

[39] K. G. Ashar, "The method of estimating delay in switching circuits and the figure of merit of a switching transistor," *IEEE Trans. Electron Devices*, vol. ED-11, pp. 497–506, Nov. 1964.

[40] W. P. Dumke, personal communication, unpublished.

[41] The concept at issue here *has* been widely discussed in the DH laser literature. See, e.g., W. Susaki, H. Namizaki, H. Kan, and A. Ito, "A new geometry double-heterostructure injection laser for room-temperature continuous operation: Junction-stripe-geometry DH lasers," *J. Appl. Phys.*, vol. 44, no. 6, pp. 2893–2894, June 1973.

[42] H. Kroemer, "Heterostructures for everything—device principle of the 1980's?," *Japan. J. Appl. Phys.*, vol. 20, suppl. 20-1, pp. 9–13, Jan. 1981 (Proc. 12th Conf. Solid-State Devices, Tokyo, Aug. 1980).

[43] Inverted transistors (with a Schottky collector) with a wide-gap emitter, but without the idea of inactivating the "uncovered" part of the emitter area, have already been reported: H. Beneking, N. Grote, W. Roth, L. M. Su, and M. N. Svilans, "Realization of a bipolar GaAs/GaAlAs Schottky-collector transistor," *Gallium Arsenide and Related Compounds 1980* (Vienna), Inst. Phys. Conf. Ser., vol. 56, pp. 385–392, 1981.

[44] W. G. Oldham and A. G. Milnes, "n-n semiconductor heterojunctions," *Solid-State Electron.*, vol. 6, no. 2, pp. 121–132, Mar./ Apr. 1963.

[45] D. T. Cheung, S. Y. Chiang, and G. L. Pearson, "A simplified model for graded-gap heterojunctions," *Solid-State Electron.*, vol. 18, no. 3, pp. 263–266, Mar. 1975.

[46] Some of the ideas on this subject were independently developed by Dr. Daniel G. Chen, to whom I owe several detailed discussions on this subject.

[47] For a discussion of those problems, see H. Kroemer, K. J. Polasko, and S. C. Wright, "On the (110) orientation as the preferred orientation for the molecular beam epitaxial growth of GaAs on Ge, GaP on Si, and similar zincblende-on-diamond systems," *Appl. Phys. Lett.*, vol. 36, no. 9, pp. 763–765, May 1980.— Unfortunately, even the switch to the (110) orientation has not solved the problems satisfactorily.

[48] See, e.g., D. Boccon-Gibod, J-P. André, P. Baudet, and J-P. Hallais, "The use of GaAs-(Ga, Al)As heterostructures for FET devices," *IEEE Trans. Electron Devices*, vol. ED-27, pp. 1141–1147, June 1980.

[49] See, e.g., S. Judaprawira, W. I. Wang, P. C. Chao, C.E.C. Wood, D. W. Woodard, and L. F. Eastman, "Modulation-doped MBE GaAs/n-Al$_x$Ga$_{1-x}$As MESFETs," *IEEE Electron Devices Lett.*, vol. EDL-2, pp. 14–15, Jan. 1981.

[50] The importance of this advantage of bipolar devices was first pointed out to me by Dr. R. C. Eden.

[51] M. Lepselter, "Integrated circuits—the new steel," Int. Electron Dev. Meeting (IEDM) 1974, *IEDM Dig.*, Dec. 1974.

Optimum emitter grading for heterojunction bipolar transistors

J. R. Hayes, F. Capasso, R. J. Malik, A. C. Gossard, and W. Wiegmann

Bell Laboratories, Murray Hill, New Jersey 07974

(Received 25 July 1983; accepted for publication 2 September 1983)

A simple procedure has been used to determine the optimum emitter grading for a heterojunction bipolar transistor (HBT). Use of this procedure allows maximum hole confinement in addition to minimum base/emitter built-in voltage, leading to a negligible collector/emitter offset voltage, both of which are necessary for high performance devices. By using a parabolic grading function at the emitter/base junction a Np^+n $Ga_{0.7}Al_{0.3}As/GaAs$ HBT has been fabricated, using molecular beam epitaxy, with a negligible collector/emitter offset voltage. A similar result can be obtained with a N-i emitter where the undoped i region is linearly graded.

PACS numbers: 85.60.Pg, 85.30.De, 73.40.Lq

A heterojunction bipolar transistor (HBT) varies significantly from a homojunction bipolar transistor because the emitter injection efficiency is determined largely by the heterojunction band-gap difference rather than the emitter doping. To obtain high current gain (β) from an Np^+n HBT it is necessary that a significant amount of the emitter/base heterojunction band-gap difference falls across the valence band. In many material systems this is not the case and the use of compositional grading has been introduced in order to achieve this.[1]

In this letter we will discuss the effect of compositional grading on the base/emitter heterojunction. Grading is by no means restricted to the base/emitter junction. By incorporating grading in the base of a transistor considerable improvement in high speed performance can be obtained.[2-5] Compositional grading of the emitter/base junction allows one to smooth out a large part of the conduction-band discontinuity so that the band-gap difference at the base emitter heterojunction is concentrated in the valence band. In a HBT where the valence-band discontinuity is small, this can greatly reduce the unwanted injection of holes from the base into the emitter. Although many coworkers have used compositional grading[1,6] in the emitter it would appear from the current literature that the grading width for a specific emitter doping has been chosen in a somewhat arbitrary fashion. It is, therefore, the intention of this letter to explain a simple procedure for optimum emitter design.

The energy-band profile of a graded heterojunction can be calculated from the generalized model of Oldham and Milnes.[7] Unfortunately, this requires extensive numerical analysis and one loses a feel for the parameters involved. A much simpler method was discussed by Cheung *et al.*[8] They showed that the conduction-band edge (E_C) can be obtained by the superposition of the grading potential φ_G and the electrostatic potential across the junction φ_{ES}; i.e.,

$$E_C = \varphi_G - \varphi_{ES}. \tag{1}$$

The valence-band edge can be obtained from

$$E_V = E_C - E_g, \tag{2}$$

where E_g is the direct band gap and E_V the valence-band edge. Results obtained using this method give exceptionally good agreement with the theory of Oldham and Milnes.[8] The

base of a HBT will always be considerably higher doped than the emitter and, hence, to a very good approximation, we can consider the emitter/base junction to be a one-sided junction (Np^+).

We will first consider three cases of linear compositional grading. This is the functional form of grading reported by other workers,[1,6] for a specific emitter doping. It will become clear later than the emitter doping determines the grading distance and for this reason we pick a specific emitter doping of 2×10^{17} cm^{-3}, for all further calculations. It will also be seen that in addition to smoothing out the conduction-band discontinuity the correct choice of grading distance also reduces the emitter/base junction built-in voltage.

This simple technique of emitter design can be used for any material system. However, as there is considerable interest in the GaAlAs/GaAs alloy system, we will consider a device with a $Ga_{0.7}Al_{0.3}As$ emitter that is linearly graded over a distance L from the p^+-GaAs base. The corresponding conduction-band discontinuity between the emitter and base is 0.35 eV.

In order to see the effect of linear grading, we calculate the conduction-band edge for three different linear compositional grading widths, namely, when the grading width is 150, 500, and 300 Å. The first case (150 Å) corresponds to a grading over a distance much smaller than the depletion layer width at the indicated voltages [Fig. 1(a)]. The second case [Fig. 1(b)] corresponds to grading over a distance approximately equal to the depletion layer width at the indicated voltages. The last case [Fig. 1(c)] corresponds to grading over about 1/3 the depletion layer width and the indicated voltages. The spikes and notches in the band structure are the result of the sum of two potentials: the electrostatic potential [which is equal to V_{BI} (the built-in potential)-V_{BE} (the base/emitter voltage)], which varies parabolically, and the copositional grading which varies linearly with distance.

Figure 1(a) shows the calculated conduction-band structure for compositional grading over a distance of 150 Å which corresponds to a near abrupt junction at these doping levels. It should be noted that the junction is not yet strongly conducting and that a forward bias greater than 1.43 V would be required for it to do so. This voltage corresponds to the band gap of the base material (GaAs) and would approximately be the built-in voltage without a heterojunction. As can be seen from Fig. 1(a), there is still a large barrier in the

Reprinted with permission from *Applied Physics Letters*, vol. 43, no. 10, pp. 949–951, 1983. ©1983
American Institute of Physics.

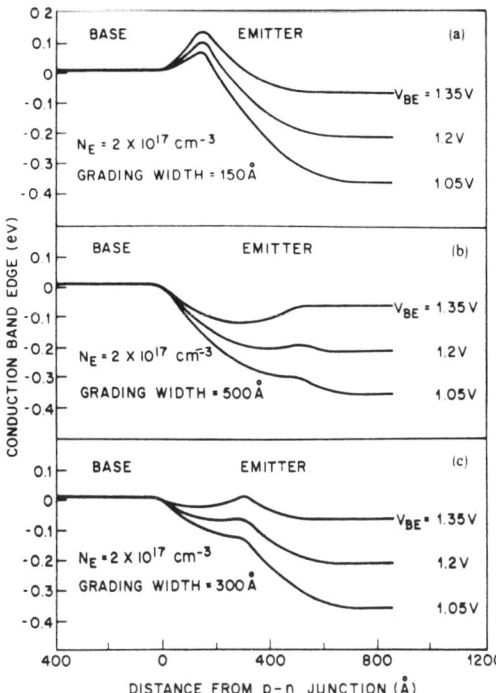

FIG. 1. Conduction-band edge vs distance from the p^+n junction for three different indicated grading width at a forward bias of 1.05, 1.2, and 1.35 V.

conduction band, and the barrier in the valence band is still of the order of ΔE_v (abrupt case), which is insufficient in preventing back injection of holes in the emitter.

The second case, to be considered, is one in which the grading width is 500 Å. Figure 1(b) shows the resulting conduction band which can be explained in terms of the functional dependence of the superimposed potentials. The dip in the conduction band will act as an electron trap and could considerably decrease the emitter injection efficiency.

The last case to be considered is illustrated in Fig. 1(c) and shows the conduction-band edge for a grading width of 300 Å. When the electrostatic and grading potentials are added there is little structure and the built-in voltage is slightly larger than that of an equivalently doped GaAs homojunction.

It appears from the previous discussion that linear grading, although it improves device performance, is not the optimal form of grading to use. It now becomes obvious that all structures can be eliminated by grading with the complementary function of the electrostatic potential of a $Ga_{0.7}Al_{0.3}As$ p^+n junction $(1 - \varphi_{ES})$ over the depletion layer

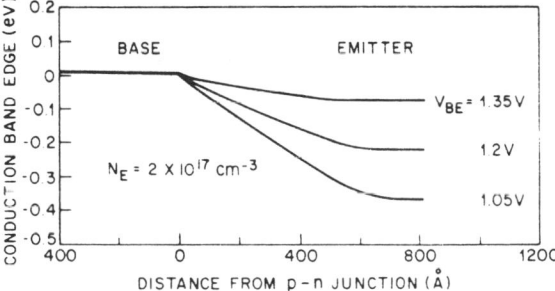

FIG. 2. Conduction-band edge vs distance from the p^+n junction, using parabolic grading over of 500 Å, for a forward bias of 1.05, 1.2 and 1.35 V.

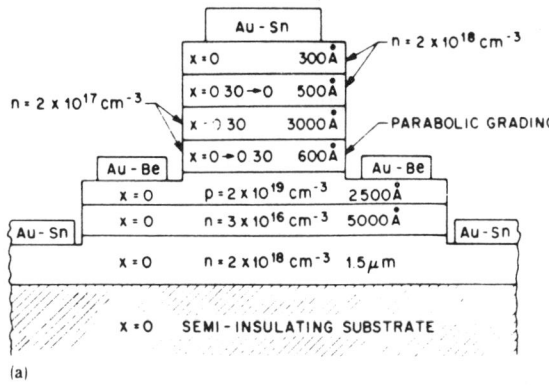

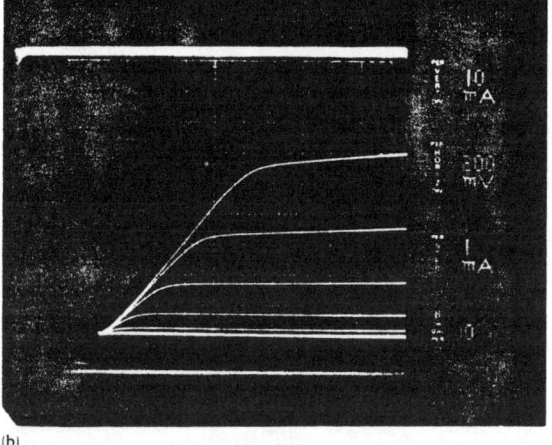

(b)

FIG. 3. (a) Schematic diagram of device structure that has a parabolic grading width of 600 Å at the emitter/base junction. (b) Common emitter characteristics of the transistor shown in (a). Note the negligible offset voltage.

width at a forward bias equivalent to the base band gap. This has been simulated and is shown in Fig. 2 for various levels of forward bias. The graded junction has a built-in voltage identical to that of a GaAs homojunction in addition to maximum hole confinement.

A HBT with such a parabolic grading has been fabricated, using MBE, with a $Ga_{0.7}Al_{0.3}As$ emitter and a GaAs base collector. A schematic diagram of the transistor structure and the common emitter characteristic are shown in Figs. 3(a) and 3(b), respectively. The emitter/base junction was graded from $x = 0$ to $x = 0.3$ on the emitter side over a distance of 600 Å, the parabolic grading function being approximated by linear grading over nine regions. It can be seen from the characteristics shown in Fig. 3(b) that the offset is very small, about 0.03 V. Virtually identical characteristics with offsets $\lesssim 0.03$ V were obtained for all devices on the wafer. Note that the offset voltage is equal to the built-in voltage of the base/emitter junction minus the built-in voltage of the base collector junction. Thus, minimizing the base/emitter junction built-in voltage also minimizes the emitter/collector offset voltage. We believe the small offset voltage was due to a difference in the theoretically designed and experimentally obtained grading and doping profiles.

Use of this grading procedure eliminates the need for a double heterojunction bipolar transistor which has been proposed in order to eliminate the transistor offset that results from unequal junctions. A few other workers[6,9] have attempted to demonstrate the absence of an offset voltage without success. This can be easily understood by considering the base collector junction. Incorrect grading in the base/collector junction does not remove the whole discontinuity from the conduction band so carriers that are injected from the emitter into the base do not have sufficient energy to surmount the collector barrier. Only when a large reverse bias is applied to the base collector junction is the remaining barrier flattened and electrons are able to be collected.

An alternative emitter structure can be made by introducing a $N - i$ emitter where the composition is linearly graded over the undoped region. Now both potentials are linear and this eliminates any conduction-band structure. In addition, due to the introduction of a graded i region the N part of the emitter doping can be significantly increased without affecting the speed of the device since the emitter capacitance is determined, to a first order by the geometric width of the graded region.

In conclusion, we have shown that compositional grading, when used properly, can considerably improve device performance not only by smoothing out the conduction-band discontinuity but also by decreasing the built-in voltage of the emitter to that of the base energy band gap. Use of this procedure has eliminated the problem of an emitter-collector offset voltage. This makes the HBT an ideal switching device and reduces the need for a DHBT unless it is necessary to have symmetrical operation, i.e., interchangeability of the collector and emitter.

[1] P. M. Asbeck, D. L.Miller, R. A. Milano, J. S. Harris, Jr., G. R. Kaelin, and R. Zucca, Proc. IEDM **273**, 629 (1981).

[2] F. Capasso, W. T. Tsang, C. G. Bethea, A. L. Hutchinson, and B. F. Levine, Appl. Phys. Lett. **42**, 93 (1983).

[3] D. L. Miller, P. M. Asbeck, R. J. Anderson, and F. H. Eisen, Electron. Lett. **19**, 367 (1983).

[4] J. R. Hayes, F. Capasso, A. C. Gossard, R. J. Malik, and W. Wiegmann, Electron. Lett. **19**, 410 (1983).

[5] B. F. Levine, C. G. Bethea, W. T. Tsang, T. Capasso, K. K. Thornber, R. C. Fulton, and D. A. Kleinman, Appl. Phys. Lett. **43**, 769 (1983).

[6] S. L. Su, O. Tejayadi, T. J. Drummond, R. Fischer, and H.Morkoç, IEEE Electron Devices Lett. **EDL-4**, 130 (1983).

[7] W. G. Oldham and A. G. Milnes, Solid State Electron. **6**, 121 (1963).

[8] D. T. Cheung, S. Y.Chiang, And G. L. Pearson **18**, 263 (1975).

[9] H. Beneking and L. M. Su, Electron Lett. **18**, 25 (1982).

Self-Consistent Particle Simulation for (AlGa)As/ GaAs HBT's with Improved Base–Collector Structures

RIICHI KATOH, MEMBER IEEE, MAMORU KURATA, FELLOW, IEEE, AND JIRO YOSHIDA, MEMBER, IEEE

Abstract—A one-dimensional self-consistent particle simulator was developed for (AlGa)As/GaAs heterojunction bipolar transistors (HBT's) to investigate how far device performance can be improved by positively utilizing nonequilibrium electron transport phenomena. Hole plasmon scattering and the screening effect on LO phonon scattering were provided to play an essential role in determining the minority-carrier temperature in a heavily doped base layer. Computation was carried out for HBT's with various types of base and collector structures. The electron transport mechanisms are discussed in detail according to the results in conjunction with the reduction in base-to-collector transit time. Intrinsic device performance is expected to improve the cutoff frequency to become over 150 GHz by adopting the graded-gap base and a p-type collector structure even under a heavily doped base condition.

I. INTRODUCTION

ULTRA-HIGH-SPEED and/or high-frequency performance have been reported for (AlGa)As/GaAs HBT's, e.g., a propagation delay time of 5.5 ps [1] and cutoff frequency f_T of 105 GHz [2], which prove their great potential for ultra-high-speed operation as well as the actual existence of the velocity overshoot effect. In view of these remarkable records, an advanced device model that includes a full account of nonequilibrium transport mechanisms, such as velocity overshoot and hot-electron transport phenomena, is indispensable both for an understanding of the device operation principle and design optimization of the device structure.

The capacity of a particle (or Monte Carlo) model excellently matches these requirements toward up-to-date progress in device development. Thus, a one-dimensional self-consistent particle simulator was developed for HBT's in order to investigate these phenomena in themselves as well as their effects on the entire device performance.

In formulating the model, hole plasmon scattering and screened LO phonon scattering were newly taken into account so that heavy doping effects corresponding to the recent tendency of increasing base doping to reduce base resistance are precisely considered.

Manuscript received February 13, 1988; revised October 20, 1988. The review of this paper was arranged by Associate Editor Y. Takeishi.
The authors are with the Research and Development Center, Toshiba Corporation, 1 Komukai Toshibacho, Saiwai-ku, Kawasaki-shi 210, Japan.
IEEE Log Number 8927033.

The validity of our model was verified by comparing computational and experimental results with respect to the minority-carrier temperature in the base region. Thereafter, a detailed discussion was made for the electron transport phenomena in the base-to-collector region, thereby choosing a wide variety of base structures and collector doping profiles.

Finally, the total device performance was discussed for HBT's with various types of base and collector structures by taking up the cutoff frequency dependence on the current density.

II. DEVICE MODEL

The basic concepts underlying the one-dimensional HBT Monte Carlo simulator are similar to that of a previous work [3], where a particle model was established for electrons and a fluid dynamic drift diffusion model [4] was established for holes.

The motion of numerous electrons is traced simultaneously through an entire device. The instantaneous electron carrier profile is calculated along the time axis, which is uniformly discretized with an interval of 5 fs. The continuity equation of holes and Poisson's equation are solved at each time step to modify the potential profile. The self-consistent steady-state solution can be obtained after continuing several thousands of time steps. Once the current density converges, further computation is carried out for a few picoseconds to obtain the average quantities such as the average electron drift velocity profile, where those quantities are averaged on the time axis in order to reduce unphysical noise.

In formulating an abrupt heterojunction, the conservation law was imposed to the total energy and the parallel component of the wave vector. The quantum mechanical reflection at the abrupt heterointerface was also taken into account, which was calculated for the step heterojunction [5].

In formulating a graded heterojunction, an electron's Hamiltonian H is described as

$$H = E + E_c(r) \tag{1a}$$

$$\frac{(\hbar k)^2}{2m_e^*(r)} = E \cdot (1 + \alpha(r) \cdot E) \tag{1b}$$

Reprinted from *IEEE Trans. Electron Devices*, vol. 36, no. 5, pp. 846–853, May 1989.

243

where E, E_c, $\hbar$, m_e^*, α, k, and r represent the electron kinetic energy, the conduction band edge, Planck's constant, the electron effective mass, the band non-parabolicity factor, the electron wavenumber vector, and the electron position vector, respectively. Since m_e^* and α depend on the position in a graded heterojunction, the following canonical equations of motion for electrons are adopted as an extension of the Bloch equation of motion.

$$\frac{dr}{dt} = \frac{1}{\hbar} \cdot \frac{\partial H}{\partial k} \tag{2a}$$

$$\frac{dk}{dt} = -\frac{1}{\hbar} \cdot \frac{\partial H}{\partial r} \tag{2b}$$

In this work, a non-parabolic Γ, L, X three-band structure was taken into consideration in order to make the model more realistic. Physical constants were extracted from [6], and [7]. Sixty-two percent of the bandgap discontinuity [8] was assumed to appear on the conduction band edge (Γ-valley minimum).

In addition to the standard scattering mechanisms for the GaAs/(AlGa)As system [9]–[11], hole plasmon scattering and screened LO phonon scattering were newly taken into account for the heavily doped base layer. Strictly speaking, a coupled system of an LO phonon/hole plasmon should be considered in such a region. However, this approximate treatment is reasonable under the present condition where the plasma frequency is larger than the LO phonon frequency [12]. Hole plasmon scattering was formulated according to a previously calculated matrix element [13], so that a detailed balance was realized for a low energy level, at least in equilibrium. The calculated scattering rate for hole plasmon $\lambda_{pl}(E)$ is as follows:

$$\lambda_{pl}(E) = \frac{(2m_e^*)^{1/2} e^2 \omega_p}{2\epsilon_0 \hbar} \frac{(1 + 2\alpha E')}{\gamma^{1/2}} \ln \left| \frac{A}{\gamma^{1/2} - \gamma'^{1/2}} \right|$$

$$\cdot \left(\frac{1}{e^{\hbar\omega_p/k_B T} - 1} + \frac{1}{2} \mp \frac{1}{2} \right),$$

$$A = \begin{cases} \hbar q_c/(2m_e^*)^{1/2}: \\ \qquad \hbar q_c < (2m_e^*)^{1/2}(\gamma^{1/2} + \gamma'^{(1/2)}) \\ \gamma^{1/2} + \gamma'^{(1/2)}: \\ \qquad \hbar q_c \geqq (2m_e^*)^{1/2}(\gamma^{1/2} + \gamma'^{(1/2)}) \end{cases}$$

$$E' = E \pm \hbar\omega_p$$

$$\gamma = E(1 + \alpha E) \tag{3}$$

where ω_p, ϵ_0, k_B, and q_c represent the hole plasmon angular frequency, the relative dielectric constant, Boltzman's constant, and the cutoff wavenumber, respectively. The cutoff wavenumber q_c is defined by the inverse of the Debye screening length in this work.

As for the LO phonon scattering, the scattering potential was considered to be statically screened with the hole plasma. Calculated scattering rate for screened LO phonon $\lambda_{sc}(E)$ is as follows:

$$\lambda_{sc}(E) = \frac{(2m_e^*)^{1/2} e^2 \omega_0}{2\hbar} \left(\frac{1}{\epsilon_\infty} - \frac{1}{\epsilon_0} \right) \gamma'^{(1/2)} (1 + 2\alpha E')$$

$$\cdot I(E, E') \left(\frac{1}{e^{\hbar\omega_0/k_B T} - 1} + \frac{1}{2} \mp \frac{1}{2} \right), \tag{4}$$

$$I(E, E') = (A + B + C)/D$$

$$A = -2\alpha\gamma^{1/2}\gamma'^{(1/2)}\{4(1 + \alpha E)(1 + \alpha E') + \alpha(\gamma + \gamma' + 2\delta)\}$$

$$B = (1/2)\Big[\{2(1 + \alpha E)(1 + \alpha E') + \alpha(\gamma + \gamma' + \delta)\}^2 + 2\alpha\delta\{2(1 + \alpha E)(1 + \alpha E') + \alpha(\gamma + \gamma' + \delta)\}\Big]$$

$$\times \ln \left| \frac{(\gamma^{1/2} + \gamma'^{(1/2)})^2 + \delta}{(\gamma^{1/2} - \gamma'^{(1/2)})^2 + \delta} \right|$$

$$C = \frac{-2\delta\gamma^{1/2}\gamma'^{(1/2)}}{\{(\gamma^{1/2} + \gamma'^{(1/2)})^2 + \delta\}\{(\gamma^{1/2} - \gamma'^{(1/2)})^2 + \delta)\}}$$

$$\cdot \{2(1 + \alpha E)(1 + \alpha E') + \alpha(\gamma + \gamma' + \delta)\}^2$$

$$D = 4\gamma^{1/2}\gamma'^{(1/2)}(1 + \alpha E)(1 + \alpha E')$$

$$\cdot (1 + 2\alpha E)(1 + 2\alpha E')$$

$$\delta = \hbar^2/(2m_e^*\lambda_D^2)$$

where ω_0, ϵ_∞, and λ_D represent the LO phonon frequency, the high-frequency dielectric constant, and the Debye screening length of hole plasma, respectively.

Fig. 1 shows the dependency of the scattering rate of the hole plasmon and that of the screened LO phonon on electron energy, where that of the unscreened LO phonon is plotted as a reference. These results show that hole plasmon scattering plays an essential role in energy relaxation in a heavily doped p-type base with 1×10^{19} cm^{-3} doping.

Simulation was carried out for a number of HBT's with the variation in the aluminum(Al) mole fraction x and doping, as specified in Table I and Fig. 2.

Throughout the whole samples, the doping concentration and thickness for the n-type emitter are fixed at 2×10^{17} cm^{-3} and 1500 Å; those for the p-type base are fixed at 10^{19} cm^{-3} and 1000 Å; the doping for the n-type subcollector is at 4×10^{17} cm^{-3}, respectively.

The first category is called "abrupt HBT's" with three variations in the emitter Al mole fraction from 0.1 to 0.3 with 0.1 steps. The second category is called a "uniform base HBT" with a graded Al mole fraction of 0.3 to 0 in the emitter, within a 500-Å range that is directly adjacent to the base. The third category, called "graded base HBT's," involves the principal parametric variations with changes in the collector doping from the standard n-type to newly adopted p-type [14]. The latter is investigated as

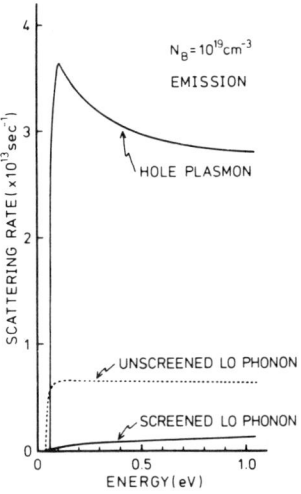

Fig. 1. Scattering rates of hole plasmon and screened LO phonon. That of unscreened LO phonon is plotted with dotted line as reference.

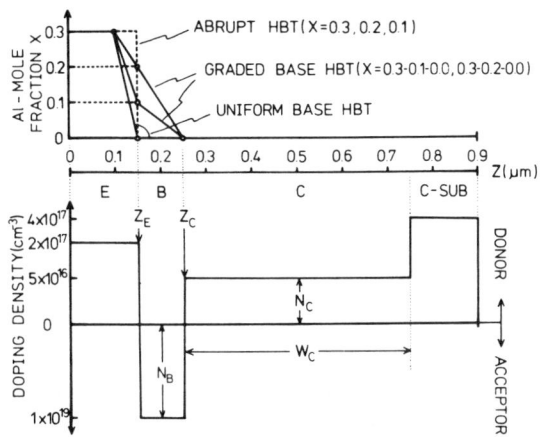

Fig. 2. Calculated HBT structures for base structure variation. Notation in graded base HBT's, e.g., $X = 0.3$-0.1-0.0, represents Al mole fractions at $Z = 1000$, 1500, and 2500 Å. Doping data are for an HBT with a standard n-type collector.

TABLE I
CALCULATED HBT STRUCTURES
(X, N_B, W_B, N_c, and W_c represent aluminum mole fraction, base doping, base thickness, collector doping, and collector thickness, respectively.)

CATEGORY	EMITTER X	BASE X, N_B(cm^{-3}) & W_B(Å)	COLLECTOR N_c(cm^{-3}) & W_c(Å)
ABRUPT HBTS	UNIFORM 0.1, 0.2, 0.3	UNIFORM 0, 10^{19} & 1000	N-TYPE 5×10^{16} & 5000
UNIFORM BASE HBT	GRADED 0.3~0	UNIFORM (SAME AS ABOVE)	(SAME AS ABOVE)
GRADED BASE HBTS	GRADED 0.3~0.1	GRADED 0.1~0, 10^{19} & 1000	N-TYPE 5×10^{16} & 5000 10^{17}, 2×10^{17} & 2000 P-TYPE 5×10^{16}, 10^{17} & 2000 P-TYPE 10^{17} & 1500
	GRADED 0.3~0.2	GRADED 0.2~0, 10^{19} & 1000	N-TYPE 5×10^{16} & 5000
	GRADED 0.3~0.05	GRADED 0.05~0, 2×10^{19} & 500	P-TYPE 10^{17} & 1500

a novel HBT structure that is especially advantageous for (AlGa)As/GaAs HBT's over the former, with the details to follow in Section III-B. Only one exceptional sample is considered for the graded-base p-collector HBT, which

is shown at the bottom of Table I. Namely, a 500-Å base thickness and 2×10^{19} cm^{-3} base doping are introduced instead of 1000 Å and 10^{19} cm^{-3}, respectively.

The collector-to-emitter bias voltage was fixed at 1.5 V and the base-to-emitter voltage was adjusted to yield a constant current density of about 2×10^4 A/cm^2 except for the calculation of cutoff frequency dependence on current density.

Every computation was carried out under 300 K operation condition.

III. COMPUTATION RESULTS

A. Base Structure Variation

Fig. 3 shows a comparison between computation and experiment [15] on minority-carrier temperature in the base for an abrupt HBT, where the Al mole fraction in the emitter was settled at 0.3. Good agreement is seen for the case including both hole plasmon scattering and screened LO phonon scattering to justify our model at high base doping, thus providing the importance of the above mentioned scattering mechanisms in the heavily doped base region.

Fig. 4(a)–(d) shows the electron energy distributions for the various base structures, where a common n-type collector structure 5000 Å in length and 5×10^{16} cm^{-3} in doping was adopted. In all cases, the majority of the electrons in the base are located near the conduction band edge (Γ-valley minimum) because of the strong energy relaxation with the hole plasmon. However, a small number of "hot" electrons are observed in the base for the abrupt HBT (b) and graded-base HBT (d). In the collector space-charge region, the majority of the electrons exists near the upper satellite valley (L, X) minima because of strong intervalley scattering.

Fig. 5 shows the average drift velocity profiles corresponding to Fig. 4, where, in spite of strong scattering, velocity overshoot is observed in the base region of the graded-base HBT's (see Fig. 4(c) and (d)). In the case of an abrupt HBT (b), the improvement in velocity is only slight in the base compared with a uniform-base HBT (a), thus providing the difficulty of ballistic or near-ballistic transport in a heavily doped base layer. Regarding electron transport in the collector region, the following three points are found: 1) a remarkable velocity overshoot is observed but is seen only within 500 Å from the base–collector junction in each case; 2) the peak overshoot velocity decreases as the electron velocity in the base increases; 3) the magnitude of the saturation velocity is about 6–8 $\times 10^6$ cm/s in each case, which agrees well with the previous results [6] calculated for a high electric field up to 10^5 V/cm.

Fig. 6 shows the base transit times for various types of base structures with variations in the aluminum mole fraction for the abrupt and graded-base HBT's. The base transit time τ_B is defined by

$$\tau_B = \int_{Z_E}^{Z_C} \frac{1}{\bar{v}(z)} \, dz \qquad (5)$$

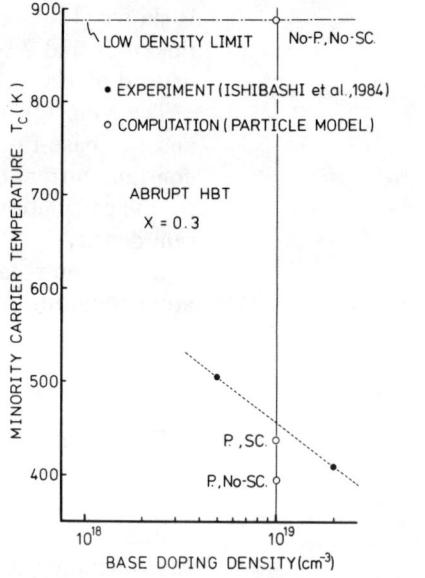

Fig. 3. Comparison between computation and experiment on minority-carrier temperature in base region for abrupt HBT. Indices "P" and "SC" represent plasmon scattering and screened LO phonon scattering, respectively. Index "NO" means that such scattering is not included.

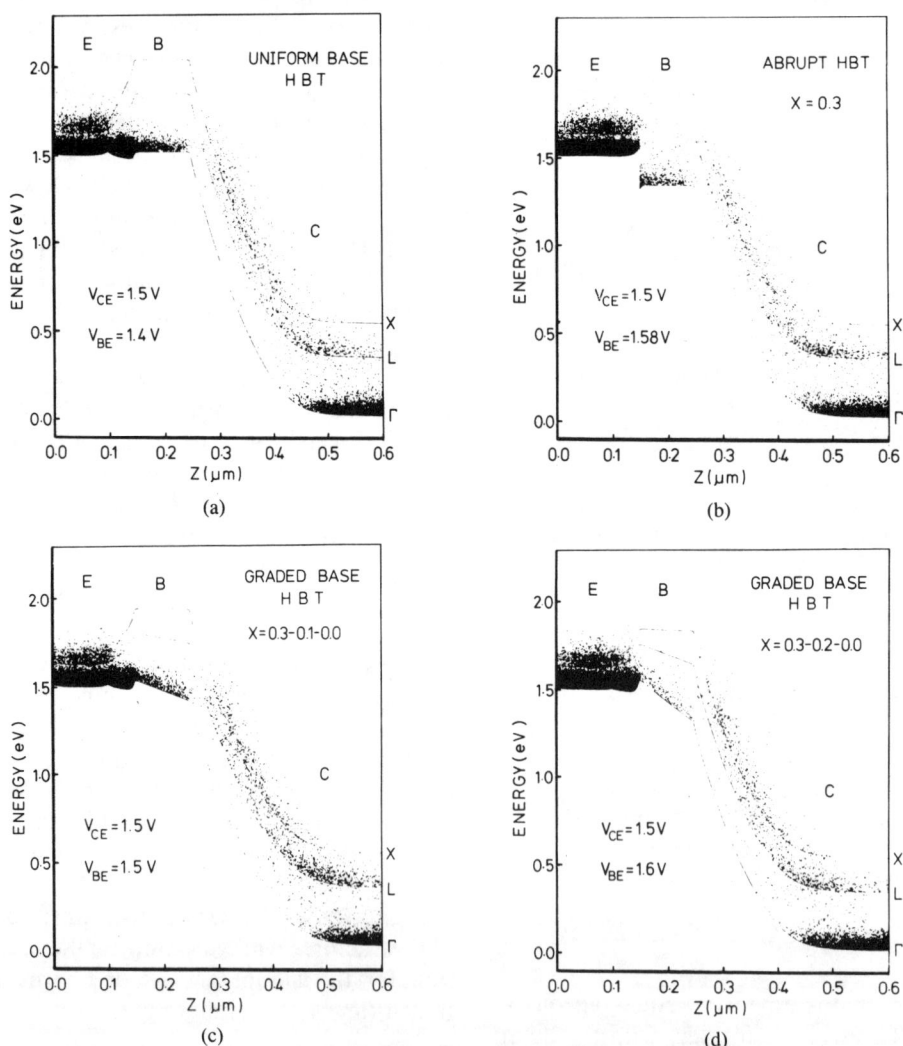

Fig. 4. Electron energy distributions for various base structures. E, B, and C represent the emitter, base, and collector, respectively. Solid lines show Γ, L, X valley minima.

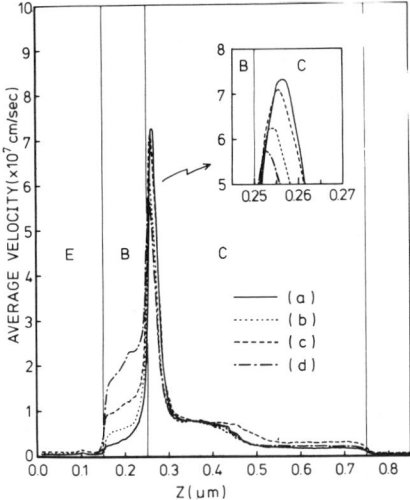

Fig. 5. Average drift velocity profiles corresponding to Fig. 4. Small window shows the magnified figure of the peak overshoot velocity profiles near the base–collector junction.

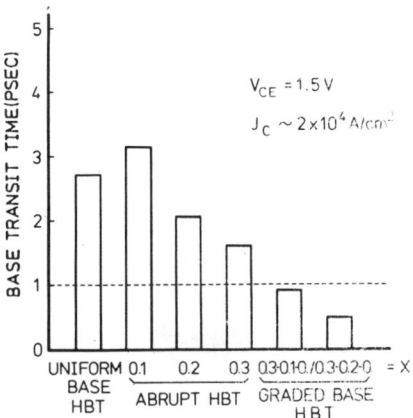

Fig. 6. Base transit times for various base structures, where base transit times for abrupt HBT's with Al mole fractions of 0.1 and 0.2 are supplemented.

where $\bar{v}(z)$ is average electron velocity at position z. Z_E and Z_C are defined at the emitter–base junction ($z = 1500$ Å) and the collector–base junction ($z = 2500$ Å), respectively.

The base transit time is shown to be improved by increasing the Al mode fraction both for the abrupt and graded-base HBT's. Transit times of less than 1 ps, however, were attainable only by graded-base HBT's under a heavily doped base condition. Especially in case (d), the base transit time was improved by a factor of 6, compared with a uniform base (a) in Figs. 4 and 5.

B. Collector Structure Variation

According to the discussion in the last section, electrons were found to drift through almost the whole range of the long depletion layer with a small saturation velocity of $6\text{–}8 \times 10^6$ cm/s. Therefore, after minimizing the base transit time, it becomes important to reduce the collector transit time. In this work, two kinds of trials were made

to reduce the collector transit time, i.e., increasing the collector doping density to decrease the depletion region length, and applying a p-type collector to make use of the velocity overshoot effect [14].

Fig. 7(a)–(d) shows the electron energy distributions in the collector for the various collector structures, where a graded-gap base structure with an aluminum mole fraction of 0.1 was applied at the base–emitter junction in each case. The collector thickness in each case is 2000 Å. For highly doped n-type collectors (Fig. 7(a) and (b)), electrons injected from the base to the collector are immediately scattered into upper valleys, to lose their kinetic energy. This is due to the steep slope of the conduction band profile in the collector space-charge region. On the contrary, for p-type collectors, (Fig. 7(c) and (d)), electrons remain in the Γ valley through a wider range of the collector, to gain large kinetic energy, where the p-layer is thoroughly depleted in Fig. 7(c), while some neutral region remains in the p-layer in Fig. 7(d). A near-ballistic transport can be seen in Fig. 7(c), while a slight energy relaxation can be observed in Fig. 7(d). The difference is attributed to the difference in the slope of the conduction band.

Fig. 8(a) and (b) shows the corresponding average drift velocity profiles for the n-collector (curves (a) and (b)), and those for the p-collector (curves (c) and (d)). The result on a longer n-collector of 5000 Å in Fig. 8(a) and that on a shorter p-collector of 1500 Å in Fig. 8(b) are plotted, each as a reference. From Fig. 8(a), the length of the saturation velocity region is found to be reduced as the doping is increased, but the saturation velocity itself is found to be simultaneously decreased, thus increasing the collector transit time. On the contrary, velocity overshoot is markedly observed in Fig. 8(b) through a wide range of the collector in each case. Comparing curve (c) with (d), the saturation velocity region length for curve (c) is found to be longer than that for curve (d). This fact indicates that a highly doped p-collector structure to avoid perfect depletion is recommended to utilize the velocity overshoot effect, rather than a perfectly depleted p-collector because a moderate electric field causes the velocity overshoot to last longer.

Furthermore, a narrower p-layer is more desirable to reduce the collector transit time because the total transit length can be reduced.

The order of superiority is determined by the collector transit time as indicated in Fig. 9, where the collector transit time τ_c is defined by

$$\tau_c = \frac{\Delta Q_c}{\Delta J_c}. \tag{6}$$

ΔQ_c and ΔJ_c show the variation in the total amount of electronic charge in the collector region and the variation in the collector current density, respectively, when the base–emitter voltage is slightly changed. They were estimated at an approximately constant current density of 2×10^4 A/cm². Obviously, the p-collector structure with

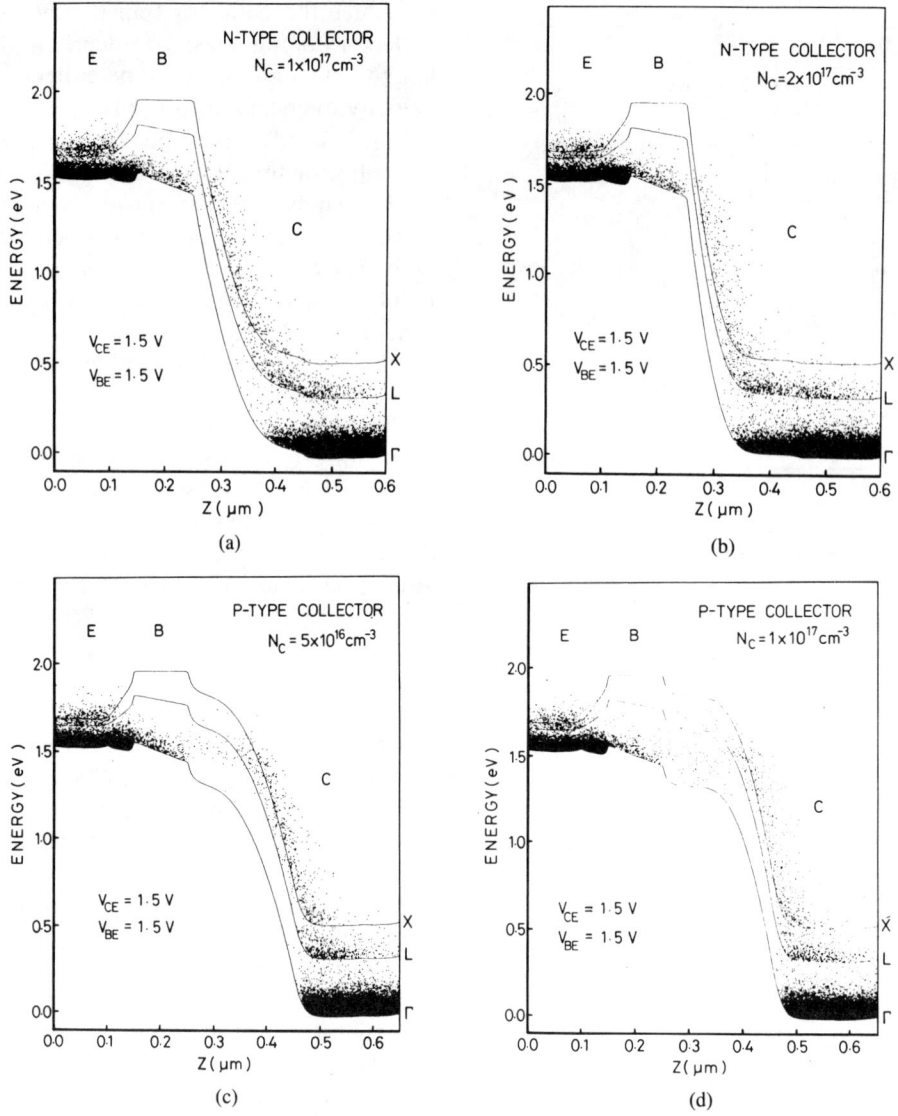

Fig. 7. Electron energy distributions for various collector structures.

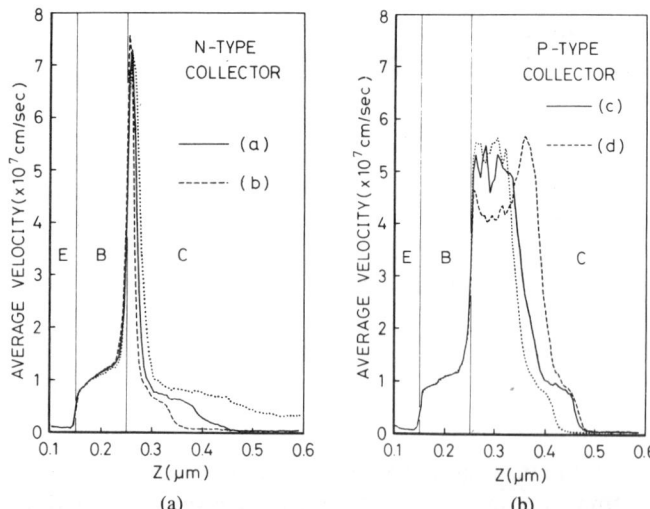

Fig. 8. Average electron drift velocity profiles corresponding to Fig. 7 n-type collector with 5×10^{16} cm^{-3} in doping and 5000 Å in length and p-type collector with 1×10^{17} cm^{-3} in doping and 1500 Å in length are plotted in (a) and (b), respectively, with a dotted line, each as a reference.

relatively high doping and narrow width exhibits the minimum collector transit time, which was smaller than that of a heavily doped n-collector, Fig. 7(b) and curve (*b*) in Fig. 8(a), by 30 percent. This fact implies that the insertion of a p-layer with the appropriate doping and width contributes to a further reduction in the collector transit time.

C. Total Device Performance

A base-to-collector transit time of less than 1 ps is obtained by simply adding the best data for base and collector transit times. However, such a simple estimate is insufficient to predict the total device performance because electron transport in the base cannot be separated from that of the collector. Furthermore, the emitter charging time τ_E should be taken into consideration. Since it is necessary to vary the base-to-emitter bias voltage in order to estimate the emitter charging time, additional computation was carried out to yield the cutoff frequency dependence on current density shown in Fig. 10. The cutoff

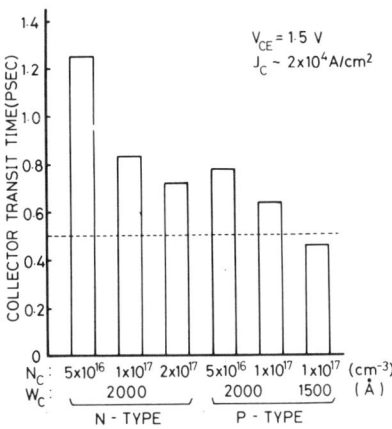

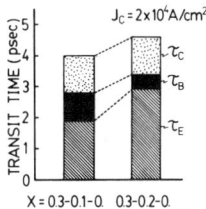

Fig. 11. Component representation of total transit times for graded base HBTs. τ_E, τ_B, and τ_c represent the emitter charging time, the base transit time, and the collector charging time, respectively.

Fig. 9. Collector transit times for various collector structures corresponding to Fig. 8.

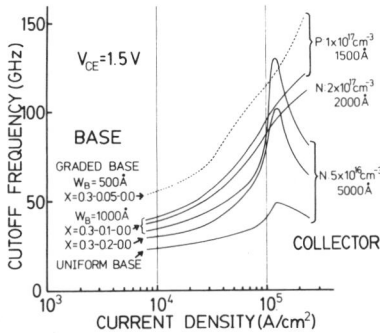

Fig. 10. Cutoff frequency dependence on current density for various base and collector structures. Cutoff frequency for a p-collector HBT with 2 × 10^19 cm^−3 in base doping and 500 Å in base thickness is plotted with dotted line as reference.

frequency f_T is defined by

$$f_T = \frac{1}{2\pi} \frac{\Delta J_c}{\Delta Q_T} \qquad (7)$$

where ΔQ_T shows the variation in the total amount of electric charge (electrons or holes) in the whole device corresponding to the slight change in bias voltage of the order of 0.01 V.

Several interesting features can be found in this figure.

Looking at the order of magnitude of the cutoff frequency for graded-base HBT's with Al mole fractions of 0.1 and 0.2, the lower Al mole fraction case is seen to have a higher cutoff frequency. This seems opposite to the result expected from the previous discussion about the base structure variations, where the base transit time became smaller as the Al mole fraction increased. In order to clarify the reason for this phenomenon, total transit times are separated into three components, τ_E, τ_B, and τ_c in Fig. 11, where they were estimated at a constant current density of 2×10^4 A/cm^2. In this figure, the increasing rate of the emitter charging time is seen to be larger than the decreasing rate of the base transit time when the Al mole fraction increases. This is because the potential barrier height in the valence band side becomes lower as

the Al mole fraction increases when it is compared under the biasing condition to yield the constant current density. This potential barrier lowering enhances the hole injection into the emitter, thus increasing the emitter charging time.

Looking at the collector variation, the order of the cutoff frequency is seen to be quite reasonable at lower current density, where the cutoff frequency for a p-collector HBT is the highest. However, at a higher current density of more than 10^5 A/cm^2, the cutoff frequency for an HBT with an n-type lightly doped collector shows a rapid increase, and the order of magnitude of the cutoff frequency becomes the opposite. This rapid increase in the cutoff frequency is found to occur at an early stage of the Kirk effect (base push-out effect), where the electric field in the collector is gradually relaxed according to the injection of a small amount of holes into the collector region. Due to this relaxation of the electric field, the electron drift velocity in the collector increases through a wide region within the collector, thus reducing the collector transit time. The successive rapid decrease in the cutoff frequency is attributed to the injection of a large amount of holes into the collector region.

The cutoff frequency for an n-type lightly doped collector HBT reaches up to 130 GHz at a high current density, while it is smaller than that for the p-collector HBT at a lower current density.

Since a high cutoff frequency for a wider range of current density is desirable for ultra-fast switching, additional computation was carried out for a p-collector HBT with a narrower graded base. In this case, the base thickness was reduced to 500 Å, while base doping was increased to 2×10^{19} cm^{-3} to keep the same base resistance. The Al mole fraction at the emitter–base junction was settled to 0.05.

The cutoff frequency is improved to more than 50 GHz at a lower current density. Furthermore, it seems that the maximum cutoff frequency of up to 150 GHz can be expected at higher current density. This result shows that the reduction in base width will become essential again for a further improvement of fast switching and high-frequency performance, even if the graded-gap base structure is utilized.

Since these data were derived under the constraint of the fabrication technology, further improvement can be expected, e.g., if a contact formation technique to an ultrathin base layer is established.

IV. Conclusions

A one-dimensional self-consistent particle simulator was developed for heavily doped base HBT's to investigate nonequilibrium electron transport dynamics in themselves, as well as their effects on device performance. A comparison of the carrier temperature between computation and experiment showed good agreement to justify our model, thus proving the importance of hole plasmon scattering and screened LO phonon scattering in the heavily doped base region. A base transit time of less than 1 ps was shown to be attainable by adopting a graded-gap base under a heavily doped base condition. The smallest collector transit time of less than 0.5 ps is attainable by adopting a p-type collector with adequate doping and width, thus proving that even for heavily doped n-type collectors, the insertion of a p-layer between the base and collector contributes to a further reduction in the collector transit time.

High-speed performance is expected to improve the cutoff frequency to become more than 150 GHz by combining the optimal base and collector structures.

Acknowledgment

The authors wish to thank Dr. T. Ozeki, Dr. A. Hojo, and M. Azuma for their continuous interest in our work.

References

[1] K. Nagata, O. Nakajima, Y. Yamauchi, H. Ito, T. Nittono, and T. Ishibashi, "High-speed performance of AlGaAs/GaAs heterojunction bipolar transistors with non-alloyed emitter contacts," presented at the 45th Ann. Dev. Res. Conf., paper IVA-2, 1987.

[2] T. Ishibashi, and Y. Yamauchi, "A novel AlGaAs/GaAs HBT structure for near-ballistic collector," presented at the 45th Ann. Dev. Res. Conf., paper IVA-6, 1987.

[3] K. Tomizawa, Y. Awano, and N. Hashizume, "Monte Carlo simulation of AlGaAs/GaAs heterojunction bipolar transistors," *IEEE Electron Device Lett.*, vol. EDL-5, pp. 362–364, Sept. 1984.

[4] M. Kurata and J. Yoshida, "Modeling and characterization for high-speed GaAlAs-GaAs n-p-n heterojunction bipolar transistors," *IEEE Trans. Electron Devices*, vol. ED-31, pp. 467–473, Apr. 1984.

[5] H. Maeda, "Electron transport across a semiconductor heterojunction," *Japan. J. Appl. Phys.*, vol. 25, pp. 1221–1226, Aug. 1986.

[6] M. A. Littlejohn, J. R. Hauser, and T. H. Glisson, "Velocity-field characteristics of GaAs with $\Gamma_6^c - L_6^c - X_6^c$ conduction-band ordering," *J. Appl. Phys.*, vol. 48, pp. 4587–4590, Nov. 1977.

[7] H. C. Casey, Jr. and M. B. Panish, *Heterostructure Lasers.* New York: Academic, 1978.

[8] M. O. Watanabe, J. Yoshida, M. Mashita, T. Nakanishi, and A. Hojo, "*C–V* profiling studies on MBE-grown GaAs/AlGaAs heterojunction interface," in *Extended Abstracts 16th Conf. Solid State Devices and Materials* (Kobe, Japan), 1984, pp. 181–184.

[9] W. Fawcett, A. D. Boardman, and S. Swain, "Monte Carlo determination of electron transport properties in gallium arsenide," *J. Phys. Chem. Solids*, vol. 31, pp. 1963–1990, 1970.

[10] J. R. Hauser, M. A. Littlejohn, and T. H. Glisson, "Velocity-field relationship of InAs-InP alloys including the effects of alloy scattering," *Appl. Phys. Lett.*, vol. 28, pp. 458–461, Apr. 1976.

[11] D. Chattopadhyay and H. J. Queisser, "Electron scattering by ionized impurities in semiconductors," *Rev. Mod. Phys.*, vol. 53, part 1, pp. 745–768, Oct. 1981.

[12] M. E. Kim, A. Das, and S. D. Senturia, "Electron scattering interaction with coupled plasmon-polar-phonon modes in degenerate semiconductors," *Phys. Rev. B*, vol. 18, pp. 6890–6899, Dec. 1978.

[13] R. A. Ferrell, "Angular dependence of the characteristic energy loss of electrons passing through metal foils," *Phys. Rev.*, vol. 101, pp. 554–563, Jan. 1956.

[14] C. M. Maziar, M. E. Klausmeier-Brown, and M. S. Lundstrom, "A proposed structure for collector transit time reduction in AlGaAs/GaAs bipolar transistors," *IEEE Electron Device Lett.*, vol. EDL-7, pp. 483–485, Aug. 1986.

[15] T. Ishibashi, H. Ito, and T. Sugeta, "Luminescence of hot carrier in the base of an AlGaAs/GaAs HBT," *Inst. Phys. Conf. Ser.*, no. 74, ch. 7, pp. 593–598, 1984.

Transit-Time Reduction in AlGaAs/GaAs HBT's Utilizing Velocity Overshoot in the p-Type Collector Region

KOUHEI MORIZUKA, RIICHI KATOH, MEMBER, IEEE, MASAYUKI ASAKA, NORIO IIZUKA, KUNIO TSUDA, AND MASAO OBARA, ASSOCIATE MEMBER, IEEE

Abstract—The effect of electron velocity overshoot in a p-type GaAs collector on the transit-time reduction of AlGaAs/GaAs HBT's has been demonstrated. An f_T improvement of about 30 percent over the conventional n-type GaAs collector was obtained in p-type collector HBT's for the same collector depletion layer width. A significant increase in electron velocity in the p-type GaAs collector layer was confirmed by a simple analysis.

I. Introduction

THOUGH AlGaAs/GaAs HBT's have been well recognized for their excellent characteristics in high-speed applications, many attempts to obtain further improvement of the intrinsic delay time have been carried out. In particular, considerable attention has been focused on how the collector transit time, which occupies a major portion of the total delay time, can be reduced by the electron velocity overshoot effect [1]–[4]. However, it has been recognized by experimental and theoretical works [1], [2] that the reduction of the transit time by velocity overshoot is relatively small in the conventional n-type collector structure. This is because the electrons injected from the base into the collector depletion region are immediately transferred from the Γ valley to upper valleys which are located at about 0.3 eV above the Γ valley. Because the electron velocity can overshoot its steady-state value until intervalley scattering occurs, it is important to design a collector potential profile in which electrons can travel a certain length without gaining enough energy for intervalley scattering.

From this point of view, several modifications of the collector structure have been proposed. Maziar *et al.* [2] have proposed replacing the n-type GaAs by a p-type GaAs as the collector layer. Katoh *et al.* [3] have also predicted the effectiveness of the p-type collector by using an elaborate particle simulator. In this p-type collector structure, a lower electric field near the base can be realized than that in the n-type collector. Because of this lower electric field, electrons can travel farther in the collector depletion region without intervalley scattering.

In this paper, the electron transport in the p-type GaAs collector was empirically investigated, and a higher electron

Manuscript received July 12, 1988; revised August 30, 1988.

The authors are with the Toshiba Research and Development Center, 1, Komukai Toshiba-cho, Saiwai-ku, Kawasaki 210, Japan.

IEEE Log Number 8824371.

velocity in the p-type collector than in the n-type collector was verified.

II. Experimental Procedure

HBT's were fabricated using MBE-grown wafers. Table I shows the epitaxial layer structure of the conventional n-type collector HBT (called HBT-A) and the newly proposed p-type collector HBT (called HBT-B). Both HBT's had the same structure in the emitter and base layers. The emitter had a 20-nm-thick composition grading region at the emitter–base interface. The base layer also had a composition grading structure to accelerate the electrons from the emitter to the collector. In HBT-A, the collector layer consisted of 500-nm-thick, 5×10^{16} cm^{-3} doped n-GaAs, while in HBT-B the layer was replaced by two stacked layers of p-GaAs (150 nm thick, 7×10^{16} cm^{-3} doped) and n-GaAs (350 nm thick, 2×10^{17} cm^{-3} doped). Both HBT's were fabricated by the same processing method [5]. Fig. 1 shows the fabricated device structure. Devices under microwave characterization had two emitter fingers of 1×8.5 μm^2 dimensions and their base–collector junction area was 6×8.5 μm^2. S-parameter measurements were carried out with an on-wafer probing instrument at 1–26 GHz. The cutoff frequency was derived from plotting h_{21} as a function of frequency and extrapolating it to a higher frequency by assuming a 6-dB/octave roll-off.

III. Results and Discussion

Fig. 2(a) and (b) shows the direct current characteristics of HBT-A and HBT-B, respectively. Almost the same characteristics were observed. Each device had the same collector off-set voltage of 0.2 V. Sufficiently high breakdown voltages under the open-base condition (BV_{CEO}) of 18 and 13 V were obtained in HBT-A and HBT-B, respectively. Though a thinner collector depletion layer is desirable for reducing the collector transit time, the breakdown voltage becomes lower. To overcome this trade-off in the collector, a two-layer structure of p- and n-GaAs was adopted in HBT-B. The collector structure, consisting of a lightly doped p-layer and a moderately doped n-layer, was effective in maintaining the breakdown voltage of HBT-B. Because the doping level was lower in the p-layer than in the n-layer, the depletion region mainly extended into the p-layer where the velocity overshoot occurred effectively.

Reprinted from *IEEE Electron Device Lett.*, vol. 9, no. 11, pp. 585–587, Nov. 1988.

TABLE I
LAYER STRUCTURE OF MBE-GROWN WAFERS

	C_{TE} (fF)	C_{TC} (fF)	W_{CB} (μm)	R_{EE} (Ω)	R_B (Ω)	R_{CC} (Ω)	τ_E (ps)	τ_{CC} (ps)	$\tau_B+\tau_C$ (ps)
HBT–A	103	29	0.19	2.6	28	11	0.68	0.39	2.3
HBT–B	107	28	0.21	4.3	32	11	0.70	0.45	1.3

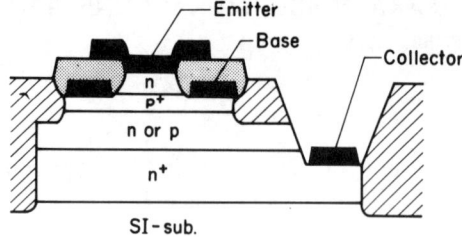

Fig. 1. Cross section of fabricated HBT.

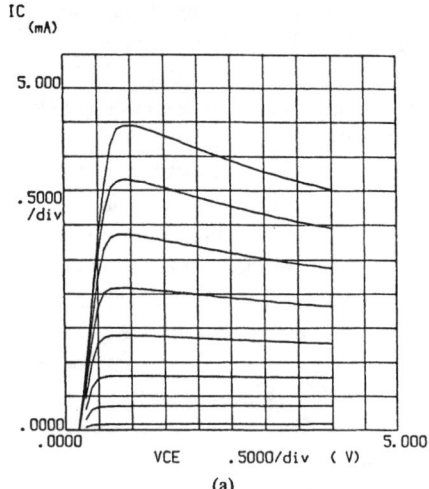

(a)

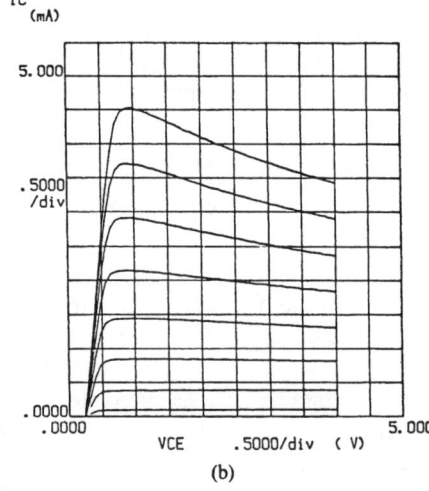

(b)

Fig. 2. Emitter-grounded direct current characteristics of (a) HBT-A, and (b) HBT-B. (Emitter area = 1.0 × 3.5 μm²; I_B = 50 μA/step.)

Fig. 3 shows the dependence of f_T on the collector current density J_C for both devices. A higher f_T was observed in HBT-B than in HBT-A. The maximum value of f_T was 64 GHz at J_C = 1.2 × 10⁵ A/cm² in HBT-A, while it was 76 GHz at J_C = 8 × 10⁴ A/cm² in HBT-B. The collector voltage V_{CE} was kept

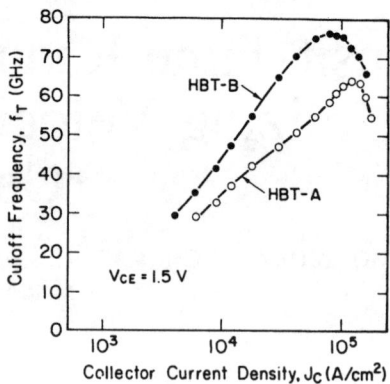

Fig. 3. Dependence of cutoff frequency on collector current density (V_{CE} = 1.5 V).

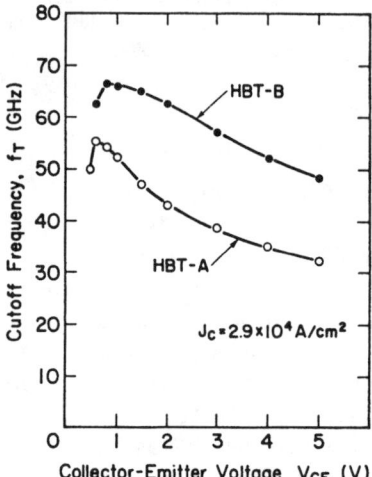

Fig. 4. Dependence of cutoff frequency on collector–emitter voltage (J_C = 2.9 × 10⁴ A/cm²).

to 1.5 V in the measurements. Under this biasing condition, both devices had nearly the same depletion layer width in the collector. Therefore, this improvement of f_T in HBT-B suggested a higher electron velocity in the p-type collector than in the n-type collector.

The dependence of f_T on the collector voltage V_{CE} was measured as shown in Fig. 4. In order to simplify the following analysis by avoiding the space-charge effect of the electrons in the collector depletion region, the collector current density was chosen to be a considerably small value of 2.9 × 10⁴ A/cm². From these measured results, the electron velocity in the collector region was derived in the following manner:

a) estimate the collector transit time τ_C by subtracting other delay components from the total delay time $\tau_{EC}(=1/2\pi f_T)$;

b) evaluate the collector depletion layer width W_{CB} by capacitance–voltage measurement on a large-area diode; and

c) calculate the average electron drift velocity in the collector region (v_d) by using the equation

$$v_d = W_{CB}/2\tau_C. \qquad (1)$$

TABLE II
EVALUATED TRANSISTOR PARAMETERS
($V_{CE} = 1.5$ V, $J_C = 2.9 \times 10^4$ A/cm^2)

	HBT-A	HBT-B
Contact	n-InGaAs 1×10^{19} cm^{-3}, 100 nm (InAs: 0-0.5)	← the same
Emitter	n-AlGaAs 5×10^{17} cm^{-3}, 140 nm (AlAs: 0.1-0.3-0)	← the same
Base	p-AlGaAs 1×10^{19} cm^{-3}, 100 nm (AlAs: 0-0.1)	← the same
Collector	n-GaAs 5×10^{16} cm^{-3} 500nm	p-GaAs 7×10^{16} cm^{-3}, 150 nm n-GaAs 2×10^{17} cm^{-3}, 350 nm
Buried collector	n-GaAs 5×10^{18} cm^{-3}, 500 nm	← the same

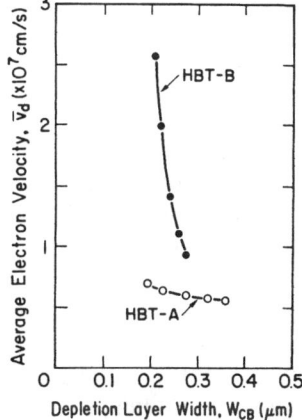

Fig. 5. Dependence of average electron drift velocity on collector depletion layer width.

According to the regional approach, the total delay time $\tau_{EC}(= 1/2\pi f_T)$ consists of the sum of the following four parts: 1) the emitter transition layer charging time τ_E; 2) the base transit time τ_B; 3) the collector transit time τ_C; and 4) the collector charging time τ_{CC}. As $\tau_E(= (kT/qI_C)\cdot(C_{TE} + C_{TC}))$ and $\tau_{CC}(= (r_{EE} + r_{CC})\cdot C_{TC})$ can be evaluated by capacitance measurement and parasitic resistance measurement, the sum of τ_B and τ_C can be derived by subtracting τ_E and τ_{CC} from τ_{EC}. Table II shows a typical set of evaluated parameters for HBT-A and HBT-B. Assuming that the base transit time is equal to the value of 0.9 ps obtained by numerical calculation [3], τ_C can be evaluated. Thus the

average electron velocity v_d can be derived from (1). Fig. 5 shows the calculated v_d as a function of the collector depletion layer width W_{CB}. In the case of the n-type collector (HBT-A), it was observed that the velocity overshoot at the base side of the collector layer slightly improved the average velocity for a narrower W_{CB}. As W_{CB} increased, v_d gradually decreased to a saturation velocity of about 6×10^6 cm/s. This value agrees well with the calculated result in [3]. On the other hand, in the case of the p-type collector (HBT-B), a significant increase in v_d was observed at a narrower W_{CB} (i.e., at a lower V_{CB}). The evaluated v_d reached 2.6×10^7 cm/s at $V_{CE} = 1.5$ V. This high electron velocity verified that the electron velocity overshoot occurs in a large portion of the p-type collector depletion region. As V_{CB} increased, the strength of the electric field in the p-type collector became high, and the region where the velocity overshoot effectively occurred became narrow. Consequently, v_d rapidly decreased as W_{CB} increased for the p-type collector.

Though the electron space-charge effect under a higher current density operation is beyond the scope of this paper, it should be noted that the electron space-charge effect appeared rather differently in HBT-A and in HBT-B because of the difference between the positive and negative ionized dopants in the n- and p-type collectors, respectively. The details of this effect will be discussed elsewhere [6].

IV. CONCLUSION

The effect of electron velocity overshoot in a p-type GaAs collector was investigated. The increased electron velocity in the p-type collector leads to a higher cutoff frequency in AlGaAs/GaAs HBT's compared with a conventional collector structure.

REFERENCES

[1] Y. Yamauchi and T. Ishibashi, "Electron velocity overshoot in the collector depletion layer of AlGaAs/GaAs HBT's," *IEEE Electron Device Lett.*, vol. EDL-7, no. 12, pp. 655–657, 1986.
[2] C. M. Maziar, M. E. Klausmeier-Brown, and M. S. Lundstrom, "A proposed structure for collector transit time reduction in AlGaAs/GaAs bipolar transistors," *IEEE Electron Device Lett.*, vol. EDL-7, no. 8, pp. 483–485, 1986.
[3] R. Katoh, M. Kurata, and J. Yoshida, "A self-consistent particle simulation for (AlGa)As/GaAs HBTs with improved base-collector structures," in *IEDM Tech. Dig.*, 1987, pp. 248–251.
[4] T. Ishibashi and Y. Yamauchi, "A possible near-ballistic collection in an AlGaAs/GaAs HBT with a modified collector structure," *IEEE Trans. Electron Devices*, vol. ED-35, no. 4, pp. 401–404, 1988.
[5] K. Morizuka, M. Asaka, N. Iizuka, K. Tsuda, and M. Obara, "AlGaAs/GaAs HBT's fabricated by a self-alignment technology using polyimide for electrode separation," *IEEE Electron Device Lett.*, vol. 9, pp. 598–600, Nov. 1988.
[6] K. Morizuka *et al.*, "Electron space-charge effects on high-frequency performance of AlGaAs/GaAs HBT's under high-current-density operation," *IEEE Electron Device Lett.*, vol. 9, no. 11, pp. 570–572, Nov. 1988.

A Model-Based Comparison of AlInAs/GaInAs and InP/GaInAs HBT's: A Monte Carlo Study

RIICHI KATOH, MEMBER, IEEE, AND MAMORU KURATA, FELLOW, IEEE

Abstract—The high-speed performances of AlInAs/GaInAs and InP/GaInAs heterojunction bipolar transistors (HBT's) have been thoroughly investigated using a one-dimensional self-consistent particle simulator. Optimum alloy compositions for a graded-gap base structure have been obtained for both transistors through the tradeoff between the emitter charging time and the base transit time. The saturation velocity in the GaInAs n-type collector has been found to be smaller than that in InP, which has been attributed to the diffusion of a large number of "hot" back-scattered Γ-valley electrons in the GaInAs collector. The difference in the collector transit time in p-type collectors has been found to be trivial, since the maximum electron velocity was restricted to below 1.2×10^8 cm/s due to a strong non-parabolicity effect. The cutoff frequency has been estimated to be twice for the former and 1.5 times for the latter higher than that for Al-GaAs/GaAs HBT's. These results have been attributed to a larger bandgap difference between the emitter and base to yield a high base built-in field, rather than a larger Γ-L band separation energy in the collector to enhance the velocity overshoot effect.

I. INTRODUCTION

AlInAs/GaInAs and InP/GaInAs heterojunction bipolar transistors (HBT's) are considered to be promising candidates as post-AlGaAs/GaAs HBT's for their capability of lower power and higher speed operation compared with the latter. Recently reported excellent data for these devices have already revealed their high potential as high-speed devices, although their research and development periods have been by far shorter than that of AlGaAs/GaAs HBT's [1], [2]. However, the effects of a large Γ-L band separation energy ($\Delta E_{\Gamma-L}$) and the base bandgap grading on nonequilibrium electron transport, as well as their high-speed performance, have been left unclear so far.

In this work, these HBT's were compared with each other using a one-dimensional self-consistent particle simulator, which had been developed for analyzing Al-GaAs/GaAs HBT's, in order to clarify the prescribed problems within the limited view of high-frequency performance. First, the alloy compositions at the emitter–base junction were optimized for both transistors to minimize the sum of the emitter charging time (τ_E) and the base transit time (τ_B).

base transit time (τ_B), from the viewpint of a tradeoff between τ_E and τ_B. Second, the effects of a large $\Delta E_{\Gamma-L}$ and a strong nonparabolicity on nonequilibrium electron transport in n-type as well as p-type collector HBT's were thoroughly analyzed in conjunction with the valley population rate, averaged energy-dependent effective mass, and electron velocity distribution. Finally, the collector transit times and the collector current density dependence of the cutoff frequency were obtained to determine the order of the high-frequency performance of these HBT's.

II. MODEL

A previously developed one-dimensional self-consistent particle (Monte Carlo) simulator [3] was applied to analyze AlInAs/GaInAs and InP/GaInAs HBT's, with several modifications in the physical parameters and the formulation of random alloy scattering. The band structure, electron affinity, effective mass, dielectric constant, and other various physical parameters appearing in the scattering rates were determined by linear interpolation of the data for binary alloy systems, i.e., GaAs, InAs, AlAs, and InP [4]–[6], so as to match the lattice constant of the InP substrate. The formulation of random alloy scattering by Littlejohn et al. [7] was adopted for the quaternary alloy systems, i.e., AlGaInAs and GaInAsP.

Fig. 1 shows the input data for the alloy composition and doping profiles, where the fractional variations in the material constants are expressed as $(Al_{\alpha}Ga_{1-\alpha})_{0.47}In_{0.53}As$ and $Ga_{0.47(1-\beta)}In_{0.53+0.47\beta}As_{1-\beta}P_{\beta}$. AlInAs/GaInAs HBT's and InP/GaInAs HBT's will be hereafter referred to as "Tr.1" and "Tr.2," respectively. A common α profile was applied to both n- and p-type collectors in Tr.1, while different β profiles were applied to n- and p-type collectors in Tr.2. Thus Tr.1 and Tr.2 had a $Ga_{0.47}In_{0.53}As$ collector and InP collector, respectively. A graded-gap base structure was adopted in order to reduce the base transit time (τ_B), which is more effective for reducing τ_B than an abrupt emitter–base heterojunction even in a heavily doped base layer [3]. The demonstrated α and β profiles were optimized data, which were derived to minimize the sum of the emitter charging time (τ_E) and the base transit time (τ_B) by the method in the following section. The doping profile was common for both Tr.1 and Tr.2, where an n-type collector with 5×10^{16} cm^{-3}

Manuscript received September 25, 1989; revised January 25, 1990. The review of this paper was arranged by Associate Editor S. Tiwari.

The authors are with the Research and Development Center, Toshiba Corporation, 1 Komukai Toshiba-cho, Saiwai-ku, Kawasaki-shi, 210 Japan.

IEEE Log Number 9034833.

Reprinted from *IEEE Trans. Electron Devices*, vol. 37, no. 5, pp. 1245–1252, May 1990.

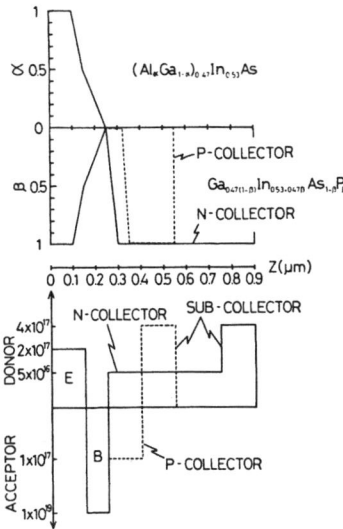

Fig. 1. Computed device structures.

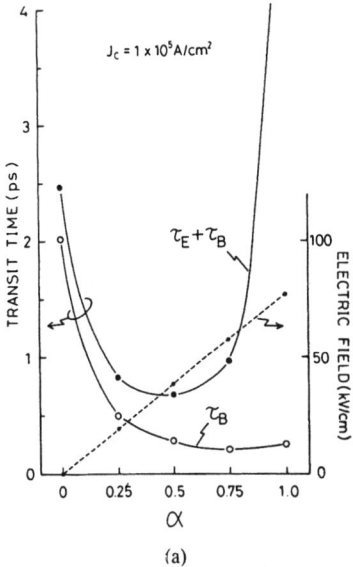

(a)

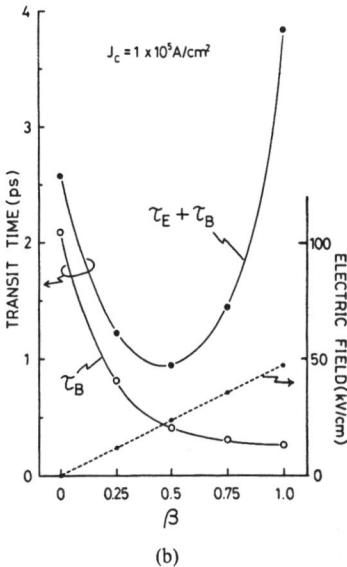

(b)

Fig. 2. Dependence of alloy compositions α and β at emitter-base junction on emitter charging time and base transit time. Dependence of α and β on base built-in field are supplemented as reference. (a) AlInAs/GaInAs HBT. (b) InP/GaInAs HBT.

in doping and 5000 Å in length and a p-type collector with 1×10^{17} cm^{-3} and 1500 Å were considered.

As for the bias condition, the collector-to-emitter bias voltage (V_{CE}) was fixed at 1.5 V throughout the paper, unless otherwise stated. Every computation was carried out under 300 K operation temperature.

III. Computational Results

A. Optimization of Emitter–Base Structures

The alloy compositions α and β at the emitter–base (E–B) junctions were first optimized from the viewpoint of a tradeoff between the emitter charging time (τ_E) and the base transit time (τ_B) [3]. Fig. 2(a) and (b) shows the dependence of ($\tau_E + \tau_B$) on compositions α and β at the E–B junctions, respectively, where the electric field strength in the graded-gap base region was supplemented as reference. The transit times were defined by $\tau_E = \Delta Q_E / \Delta J_C$ and $\tau_B = \Delta Q_B / \Delta J_C$ at a collector current density (J_C) of around 1×10^5 A/cm^2. It should be noted that τ_E's were obtained from the conventional drift-diffusion model, and τ_B's from the particle model.

τ_B's were seen to decrease monotonically as α and β increase, because of the enhancement of velocity overshoot corresponding to the increase in built-in field strength. On the other hand, τ_E's were seen to increase as α and β increase, because of the increase in emitter capacitance. Since the turn-on voltage becomes larger as α and β increase, the potential barrier for holes becomes lower if compared under a fixed current density condition, thus enhancing the injection of holes from the base into the emitter. Consequently, there existed minima in ($\tau_E + \tau_B$) at around $\alpha = \beta = 0.5$ for both transistors. In Tr.1, τ_B seemed to slightly increase at $\alpha \geq 0.75$. This was attributed to the reduction in electron velocity due to the upper valley transition.

Hereafter, $\alpha = \beta = 0.5$ will be chosen at the E–B junctions to minimize ($\tau_E + \tau_B$). Under this condition, the $\tau_B = 0.29$ ps obtained for Tr.1 and 0.41 ps for Tr.2 were

1/3 and 2/5 of τ_B for AlGaAs/GaAs HBT's, respectively. It should be noted that these small τ_B's were obtained as a consequence of the large bandgap difference between the emitter and base layers to increase the base built-in field.

Fig. 3 shows the J_C versus V_{BE} characteristics for the above optimized HBT's. There was a significant difference in the turn-on voltage (V_{on}) of about 0.14 V between Tr.1 and Tr.2. Since V_{on} for Tr.2 was smaller than that for Tr.1, Tr.2 was considered to be more suitable for lower power consumption operation. Furthermore, it is noteworthy that V_{on} for Tr.2 was fairly similar to that of Si bipolar transistors.

B. n-Type Collector HBT's

Fig. 4(a) and (b) shows the electron energy distributions for Tr.1 and Tr.2 with an n-type collector, respec-

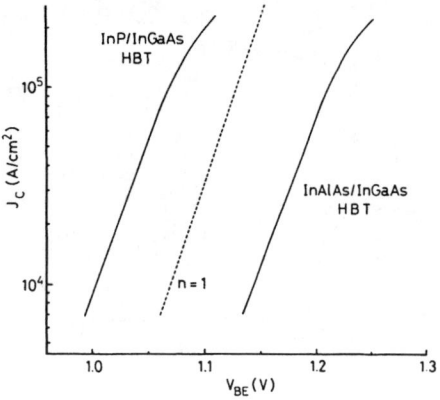

Fig. 3. Collector current density versus base-to-emitter bias voltage characteristics. Ideality factor $n = 1$ is supplemented as reference.

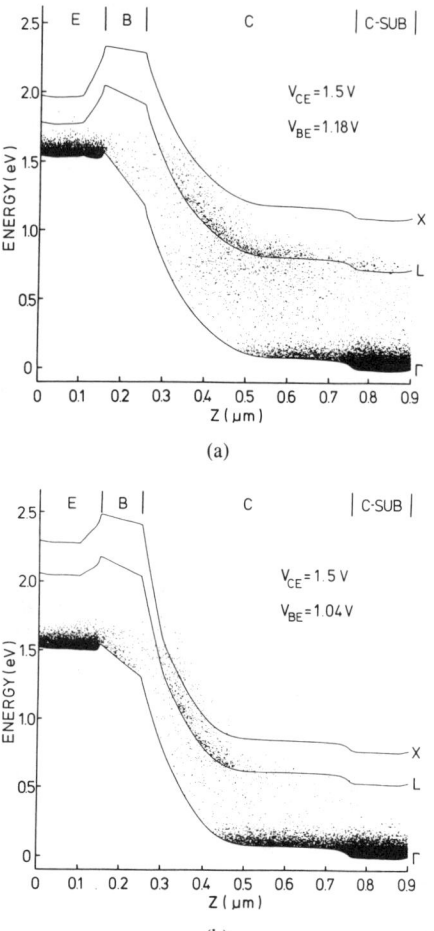

Fig. 4. Electron energy distributions of n-type collector HBT's. Solid line curves Γ, L, and X are corresponding valley minima. (a) AlInAs/GaInAs HBT. (b) InP/GaInAs HBT.

tively, where the base-to-emitter bias voltage (V_{BE}) was chosen to yield a current density of about 5×10^4 A/cm^2. In the emitter regions, since the Γ–L band separation energy ($\Delta E_{\Gamma-L}$) for Tr.2 was by far larger than that for Tr.1, there were no electrons in the upper valley of Tr.2. In the base regions, the inclination of the conduction band for Tr.1 was steeper than that for Tr.2, because of a larger

bandgap difference of the AlInAs/GaInAs heterosystem compared with the InP/GaInAs heterosystem. Therefore, the electrons in Tr.1 gained a higher energy within the base region that that in Tr.2. In the collector regions, there can be seen a large amount of L-valley electrons for both transistors, thus reducing the average electron velocity, even though GaInAs and InP have larger $\Delta E_{\Gamma-L}$ values than GaAs. It should be noted that there was much less X-valley electrons in these collectors than in a GaAs collector [3]. In the collector depletion regions, i.e., $0.25 \leq Z \leq 0.5$ for Tr.1 and $0.25 \leq Z \leq 0.45$ for Tr.2, a larger number of Γ-valley electrons can be seen in Tr.1 compared with Tr.2, because of a larger $\Delta E_{\Gamma-L}$ of GaInAs compared with that of InP. These Γ-valley electrons were transferred from the L-valleys via intervalley scattering, thus having a large energy with a random direction wavenumber vector. In the collector neutral regions, i.e., $Z \geq 0.5$ for Tr.1 and $Z \geq 0.45$ for Tr.2, there still existed numerous L-valley electrons for Tr.1, while only a small number of L-valley electrons was found for Tr.2. This difference in the L-valley population was attributed to the difference in the magnitude of the transfer rate by intervalley scattering from the L-valley to the Γ-valley. Since the intervalley scattering rate is proportional to the power of the electron effective mass in the transferred valley (Γ-valley in this case), the transfer rate in the InP collector is larger than that in GaInAs.

Fig. 5(a) and (b) shows the average electron velocity (V_d) profiles for Tr.1 and Tr.2 with an n-type collector, respectively, with V_{BE} as a parameter. In the base regions, V_d for Tr.1 was about 1.5 times as large as that for Tr.2, resulting in a smaller τ_B for Tr.1. In the collector regions, the peak overshoot velocity for Tr.2 was a little larger than that for Tr.1, because of the smaller velocity for Tr.2 in the base region [3]. The overshoot distance at a lower V_{BE} was about 750 Å for both HBT's, which was a little larger than that for GaAs. At higher V_{BE} values, V_d of both HBT's began to increase in a wider range of the collector region, thus decreasing the collector transit time τ_C. This phenomenon was attributed to the relaxation of the electric field at the onset of the collector high-injection effect (the Kirk effect) [3], [8], [9]. Though the peak overshoot velocity decreased markedly as V_{BE} increased, it was insensitive to the magnitude of τ_C since the overshoot velocity was inherently large.

Another interesting feature was found in the collector depletion region at low V_{BE} values, i.e., the saturation velocity (V_s) for Tr.1 (6×10^6 cm/s) was unexpectedly smaller than that for Tr.2 (1×10^7 cm/s), although $\Delta E_{\Gamma-L}$ in the collector for Tr.1 was larger than that for Tr.2, and the effective mass in the collector for Tr.1 was smaller than that for Tr.2. In order to clarify this small V_s for Tr.1, the Γ-valley population rates were plotted for both transistors in Fig. 6, with the bias conditions corresponding to Fig. 4(a) and (b). In the collector depletion region, i.e., $0.25 \leq Z \leq 0.45$, the Γ-valley population rate for Tr.1 was larger than that for Tr.2 because of the larger $\Delta E_{\Gamma-L}$ for Tr.1. Therefore, upper valley transition

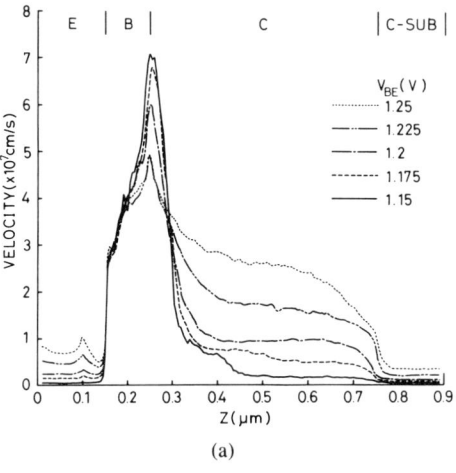

(a)

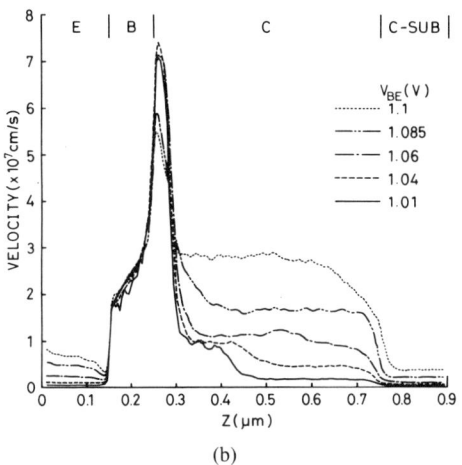

(b)

Fig. 5. Average electron velocity profiles of n-type collector HBT's for various bias conditions. (a) AlInAs/GaInAs HBT. (b) InP/GaInAs HBT.

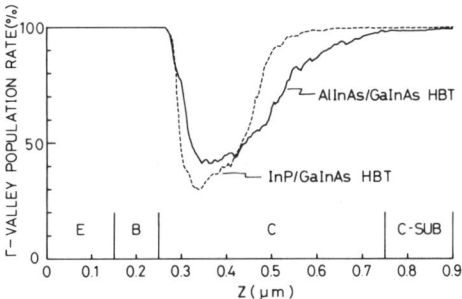

Fig. 6. Γ-valley population rate for n-type collector HBT's corresponding to Fig. 4.

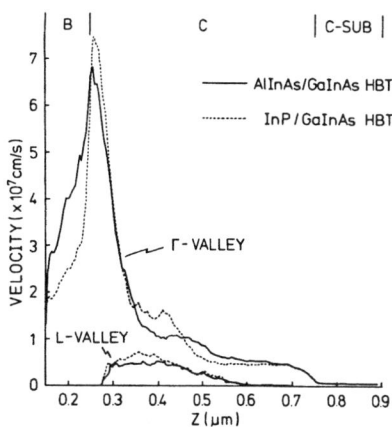

Fig. 7. Γ- and L-valley average electron velocity profiles in base–collector regions for n-type collector HBT's corresponding to Fig. 4.

was not responsible for the small V_s for Tr.1. The inverted profiles of the Γ-valley population rate at the collector neutral region, i.e., $Z \geqq 0.45$, were attributed to the smaller transition rate of L-to-Γ intervalley scattering for Tr.1, as was mentioned in the explanation of Fig. 4.

Fig. 7 shows the Γ- and L-valley electrons' average velocity profiles for both transistors corresponding to Fig. 4(a) and (b). It can be clearly seen that the small V_s for Tr.1 is mainly attributed to the small Γ-valley electron velocity for Tr.1. Since narrow-gap materials, such as GaInAs, have large nonparabolicity, there is some pos-

sibility for the Γ-valley electron velocity to be deteriorated by a large effective mass due to this strong nonparabolicity. Fig. 8 shows the averaged Γ-valley electron effective mass profiles for both transistors corresponding to Fig. 5. The energy-dependent effective mass $m_e^* = m_{e0}^* (1 + 2 \alpha E)$ was averaged over every particle, where m_{e0}^* and α represent the electron effective mass at the Γ-valley minimum and the nonparabolicity factor, respectively. The thin solid and dashed lines correspond to the effective masses of the Γ-valley minima for Tr.1 and Tr.2, respectively. Looking at the collector depletion region, i.e., $0.25 \leqq Z \leqq 0.45$, although the increased rate of the effective mass from m_{e0}^* for Tr.1 was larger than that for Tr.2, the magnitude of the effective mass for Tr.2 was larger than that for Tr.1, because m_{e0}^* for Tr.2 was inherently larger than that for Tr.1. Thus the large nonparabolicity was found to be not responsible for the small Γ-valley electron velocity for Tr.1.

In order to investigate the reason for the small Γ-valley electron velocity, the Γ-valley electron velocity distributions in the Z-direction for Tr.1 and Tr.2 are presented in Fig. 9(a) and (b), respectively. Looking at the left-half portion of the collector regions, a larger number of back-scattered electrons, having a large velocity with a negative component, can be found in Tr.1, compared with Tr.2.

The reason for this phenomenon is as follows. As can be understood from Fig. 4, the distance from the edge of the collector depletion region in the subcollector side to the intersecting point of the conduction band edge (Γ-valley minimum) and the extended line of the L-valley minimum in the collector neutral region, which is parallel to the Z-axis, is larger for Tr.1 than for Tr.2. Furthermore, the time required to lose the energy $\Delta E_{\Gamma-L}$ by emitting LO phonons from the L-valley minimum down to the Γ-valley minimum is longer for Tr.1 than for Tr.2. Therefore, the back-scattered electrons in the InGaAs collector have a larger probability to travel far into the collector depletion region toward the base layer than in the InP collector. Of course, a lighter effective mass for InGaAs compared with InP is also considered to be another reason for the above

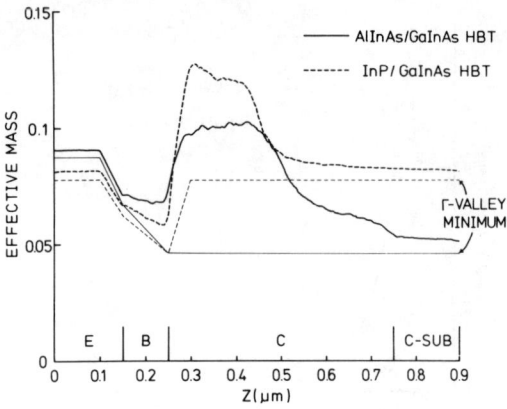

Fig. 8. Average Γ-valley electron effective mass for n-type collector HBT's corresponding to Fig. 4. Thin solid and dashed lines represent electron effective mass at Γ-valley minima.

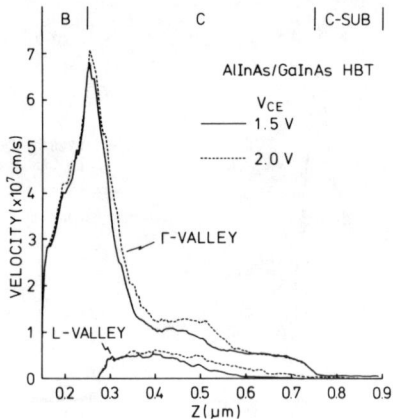

Fig. 10. Γ- and L-valley average electron velocity profiles in base–collector regions for AlInAs/GaInAs n-type collector HBT's with variation in collector-to-emitter bias voltage.

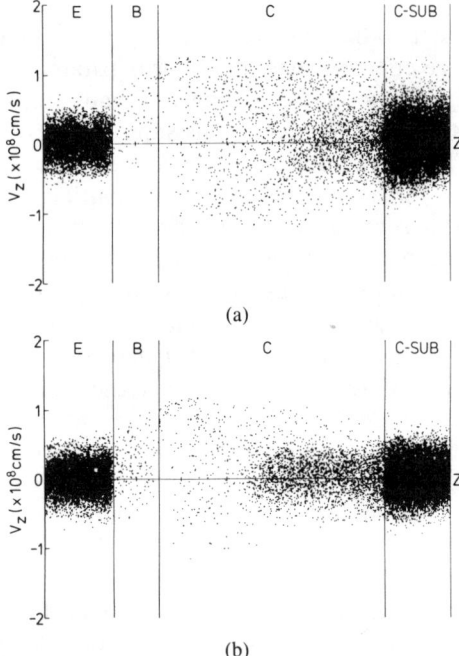

(a)

(b)

Fig. 9. Z-direction electron velocity distribution for n-type collector HBT's corresponding to Fig. 4. (a) AlInAs/GaInAs HBT. (b) InP/GaInAs HBT.

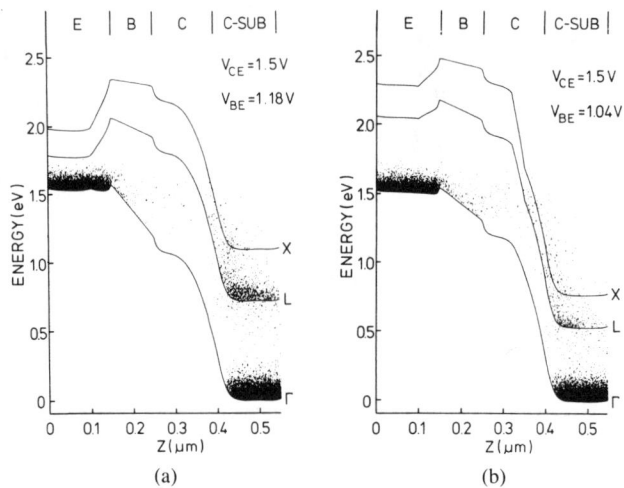

(a)

(b)

Fig. 11. Electron energy distributions of p-type collector HBT's. Solid line curves Γ, L, and X are corresponding valley minima. The same bias conditions were applied as in Fig. 4. (a) AlInAs/GaInAs HBT. (b) InP/GaInAs HBT.

phenomenon. These "hot" back-scattered electrons diffuse to the base–collector junction with a large diffusion constant, thus reducing the Γ-valley avarage electron velocity. The collector-to-emitter bias (V_{CE}) dependences of the Γ- and L-valley electrons' average velocity profiles for Tr.1 are further demonstrated in Fig. 10, where V_{CE} was increased to 2.0 V. At $V_{CE} = 2.0$ V, the Γ- as well as L-valley electrons' average velocity increased compared with those at $V_{CE} = 1.5$ V, because the increased electric field suppressed the Γ-valley "hot" electron diffusion current in the collector depletion region, thus verifying the above-mentioned reason for the small Γ-valley electron velocity.

C. p-Type Collector HBT's

Fig. 11(a) and (b) shows the electron energy distributions for Tr.1 and Tr.2 with a p-type collector, respec-

tively, where the same bias conditions were chosen as in Fig. 4(a) and (b). In the emitter and base regions, almost no difference was found in the energy distribution compared with n-type collector HBT's, since a common emitter–base structure was adopted for both n- and p-type collector HBT's. In the p-type collector regions, electrons remained in the Γ-valley through a wider range of the collector, with their kinetic energy far beyond the conduction band edges. The main difference between Tr.1 and Tr.2 was the magnitude of electron kinetic energy in the p-type collectors, i.e., the electron energy for Tr.1 was higher than that for Tr.2 because of the difference in the base built-in field.

Fig. 12(a) and (b) shows the average electron velocity profiles for Tr.1 and Tr.2, respectively, with V_{BE} as a parameter. Compared with the n-type collector cases, the difference between Tr.1 and Tr.2 was very slight. Both the peak overshoot velocity and the overshoot distance for Tr.1 were only a little larger than those for Tr.2, although

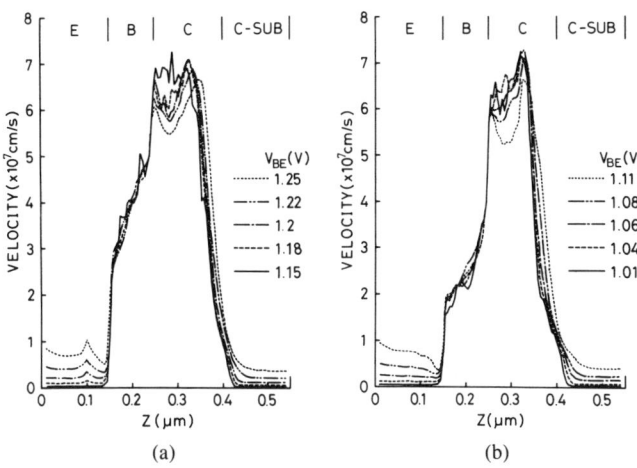

Fig. 12. Average electron velocity profiles of p-type collector HBT's for various bias conditions. (a) AlInAs/GaInAs HBT. (b) InP/GaInAs HBT.

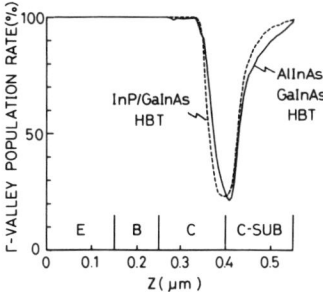

Fig. 13. Γ-valley population rate for p-type collector HBT's corresponding to Fig. 11.

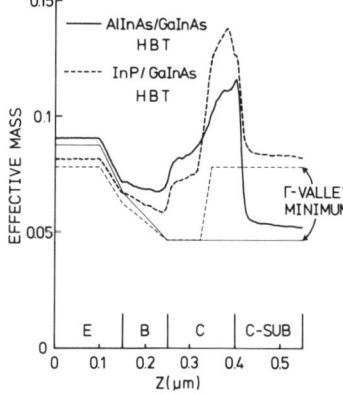

Fig. 14. Average Γ-valley electron effective mass for p-type collector HBT's corresponding to Fig. 11. Thin solid and dashed lines represent electron effective mass at Γ-valley minima.

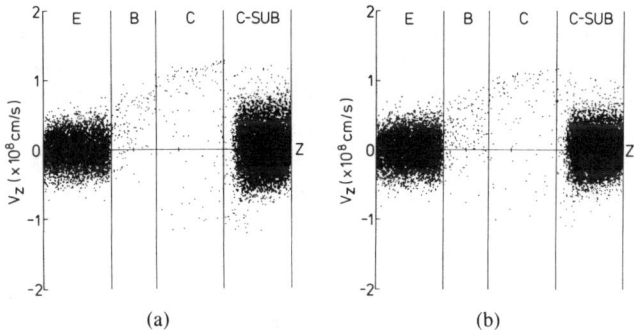

Fig. 15. Z-direction electron velocity distribution for p-type collector HBT's corresponding to Fig. 11. (a) AlInAs/GaInAs HBT. (b) InP/GaInAs HBT.

the electron kinetic energy and $\Delta E_{\Gamma-L}$ for Tr.1 were sufficiently larger than those for Tr.2 in the collector region.

In order to further investigate the reason for these slight differences, the Γ-valley population rates and the averaged Γ-valley electron effective mass profiles were demonstrated in Figs. 13 and 14, respectively, where the averaged effective mass was defined in the same manner as in Fig. 8 and the bias conditions corresponded to those in Fig. 11. In Fig. 13, the Γ-valley population rate for Tr.1 is seen to be slightly larger than that for Tr.2 in the p-type collector regions, which corresponded to the slightly longer overshoot distance for Tr.1. Since the inclination of the conduction band edge changes steeply at the p-n collector-to-subcollector junction, the merit of a large $\Delta E_{\Gamma-L}$ material is lost in such regions even for the p-type collector structure. In Fig. 14, the effective mass for Tr.1 is seen to be larger than that for Tr.2 in the left-half portion of the collector, while their order of magnitude is inverted in the right-half portion of the collector. Since the left-half portion of the collector was composed of GaInAs for both transistors, the magnitude of the averaged effective mass could be compared with each other. The larger effective mass for Tr.1 is considered to be a direct reflection of the broad electron energy distribution, since higher energy electrons have a larger effective mass due to the strong nonparabolicity in a narrow-bandgap material. Therefore, the unexpectedly small average velocity for Tr.1 was considered to be due to the larger average effective mass of higher energy electrons.

Since it is significant to confirm the above reasoning by taking up microscopic data, the Γ-valley electron velocity distributions in the Z-direction for Tr.1 and Tr.2 are demonstrated in Fig. 15(a) and (b), respectively. A quite similar velocity distribution is seen in the collector regions for both transistors. It is noteworthy that the maximum electron velocity V_{max} of 1.2×10^8 cm/s for Tr.1 was almost as large as that for Tr.2, though the electron energy for Tr.1 was larger than that for Tr.2 as was seen in Fig. 11. Inherently, there exists an upper limit for the electron group velocity at the point of inflection in the

band diagram. The obtained V_{max}, however, is not equal to this limited velocity, because such a detailed band structure has not been considered in our formulation as well as because the electron energy will never reach the point of inflection in these transistors if used under moderate bias conditions. Therefore, it can be said that the Γ-valley electron velocity is suppressed to at most the order of 10^8 cm/s by strong nonparabolicity even if a large $\Delta E_{\Gamma-L}$ material is utilized.

D. Collector Transit Time and Cutoff Frequency

Fig. 16 shows the collector transit times (τ_C's) for Tr.1 and Tr.2 with n- and p-type collectors, with correspond-

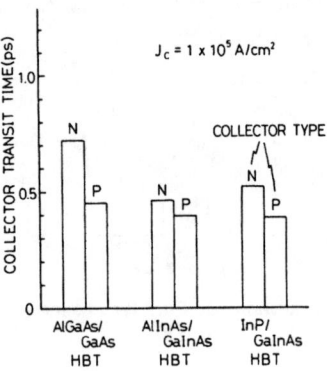

Fig. 16. Collector transit times for AlInAs/GaInAs and InP/GaInAs HBT's with n- and p-type collectors. Those for AlGaAs/GaAs HBT's are supplemented as reference.

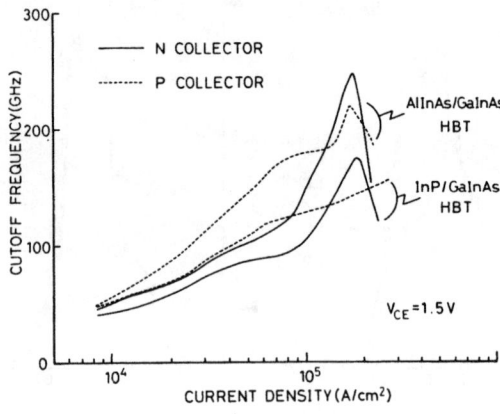

Fig. 17. Cutoff frequency versus collector current density characteristics for AlInAs/GaInAs and InP/GaInAs HBT's with n- and p-type collectors.

ing data for AlGaAs/GaAs HBT's as reference, where τ_C was obtained by $\Delta Q_C/\Delta J_C$ at around $J_C = 1 \times 10^5$ A/cm². With n-type collector HBT's, Tr.1 exhibited the smallest τ_C, while the difference between Tr.1 and Tr.2 was less than 0.1 ps. Since τ_C for an n-type collector transistor is known to decrease dramatically under a high-injection condition at the upper half of the 10^4 A/cm² order, τ_C became sensitive to the bias condition in such a high current-density region. In this case, however, smaller τ_C's for Tr.1 and Tr.2 compared with the AlGaAs/GaAs HBT were considered to be a consequence of their larger $\Delta E_{\Gamma-L}$. It should be noted that the reduction rate of τ_C by utilizing large $\Delta E_{\Gamma-L}$ materials was not so large as expected, since the improvement rate in τ_C was at most 30% compared with GaAs HBT, while that for τ_B was at least 60%. Contrary to the n-type collector cases, little difference was observed in the τ_C's for p-type collectors, where the improvement rate in τ_C was at most 10% compared with GaAs HBT. This is because a high average electron velocity has already been achieved by introducing a p-type collector, as has been discussed in Section III-C.

In order to investigate the high-frequency performance of these HBT's, the cutoff frequency (f_T) versus collector current density (J_C) characteristic is demonstrated for Tr.1 and Tr.2 with n- and p-type collectors in Fig. 17. At J_C of less than 10^5 A/cm², p-type collector HBT's exhibited higher f_T values than n-type collector HBT's. On the other hand, under a higher J_C condition, f_T for n-type collector HBT's became higher than those for p-type collector HBT's. This rapid increase in f_T's for n-type collector HBT's has been attributed to the rapid reduction in τ_C's corresponding to the collector high-injection effect as has been mentioned in the explanation of Fig. 5. The maximum f_T's were 250 and 220 GHz for Tr.1 with n- and p-type collectors, and 180 and 160 GHz for Tr.2 with n- and p-type collectors, respectively. Therefore, Tr.1 and Tr.2 were concluded to be twice and 1.5 times faster than AlGaAs/GaAs HBT's, respectively, since the previously obtained maximum f_T's for AlGaAs/GaAs HBT's were 120 GHz for an n-type collector HBT and 110 GHz for a p-type collector HBT [3].

Since the high-speed performance is considered to be mainly attributed to the larger reduction rate in the base transit times than that in the collector transit times, a combination of AlInAs/GaInAs/InP is recommended as an emitter/base/collector material system, where the emitter–base materials are chosen to yield a higher base built-in field and the collector material to increase the breakdown voltage as high as possible.

IV. CONCLUSIONS

The high-speed performances of AlInAs/GaInAs and InP/GaInAs HBT's have been thoroughly investigated using a one-dimensional self-consistent particle simulator. Optimum alloy compositions have been derived for both transistors with a graded-gap base structure through the tradeoff between the emitter charging time and the base transit time. Under these optimized conditions, the base transit times for AlInAs/GaInAs HBT and InP/GaInAs HBT have been estimated to be 1/3 and 2/5 of that for an AlGaAs/GaAs HBT, respectively.

The saturation velocity for an AlInAs/GaInAs HBT (6×10^6 cm/s) has been found to be smaller than that for an InP/GaInAs HBT (1×10^7 cm/s), where this small saturation velocity has been attributed to the reduction in the average electron velocity by the reverse-direction diffusion of back-scattered "hot" Γ-valley electrons. Nonparabolicity was found to have the function of suppressing the maximum electron velocity by increasing the electron effective mass when the electron kinetic energy increased. Thus the peak overshoot velocity in the p-type collector has been found to be almost independent of the collector material. A large $\Gamma-L$ band separation energy has been found to be not so effective in reducing the collector transit times, since their improvement rates were at most 30% for an n-type collector and 10% for a p-type collector compared with AlGaAs/Gas HBT's.

The cutoff frequencies have been estimated to be twice for AlInAs/GaAs HBT's and 1.5 times for InP/GaInAs HBT's higher than that for AlGaAs/GaAs HBT's. These results have been attributed to a larger bandgap difference between the emitter and base to yield a high base built-in field, rather than a larger Γ–L band separation energy in the collector to enhance the velocity overshoot effect.

ACKNOWLEDGMENT

The authors wish to thank K. Morizuka for his stimulating discussions and also Dr. A. Hojo for offering the authors an opportunity to continue this work.

REFERENCES

[1] U. K. Mishra, J. F. Jensen, D. B. Rensch, A. S. Brown, M. W. Pierce, L. G. McCray, T. V. Kargodorian, W. S. Hoefer, and R. E. Kastris, "48 GHz AlInAs/GaInAs heterojunction bipolar transistors," in *IEDM Tech. Dig.*, pp. 873–875, 1988.

[2] Y-K. Chen, R. N. Nottenburg, M. B. Panish, R. A. Mamm, and D. A. Humphrey, "Subpicosecond InP/InGaAs heterostructure bipolar transistors," *IEEE Electron Device Lett.*, vol. 10, no. 6, pp. 267–269, 1989.

[3] R. Katoh, M. Kurata, and J. Yoshida, "Self-consistent particle simulation for (AlGa)As/GaAs HBT's with improved base-collector structures," *IEEE Trans. Electron Devices*, vol. 36, no. 5 pp. 846–853, 1989.

[4] S. Adachi, "Material parameters of $In_{1-x}Ga_xAs_yP_{1-y}$ and related binaries," *J. Appl. Phys.*, vol. 53, no. 12, pp. 8775–8792, 1982.

[5] ——, "GaAs, AlAs, and $Al_xGa_{1-x}As$: Material parameters for use in research and device applications," *J. Appl. Phys.*, vol. 58, no. 3, pp. R1–R29, 1985.

[6] K. Brennan and K. Hess, "High field transport in GaAs, InP and InAs," *Solid-State Electron.*, vol. 27, no. 4, pp. 347–357, 1984.

[7] M. A. Littlejohn, J. R. Hauser, and T. H. Glisson, "Alloy scattering and high field transport in ternary and quaternary III-V semiconductors," *Solid-State Electron.*, vol. 21, pp. 107–114, 1978.

[8] K. Morizuka, R. Katoh, T. Tsuda, M. Asaka, N. Iizuka, and M. Obara, "Electron space charge effects on high-frequency performance of AlGaAs/GaAs HBT's under high-current-density operation," *IEEE Electron Device Lett.*, vol. 9, no. 11, pp. 570–572, 1988.

[9] R. Katoh and M. Kurata, "Self-consistent particle simulation for (AlGa)As/GaAs HBT's under high bias conditions," *IEEE Trans. Electron Devices*, vol. 36, no. 10, pp. 2122–2128, 1989.

Dramatic enhancement in the gain of a GaAs/AlGaAs heterostructure bipolar transistor by surface chemical passivation

C. J. Sandroff, R. N. Nottenburg, J.-C. Bischoff,[a] and R. Bhat

Bell Communications Research, Inc., 331 Newman Springs Road, Red Bank, New Jersey 07701

(Received 2 March 1987; accepted for publication 11 May 1987)

With a simple chemical treatment we have passivated nonradiative recombination centers at the periphery of a GaAs/AlGaAs heterostructure bipolar transistor, resulting in a 60-fold increase in the current gain of the device at low collector currents. This large enhancement in gain was achieved by spin coating thin films of $Na_2S \cdot 9H_2O$ onto the devices after their fabrication. We briefly discuss the passivation mechanism and the implications for other III-V optoelectronic devices.

It is well known that the free GaAs surface is plagued with both a large surface recombination velocity ($S_0 \sim 10^6$ cm/s) and a high surface state density ($N_s \sim 10^{12}$ cm^{-2} eV^{-1}). These characteristics continue to impede the development of a metal-insulator-semiconductor technology,[1,2] and limit the scale down of minority-carrier devices such as bipolar transistors.[3] Though much progress has been made in understanding the poor electronic quality of the GaAs surface,[4-6] no completely successful scheme has been devised which passivates the GaAs interface both electronically and chemically while permitting the growth of an insulating barrier. In this letter we report on a simple chemical treatment having some favorable attributes as a passivating scheme for compound semiconductor surfaces. A detailed discussion of the passivating mechanism will be published elsewhere.[7] Here we concentrate on what we believe is the first example of chemical passivation leading to the improved performance of a three-terminal device.

The heterostructure bipolar transistor (HBT) employed in this work was grown by organometallic chemical vapor deposition and a schematic representation of the device structure is shown in Fig. 1. Mesa transistors were fabricated with emitter sizes of 50×150, 40×100, 18×40, and 6×20 μm^2. In the device fabrication the emitter-base junction was first formed by sequential growth and removal of an anodic oxide. The base mesa was then defined using 5:1:1 H_2SO_4:H_2O_2:H_2O and alloyed ohmic contacts were realized from evaporated Au/Ge and Au/Be.

The common-emitter characteristics for an as-processed (unpassivated) transistor are shown in Fig. 2. A maximum differential current gain of ~ 3000 is measured at a collector current of $I_C = 18$ mA, corresponding to a collector current density of 720 A/cm^2. The high current gain of these HBT devices is due to the long electron diffusion length in the base together with a defect-free emitter-base heterointerface. The current gain β for these transistors, however, was found to decrease markedly at low current levels and to depend on the perimeter-to-area (P/A) ratio of the emitter junction. These observations indicate that β is limited primarily by the nonradiative recombination of mi-

nority carriers at the device perimeter as noted previously,[3] for AlGaAs/GaAs HBT's. With the current gain so strongly dependent on the P/A ratio, these transistors are ideal tools for studying the effects of any surface passivation scheme, since any accompanying decrease in the surface recombination velocity at the emitter junction perimeter should increase β particularly at low emitter junction bias.

In our passivating scheme, fabricated device structures were first immersed in a mild etch (1:8:500 H_2SO_4: H_2O_2:H_2O) for 5 s and rinsed in distilled water. An aqueous solution of $Na_2S \cdot 9H_2O$ (Aldrich Chemical Co.) was then deposited on the device substrate which was spun at 5000 rpm for 60 s. This procedure left a colorless, crystalline film whose passivating effects persisted for several days. The film thickness varied substantially across a given substrate and was a sensitive function of the $Na_2S \cdot 9H_2O$ concentration in solution. In general, we achieved the best results with $Na_2S \cdot 9H_2O$ solutions ranging in concentration from 0.5 to 1.0 M. At the high end of this concentration range it was sometimes found that a thick $Na_2S \cdot 9H_2O$ film covered the ohmic contacts of most of the devices. Under these circumstances no meaningful common-emitter characteristics could be obtained because of parallel conduction pathways in the ionic $Na_2S \cdot H_2O$ film. When such thick films were deposited, they were rinsed off in distilled water and $Na_2S \cdot 9H_2O$ was reapplied from a more dilute solution. The most substantial improvements in current gain were

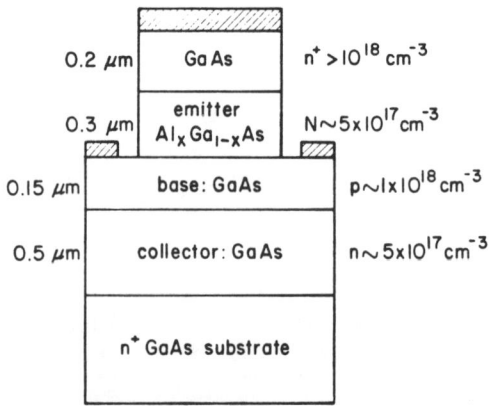

FIG. 1. Schematic representation of the AlGaAs/GaAs HBT device.

[a] On leave from the Swiss Federal Institute of Technology, Lausanne, Switzerland.

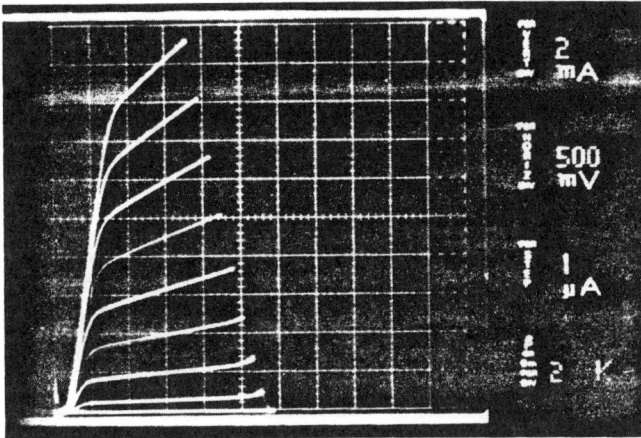

FIG. 2. Common-emitter characteristics for an unpassivated HBT at high collector current levels (collector area $\sim 2.5 \times 10^{-5}$ cm^{-2}).

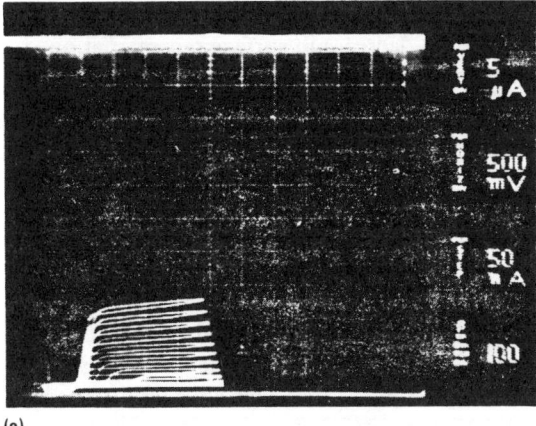

(a)

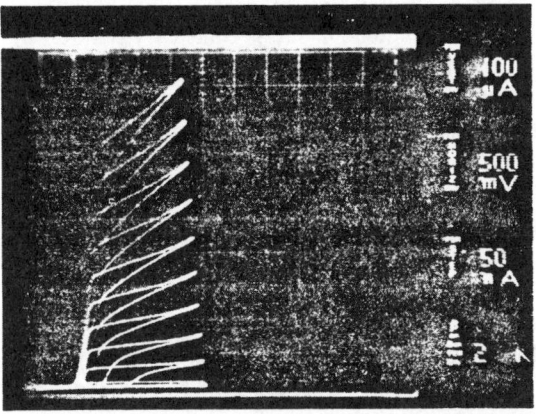

(b)

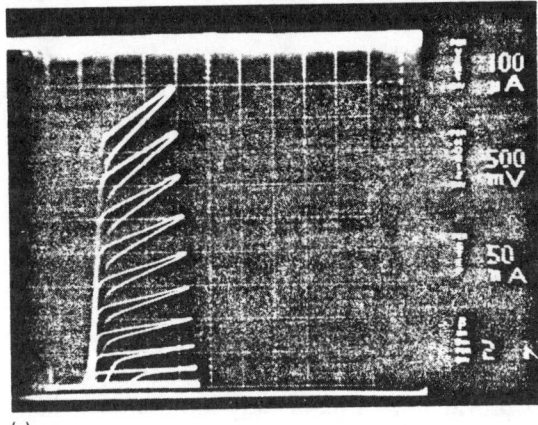

(c)

FIG. 3. Common-emitter characteristics at low collector current levels for (a) an unpassivated device with $\beta \sim 30$. (b) The device after Na$_2$S·9H$_2$O treatment, $\beta \sim 2000$. (c) The same device with the passivating film removed, then reapplied. Again $\beta \sim 2000$. (Collector area same as in Fig. 2.)

achieved when the entire emitter periphery was surrounded by the passivating film while the ohmic contacts remained uncovered.

In Fig. 3 we compare the common-emitter characteristics at low base current for a transistor both before and after the sulfide treatment. The current gain for the untreated device in Fig. 3(a) is ~ 30. After a single passivation step, Fig. 3(b) reveals a more than a 60-fold increase in β to a value of ~ 2000. To test if the effects of passivation could be reversed, the treated wafer was rinsed in distilled water and re-etched in 1:8:500 H$_2$SO$_4$:H$_2$O$_2$:H$_2$O. After this procedure, the current gain returned to its same initial low value of ~ 30. The characteristic for the same transistor after the Na$_2$S·9H$_2$O solution was reapplied appears in Fig. 3(c). Not only does β increase but the same high value is once again obtained. This sequence could be repeated indefinitely without any degradation in β, indicating that only the device interface is affected by the passivating process with no damage occurring in the bulk HBT structure.

The increase in β associated with the passivation treatment is directly related to a decrease in surface recombination velocity. By neglecting the emitter hole diffusion current and assuming a base transport factor close to unity, we find that

$$\beta \approx \frac{i_n}{i_s} \sim \left(\frac{v_n I_c}{S_0^2}\right)^{1/2} \frac{A_E^{1/2}}{P_E},$$

where i_n is the electron diffusion current, i_s is the surface recombination current, v_n is the electron diffusion velocity, and A_E and P_E are the emitter junction area and perimeter, respectively. Thus the increase in current gain by ~ 60 due to the passivation process implies a decrease in S_0 by the same factor. Furthermore, by passivation considerable device scale down can be realized since a treated transistor could have a smaller P/A ratio while yielding the same current gain.

On the microscopic scale, the 60-fold decrease in S_0 implies a reduction by the same factor in the number of active nonradiative recombination centers on the device perimeter.[8] To provide further evidence for a decrease in the number of electronically active surface states luminescence stud-

ies were also performed on passivated GaAs surfaces.[9] For photoluminescence experiments, the sulfide films were deposited on n-type GaAs epilayers. At room temperature a 250-fold increase in photoluminescence intensity was observed relative to the untreated surface. This marked improvement in radiative efficiency could also be seen directly by imaging the HBT using an infrared camera and monitoring the injection luminescence arising from recombination in the base region. Only in passivated transistors were we able to record bright images from the emitter-base periphery. Thus, both photoluminescence and injection luminescence verify the significant decrease in surface recombination velocity which accompanies the deposition of our sulfide films on GaAs.

For many technological applications, it is essential to fabricate GaAs interfaces which not only possess a low density of surface states (i.e., low S_0) but also remain chemically stable in air for long periods of time. A recent photochemical procedure has gone far towards this goal showing that a relatively simple treatment could result in a lowering of S_0 persisting for several hours.[10] We have found that HBT structures electronically passivated by films of $Na_2S \cdot 9H_2O$ showed enhanced gain for several days after treatment. Besides offering the possibility of long term electronic passivation, a $Na_2S \cdot 9H_2O$ film possesses several other favorable material properties. Its ionic nature permits easy removal by water rinse, and its high melting point ($\sim 920 °C$) implies that films could be heated to high temperatures without any significant chemical degradation. To explore this idea we submitted passivated HBT's to a 30 min heat treatment at $200 °C$ in a N_2 atmosphere. After rinsing off the bulk $Na_2S \cdot 9H_2O$ film we could still observe a significant degree of surface passivation, with β a factor of 3 to 10 greater than in the unpassivated devices.

We believe that the chemical reaction mechanism responsible for the passivation can be described by a two-step process where the native oxide and elemental arsenic are etched away exposing a pristine surface to which sulfur can strongly bond. Indeed, sulfur forms many stable binary compounds with both Ga and As; GaS and As_2S_3 for example, are well known layered semiconductors, the latter having excellent glass forming properties. The importance of covalently bound sulfur in the passivation mechanism was suggested by Auger analysis of GaAs substrates which had been treated with the sulfide at room temperature. Even after a thorough water rinse roughly 1/2 monolayer of sulfur could be detected on the surface. At this point it would be difficult to make any definitive correlation between surface structure and the efficient electronic termination of the GaAs. However, one plausible way to rid the surface of electronically active sites entails the bridging of pairs of Ga atoms by sulfur. Besides providing an efficient way to tie up dangling bonds such a termination could account for the 1/2 monolayer of sulfur remaining after the water rinse, since hydrolysis of Ga–S–Ga bonds could result in one bond to sulfur broken and replaced by a bond to oxygen. Finally, we note that previous work conducted in high vacuum[11] and electrochemical[12] environments also concluded that a higher quality electronic interface resulted when GaAs was exposed to chalcogenide compounds.

In summary, we have shown that a simple chemical treatment of a GaAs/AlGaAs HBT results in a significant improvement in the current gain of the device. Very efficient passivation of nonradiative recombination centers was achieved by the deposition of $Na_2S \cdot 9H_2O$ films onto the transistor perimeter. The high solubility of $Na_2S \cdot 9H_2O$ and its elevated melting point combine to make it a versatile, chemically stable passivating material. The beneficial effects of the treatment should be quite general, and we believe that similar improvements in performance could be obtained in any number of compound semiconductor devices. Specifically, it seems plausible that conditions could be found under which a thin sulfide layer would be formed which would be fully passivating and chemically robust. On such an interface it might be possible to deposit thick insulating layers from which metal-insulator-semiconductor devices could be fabricated.

We thank C. C. Chang for his Auger analysis and D. Humphrey for technical assistance.

[1]C. R. Zeisse, L. J. Messick, and D. L. Lile, J. Vac. Sci. Technol. **14**, 957 (1977).

[2]L. G. Meiners, J. Vac. Sci. Technol. **15**, 1402 (1978).

[3]T. Azawa, T. Ishibashi, and T. Sugeta, in IEEE Tech. Dig. p. 328, Washington D.C., December, 1985.

[4]H. H. Wieder, J. Vac. Sci. Technol. **15**, 1498 (1978).

[5]J. M. Woodall and J. L. Freeouf, J. Vac. Sci. Technol. **19**, 794 (1981).

[6]H. Hasegawa and T. Sawada, Thin Solid Films **103**, 119 (1983).

[7]E. Yablonovitch, C. J. Sandroff, R. Bhat, and T. G. Gmitter (unpublished results).

[8]A. S. Grove, in *Physics and Technology of Semiconductor Devices* (Wiley, New York, 1967), p. 136.

[9]B. J. Skromme, C. J. Sandroff, E. Yablonovitch, R. Bhat, and T. J. Gmitter (unpublished results).

[10]S. D. Offsey, J. M. Woodall, A. C. Warren, P. D. Kirchner, T. I. Chappell, and G. D. Pettit, Appl. Phys. Lett. **48**, 475 (1986).

[11]J. Massies, J. Chaplart, M. Laviron, and N. T. Linh, Appl. Phys. Lett. **38**, 693 (1981).

[12]R. J. Nelson, J. S. Williams, H. J. Leamy, B. Miller, H. C. Casey, Jr., B. A. Parkinson, and A. Heller, Appl. Phys. Lett. **36**, 76 (1980).

Current Gain Enhancement in Graded Base AlGaAs/GaAs HBTs
Associated with Electron Drift Motion

Hiroshi Ito, Tadao Ishibashi and Takayuki Sugeta

Atsugi Electrical Communication Laboratory,
Nippon Telegraph and Telephone Public Corporation,
1839 Ono, Atsugi-shi, Kanagawa 243–01

(Received January 23, 1985; accepted for publication March 23, 1985)

The temperature dependence of current gain in AlGaAs/GaAs heterojunction bipolar transistors with a graded bandgap base (GB-HBTs) is described. High current gain up to 1100 has been achieved at a collector current density of 6×10^3 A/cm^2 in MBE grown GB-HBTs with a 1×10^{19} cm^{-3} doped base. The current gain in the low temperature region does not depend on temperature. This tendency is clearly explained by a nearly constant base transit time due to electron drift motion caused by the built-in field, which contrasts with the electron diffusion.

It was first proposed by Kroemer[1] that the built-in field in the compositionally graded base region of heterojunction bipolar transistors (HBTs) would give a shorter base transit time due to the effect of electron acceleration. It is also important for high-speed operation that thicker base width is used for base resistance reduction without much increase in the base transit time. Promising experimental results for this device structure have already been reported.[2~4] In addition to these advantages, the current gain enhancement with this structure is expected not only at room temperature but also at lower temperatures. In a conventional uniform base structure, current gain reduction is inevitable due to a diffusion velocity decrement.[5] Therefore, the differences in temperature dependence of electron transport in the base region between graded base HBTs (GB-HBTs) and uniform base HBTs (UB-HBTs) are of interest. This paper presents evidence of electron drift field transport in the base region by measuring the temperature dependence of current gain in GB-HBTs.

The epitaxial layer structure parameters of HBTs grown by MBE on (100) oriented semi-insulating GaAs substrates are shown in Table I. The base layer, with a thickness of 1000 Å, was graded linearly from $x=0$ to $x=0.1$. The emitter-base junction was graded from $x=0.1$ to $x=0.3$ on the emitter side over a distance of 300 Å, where parabolic grading[6] was utilized to provide a smooth conduction band edge at the interface. The contact side of the emitter layer was also parabolically graded from $x=0.3$ to $x=0$ to reduce the emitter resistance. Each grading growth procedure was performed by a com-

puter-controlled Al effusion cell. The built-in field, E_{bi}, in the compositionally graded base was calculated to be 12 kV/cm. The base layer was heavily doped with Be to 1×10^{19} cm^{-3} to reduce base resistance. The n type dopant was Si and the substrate temperature during the growth was 680°C. An undoped spacer layer[5] of 200 Å was inserted between the emitter and the base layer to prevent shift of the pn junction towards the upper layer due to Be diffusion.

Mesa structure devices were fabricated by wet chemical etching and lift-off processes. SiN film was deposited for surface passivation. The ohmic metals used for the n and p type layers were AuGe/Ni/Ti/Au[7] and Cr/Au, respectively. The transistor had a 48×48 μm^2 emitter area.

The DC characteristics of the transistor are shown in Fig. 1. The device exhibited current gain up to 1100 for a collector current of 140 mA, which corresponds to a current density of 6×10^3 A/cm^2. Thermal influences caused negative differential collector resistance to appear, providing the collector current limitation in this large device. Higher current gain will be obtained at a higher collector current over $\sim 1 \times 10^4$ A/cm^2 in smaller devices. A current gain value as high as that observed has not previous-

Table I. Epitaxial layer structure parameters for graded base HBT.

layer	thickness (Å)	doping (cm^{-3})	Al composition	
n$^+$GaAs	1000	5×10^{18}		
nGaAs	2000	2×10^{17}		
NAlGaAs	300	5×10^{17}	0–0.3	parabolic
NAlGaAs	900	5×10^{17}	0.3	
NAlGaAs	300	5×10^{17}	0.3–0.1	parabolic
P$^+$AlGaAs	1000	1×10^{19}	0.1–0	linear
nGaAs	4000	5×10^{16}		
n$^+$GaAs	7000	3×10^{18}		

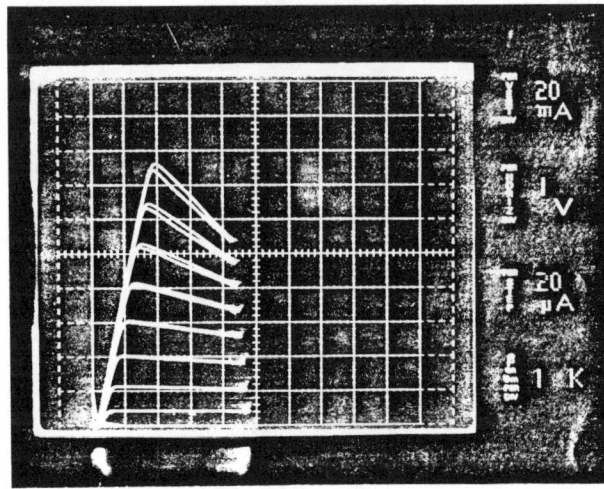

Fig. 1. Common emitter collector *I-V* characteristics of graded base HBT with emitter area of 2.3×10^{-5} cm^2.

Reprinted from *Japan J. Appl. Phys.*, vol. 24, no. 4, pp. L241–L243, April 1985.

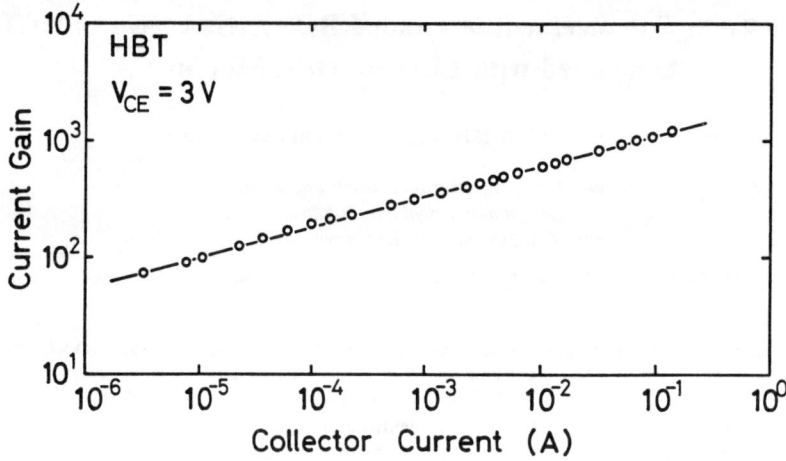

Fig. 2.   Collector current dependence of current gain.

ly been obtained by nuiform base HBTs with similar high base doping density.[5]

The collector current dependence of the current gain is shown in Fig. 2. The ideality factor, n, deduced from the slope of the I-β curve, was 1.3. This implies that the influence of g-r current in the emitter-base depletion region was very small.

The temperature dependence of the current gain in a GB-HBT is shown in Fig. 3; the gain increases with decreasing temperature and then saturates below 150 K. The saturated current gain value is 870. For comparison, the typical current gain characteristic for a 1×10^{19} cm^{-3} base doped UB-HBT[5] is also shown in this figure. The current gain measurement was performed at a low collector current of 2×10^{-2} A to avoid temperature rise in a device. It has already been experimentally confirmed that the current gain of a UB-HBT has a linear relationship to temperature below 150 K. This characteristic has been explained by a current gain relation based on a carrier diffusion model with a constant lifetime and a unity emitter in-

jection efficiency, and is expressed as

$$\beta \cong \frac{2\,\mu k T \tau}{q W_{\mathrm{B}}^{2}} \qquad (1)$$

where μ is the electron mobility; k, Boltzmann constant; T, temperature; τ, electron lifetime; q, electron charge; and W_{B}, base thickness. When the electron transport in the base region is dominated by drift motion due to the intense electric field, β is expressed, assuming a unity emitter injection efficiency, as

$$\beta \cong \frac{1 - \dfrac{\tau_{\mathrm{B}}}{\tau}}{\dfrac{\tau_{\mathrm{B}}}{\tau}} \cong \frac{\tau \cdot v_{\mathrm{d}}}{W_{\mathrm{B}}} \qquad (2)$$

where τ_{B} is the base transit time and $v_{\mathrm{d}} = \mu \cdot E_{\mathrm{bi}}$ is the electron drift velocity. The effect of the momentum relaxation time on τ_{B} can be neglected here. When τ is a radiative recombination lifetime, which is independent of temperature, β is independent of temperature. Therefore, the observed behavior of current gain saturation below 150 K is explained well by equation (2) with a constant electron lifetime. If we account for the current gain behavior in a GB-HBT with the diffusion model, an unacceptable electron lifetime as long as 5.5 ns must be assumed.

The decreasing tendency of β for GB-HBTs at higher temperatures is considered to be due to the nonradiative recombination process, which is dominant at the higher temperature region. This is because nonradiative recombination lifetime decreases as temperature increases, a situation similar to that of the UB-HBTs. If we assume the electron mobility in a 1×10^{19} cm^{-3} doped p^{+} GaAs region to be 1200 cm^{2}/V·s,[8] v_{d} is 1.4×10^{7} cm/s. Thus, the electron lifetime in the base of GB-HBTs is estimated from equation (2) to be 0.50 ns at RT and 0.62 ns below 150 K. The electron lifetimes for GB-HBTs at RT are slightly longer than those of UB-HBTs (0.33 ns at RT and 0.64 ns at 77 K); this is probably due to the higher substrate temperature during the growth.

GB-HBTs can be operated as an ultrahigh-speed device not only at RT but also at 77 K, because their base transit time as short as 1 ps is independent of temperature, while the base transit time of UB-HBTs at 77 K is 4 times larger

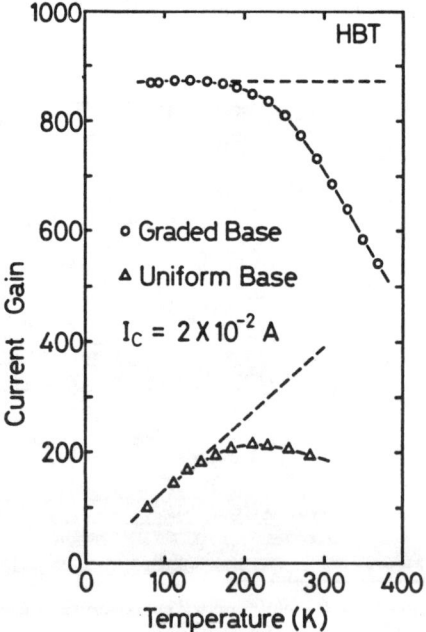

Fig. 3.   Temperature dependence of current gain in graded base HBT and uniform base HBT[5] for $I_{\mathrm{c}} = 2 \times 10^{-2}$ A.

than that at 300 K. It was also found that this structure restrains the current gain reduction effect in small emitter devices. This will be presented elsewhere.

In conclusion, an HBT with a linearly graded base was grown using the MBE method. Current gain of up to 1100 is achieved around a current density of 6×10^3 A/cm^2. Current gain is independent of temperature below 150 K. These characteristics strongly suggest the electron drift motion in the base due to the built-in field of graded base HBTs.

We wish to thank O. Nakajima, K. Nagata, and S. Adachi for valuable discussions.

References

1) H. Kroemer: RCA Rev. **18** (1957) 332.
2) F. Capasso, W. T. Tsang, C. G. Bethea, A. L. Hutchinson and B. F. Levine: Appl. Phys. Lett. **42** (1983) 93.
3) D. L. Miller, P. M. Asbeck, R. J. Anderson and F. H. Eisen: Electron. Lett. **19** (1983) 367.
4) J. R. Hayes, F. Capasso, A. C. Gossard, R. J. Malik and W. Wiegmann: Electron. Lett. **19** (1983) 410.
5) H. Ito, T. Ishibashi and T. Sugeta: *Extended Abst. 16th Int. Conf. Solid State Device & Materials, Kobe 1984* (Business Center for Academic Societies Japan, Tokyo, 1984) p. 351.
6) R. J. Hayes, F. Capasso, R. J. Malik, A. C. Gossard and W. Wiegmann: Appl. Phys. Lett. **43** (1983) 949.
7) H. Ito, T. Ishibashi and T. Sugeta: Jpn. J. Appl. Phys. **23** (1984) L635.
8) S. M. Sze: *Physics of Semiconductor Devices*, 2nd ed. (Wiley, New York, 1981).

A New Effect at High Currents in Heterostructure Bipolar Transistors

SANDIP TIWARI, MEMBER, IEEE

Abstract—Large current densities in heterostructure bipolar transistors with heterostructure collectors are shown to cause an excess electron barrier leading to an increase in minority-carrier charge storage in the base and a decrease in current gain of the device. This effect occurs at current densities where the mobile charge in the collector depletion region significantly reduces the electrostatic field, thus exposing an electron chemical potential barrier due to bandgap grading at the junction. The effect appears at lower current densities than the Kirk effect and should occur in wide-gap heterostructure collector devices. The effect is demonstrated using experimental data, analyzed using device modeling, and solutions are suggested for its elimination.

HETEROSTRUCTURE bipolar transistors rely on a difference in bandgap to obtain a selective suppression of carriers from the base into the emitter. Double heterostructure devices using heterostructure collectors are of interest because they are a more general device allowing both emitter-down and emitter-up operation [1], and because certain heterostructures that rely on special substrates or pseudomorphic growth can, in practice, be obtained only as double heterostructures. Examples of the latter are GaInAs/GaAs [2], GaInAsP/InP [3], GaInAs/GaAlAs [4], and SiGe/Si [5]. This paper describes an effect unique to these devices that utilizes a larger bandgap collector, and that takes place at high current densities. We use GaAs/GaAlAs devices to demonstrate this.

Fig. 1 shows measured collector and base current as a function of base–emitter bias at various base–collector voltages for a $4 \times 12 \ \mu m^2$ emitter device. This device employed a base doping of $5 \times 10^{18} \ cm^{-3}$ and thickness of 1000 Å, an emitter doping of $4 \times 10^{17} \ cm^{-3}$, a collector doping of $1 \times 10^{17} \ cm^{-3}$, and an aluminum mole fraction of 0.3. The device structure and the technology are summarized in [6]. The built-in voltage for electrons of the base–emitter and the base–collector junctions, measured in the regions of ideal collector current behavior, are identical and independent of collector bias, indicating the absence of excess electron barrier in that bias range. In Fig. 1, the base current shows a rapid rise at high current densities, with the point of rapid increase dependent on the applied base–collector voltage. The onset of this effect is pushed toward higher current densities or base–emitter voltages with increased reverse biasing of the base–collector junction. The rise is accompanied by a saturation in the collector current, which is shown clearly in the bottom part of the figure.

Manuscript received November 30, 1987.

The author is with the IBM T. J. Watson Research Center, Yorktown Heights, NY 10598.

IEEE Log Number 8819752.

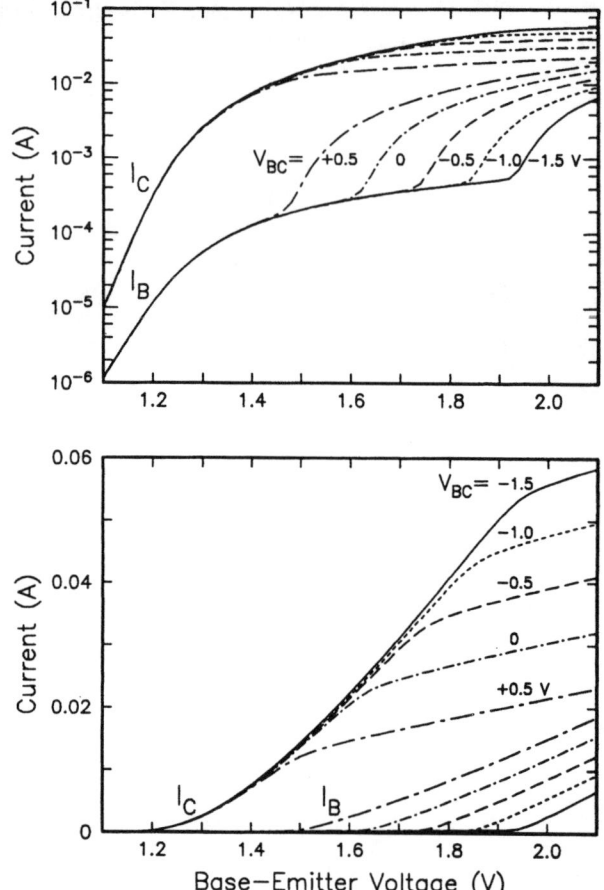

Fig. 1. Gummel plots (collector current I_C and base current I_B) of a double heterostructure bipolar transistor in the high current region for base-collector voltages of 0.5, −0.5, −1.0, and −1.5 V are shown in the top part of the figure. The bottom part shows the same characteristics with linear ordinate to show onset of saturation in the collector current. The emitter dimensions are $4 \times 12 \ \mu m^2$.

In order to understand this, consider a base–collector heterostructure design (a parabolically graded [7] GaAs/GaAlAs n-p-n transistor) with grading over distance x_0 and a final mole fraction of u_f, permittivity of ϵ_s, electron charge q, collector doping of N_D, a carrier saturated velocity v in the collector, a current density of J, and a depletion width of W. Under low injection conditions, the alloy grading results in a retarding chemical field for holes from the base and none for electrons because most of the alloy grading potential appears in the valence band. As injection increases, the decreasing net charge density in the depletion region and hence reduced electrostatic field, leads to the appearance of some of this potential as an electron chemical barrier that retards the flow

Reprinted from *IEEE Electron Device Lett.*, vol. 9, no. 3, pp. 142–144, March 1988.

of electrons also. An electron barrier just begins to appear when the electrostatic field is the same as the electron chemical field resulting from alloy grading. The maximum of the latter is derived from the grading potential. This condition is related by

$$\int \frac{q}{\epsilon_s} \left(N_D - \frac{J}{qv} \right) dx = -\frac{dV_G}{dx}. \qquad (1)$$

Here, V_G is the grading potential, and the limits of integration are from the collector depletion edge to the point x in the depletion region. For a grading of the form

$$u(x) = u_f - u_f \left(\frac{x_0 - x}{x_0} \right)^2$$

for $x < x_0$, with the proportionality factor between bandgap and mole fraction of k, this implies that

$$N_D = \frac{2\epsilon_s k u_f}{q x_0 W} + \frac{J}{qv}. \qquad (2)$$

For a current density $J = 1 \times 10^5$ A·cm^{-2}, a saturated velocity of 6×10^6 cm·s^{-1}, a parabolic grading length $x_0 = 300$ Å, and a depletion width of 1800 Å (ideally this parameter should be applied self-consistently), the effect appears at all doping levels below 1.3×10^{17} cm^{-3}. Of this, the first part to compensate for electron chemical potential is 7×10^{16} cm^{-3}, and the second part to compensate for the mobile charge (Kirk effect [8] or the base push-out effect) is 6×10^{16} cm^{-3}. This effect, thus, appears at a doping level in the collector that is significantly below that required for the Kirk effect and should be present for most conventional designs [2]–[5], [9].

We now show device simulations for design parameters similar to that of the measured device in order to analyze the effect more accurately. These simulations are performed using the two-dimensional drift-diffusion-thermionic emission heterostructure modeling program MONTE [10]. The device simulations are for devices with 1.0-μm emitters, a spacing of 0.5 μm between base-contact and the emitter edge, a base doping of 5×10^{18} cm^{-3}, and a base width of 0.1 μm. The emitter doping is 8×10^{17} cm^{-3} and the other device dimensions are fairly conventional. The parabolic alloy grading parameters are $x_0 = 300$ Å and $u_f = 0.3$ for both junctions.

Fig. 2 shows the conduction-band edge and quasi-Fermi-level for electrons at the base–collector junction for base–emitter voltages of 1.3, 1.4, and 1.5 V at a collector doping of 1×10^{17} cm^{-3} and the dot-dashed curves of Fig. 3 show the Gummel plots. A mole fraction of 0.3 implies a grading potential of ~ 0.3 V in this structure. At high current densities, the mobile charge concentration becomes sufficiently large (low to high 10^{16} cm^{-3} at these biases) resulting in the appearance of the bandgap grading related electron chemical potential barrier. Fig. 2 shows this as a barrier of approximately 0.07 eV, at a forward voltage of 1.5 V at the base–emitter junction (an emitter current density of 8×10^4 A·cm^{-2}). Simulations show that when a reverse bias is

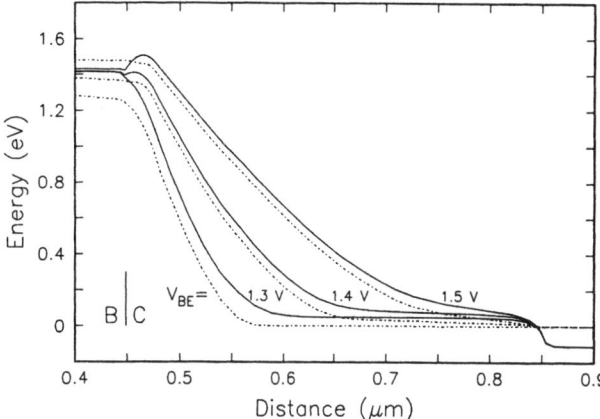

Fig. 2. Conduction-band edge (E_C) and the electron quasi-Fermi level (ϕ_n) plotted as a function of base–emitter voltage (1.3 V, 1.4 V, 1.5 V) for a simulated device with collector doping of 1×10^{17} cm^{-3} and $V_{BC} = 0$ V. The electron barrier is seen to form after the forward bias of 1.3 V, and increases with forward biasing of the emitter-base junction.

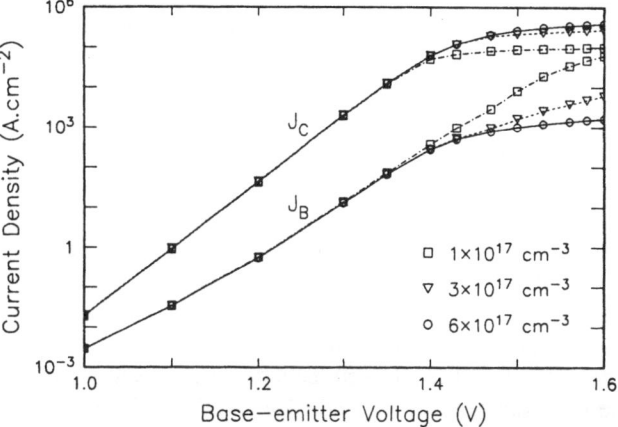

Fig. 3. Gummel plots (collector current density J_C and base current density J_B) for the simulated devices. The dot-dashed curves are J_C and J_B for the device with 1×10^{17} cm^{-3} doping in the collector, the dashed curves are J_C and J_B for the device with 3×10^{17} cm^{-3} doping in the collector, and the solid curves are J_C and J_B for the device with 6×10^{17} cm^{-3} doping in the collector.

applied at the base–collector junction, the barrier is reduced because of an increase in the compensating electrostatic field, thus reducing the magnitude of this effect.

The negative feedback associated with an increase in barrier that impedes the flow of carriers leads to quasi-saturation of the collector current seen in the dot-dashed curve of Fig. 3 and in the curves of Fig. 1. The barrier also leads to large electron accumulation in the base region. The quasi-Fermi levels of Fig. 2 show the accumulation phenomenon of electrons in the base region. A similar saturation and accumulation behavior occurs even at low currents in abrupt collector devices because of the built-in electron chemical barrier, in improperly fabricated transistors that have a barrier due to dopant diffusion into the collector [11] and when the base–collector voltage is forward biased in a normal alloy graded transistor. The important consequence of increased charge in the base is a decrease in the gain and speed of the device. The Gummel curves of Fig. 3 show the effect related to gain. The current densities where this effect becomes very clearly obvious are

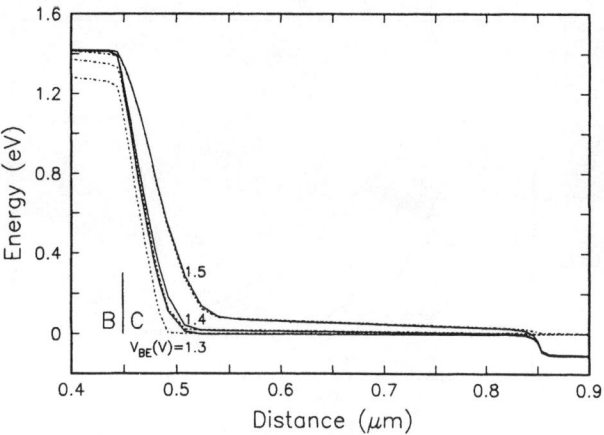

Fig. 4. Conduction-band edge (E_C, solid curves) and quasi-Fermi level for electrons (ϕ_n, dashed curves) for the device with 6×10^{17} cm^{-3} doping in collector. The three sets of curves are for an applied base–emitter voltage of 1.3, 1.4, and 1.5 V, and $V_{BC} = 0$ V.

high ($> 10^4$ A·cm^{-2}), therefore, the phenomenon shows itself only in experimental devices where parasitic resistances are low. These parasitic resistances are also responsible for the large shifts in the onset voltage for this effect for the experimental curves of Fig. 1.

Equation (2) indicates that at sufficiently high doping levels this effect should disappear. This is demonstrated by simulating the device with collector doping increased to 3×10^{17} cm^{-3} and 6×10^{17} cm^{-3} and results are shown in the dashed and solid curves of Fig. 3. The figure shows that at 3×10^{17} cm^{-3} collector doping this effect appears at a higher current density, and at 6×10^{17} cm^{-3} collector doping it is absent. The absence of the effect in the latter is more clearly seen in the plots for conduction-band edge and the quasi-Fermi level for electrons in Fig. 4. The barrier does not appear at base–emitter biases up to 1.5 V, increasing the usable emitter current density of the device beyond 1×10^5 A·cm^{-2}. The influence of mobile charge in pushing the collector depletion region is also significantly reduced.

It is interesting to observe that the appearance of this barrier slows the carriers leading to increased charge at the junction. Two-dimensional effects also become more important because of increased base charge and hence increased lateral diffusion effects. We showed that the effect is avoided by going to a higher doping level in the collector. The upper limit on this doping is placed by considerations of breakdown voltage at the junction, and speed-related effects of a higher base–collector capacitance. Alternately, this effect can also be avoided by using a much larger grading region, or a smaller aluminum mole fraction. However, a consequence of these is loss of symmetry and speed in the reverse operation of the device. Finally, we emphasize that this alloy-grading effect has a counterpart in the base–emitter junction also. The compensa-

tion of donor space charge at high current densities by the mobile charge [12] leads to a current limiting barrier at the injecting junction if sufficiently high doping levels are not used [13].

In conclusion, we have demonstrated and explained in this letter a unique effect in wide-gap heterostructure collector bipolar transistors that appears at high current densities. At these current densities, partial compensation of the collector space charge reveals the electron chemical potential barrier. This results in an increase in charge storage in the base, an increase in the base current, and a decrease in device speed. This effect can be prevented by employing a higher-doped collector than is employed for preventing the Kirk effect, by an increase in the width of the alloy-grading region, or by a decrease in the alloy mole fraction.

ACKNOWLEDGMENT

The author thanks D. Frank, A. Ginzberg, R. Marks, S. Wright, and P. Hoh for help during the course of this work.

REFERENCES

[1] H. Kroemer, "Heterostructure bipolar transistors and integrated circuits," *Proc. IEEE,* vol. 70, no. 1, p. 13, 1982.

[2] G. J. Sullivan, P. M. Asbeck, M. F. Chang, D. L. Miller, and K. C. Wang, "AlGaAs/InGaAs/GaAs strained-layer heterojunction bipolar transistors by molecular beam epitaxy," *Electron. Lett.,* vol. 22, no. 8, p. 419, 1986.

[3] R. N. Nottenburg, J. C. Bischoff, J. H. Abeles, M. B. Panish, and H. Temkin, "Base doping effects in InGaAs/InP double heterostructure bipolar transistors," in *IEDM Tech. Dig.,* Dec. 7-10, 1986, p. 278.

[4] W. Lee and C. G. Fonstad, "Application of O$^+$ implantation in inverted InGaAs/InAlAs heterojunction bipolar transistors," *IEEE Electron Device Lett.,* vol. EDL-8, no. 5, p. 217, 1987.

[5] S. S. Iyer, G. L. Patton, S. L. Delage, S. Tiwari, and J. M. C. Stork, "Silicon-germanium heterojunction bipolar transistors by molecular beam epitaxy," in *IEDM Tech. Dig.,* Dec. 6-9, 1987.

[6] S. Tiwari and S. L. Wright, "Symmetric-gain, zero offset, self-aligned, and refractory-contact double HBTs," *IEEE Electron Device Lett.,* vol. EDL-8, no. 9, p. 417, 1987.

[7] J. R. Hayes, F. Capasso, R. J. Malik, A. C. Gossard, and W. Wiegmann, "Optimum emitter grading for heterojunction bipolar transistors," *Appl. Phys. Lett.,* vol. 43, no. 10, p. 949, 1983.

[8] C. T. Kirk, "Theory of transistor cutoff frequency fall off at high current densities," *IRE Trans. Electron Devices,* vol. ED-3, no. 3, p. 164, 1962.

[9] S. Tiwari, "Circuit performance of the GaAlAs/GaAs heterostructure bipolar junction transistors," in *Tech. Dig. GaAs IC Symp.,* Nov. 12-14, 1985, p. 95.

[10] J. Y.-F. Tang and S. E. Laux, "MONTE: A program to simulate the heterojunction devices in two dimensions," *IEEE Trans. Computer-Aided Des.,* vol. CAD-5, no. 4, p. 645, 1986.

[11] S. Tiwari, S. L. Wright, and A. Kleinsasser, "Transport and related properties of (Ga,Al)As/GaAs double heterostructure bipolar junction transistors," *IEEE Trans. Electron Devices,* vol. ED-34, no. 2, p. 185, Feb. 1987.

[12] S. Tiwari and D. J. Frank, "Simulation of recombination and grading effects in heterostructure bipolar transistors," presented at the IEEE Dev. Res. Conf., Amherst, MA, June 21-23, 1986.

[13] S. Tiwari and D. J. Frank, "Barrier and recombination effects in the base-emitter junction of heterostructure bipolar transistors," submitted for publication, 1987.

Analysis of the Operation of GaAlAs/GaAs HBT's

SANDIP TIWARI, SENIOR MEMBER, IEEE, AND DAVID J. FRANK

Abstract—Drift-diffusion modeling in two dimensions has been employed to characterize and analyze storage, transport, and recombination effects in GaAlAs/GaAs heterostructure bipolar transistors. Both intrinsic and parasitic effects have been studied, and their relationship to the design of the device is discussed. For conventional dopings and high current densities, the heterojunction grading potential causes a barrier in the base–emitter junction, which results in a large increase in the dynamic resistance. In heterojunction collectors, a similar barrier leads to a large increase in base charge storage and to spreading of collector current. It is shown that increased doping levels can successfully suppress these barrier effects. The capacitance and transport phenomena at the base–emitter junction are also analyzed under conditions of large forward bias, where the junction space-charge region is shorter than the alloy grading length. Recombination is analyzed in the limit of high surface recombination velocities using Shockley–Read–Hall theory in the presence of Fermi-level pinning due to surface states. The pinning results in a potential energy saddle point, at the edge of the base–emitter junction, which largely determines the surface recombination behavior of the transistor when the recombination velocity is high. The characteristics of this behavior are found to agree with the experimentally observed recombination behavior of conventional devices. Finally, the design and scaling of heterojunction bipolars is discussed in the light of these results.

I. INTRODUCTION

COMPOUND semiconductor heterojunction bipolar transistors (HBT's) exhibit several phenomena that differ from those seen in silicon homojunction bipolar transistors. The primary operational differences are in the bias dependence of charge transport and storage. These differences are caused by several physical processes: the injection of carriers at a varying-bandgap heterostructure [1]–[3], the collection of carriers at a varying-bandgap heterostructure [4], and hot-carrier and quasi-drift field effects in the base [5] and collector [6]–[9]. In addition, recombination at the surface [10]–[12] and in the bulk [10], [13] is important as a parasitic effect, while it is virtually nonexistent in silicon bipolar transistors.

In this paper, we provide new insights into, and analyze the magnitudes of, these effects in an n-p-n HBT by employing a conventional drift-diffusion approach. We use a two-dimensional model to study effects related to storage, recombination, and collector and base transport. Although this model does not include ballistic carrier effects, it does adequately describe most of the important effects in HBT's. Recently, it has been demonstrated that ballistic effects reduce the collector transit time in suitably designed homojunction collectors [9]. To fully take advantage of such effects, one must also deal with the effects described here.

Manuscript received January 13, 1989; revised May 22, 1989.
The authors are with the IBM Thomas J. Watson Research Center, Yorktown Heights, NY 10598.
IEEE Log Number 8930057.

The injection and collection of carriers at varying-bandgap heterojunctions is strongly influenced by the development, at high current density and/or high forward bias, of a potential barrier due to the alloy grading potential of the heterojunction. We demonstrate the importance of this barrier by analyzing its behavior and consequences under various conditions. In the base–emitter junction it causes a large increase in the dynamic resistance at nominal dopings. In double-heterojunction devices, it leads to an increase in base storage, a decrease in current gain, and a significant collector current spreading at high current densities. For single- and double-heterojunction devices, we derive the behavior of the quasi-static capacitance of the base–emitter junction and the behavior of the charge storage and the forward time constant in the base at the highest current densities to elucidate the influence of the heterojunctions on these characteristics. We analyze surface recombination in the conventional HBT [11] using the Shockley–Read–Hall (SRH) theory of recombination in the presence of Fermi-level pinning due to surface states. Fermi-level pinning leads to the formation of a saddle point at the surface through which electrons are injected to the p-GaAs surface. The flow of these electrons along the extrinsic base surface is strongly channeled, also due to the pinning. This injection, together with the large surface recombination velocity of GaAs, leads to the substantial surface recombination current observed at moderate current densities. This current has an exponential characteristic energy that changes from $\sim 2kT$ to $\sim kT$ over the operating current densities. Finally, we discuss how this analysis of the operation of the device can be applied to designing devices for high-speed and high-frequency operation.

The organization of the paper is as follows. Section II describes the model and its assumptions. Section III describes the transport and storage characteristics of the base–emitter junction and relates them to the design of the junction. Section IV describes effects related to the base–collector junction. Section V describes the behavior of the bulk and surface recombination and relates them to the effects described in the earlier sections and to the perimeter dependence observed in the conventional HBT. Finally, Section VI relates all of these to the scaling of the HBT.

II. MODEL DESCRIPTION

The two-dimensional device simulator [14] uses drift diffusion as the basis for electron and hole transport. The electron velocity field relationship in this model is a complex function of doping and the electron quasi-electric

Reprinted from *IEEE Trans. Electron Devices*, vol. 36, no. 10, pp. 2105–2121, Oct. 1989.

field that reproduces the peak velocities and negative resistance features of low doping as well as the monotonically increasing velocity field characteristics of high doping. "Electron quasi-electric field" refers to the gradient of the conduction band edge, $\vec{\nabla}(E_c/q)$, where $E_c = -q\psi + \phi_C^A$, with q being the electron charge, ψ being the electrostatic potential, and ϕ_C^A being the conduction band alloy potential energy,[1] i.e., the extra conduction band edge energy associated with the alloy composition of the semiconductor. Similar definitions apply for the valence band.

This model does not include bandgap shrinkage ($\Delta E_G \propto 1.6 \times 10^{-8} p^{1/3}$ eV where p is the hole concentration [15], and $5.8 \times 10^{-13} p^{0.58}$ according to our experiments). At a background hole concentration of 5×10^{18} cm^{-3}, a bandgap shrinkage of ≥ 27 meV would be expected, which would improve the electron injection behavior of the transistor. In view of this and other assumptions in the simulations, such as the magnitudes of lifetimes, recombination velocities, bulk mobilities and diffusivities, and the validity of the Einstein relationship (generalized to Fermi–Dirac statistics, which have been used in all calculations), the model should be considered as a guide that provides excellent qualitative understanding and fair quantitative agreements. We have checked the model against experiments and found it to agree within a few percent with experimental data for the collector current at moderate current densities in graded and abrupt structures of varying doping. It also qualitatively agrees with the behavior of base current for structures of varying size, both with and without surface passivation. Quantitative agreement with this behavior requires accurate characterization of both surface and bulk parameters—all of which require intelligent guesses in the absence of independent measurements.

The model includes deep levels and SRH recombination resulting from such levels. This allows the model to simulate, self-consistently with the charge, the conditions resulting from Fermi level pinning and SRH recombination. Inclusion of the charge is essential to the statistics of surface recombination. Fermi-level pinning is simulated by using equal donor and acceptor trap concentrations of 1×10^{13} cm^{-2} at 0.8 eV [16]. This yields a Fermi level still pinned at 0.8 eV for 5×10^{18} cm^{-3} doping in the base. The real distribution of surface states and bulk generation-recombination (g-r) centers is nonuniform, and recombination occurs through only those levels that capture electrons and holes efficiently, i.e., those levels that lie between the quasi-Fermi levels and have sufficiently large capture cross sections. In our model, by choosing a position near mid-gap and suitable lifetimes for the surface states and the bulk g-r centers, we lump the effect of capture cross section and distribution into a single lifetime. Provided the electron-to-hole capture cross section

ratios of the dominant g-r centers in the real distribution are not wildly different from unity, and provided the dominant g-r centers are in fact in the vicinity of mid-gap, the use of a single acceptor/donor pair g-r center near midgap should qualitatively reproduce the characteristics of the real distribution. These assumptions are thought to be valid for GaAs and GaAlAs, but for a material in which these assumptions are seriously violated, our recombination results would need to be re-evaluated. Surface recombination is simulated by a decrease in electron and hole lifetime τ in a thin surface layer determined by the mesh spacing Δy at the surface. The surface recombination velocity definition, $S = \Delta y/\tau$, appropriate to this approach differs from the conventional non-local definition [17] that relates the surface recombination current to the bulk minority-carrier population, but agrees with the definition used by Henry et al. [18]. The schematic cross section of a typical simulated device is shown in Fig. 1, and the values of the parameters used for the calculations are summarized in Table I.

III. CHARGE TRANSPORT AND STORAGE IN THE BASE–EMITTER JUNCTION

The current flow through an n-GaAlAs/p-GaAs heterojunction is much more complicated than the flow through a homojunction. It depends on whether the heterojunction is graded or abrupt, and if it is graded, it depends on the details of the grading because the changing alloy potential causes alloy fields, $\vec{\nabla}(\phi^A/q)$, which add to the usual electrostatic fields and modify the current flow.

At low currents, with appropriate heterostructure grading, the built-in quasi-electric fields for electron transport remain the same as in the homojunction transistor except for minor second-order effects related to the density of states, donor ionization energy, etc. The built-in fields for holes change such that the hole potential barrier increases by the difference in bandgap, adjusted by the second-order effects mentioned above. The resulting suppression of hole injection allows the unique design advantage [5] of higher doping in the base while still maintaining good injection efficiency. In addition, recombination in the depletion region of the base–emitter junction is reduced in proportion to the hole suppression. This is important, since compound semiconductors have low carrier lifetime due to SRH recombination.

At large forward bias and current, the electrostatic potential and the space-charge region $(SCR)^2$ width are reduced, leading to anomalies in transport and storage. These occur because the relative importance of the alloy field in the conduction band increases as the electrostatic field decreases. Furthermore, the hole suppression property of the alloy grading region becomes less effective as the SCR width becomes shorter than the alloy grading width.

---

[1] In our previous papers [3], [4] this was called a "chemical" potential; we have adopted the new name here to avoid conflict with the thermodynamic chemical potential.

[2] The space-charge region, which we also sometimes call the depletion region, refers to the region in which electrostatic fields exist due to the absence of quasi-neutrality.

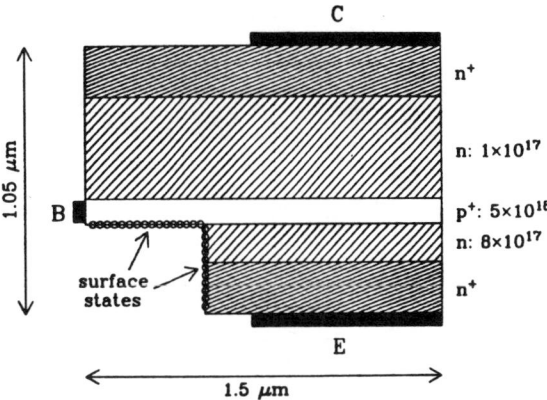

Fig. 1. Schematic cross section of a typical simulated HBT. The 8×10^{17} cm^{-3} doped area is the wide-gap emitter region, and the 1×10^{17} cm^{-3} area is the wide-gap collector region for those simulations in which the collector is a heterojunction. Most of the dimensions are given in Table I. The extrinsic base length is 0.5 μm.

TABLE I
SELECTED PARAMETERS USED IN COMPUTATION

Emitter:	
Contact Layer Doping	5×10^{18} cm^{-3}
Contact Layer Thickness	2300Å
Wide Gap Emitter Doping	$2 - 8 \times 10^{17}$ cm^{-3}
Total Wide Gap Emitter Thickness	1200Å
GaAlAs Alloy Composition	30%
Parabolic Grading Length	300Å
Base:	
Base Doping	5×10^{18} cm^{-3}
Base Thickness	1000Å
Collector:	
Parabolic Grading Length	300Å
GaAlAs Alloy Composition	30%
Collector Thickness	3500Å
Collector Doping	1×10^{17} cm-3
Sub-collector Doping	5×10^{18} cm^{-3}
Bulk Recombination Parameters:	
Donor Trap:	
τ_n	1×10^{-9} s
τ_p	1×10^{-9} s
Energy Level:	0.8 eV below Conduction Band
Acceptor Trap:	
τ_n	1×10^{-9} s
τ_p	1×10^{-9} s
Energy Level:	0.8 eV below Conduction Band
Surface Recombination Parameters†:	
Donor Trap:	
Trap Density	1×10^{13} cm^{-2}
S_n	$0 - 1 \times 10^{6}$ cm/s
S_p	$0 - 1 \times 10^{6}$ cm/s
Energy Level:	0.8 eV below Conduction Band
Acceptor Trap:	
Trap Density	1×10^{13} cm^{-2}
S_n	$0 - 1 \times 10^{6}$ cm/s
S_p	$0 - 1 \times 10^{6}$ cm/s
Energy Level:	0.8 eV below Conduction Band

† - Surface recombination velocities derived from lifetime specified at the surface.

A. Alloy Grading, Doping Design and Transport at the Base-Emitter Junction

The central requirements of the design of the alloy grading at a heterojunction are maximizing the current drive capability, maintaining the device gain, and minimizing the delays associated with the resistance, capaci-tance, and transit time. The transit time for the base-emitter junction is generally much smaller than the delays associated with the base-collector junction. Similarly, except at the largest forward bias, the time constant for the charging the base-emitter depletion capacitance is smaller than the base-collector delays because of the relatively large collector area. Thus, maximization of injection current becomes a more important design criterion at the base-emitter junction.

In homojunction transistors, high-level injection effects at the base-emitter junction are primarily caused by reduction in the forward-bias voltage due to lateral ohmic drop in the base, vertical ohmic drop in the emitter, and base push-out effect in the collector. A secondary effect due to the majority-carrier-induced drift field is kept low by choosing an adequate base doping level. This secondary effect is even less important in HBT's because the doping levels are higher than in homojunction transistors. Homojunction transistors are designed with emitter dopings in excess of $\sim 10^{20}$ cm^{-3} and base dopings exceeding $\sim 10^{18}$ cm^{-3} in order to maintain sufficient injection efficiency. Since HBT's have a much higher injection efficiency, they are usually designed with much smaller emitter dopings to reduce emitter capacitance and tunneling effects associated with heavier doping in both the emitter and the base. However, if the emitter doping is too small, the injection of minority carriers into the base becomes limited [3] by space-charge effects and by the opposing effect of the alloy potential at small electrostatic fields.

This alloy potential effect is also significant at the collector heterojunction—which is where it was first observed and explained [4]. At high electrostatic fields, the barrier due to the difference in the bandgaps appears in the valence band; but at low electrostatic fields (i.e., high biases), it also begins to appear in the conduction band. Hence, at high bias, the alloy potential causes an "alloy barrier" to the flow of carriers.

This effect, which does not exist in the homojunction case (the smaller bandgap in the emitter causes an alloy field in the opposite direction from that in the heterojunction), must be suppressed to obtain high current density operation. For a current density J_E in the base-emitter junction, suppression of this alloy barrier can be assured by having a monotonic increase in the total electron potential (electrostatic and alloy) from the emitter to the base. This implies that the alloy field should be maintained lower than the electrostatic field throughout the grading region [3]. Using Poisson's equation and the modeling observation that the hole space charge is almost always negligible, this condition can be written as

$$\int_{y_E}^{y} \frac{e}{\epsilon_s} \left(N_D - \frac{J_E}{qv} \right) dy \geq - \frac{d\phi_C^A(y)}{q \, dy} \qquad (1)$$

where y refers to the vertical, or depth, direction in the device, ϵ_s is the permittivity, N_D is the emitter doping, e

is the elementary charge,[3] and $v(y)$ is the electron velocity. The limits of integration are from y_E, the emitter edge of the SCR or the emitter edge of the graded region, whichever comes first, to the point y in the SCR. Alloy barrier suppression requires that this inequality hold for all points y in the junction region. Note that the sense of y is such that y increases from the emitter toward the base.

By differentiating and solving for N_D, one obtains the following upper bound on the minimum N_D needed to guaranty that (1) is satisfied:

$$N_D(y) \geq -\frac{\epsilon_s}{e} \frac{d^2\phi_C^A(y)}{q \, dy^2} + \frac{J_E}{qv(y)}. \tag{2}$$

Lower values for N_D can in fact be sufficient if the SCR is wider than the grading length, as will be shown later for the base–collector junction. On the right-hand side of this inequality, the first term is associated with the alloy barrier, and the second term is simply the concentration of mobile carriers in the depletion region. Although the first term is readily determined, the carrier concentration in a depletion region does not yield to analytic solution except in special cases such as the Schottky diode [19]. In the situation of interest—high current density and substantial degeneracy—the carrier mobility and diffusivity are nonlinearly dependent on the carrier density, the alloy composition, the quasi-electric field, and the doping concentration. Hence, it appears unlikely that good analytic formulas for N_D can be found.

Applying (2) to a specific case, we consider a parabolically graded, uniformly doped junction, where the mole fraction $u(y)$ is given by

$$u(y) = u_f - u_f \left(\frac{y + y_0}{y_0}\right)^2, \quad \text{for} \quad -y_0 \leq y \leq 0. \tag{3}$$

Here, y_0 is the grading length, and u_f is the final mole fraction in the parabolic grading. Note that the largest alloy field for this grading occurs at the junction ($y = 0$). Use of $\phi_C^A(y) = \gamma u(y)$ for the alloy potential of GaAlAs relative to GaAs, and this grading, gives

$$N_D^{\text{uniform}} \geq \frac{\epsilon_s}{e} \frac{2\gamma u_f}{qy_0^2} + \frac{J_E}{qv_{\min}} \tag{4}$$

as the minimum uniform doping density needed to maintain proper heterojunction operation at high current density, where $v_{\min}$ is the minimum carrier velocity in the graded region. Since the $J_E/v_{\min}$ term is strongly dependent on the base–emitter junction bias, numerical simulations must be used to clarify the importance of the individual terms in (4). The simulations show that, for a GaAlAs/GaAs junction with parabolic alloy grading of

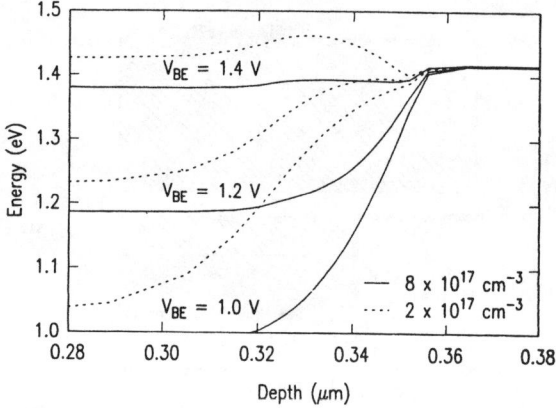

Fig. 2. Conduction band edge (E_C) versus position at the base–emitter junction for 1.0-, 1.2-, and 1.4-V forward bias at the junction. The emitter doping is 2×10^{17} cm^{-3}, and 8×10^{17} cm^{-3}. The current limiting barrier appears at the highest bias for the lower doped device. The metallurgical and grading junction is located at 0.35 μm.

0.3 over 300 Å, the contributions of the two terms are approximately equal at a current density of 10^5 A/cm^2. Because of the low velocity of injected electrons in the junction, emitter dopings exceeding 8×10^{17} cm^{-3} are needed to assure operation in the 10^5 A/cm^2 range.

Fig. 2 shows the behavior of a 2×10^{17} and a 8×10^{17} cm^{-3} doped emitter junction as a function of bias. The lower doped device shows an alloy barrier at 1.4-V bias, while the higher doped device does not. Clearly, such an alloy barrier reduces the current that can be injected into the base. In this example, the current density in the 2×10^{17} cm^{-3} emitter doped device is only $\sim 1.6 \times 10^4$ A/cm^2 at 1.4 V, while it is $\sim 6 \times 10^4$ A/cm^2 in the higher doped device.

The effects of this alloy barrier and its emitter doping dependence are shown more completely in Fig. 3(a) and (b), where Fig. 3(a) shows the Gummel plots for the two emitter dopings, and Fig. 3(b) plots the excess voltage (i.e., the forward-bias voltage needed in excess of the extrapolated low current characteristics) and the current gain in the absence of surface recombination. Note the larger deviation from the low current extrapolation in the lower doped device. This device also has smaller current-handling capability, as can be seen in Fig. 3(b), where the rapid increase in excess voltage occurs at $\sim 5 \times$ lower current densities. The higher current gain of the lower doped device at lower current densities is due to lower SCR recombination [3]. This overall behavior is in agreement with our observations, even for single-heterojunction bipolar transistors, of saturation of collector current and lack of correlation between parasitic resistances estimated from device I–V curves (at high current densities) and alternate estimates of those resistances based on independent (low current density) test sites.

A lower alloy field can be achieved at the junction by employing linear grading instead of parabolic grading. However, in this case the alloy field is larger at the emitter end of the grading region where the electrostatic fields are even lower. The barrier now appears at the emitter edge

---

[3]The distinction between e and q is maintained so that the equations will be valid for both the absolute convention, where $q = -e$, and the common convention for band diagrams, etc., which is equivalent to $q = +e$. Note that this latter convention also includes the treatment of electron current density and velocity as having the same sign.

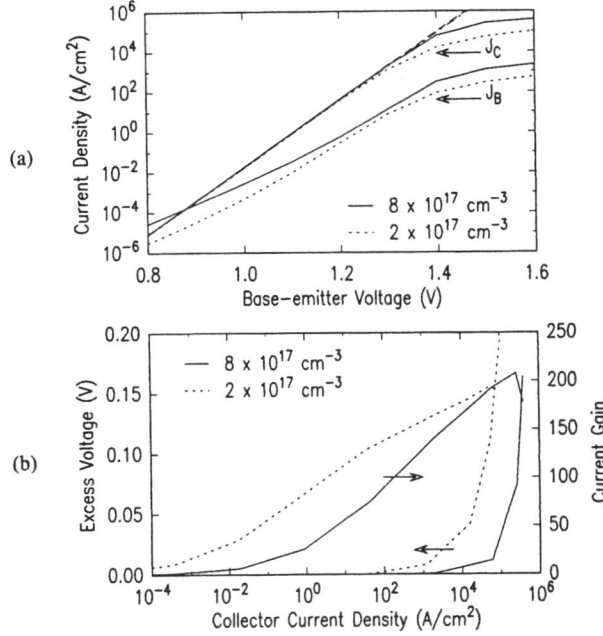

Fig. 3. (a) Gummel plots for parabolically graded single heterojunction devices with emitter doping of 8×10^{17} cm^{-3} and 2×10^{17} cm^{-3}. The collector current density is identified by J_C and the base current density by J_B. The figure also shows a low current fit to the collector current density plot. This fit is used to derive the excess voltage—the extra base-emitter voltage that must be added to the low-current extrapolation in order to sustain a large current density J_C. The fit shows the current density varying at ~ 1 decade/60 mV. (b) Excess voltage and current gain of these devices as a function of the collector current density. Surface recombination is set to zero in this calculation. The rapid increase in excess voltage shows the onset of current saturation.

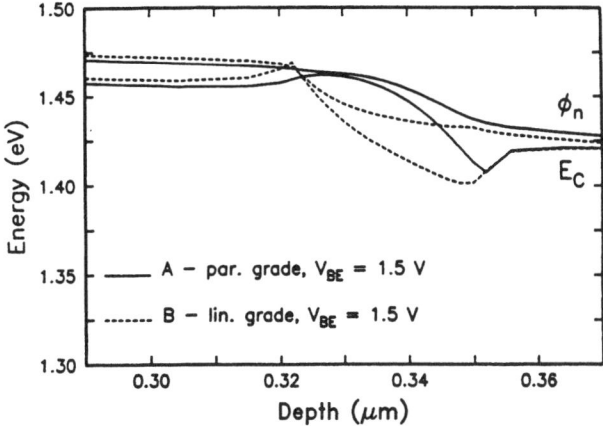

Fig. 4. Conduction band edge (E_C) and quasi-Fermi level for electrons (ϕ_n) at a base-emitter bias of 1.5 V for parabolic and linear grading in a single heterojunction transistor. The junction is located at 0.35 μm. The current density is $\sim 3 \times 10^5$ A/cm^2 through the parabolically graded device and $\sim 2.3 \times 10^5$ A/cm^2 through the linearly graded device.

of the grading region. Fig. 4 demonstrates this by plotting the conduction band edge and the quasi-Fermi level at V_{BE} = 1.5 V for a linearly graded junction and a parabolically graded junction for 8×10^{17} cm^{-3} emitter doping. The linearly graded junction shows a barrier to injection at the emitter end of the grading region that persists down to lower bias currents; an example at 1×10^4 A/cm^2 is shown in [3]. One consequence of this barrier is that less current is carried by the linearly graded device than by the parabolically graded device at the same bias. The limiting current is also lower—by about a factor of three in our simulations.

B. Base-Emitter Capacitance

Compared to Si, the high diffusivity of electrons in the base of the GaAs HBT leads to a smaller base diffusion capacitance, which increases the relative importance of the emitter capacitance. Since the devices are operated at high currents and large forward bias, this capacitance has to include both the immobile charge component (i.e., the textbook depletion capacitance) and the mobile charge component (i.e., that due to the charges carrying the current). We calculate the capacitances from the change in total charge on the emitter side of the junction due to a small perturbation (~ 10 mV) in applied bias. We do not separate the mobile and immobile charge contributions because they cannot be independently determined. The

injection hole charge (which can be separately considered) is generally quite small and gives rise to a negligible diffusion capacitance. We can and do separate the depletion and diffusion capacitance (due to electron storage) in the base. The depletion region charge is the same as the net charge on the emitter side, and the storage charge is calculated by integrating the increase in electron density in the base region. In calculating these capacitances in the bipolar transistor, the partitioning of charge occurs quite naturally because of the p-n junctions, and hence is not subject to the inherent inaccuracies of this approach in more general problems [20].

Fig. 5 plots the total capacitance (depletion and diffusion) of the base-emitter junction for a single- and double-heterojunction transistor. Since hole injection is much smaller than in the homojunction transistor, this capacitance is primarily determined by the depletion region thickness and the electron space charge in the depletion region. Both of these are in turn sensitive to the behavior of the potential in the presence of this injected charge. These junction characteristics, for the single-heterojunction device, are illustrated in Fig. 6(a)-(d), which shows the conduction band edge, the electron density, the hole density, and the total charge density versus depth for bias voltages of 0.8, 1.1, and 1.4 V. Note that the hole density is, indeed, very small because of the high injection efficiency. Even at 1.4 V, where the depletion region is smaller than the grading region, the contribution of hole density remains small except in a region a few 10's of angstroms wide at the junction. Only at extremely low emitter dopings does the alloy barrier effect become large enough that hole injection is important.

The behavior of the capacitance—a rapid increase followed by a decrease at high bias—is similar to that observed in homojunction transistors [21] with the maximum capacitance limited by Debye lengths at near flat-band conditions. At high bias the junction reaches or exceeds flat band, and the ability of the bias to modulate the current decreases, leading to less charge modulation and,

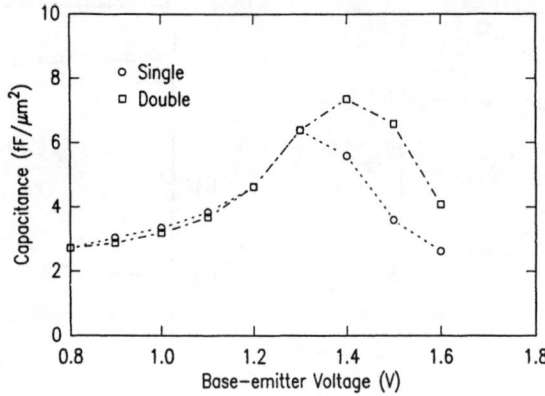

Fig. 5. The total capacitance in femtofarads per square micrometer at the base–emitter junction as a function of the applied base–emitter voltage. The base–collector voltage is 0 V. The alloy grading at the junction is parabolic, the emitter doping is 8×10^{17} cm^{-3}, and the base doping is 5×10^{18} cm^{-3}. Capacitances for both single- and double-heterojunction devices are plotted.

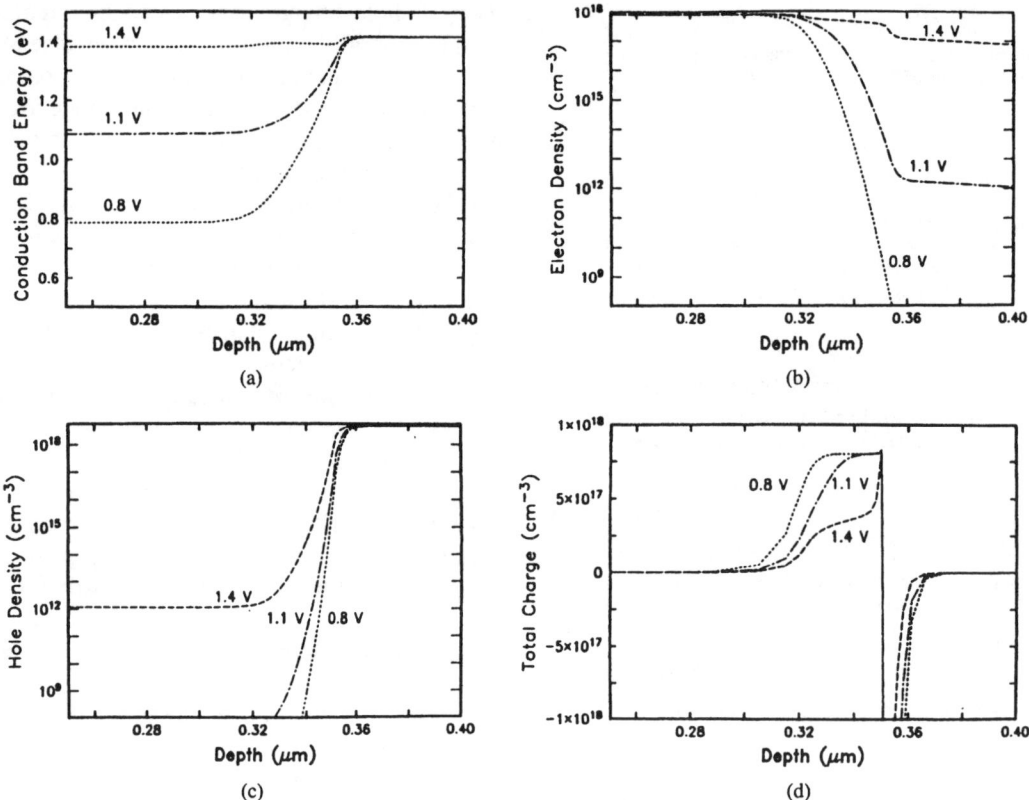

Fig. 6. Plots of the conduction band energy, electron density, hole density, and the total charge density at the base–emitter junction (located at 0.35 μm) for a single-heterojunction device, with bias varying from low-level injection conditions (0.8 V) to moderately high-level injection conditions.

hence, decreased capacitance. As in the homojunction transistor, large mobile charge in the space-charge region and voltage drops in the quasi-neutral regions diminish the effect of increasing bias. In the HBT, the alloy barrier also works to decrease the capacitance at high biases, by decreasing the modulation of the depletion region width. The increasing voltage is accommodated by the relatively poor injection characteristics of the alloy barrier and does not result in much change in depletion region charge.

HBT's employ lower emitter doping than homojunction

transistors because of their higher injection efficiency. Lower doping gives rise to a lower emitter capacitance, and the alloy barrier at high bias reduces the peak capacitance even further. Increasing emitter doping leads to higher current drive capability, but causes higher capacitance throughout the bias range. In particular, the peak capacitance increases with doping, and the peak occurs at higher bias voltage because of a larger built-in voltage and reduced effect of the alloy barrier.

The double-heterojunction device C–V curve in Fig. 5

has a larger peak capacitance than the single-heterojunction device. At these high bias levels—as we will show later, a large storage of electrons takes place in the quasi-neutral base. The increase in base charge influences the emitter injection behavior because a larger charge at the base–emitter junction also implies a larger space charge in the emitter depletion region. This increase in space charge causes most of the increase in the base–emitter capacitance at large voltages, and also makes the barrier effect more prominent in the base–emitter depletion region. The barrier modulation and the increase in majority-carrier concentration due to the charge storage both lead to a larger hole diffusion capacitance for the double-heterojunction device, which also contributes to the observed increase at sufficiently high bias.

C. Electron Quasi-Electric Fields in Single-Heterojunction Bipolars

Fig. 7 shows the negative of the electron quasi-electric field versus position in the device as a function of base–emitter bias voltage for the same HBT as in Fig. 6. In this figure, the base region has near-zero electric field; the region of high field on the left is associated with the emitter, and the one on the right with the collector. The base–emitter junction field distribution should be compared to that of a homojunction transistor (e.g., field distributions drawn in Poon's work [22] and reproduced in Sze's book [23]). At low biases, the quasi-electric field varies nearly linearly with distance and is dominated by electrostatic considerations in the depletion approximation. As the bias increases, the quasi-field decreases and the depletion region shrinks, until at a bias of about 1.3 V most of the region lies inside the base–emitter alloy grading region. At this bias, the alloy field $d\phi_C^A/qdy$ acts to substantially reduce the electrostatic field component. At higher biases the quasi-electric field, which is the sum of the electrostatic and alloy fields, decreases and even changes direction due to the influence of the alloy field (this is quite large at the 1.5-V bias). The negative spikes in the quasi-field at the beginning and end of the grading region for 1.3-V bias and above are due to the abrupt changes in $d^2\phi_C^A/qdy^2$ at the grading region edges, which cause short-range breakdown of quasi-neutrality.

Although the quantitative magnitudes of the emitter-base effects we have described can be changed by varying the modeling parameters (diffusivity, mobility, etc.) and their dependence on alloy composition, the qualitative variations remain the same. The largest difference in base-emitter junction behavior between HBT's and homojunction bipolar transistors is the influence of the alloy field.

IV. HIGH CURRENT CONSIDERATIONS OF THE BASE–COLLECTOR JUNCTION

Having considered the high current and high forward bias effects in the base–emitter junction, we next consider the base–collector junction. Many bipolar logic families—including I²L and ECL—involve either a significant forward bias of the base–collector junction or a large for-

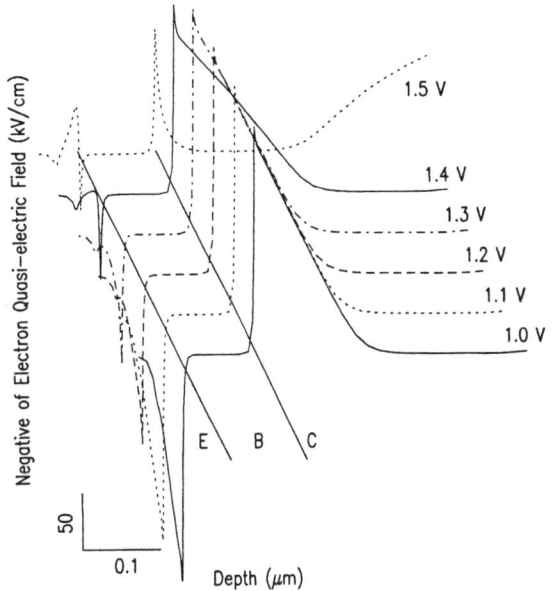

Fig. 7. Electron quasi-electric field at the base–emitter junction, in the base, and at the base–collector junction as a function of base–emitter junction bias for the single-heterojunction HBT. The base is 0.1 μm wide (parameters in Table I), and the different plots are shifted to allow a perspective view. E, B, and C identify the emitter, base, and collector regions. The scales of distance and field are located in the lower left corner. The field in the base region is essentially zero in each plot.

ward bias current that tends to debias the base–collector junction because of the increased minority-carrier charge. Our discussion of the base–emitter junction dwelt on designing a junction capable of injecting large currents. We also need to assure that the device is capable of collecting large currents with low storage and transit time to assure the desired high-speed and high-frequency characteristics. The following analyzes these properties for heterojunction collectors and provides comparisons with homojunction collectors. In particular, we describe the alloy barrier effects that occur in heterojunction collectors at high current densities and the resulting two-dimensional storage and transport characteristics.

A. Barriers and Their Influence in Heterojunction Collectors

Under low-current operation, the homojunction theory of collector transport applies quite well to the heterojunction transistor. Since hole transport at the collector is insignificant—the junction being usually reverse biased—a heterojunction collector has little effect on the transport physics, and the collector efficiently transports the electron current. At medium and high current densities, various effects are known to become important in homojunction transport, e.g., Kirk effect (base push out), and Webster effect (base field and conductivity modulation). In HBTs, the first of these two effects remains equally important but the second is relatively insignificant because of the large base doping densities. In an HBT with a heterojunction collector, an alloy barrier effect [4] also appears, which can be more important than the Kirk effect.

The nature and origin of the alloy barrier effect in the collector are essentially the same as in the emitter: at low currents the alloy potential difference appears in the valence band resulting in the usual hole retarding barrier and field, while at higher currents it appears partially in the conduction band as a barrier to the flow of the minority carriers. The consequences of this barrier are excess charge storage in the base and an increase in the associated capacitance, a decrease in device current gain due to recombination, and a saturation of collector current. All three effects worsen as the base–collector forward bias increases, and thus limit the minimum usable value of V_{CE} in circuit applications.

Fig. 8 shows two perspective plots of the conduction band edge energy in the base region of a double-heterojunction transistor at $V_{BE} = 1.3$ and 1.5 V, with $V_{BC} = 0$ V. The region plotted extends 500 Å into both the emitter and the collector. At the lower bias, the potential profile is similar to that of a well behaved homojunction bipolar transistor. However, at the higher voltage, a barrier due to both the alloy potential and the electron space charge appears at the collector junction. This barrier is largest where the current density is largest, i.e., opposite the emitter junction. The barrier is lower toward and in the extrinsic part of the device because the current density is lower there. Significant spreading of the electron flow toward the extrinsic base–collector junction area occurs in this device because the alloy barrier varies proportionally with current density.

We will refer to this consequence of the alloy barrier as the "collector current spreading effect." (Emitter crowding remains low in HBT's because the base doping is high.) Spreading of the current to the extrinsic part of the device causes increased storage of carriers there. Before quantifying this, we point out its serious consequences in a double-heterojunction bipolar where a wider gap junction is buried in the extrinsic base–collector region in order to achieve symmetrical operation of the transistor [5]. Fig. 9 shows the same two-dimensional band edge plots as in Fig. 8, except that there is a buried extrinsic wider gap base–collector junction. The large barrier on the left is a result of the extrinsic p-type base region extending into the collector. This barrier, with a height approximately equal to the bandgap difference, is much larger than the alloy barrier in Fig. 8. The current is forced to transport to the collector in the intrinsic region, leading to larger storage in both the intrinsic and the extrinsic base regions, due to diffusion of carriers. This transistor has even worse minority-carrier storage and current saturation than the simpler double-heterojunction bipolar transistor.

Fig. 10(a) and (b) provides a comparison of the base storage effects in hetero- and homojunction collector devices. Each part of the figure shows the electron concentration along a lateral cross section at the middle of the base for applied biases varying from 1.0 to 1.6 V. The storage behavior is similar for the two devices at low biases (up to 1.4 V), with an exponential decay length of ~ 850 Å which is determined primarily by the base width

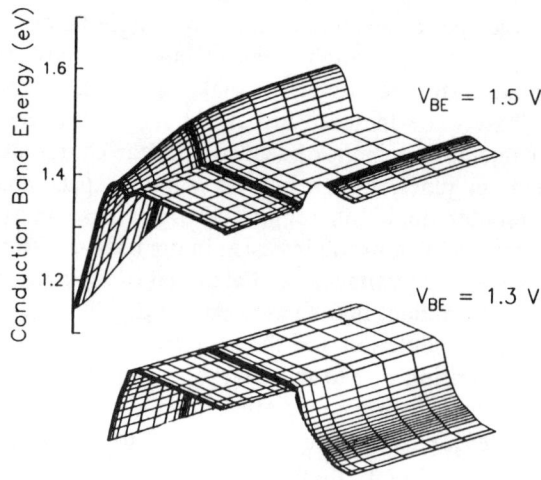

Fig. 8. Surface plots of the conduction band edge in the base and at the emitter and collector junctions for forward-bias voltages of 1.3 and 1.5 V for a double-heterojunction HBT. The 1.3-V surface is shifted downward for clarity. The perspective is similar to that of Fig. 15. The emitter junction is located in the lower right, and the collector is the broader region in the background. The base is the constant energy region. Both heterojunctions are parabolically graded.

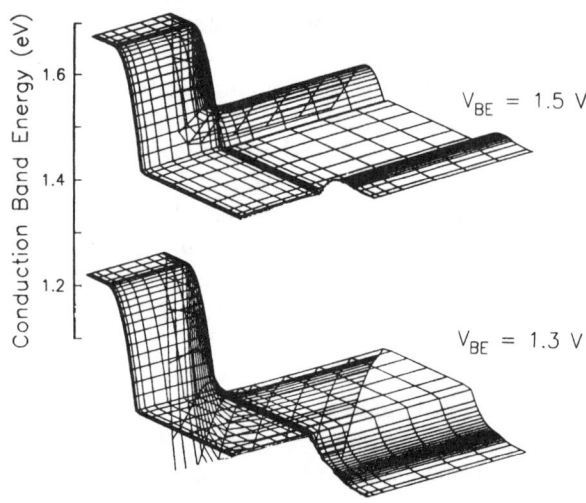

Fig. 9. Surface plots of the conduction band edge in the base and at the emitter and collector junctions for forward-bias voltages of 1.3 and 1.5 V for a double-heterojunction HBT with a buried wide-gap p-type region in the extrinsic region. The perspective is identical to that of Fig. 8. Note that compared to Fig. 8 there is a decrease in the effective collector area because electrons that diffuse out into the extrinsic base region cannot be collected.

of the device (1000 Å). At high biases, the homojunction collector device shows only the limitations of the injection process and maintains the exponential decay length. The heterojunction device, however, develops a much longer, nonexponential decay characteristic, which appears limited by the designed spacing of 0.5 μm between the base ohmic contact and the intrinsic base region. Thus, the increased storage in the heterojunction collector bipolar occurs both in the intrinsic and in the extrinsic part of the device.

The magnitude of this storage depends exponentially on the height of the barrier between the base and the collector. As in the emitter, the height of this barrier can be

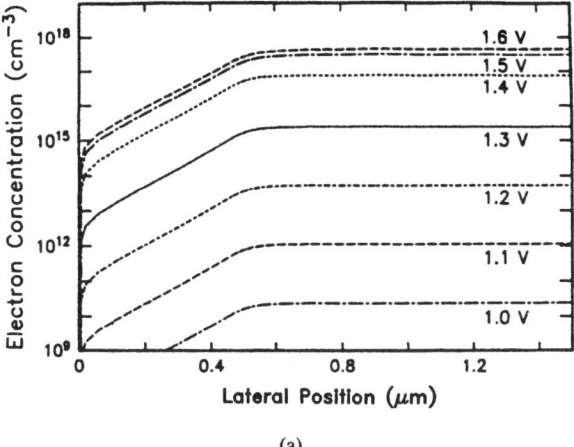

(a)

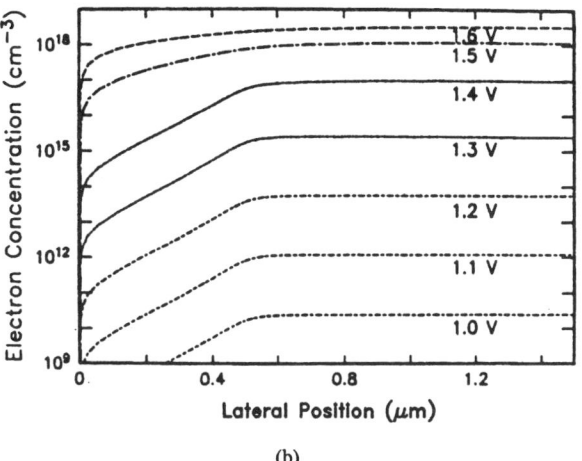

(b)

Fig. 10. Electron concentration along a cross section in the middle of the base, as a function of bias for (a) a single-heterojunction and (b) a double-heterojunction device. The figures differ at the highest biases where the extrinsic decay is much slower in the double-heterojunction device (~ diffusion length) than in the single-heterojunction device (~ base width).

controlled by the use of doping, as discussed below. Alternately, since a large portion of the barrier is due to the alloy potential, one could reduce the base charge storage by reducing the Al mole fraction u_f, and hence the alloy potential. Unfortunately, this would result in more hole injection. This tradeoff could be improved by finding and using a different semiconductor alloy system with a smaller alloy potential conduction band-to-valence band ratio ϕ_C^A / ϕ_V^A.

A comparison between the size of the alloy barrier effect and the size of the Kirk effect (electron space-charge barrier) can be obtained by considering the collector junction version of (4). If one assumes a base doping much higher than the collector doping, constant collector doping, parabolic grading, and an SCR thickness W greater than the grading length y_0, then one can approximately solve the collector version of (1) for N_D (rather than differentiating it as in (2)) to obtain [4].

$$N_D \geq \frac{\epsilon_s}{e} \frac{2\gamma u_f}{q y_0 W(N_D, V_{BC})} + \frac{J_C}{q\bar{v}}. \quad (5)$$

Here, $\bar{v} = \langle 1/v_y \rangle^{-1}$ is the harmonic mean of the carrier velocity in the collector, and W depends on the doping and the bias voltage. This is a more relaxed requirement than for the emitter in (4), especially when the depletion region extends substantially farther than the grading region into the collector. If one further assumes $J_C/\bar{v}$ is constant, one can evaluate W and solve (5) for N_D explicitly

$$N_D \geq \frac{\epsilon_s}{e} \frac{2\gamma^2 u_f^2}{q^2 y_0^2 \left(-V_{BC} + V_{Bi} + \dfrac{\gamma u_f}{q}\right)} + \frac{J_C}{q\bar{v}} \quad (6)$$

where V_{Bi} is the built-in potential. To illustrate the characteristics of this constraint, we consider a specific example. For a current density of $J_c = 1 \times 10^5$ A/cm^2, a mean velocity equal to the saturated velocity of 1×10^7 cm/s, $\gamma u_f = 0.24$ eV, $V_{BC} = 0$ V, $V_{Bi} = 1.4$ V, and a parabolic grading length $y_0 = 300$ Å, at least some barrier appears at all doping levels below ~ 1.2×10^{17} cm^{-3}. The second term of (6), which is 6.2×10^{16} cm^{-3}, is due to the electrostatic effect from the mobile charge (Kirk effect). The first term of (6) is due to the alloy grading and requires a compensation of at least 5.4×10^{16} cm^{-3} for the bias. Thus, the two effects, Kirk and alloy barrier, are approximately equal for this set of conditions. Since the doping requirement for a homojunction collector omits the first term in (6), it would be about a factor of two lower for these conditions. On the other hand, for a heterojunction collector, the doping required by the first term increases as the maximum forward collector-base bias at which the barrier must be suppressed increases, until, when $W \leq y_0$, (4) begins to apply.

Choosing the grading in the collector requires slightly different considerations than those in the base-emitter junction. In most high-speed circuit designs, the base-collector junction is not forward biased far enough that $W < y_0$ during operation. Hence, the electrostatic fields at and near the junction are larger than in the base-emitter junction, and the mobile charge moves faster than in the emitter. As a result, we obtain the lower doping requirement in this case. Thus, designing junctions that do not show barrier effects is easier at the collector than at the emitter. Linearly graded heterojunction collectors without alloy barriers are also possible, since their grading field constraint occurs at the collector edge of the grading region. In the same manner as (5) was obtained, one may obtain the following relation for a linearly graded region [4]:

$$N_D \geq \frac{\epsilon_s}{e} \frac{\gamma u_f}{q y_0 (W(N_D, V_{BC}) - y_0)} + \frac{J_C}{q\bar{v}}. \quad (7)$$

Solving for the doping typically yields only a limited range of N_D for which (7) can be satisfied. As the forward bias increases, that range decreases until it vanishes and the inequality can no longer be satisfied. Typically, also, the minimum N_D that satisfies (7) (when there is one) is somewhat larger than for parabolic grading.

B. Collector Electron Quasi-Electric Fields

In analyzing the electron quasi-electric fields in the collector, we first consider the homojunction collector HBT. Fig. 7 shows that in such a device the peak quasi-electric field at the base–collector junction decreases as the current density increases (due to the space charge of the electrons in the collector SCR), until finally base push-out occurs (the Kirk effect). Thus, at very high current density, the single heterojunction device has an aiding field at the collector junction due to electrostatics, followed by a decrease to low values because of the Kirk effect. Finally the field increases again near the subcollector where the donor density is higher. The push-out can also be seen in the hole concentration, which is shown in Fig. 11 for both the single and the double-heterojunction devices at 1.4 and 1.5-V base–emitter bias. In the double-heterojunction device there is no significant hole injection into the collector, but in the single-heterojunction device there is substantial hole injection at the higher current density (1.5 V) because of junction debiasing and the absence of a hole barrier.

Fig. 12 shows, in a figure similar to Fig. 7, the variation in the electron quasi-electric field for a double-heterojunction HBT. The collector junction behavior is very different from that of the single-heterojunction device. Here, at high base–emitter bias, the quasi-electric field goes through a reversal in direction, and there is no high field region at the subcollector. There are two major differences. First, the valence band alloy barrier prevents injection of holes into the collector, allowing a rapid decrease in electrostatic field at the junction in the presence of large electron charge. Second, the conduction band alloy field, together with the decrease in the electrostatic field, results in a retarding field at the base–collector junction, as seen in Fig. 12 for biases of 1.4 and 1.5 V. Unlike the situation in the homojunction collector, base push-out does not occur in this device, even at the highest current density, because hole injection is blocked and the electron alloy barrier diminishes the numbers of electrons entering the collector depletion region by a factor between 1.5 and 2.

C. Diffusion Capacitances

Fig. 10 showed the electron concentration in the base for a range of biases for both hetero- and homojunction collectors. As a further comparison, Fig. 13(a) and (b) shows the total stored electron density (in reciprocal square centimeters) and the base time constant (obtained by dividing total stored base charge by collector current). The stored charge in the base is substantially higher for the heterojunction collector case at biases exceeding 1.4 V (corresponding to a current density of 6×10^4 A/cm^2), and can be as much as a factor of two higher at the highest biases. The time constant associated with base charge storage in the heterojunction collector is $\sim$ 3 ps at low currents, decreases with increasing bias because of minute aiding drift fields (Webster effect), and then in-

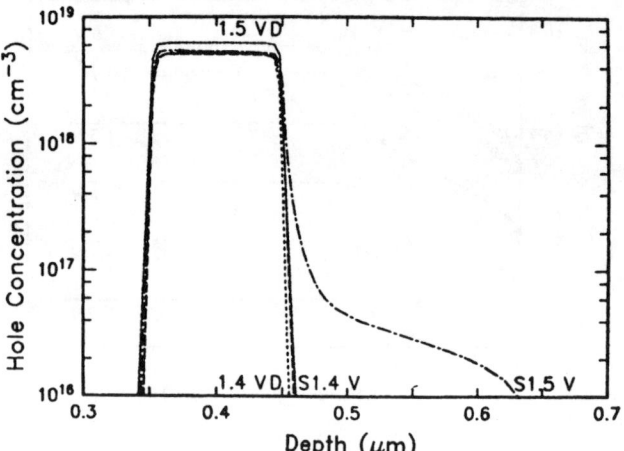

Fig. 11. Hole concentration in the base and into the emitter and collector region for applied biases of 1.4 and 1.5 V, for a single-heterojunction (S) and double-heterojunction (D) device. The base extends from 0.35 to 0.45 μm. The single-heterojunction HBT shows additional hole storage in the collector depletion region.

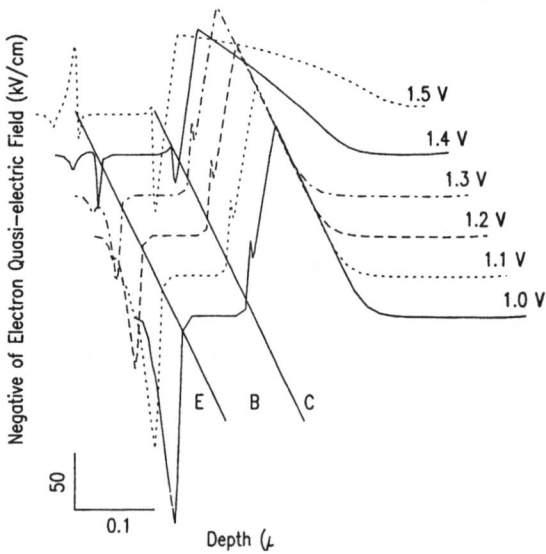

Fig. 12. Electron quasi-electric field for an electron at the base–emitter junction, in the base, and at the base–collector junction as a function of base–emitter bias for the double-heterojunction HBT. Field for a single-heterojunction HBT was plotted in Fig. 7 for identical conditions. The base is 0.1 μm wide (parameters in Table I), and the different plots are shifted to allow a perspective view. E, B, and C identify the emitter, base, and collector regions. The scales of distance and field are located in the lower left corner. The field in the base region is essentially zero in each plot.

creases to greater than 12 ps for extreme forward bias where current densities reach $\sim 1 \times 10^5$ A/cm^2. Under these conditions the base time constant is a significant portion of the total time constant of the device. In the homojunction collector device, the low and medium current behavior is similar to that in the heterojunction collector device but only increases to $\sim$ 6 ps at high currents ($\sim 4 \times 10^5$ A/cm^2) due to the Kirk effect.

Although the heterojunction collector leads to an increase in base storage, it continues to suppress hole injection. The homojunction collector does not have a hole alloy barrier, and hole injection into the collector is

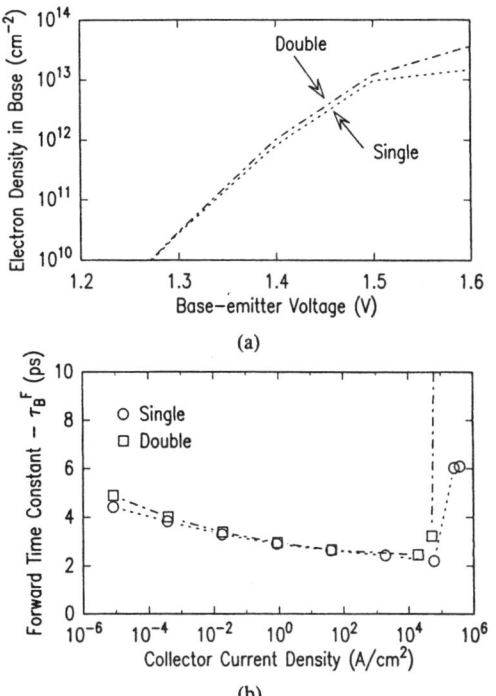

(a)

(b)

Fig. 13. (a) Total electron density in the base versus applied bias, and (b) the base time constant as a function of collector current density, for single- and double-heterojunction devices.

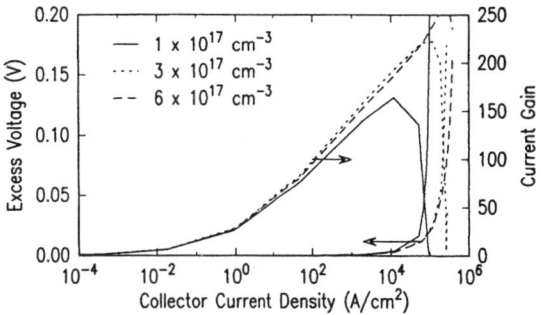

Fig. 14. Excess voltage and current gain derived from Gummel plot for a parabolically graded double-heterojunction transistor with an emitter doping of 8×10^{17} cm^{-3} and collector dopings of 1×10^{17} cm^{-3}, 3×10^{17} cm^{-3}, and 6×10^{17} cm^{-3}. This figure is analogous to Fig. 3(b). The rapid increase in excess voltage shows the onset of current saturation. It occurs together with a decrease in gain and an increase in storage in the quasi-neutral base.

important in these devices. Fig. 11 shows $\sim 4 \times 10^{11}$ cm^{-2} holes stored in the homojunction collector to sustain a current density of 2.5×10^5 A/cm^2 (an additional time constant of $\tau \sim 0.3$ ps). Thus, while the heterojunction collector bipolar has an increase in diffusion capacitance associated with electron storage in the base, the homojunction collector device has an increased capacitance associated with hole storage in the collector.

D. Current Gain Effects

Increased base charge storage reduces the gain of HBT's because of increased base recombination current. Thus, in addition to surface and bulk recombination effects and their voltage-dependent features, the HBT shows an anomalous decrease in gain at high current density when heterojunction collectors are used. This decrease is mostly due to quasi-neutral base recombination. However, the base charge storage also affects the transport of the carriers to the surface. Hence, surface recombination should also be slightly increased at high current densities by the presence of a heterojunction collector. In the absence of excess storage, the device gain continues to improve with current.

Fig. 14 shows the current gain and the excess voltage at several collector dopings in a form similar to Fig. 3(b). The decrease in gain at ~ 1.4 V for the 1×10^{17} cm^{-3} doped collector occurs because of increased base charge storage. Along with the increase in base current this implies, there is a saturation of the collector current, which is reflected in the rapid increase in the excess voltage as the current gain drops. This effect does not occur at 6×10^{17} cm^{-3} due to the suppression of the alloy barrier effect

[4]. The collector saturation current density is a factor of six higher in the higher doped collector structure.

V. GENERATION AND RECOMBINATION EFFECTS

Recombination is a more important parasitic effect in GaAs HBT's than in silicon bipolars because of the much shorter bulk recombination lifetime [13] and higher surface recombination velocity of GaAs. At moderate and low doping levels, the bulk recombination lifetime is dominated by deep trap recombination centers (centers that can capture electrons and holes nearly equally efficiently) and the lifetime can be as much as three orders of magnitude lower than in silicon at similar doping levels. At higher doping levels, the radiative lifetime may predominate—since GaAs is a direct bandgap material together with that due to Auger recombination—even reaching hundreds of picoseconds. In silicon the lifetime is also low at high doping levels because of Auger recombination. Surface recombination is higher for GaAs than for Si because the unpassivated GaAs surface contains a large number of interface states (10^{13} cm^{-2} and higher), leading to mid-gap pinning of the Fermi level at the surface, as well as recombination. As a result of the poor lifetime and surface pinning, the current–voltage behavior of p-n junctions and HBT's deviates substantially from the ideal due to recombination. Reference [2] has discussed the bulk recombination phenomena for abrupt junction devices. Reference [12] has discussed surface recombination at moderate surface recombination velocities, and [24] discusses the origin of ``kT''-like surface recombination current. We emphasize here the bulk and surface recombination behavior of graded junction devices at high surface recombination velocities.

The nonidealities in current–voltage behavior are introduced by recombination at SRH centers in the space-charge region of the p-n heterojunctions and at the surface interface states. Fig. 15 shows the relative recombination rates of these regions compared to the quasi-neutral recombination by plotting the volume recombination density in a realistic HBT whose parameters were summarized in Table I. This figure shows that the electrons

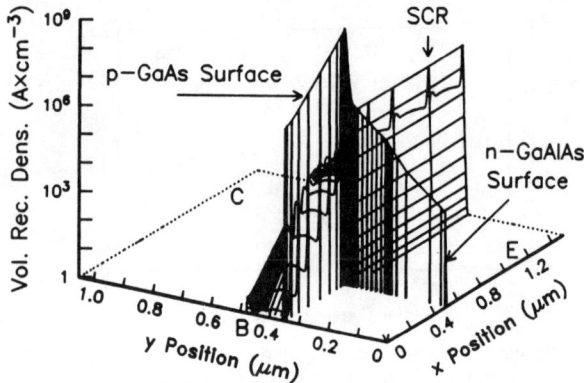

Fig. 15. Perspective plot of volume recombination current density for a surface recombination velocity S of 2×10^5 cm/s, and other parameters as specified in Table I. (Taken from [24]).

injected into the base recombine copiously at the surface of the GaAs base and at the base–emitter junction and that holes injected from the base recombine at the GaAlAs emitter surface. Of these, the base surface and the p-n junction recombination are the dominant components. The textbook component constituting the base transport factor—the quasi-neutral base recombination—is the smallest and is generally dominated by the surface recombination. The bulk recombination in the base–emitter SCR contributes to degrading the injection efficiency. The nonidealities corresponding to these recombinations appear in the form of multiple exponential regions in the forward characteristics, and as nonsaturating current in the reverse characteristics. The forward characteristics also have nonidealities due to the heterojunction injection phenomena, which we have considered, but these occur only at high currents. At low currents the generation-recombination effects dominate.

A. Bulk Effects

We first consider the bulk effect—the recombination taking place in the SCR and in the quasi-neutral region. The former has been studied extensively for homojunctions. Simple theories taking into account the SRH recombination in the junction [25] show that it results in a "$2kT$" exponential dependence. This dependence arises because recombination requires each of the carriers to surmount only half of the barrier on average, even though current flow across the junction requires the carriers to cross the whole barrier. Thus, the exponential dependence varies as $V/2$, which is "$2kT$." In practice, even for homojunction devices such recombination is rarely exactly "$2kT$" and has been shown [26] to result from junction asymmetry, asymmetric capture statistics, position-dependent trap distributions, and electric field dependence of capture statistics. This results in a recombination dependence that is "$2kT$"-like instead of being exactly "$2kT$." In heterojunctions, in addition to these effects, we have to consider the suppression of hole injection into the emitter, which is crucial to limiting SCR recombination. Homojunctions in GaAs have orders of magnitude higher SCR recombination than heterojunctions.

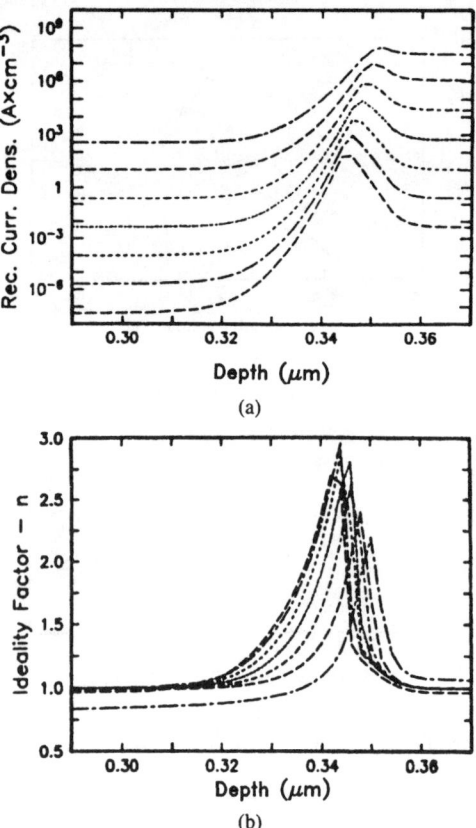

Fig. 16. (a) Recombination current density and (b) ideality factor of the recombination current density versus depth for biases varying from 0.8 to 1.4 V. The base–emitter junction is located at 0.35 μm.

Fig. 16(a) shows the volume recombination current density in the junction SCR for the parabolic graded junction, and Fig. 16(b) shows the ideality of this volume recombination current density as a function of position. The ideality is derived from the rate of exponential increase in the local volume recombination density, as determined by steady-state analysis of a small perturbation at the bias point. The junction grading extends from 0.32 to 0.35 μm. At these biases, the peak in the volume recombination density occurs in the junction alloy grading region, within 100 Å of the junction—a region where the effective barrier is changing with bias because of the varying bandgap. The ideality of the volume recombination current density peaks in this junction region but not at the same position as the peak of the volume recombination density. This is a result of the recombination process, which drives the system toward equilibrium (i.e., reducing the np product to n_i^2—a position-varying quantity). The np product does not peak at the junction itself in a heterojunction but close to it because of the suppression of holes. The ideality of the recombination current depends on the carrier transport to the position of recombination. Although the ideality approaches kT-like dependence outside the SCR, within the junction SCR it rapidly increases. It can exceed 2 because of the heterojunction effect and junction asymmetry. Note that at the highest biases in this figure the quasi-neutral base recombination nearly reaches the magnitude of the volume recombination density in the SCR, and has a faster exponential dependence. Thus, at sufficiently high biases the bulk recombination component will begin to dominate the SCR component, resulting in a "kT"-like dependence.

Generally, the bulk base transport factor only becomes truly important in determining the device gain in the high current regime of a double-heterojunction HBT. At low currents, the dominant factor is recombination in the SCR and at the surface, and in the medium current range it is mostly surface recombination.

B. Surface Effects

Surface effects have been a dominant source of gain degradation in GaAs HBT's and have not yet been studied extensively [10], [12], [24]. Experimental evidence shows that they give rise to a "$2kT$"-like dependence going towards a "kT"-like dependence at high-bias conditions [10], [27], [24]. The surface is a dominant recombination source in most common designs [28] due to a high rate of recombination through surface states and to Fermi-level pinning at the surface. The first by itself would give rise to recombination-dominated transport only in the low current regime, but Fermi-level pinning attracts more electrons to the surface, leading to much higher recombination. Surface recombination can be minimized for the operating current densities of interest by various techniques including the use of surface chemical passivation [29] or the use of a wide-gap material at the surface of the extrinsic base. This could be a p-type GaAlAs obtained by converting the polarity of the injecting emitter [10], or a depleted GaAlAs emitter region (by thinning of this material) [30]. (Minimum lateral dimensions for such layers are discussed in Section V-C.) The use of a graded base layer can also reduce the surface recombination [28]. In the following we analyze surface recombination in the absence of any of these techniques.

Henry et al. [18] studied this phenomenon both theoretically and experimentally. Their experiments on p-n junctions showed a "$2kT$" current that could not be explained as bulk recombination in the SCR. They showed theoretically that the presence of Fermi-level pinning leads to a "$2kT$" dependence for recombination current. The argument is in two parts. First, a high density of surface states (as required for surface pinning) causes the electron-to-hole ratio at the surface to remain constant and close to unity. Second, if there is essentially equilibrium between the surface and the bulk and only a small charge flow to the surface, such that the quasi-Fermi levels remain essentially flat, then the np product remains constant. Putting these two parts together yields a surface carrier concentration varying as the square root of the bulk minority-carrier density, $n_s \propto \sqrt{n_{\text{Bulk}} \times p_{\text{Bulk}}}$ and hence a "$2kT$" dependence for the surface recombination. At moderate recombination velocity, the rate-limiting process is the recombination velocity, and the assumptions used in deriving the "$2kT$" dependence are valid. At high recombination velocity, however, large deviations from the assumptions can occur. In particular, our numerical

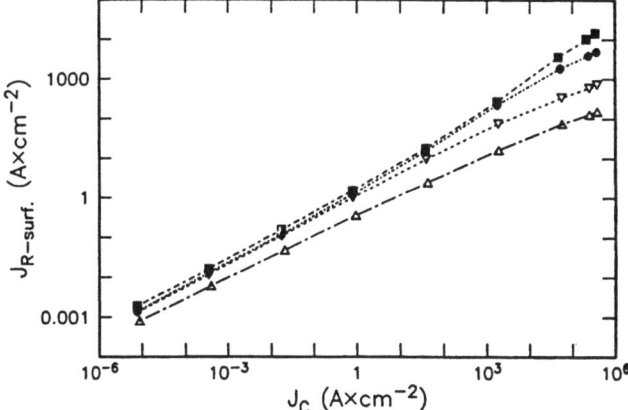

Fig. 17. Surface recombination current density (normalized to the emitter area) as a function of collector density, with S as a parameter varying from 2×10^3 cm/s^2 for the lowest curve to 2×10^6 cm/s for the highest curve in steps of a decade. The low-current behavior in this plot has a "$2kT$"-like dependence, and the high-current behavior for the higher surface recombination velocities has a "kT"-like dependence.

simulation of this problem in the limit of high surface recombination velocity shows that the electron quasi-Fermi level does not remain flat between the surface and the bulk. For example, at $S = 2 \times 10^6$ cm/s the quasi-Fermi level bends by more than 200 meV near the injecting junction for a bias of 1.2 V.

We have numerically analyzed the surface recombination characteristics of a variety of cases using the single acceptor/donor model described in Section II. Fig. 17 shows the total surface recombination current density as a function of device current density for moderate to high surface recombination velocity. At low currents, the surface recombination current density follows a "$1.8kT$" behavior. The higher current behavior is a function of S. At high S, the ideality shows a deviation toward a "$1.2kT$" dependence. Note that the high S characteristics are in general agreement with the previously mentioned experimental observations, while the lower S characteristics are in general agreement with the simple analytic theory, as they should be. The high S behavior occurs because recombination depends on the rate at which carriers are provided to the surface, and this depends on device geometry, surface conditions, and the design of the injecting junction. The Fermi level of electrons is no longer flat between the surface and the bulk—and there is no simple relationship between n_s and n_{Bulk} and p_{Bulk}.

To better understand the high S behavior, we have studied how carriers reach the surface to recombine. Fig. 18 plots the electron and hole current flow lines in the vicinity of the extrinsic base, showing the paths carriers follow to get to the surface. Note that the electron current to the surface is almost entirely due to injection of carriers into a surface channel at the base–emitter junction. This surface electron channel is caused by the surface Fermi-level pinning. Only a very small flux of electrons into this channel from the quasi-neutral base region is observed in our modeling—and it occurs at the intersection of the surface with the base–emitter junction—where two-dimen-

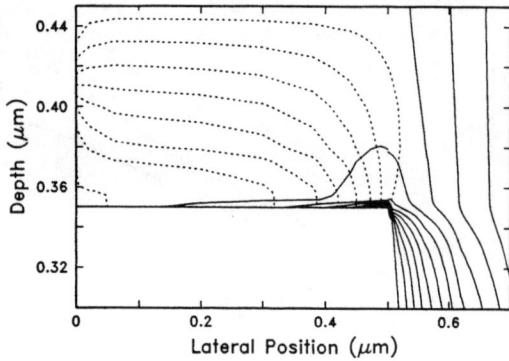

Fig. 18. Current stream lines in an HBT, showing the way in which carriers reach the surface of the extrinsic base. The solid lines are electron flow lines, and the dashed lines are hole flow lines. The base-collector junction is at the top of the figure, and the emitter–base junction is at a depth of 0.35 μm. Surface recombination occurs along the exposed surface of the extrinsic base, which stretches between 0.0 and 0.5 μm laterally and is at the same depth, 0.35 μm. The base contact is at the left side of the figure where the hole stream lines converge, while the emitter and collector contacts are outside the range of the figure.

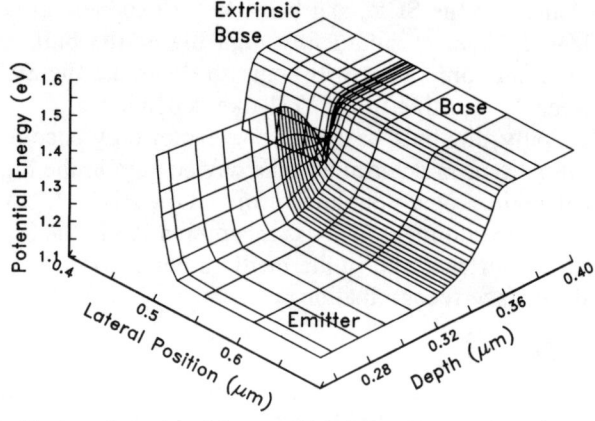

Fig. 19. A surface plot of the negative of the conduction band energy at the intersection of the base–emitter junction with the base surface, showing the saddle point that allows easier flow of the injected electrons into the surface depletion region. Note that this view is rotated about 90 degrees clockwise relative to the views in Figs. 8, 9, and 15.

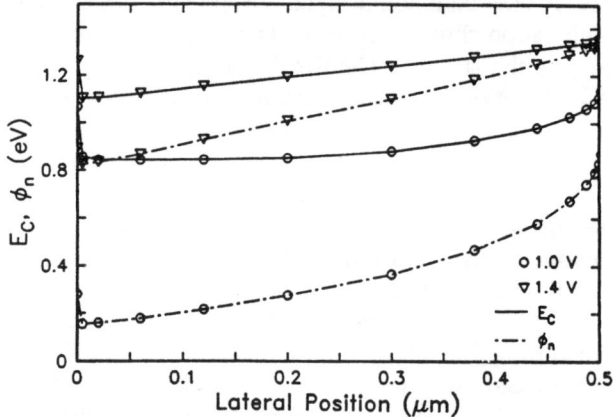

Fig. 20. Conduction band edge energy and electron quasi-Fermi level along the base surface, at 1.0- and 1.4-V forward bias, with the base–emitter junction intersecting at 0.5 μm on the horizontal scale.

sional effects are strong. Note also that the hole current for recombination is mostly perpendicular to the surface and originates in the quasi-neutral base region.

Because of its close proximity to the surface, this electron injection phenomenon is a function of the chosen boundary conditions. Entry into the channel is through a saddle point in the conduction band edge potential, as shown in Fig. 19. This complex barrier shape is the result of the surface interface state distribution that caused the Fermi-level pinning, the choice of bandgap grading, and the choice of bias. This barrier causes, at low bias conditions, a narrow constricted region through which it is energetically favorable for the carriers to stream into the surface channel. At high S, this constriction is the rate-limiting step in the surface recombination, and the magnitude of current transported to the surface is determined by the barrier height and the cross section of this saddle point, which are nonlinearly dependent on the bias because of the changing junction depletion thickness.

Once the carriers reach the surface, the SRH statistics determine the spatial dependence of the recombination of these carriers. At high surface recombination velocities, these carriers recombine rapidly in a very short region. At moderate surface recombination velocities, many carriers drift-diffuse down the channel before they recombine. Fig. 20 shows the conduction band edge energy and the electron quasi-Fermi level as a function of position along the surface at two bias conditions: 10 and 1.4 V. At the lower bias, further away from the intersection of the surface with the base–emitter junction (at $x = 0.5$ μm), the electron current is mostly diffusive, with drift only becoming important close to the junction. At the higher bias, however, drift and diffusion contribute approximately equally to the transport process. The rapid fall-off of the Fermi level reflects the rapid depletion of electrons by recombination.

Just as the electron concentration rapidly diminishes with distance from the junction, so also the surface recombination density is highest at the injecting point and

rapidly decreases thereafter. The length scale of this fall off is ~ 1250 Å at $S = 2 \times 10^5$ cm/s and ~ 500 Å at $S = 2 \times 10^6$ cm/s, giving between a square and cube root dependence on S. This is the approximate length of the thinned wide-gap emitter region [30] necessary to suppress surface recombination. The effective ideality of the surface recombination current density (determined as for Fig. 16) depends on recombination velocity, bias, and (weakly) position. At $S = 2 \times 10^5$ cm/s, the ideality varies from ~ 1.7 at 0.95 V forward bias to ~ 1.3 at 1.25 V bias. At $S = 2 \times 10^6$ cm/s, the idealities behave similarly, but are somewhat lower. (A figure similar to Fig. 16, showing the surface recombination density and its effective ideality as a function of position, can be found in [24].)

These characteristics of the ideality and the magnitude of the surface recombination current depend not only on the value of S, but also on the geometry, surface conditions, and the design of the device because of the two-dimensional nature of the phenomenon. Thus, surface recombination is quite variable. Thick bases allow a larger flux of carriers to the surface, quasi-fields which pull the carriers away from the junction yield a smaller flux of carriers, and abrupt junctions cause a larger voltage barrier at the surface and hence result in a smaller flux of carriers. The total surface recombination current is proportional to the flux of electrons; hence, these devices show different surface recombination.

C. Current Gain Behavior

Current gain behavior in the absence and presence of surface recombination is plotted in Fig. 21(a) and (b) for single heterojunction, graded-base (an electron quasi-field of 10 kV/cm created by grading the aluminum mole fraction to GaAs at the collector), and double-heterojunction bipolar transistors. In the absence of surface recombination, Fig. 21(a) shows, at large current densities, the largest current gains in the graded-base device followed by the single-heterojunction device and then the double-heterojunction device. The larger current gain in the graded-base device is due to the aiding quasi-electric field, which reduces the storage of electrons in the base and, hence, the neutral base recombination. The double-heterojunction device shows the drop in gain at high currents due to the increased base storage, as discussed previously. Although the graded-base device has a lower barrier-to-hole injection into the emitter, its gain actually continues to be comparable to that of the other devices at low current densities. The increased hole injection does not result in increased recombination in the base–emitter SCR because it is compensated by a decrease in electron density due to the quasi-electric field in the neutral base. Fig. 21(b) shows the behavior of these devices with surface recombination for a perimeter-to-area ratio of 1×10^4 cm^{-1}. The general trends of Fig. 21(a) still apply, with the largest gains at the highest currents in the graded-base device and a decrease in gain of the double-heterojunction device due to excess storage at high currents. The current gain of the graded-base device is substantially higher than the single-heterojunction device because the quasi-drift field results in fewer electrons to inject into the surface recombination region.

The dependence of current gain on perimeter-to-area ratio—the device size effect—is shown in Fig. 22(a) and (b) at biases of 1.5 and 1.4 V, respectively, for the graded-base and single-heterojunction device. The current flowing through the graded base device is $\sim 10 \times$ smaller at 1.4 V, and $2 \times$ smaller at 1.5 V due to the different injection characteristics of the two devices. At 1.4 V, where current saturation effects are not large, the graded device (which is operating at a lower current density) shows less sensitivity to surface recombination velocity. The current gain behavior is in good agreement with the published current gains of large-area (small perimeter-to-area ratio) and small-area (large perimeter-to-area ratio) devices without surface passivation. For small digital devices, with perimeter-to-area ratios of $\sim 2 \times 10^4$ cm^{-1}, the figure indicates that current gains should be in the range of 10 to 40. Using a one-dimensional model, the highest current gain in these structures, if it were determined only by neutral base recombination, would be ~ 380 ($D_n \sim$

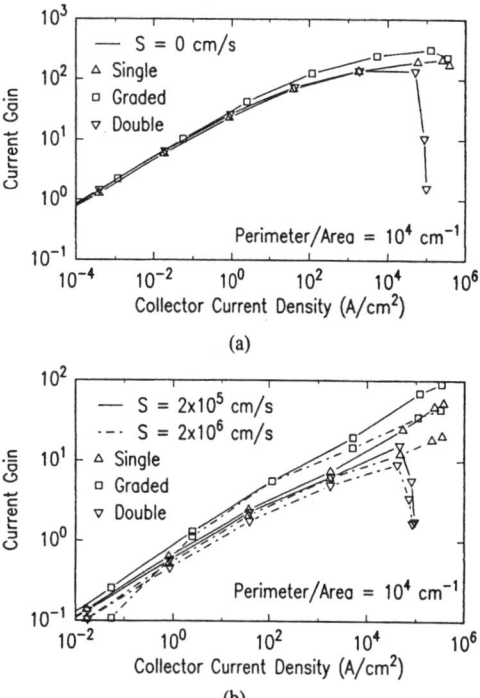

(a)

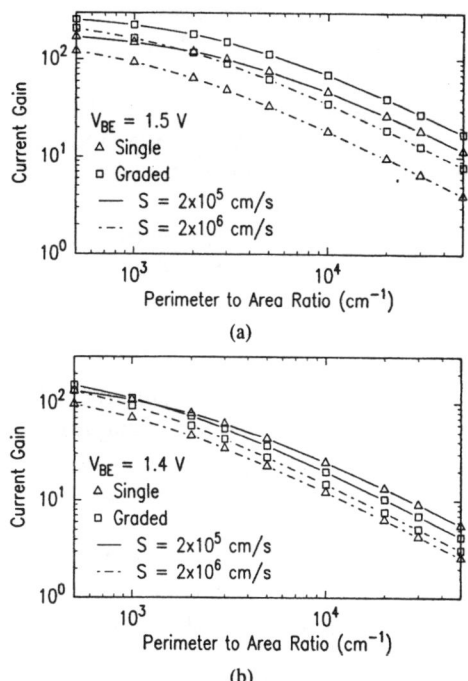

(b)

Fig. 21. (a) Current gain dependence in the absence of surface recombination, for single-heterojunction, graded-base, and double-heterojunction bipolar transistors. The perimeter-to-area ratio is 10^4 cm^{-1}. (b) Gain for the same devices at $S = 2 \times 10^5$ cm/s and $S = 2 \times 10^6$ cm/s.

Fig. 22. Current gain plotted as a function of perimeter-to-area ratio for $S = 2 \times 10^5$ cm/s and $S = 2 \times 10^6$ cm/s. (a) is for a base–emitter forward bias of 1.5 V and (b) is for 1.4 V.

38 cm^2/s, $\tau_n \sim 0.5$ ns, and hence the diffusion length $L_n \sim 1.38$ μm). The current gains in Fig. 21(a) approach this value but are lower and depend on device size because the full two-dimensional base transport factor includes recombination in the extrinsic region. The lateral

extent of the extrinsic base that must be included is of the order of the base width in a single-heterojunction device and the diffusion length in a double-heterojunction device. Hence, the gain of HBT's shows perimeter dependence even in the absence of surface recombination.

VI. Device Design Overview

Having viewed the behavior and effect of grading, doping, current drive, recombination, forward base time constants, etc., we now consider design strategy for an HBT in a compound semiconductor system that employs surface passivation. Parabolic grading has been shown to give better current drive capability than linear grading and hence is the preferred technique for the base–emitter junction. Although SCR recombination is lower for linear grading [3], quasi-neutral recombination dominates at high current densities for all grading types anyway, and hence this is of secondary concern. The absence of doping in the grading regions gives lower SCR recombination, but for the same reason as in linear grading this is also not a preferable design technique. For heterojunction collector devices, linear collector alloy grading is attractive because it reduces the alloy field near the junction compared to parabolic grading, and the larger space-charge width allows a design wherein the barrier at the end of the linear grading can be suppressed by a sufficiently large electrostatic field. For both junctions it is important to use dopings that are high enough to avoid both the conventional bipolar effects and the HBT alloy barrier.

We emphasize the digital aspects of the design—the high-frequency aspects follow in a similar way. Digital operation requires high current drive, low capacitances and time constants, and sufficiently large current gains. HBT scaling therefore differs in two important respects from scaling of homojunction silicon transistors. First, the device design must compensate for the barrier effects. This requires the emitter and the collector doping to increase with current density. In silicon bipolars, only the collector doping must be increased; the emitter is already highly doped to maintain good injection efficiency. Second, the gain of the device is maintained by suppressing both the surface and the SCR recombination and by ensuring adequate base transport factor. The surface recombination can be minimized using the techniques described in Section V-B. The SCR recombination is sufficiently small at current densities of interest, as we have seen, and remains so even with increasing emitter and base dopings.

Maintaining adequate base transport factor at high current densities requires considering the dependence of lifetime on doping. While injection efficiency is a constraint in silicon bipolars, the base transport factor is a constraint in GaAs. The base doping can be much higher than in the homojunction bipolar. At base dopings of interest ($N_B > 10^{18}$ cm^{-3}), the lifetime varies inversely with the doping ($\tau \propto N_B^{-1}$) [31]. Ignoring two-dimensional effects, the gain varies as $2L_n^2/W_B^2$, where L_n is the electron diffusion length and W_B is the base width. Thus, current gain $\beta \propto D_n/N_B W_B^2$, where D_n is the diffusion coefficient. The base

TABLE II
VERTICAL SCALING FOR HORIZONTAL SCALING OF λ

Current Density (J)	$1/\lambda^2$
Emitter Doping (N_E)	$J \sim 1/\lambda^2$
Emitter Thickness (W_E)	λ
Base Thickness (W_B)	λ
Base Doping (N_B)	$1/\lambda$
Collector Doping (N_E)	$J \sim 1/\lambda^2$
Collector Thickness (W_C)	λ

time constant τ_B varies approximately as $W_B^2/2D_n$. D_n varies inversely with the doping N_B because of ionized impurity scattering. (The data on this are scarce, but in the regime dominated by ionized impurity scattering, they vary for majority carriers as $N_B^{-0.5}$ [23, p. 29] to $N_B^{-1.0}$).

A self-consistent method for scaling the base time constant and maintaining gain requires varying the base width as the scaling factor λ, while varying the base doping between $1/\lambda^{1.33}$ and $1/\lambda$, depending on the dependence of D_n on N_B. This allows the base time constant to vary between $\lambda^{1.33}$ and λ (slightly faster than the scaling factor) while maintaining the current gain in device structure. This increase in base doping is not as rapid as in silicon bipolars where the doping levels are lower. For constant voltage swing (usually chosen as the minimum that satisfies noise and power constraints), and constant power supply voltages (determined by the bandgap of the base material and the logic swing), the currents remain constant and the current density increases as λ^{-2}. The collector and the emitter dopings also increase at this rate. This results in the total delay due to intrinsic components scaling as λ. The scaling factors of the important epitaxial parameters are summarized in Table II.

The minimum thicknesses of the emitter and collector regions depend on the doping, the junction voltage swing, and the minimization of the parasitic resistance. For the collector, the breakdown voltage must also be considered, and for the emitter, tunneling currents at low bias due to degenerate doping in the emitter and the base must be taken into account. While DX centers play a negligible role in the operation of HBT's with regard to transient effects in current transport, capacitance [33], and recombination current [34], their inefficient ionization leads to larger emitter resistance, which may become important at current densities about 10^5 A/cm^2. Thus, although the HBT has generally employed mole fractions close to 0.3, scaled submicrometer devices operating at high current densities may require lower aluminum mole fractions.

VII. Conclusions

This paper has employed two-dimensional drift-diffusion modeling to study some of the important properties of HBT's—transport across heterojunctions, the effect of transport characteristics on storage, and the recombination mechanisms. Using the qualitative understanding and the design principles derived here, it should be possible to make preliminary designs for HBT's that can operate at specified current densities, with minimal capacitances, resistances, and recombination effects. Although the operation and design of HBT's have been discussed here solely in the context of the GaAlAs/GaAs material system, the extension of these principles to other materials systems should be straightforward.

REFERENCES

[1] J. Yoshida, M. Kurata, K. Morizuka, and A. Hojo, "Emitter-base bandgap grading effects on GaAlAs/GaAs heterojunction bipolar transistor characteristics," *IEEE Trans. Electron Devices*, vol. ED-32, no. 9, p. 1714, 1985.

[2] A. Das and M. S. Lundstrom, "Numerical study of emitter-base for AlGaAs/GaAs heterojunction bipolar transistors," *IEEE Trans. Electron Devices*, vol. 35, no. 7, p. 863, 1988.

[3] S. Tiwari and D. J. Frank, "Barrier and recombination effects in the base-emitter junction of heterostructure bipolar transistors," *Appl. Phys. Lett.*, vol. 52, no. 12, p. 993, 1988.

[4] S. Tiwari, "A new effect in heterostructure bipolar transistors," *IEEE Electron Device Lett.*, vol. 9, no. 3, p. 142, 1988.

[5] H. Kroemer, "Heterostructure bipolar transistors and integrated circuits," *Proc. IEEE*, vol. 70, no. 1, p. 30, 1982.

[6] C. M. Maziar, M. E. Klausmeier-Brown, and M. S. Lundstrom, "A proposed structure for collector transit time reduction in AlGaAs/GaAs bipolar transistors," *IEEE Electron Device Lett.*, vol. EDL-7, no. 8, p. 483, 1986.

[7] Y. Yamauchi and T. Ishibashi, "Electron velocity overshoot in the collector depletion layer of AlGaAs/GaAs HBT's," *IEEE Electron Device Lett.*, vol. EDL-7, no. 12, p. 655, 1986.

[8] R. Katoh, M. Kurata, and J. Yoshida, "A self-consistent particle simulation for (AlGa)As/GaAs HBTs with improved base-collector structures," in *IEDM Tech. Dig.*, 1987, p. 248.

[9] K. Morizuka et al., "Transit-time reduction in AlGaAs/GaAs HBT's utilizing velocity overshoot in the p-type collector region," *IEEE Electron Device Lett.*, vol. 9, no. 11, p. 585, 1988.

[10] S. Tiwari, "GaAlAs/GaAs heterostructure bipolar transistors: experiment and theory," in *IEDM Tech. Dig.*, 1986, p. 262.

[11] P. M. Asbeck et al., "Heterojunction bipolar transistors for microwave and millimeter-wave integrated circuits," *IEEE Trans. Electron Devices*, vol. ED-34, no. 12, p. 2571, 1987.

[12] Y. S. Hiraoka and J. Yoshida, "Two-dimensional analysis of the surface recombination effect on current gain for GaAlAs/GaAs HBT's," *IEEE Trans. Electron Devices*, vol. 35, no. 7, p. 857, 1988.

[13] Y. S. Hiroaka, J. Yoshida, and M. Azuma, "Two dimensional analysis of emitter-size effect on current gain for GaAlAs/GaAs HBT's," *IEEE Trans. Electron Devices*, vol. ED-34, no. 4, p. 721, 1987.

[14] J. Y.-F. Tang and S. E. Laux, "MONTE—A program to simulate the heterojunction devices in two dimensions," *IEEE Trans. Computer-Aided Design*, vol. CAD-5, no. 4, p. 645, 1986.

[15] H. C. Casey and F. Stern, "Concentration-dependent absorption and spontaneous emission of heavily doped GaAs," *J. Appl. Phys.*, vol. 47, no. 2, p. 631, 1976.

[16] J. Y.-F. Tang and J. L. Freeouf, "Non-uniform surface potentials and their observation by surface sensitive techniques," *J. Vac. Sci. Technol.*, vol. B2, p. 459, 1984.

[17] J. P. McKelvey, *Solid State and Semiconductor Physics.* Melbourne: R. E. Krieger, 1986, p. 350.

[18] C. H. Henry, R. A. Logan, and F. R. Merritt, "The effect of surface recombination on current in AlGaAs heterojunctions," *J. Appl. Phys.*, vol. 49, no. 6, p. 3530, 1978.

[19] A. van der Ziel, *Solid State Physical Electronics.* Englewood Cliffs, NJ: Prentice-Hall, 1971, p. 273.

[20] S. E. Laux, "Techniques for small-signal analysis of semiconductor devices," *IEEE Trans. Electron Devices*, vol. ED-32, no. 10, p. 2028, 1985.

[21] H. C. Poon and H. K. Gummel, "Modeling of emitter capacitance," *Proc. IEEE*, vol. 57, no. 12, p. 2181, 1969.

[22] H. C. Poon, H. K. Gummel, and D. L. Scharfetter, "High injection in epitaxial transistors," *IEEE Trans. Electron Devices*, vol. ED-16, no. 5, p. 455, 1969.

[23] S. M. Sze, *Physics of Semiconductor Devices.* New York: Wiley, 1981, p. 147.

[24] S. Tiwari, D. J. Frank, and S. L. Wright, "Surface recombination in GaAlAs/GaAs heterostructure bipolar transistors," *J. Appl Phys.*, vol. 64, no. 10, p. 5009, 1988.

[25] C.-T. Sah, R. N. Noyce, and W. Shockley, "Carrier generation and recombination in p-n junctions and p-n junction characteristics," *Proc. IRE*, vol. 45, no. 9, p. 1228, 1957.

[26] K. Lee and A. Nussbaum, "The influence of traps on the generation-recombination current in silicon diodes," *Solid-State Electron.*, vol. 23, p. 655, 1980.

[27] S. Tiwari and S. L. Wright, "Symmetric-gain, zero-offset, self-aligned, and refractory-contact double HBTs," *IEEE Electron Device Lett.*, vol. EDL-8, no. 9, p. 417, 1987.

[28] H. Ito, T. Ishibashi, and T. Sugeta, "Current gain enhancement in graded base AlGaAs/GaAs HBT's associated with electron drift motion," *Japan. J. Appl. Phys.*, vol. 24, no. 4, p. L241, 1985.

[29] C. J. Sandroff, R. N. Nottenburg, J.-C. Bischoff, and R. Bhat, "Dramatic enhancement in the gain of a GaAs/AlGaAs heterostructure bipolar transistor by surface chemical passivation," *Appl. Phys. Lett.*, vol. 51, no. 1, p. 33, 1987.

[30] H. H. Lin and S. C. Lee, "Super-gain AlGaAs/GaAs heterojunction bipolar transistors using an emitter edge thinning design," *Appl. Phys. Lett.*, vol. 47, p. 839, 1985.

[31] H. C. Casey and M. B. Panish, *Heterostructure Laser—Part A.* New York: Academic, 1978, p. 161.

[32] D. D. Tang and P. Solomon, "Bipolar transistor design for optimized power-delay logic circuits," *IEEE J. Solid-State Circuits*, vol. SC-14, no. 4, p. 679, 1979.

[33] M. I. Nathan, S. Tiwari, P. M. Mooney, and S. L. Wright, "DX centers in AlGaAs p-n heterojunctions and heterojunction bipolar transistors," *J. Appl. Phys.*, vol. 62, no. 8, p. 3234, 1987.

[34] S. Tiwari, S. L. Wright, and A. Kleinsasser, "Transport and related properties of (Ga,Al)As/GaAs double heterostructure bipolar junction transistors," *IEEE Trans. Electron Devices*, vol. ED-34, no. 2, p. 185, 1987.

Subpicosecond InP/InGaAs Heterostructure Bipolar Transistors

YOUNG-KAI CHEN, MEMBER, IEEE, RICHARD N. NOTTENBURG, MEMBER, IEEE,
MORTON B. PANISH, SENIOR MEMBER, IEEE, R. A. HAMM, AND
D. A. HUMPHREY

Abstract—We demonstrate the first bipolar transistors with subpicosecond extrinsic delay. These InP/InGaAs heterostructure transistors show a unity current gain cutoff frequency f_T = 165 GHz and maximum oscillation frequency f_{MAX} = 100 GHz at room temperature. Our model shows that an f_T beyond 386 GHz is obtainable by further vertical scaling. Ring oscillators implemented with nonthreshold logic (NTL) and transistors having f_{MAX} = 71 GHz show a propagation delay of 14.7 ps and 5.4 mW average power consumption per stage.

BIPOLAR transistors are one of the dominant semiconductor devices for today's high-speed integrated circuits because of their high cutoff frequency, exceptional threshold voltage control, large transconductance, and high current driving capability for interconnection parasitics [1], [2]. Interest in InP-based heterostructure bipolar transistors (HBT's) has increased because of the excellent transport properties of InGaAs. The use of an InP/InGaAs heterostructure emitter permits ballistic injection into InGaAs [3]. Good lateral scaling has been demonstrated in submicrometer InP/InGaAs HBT's because of the low surface recombination velocity and high electron velocity ($v_e > 5 \times 10^7$ cm/s) in the thin InGaAs base [4]. Recently, extremely high doping (p > 10^{20} cm^{-3}) has been achieved in InGaAs using gas source molecular beam epitaxy (GSMBE) [5]. The high base doping in these heterostructures eliminates current crowding in the thin base layer while maintaining high current gain. Because of the large conduction-band intervalley separations (Γ–L and Γ–X) in InGaAs and ballistic injection, transistors with intrinsic transit time less than 0.5 ps are realized [6]. In this letter, we report the first bipolar transistor with subpicosecond extrinsic delay. Ring oscillators using nonthreshold logic (NTL) are illustrated with a propagation delay as short as 14.7 ps.

The heterostructure was grown lattice matched to Fe-doped semi-insulating ⟨100⟩ InP substrates by GSMBE in the following sequence: 2500-Å InGaAs subcollector (n = 3 × 10^{19} cm^{-3}), 3000-Å InGaAs collector (n = 2 × 10^{16} cm^{-3}), 500-Å InGaAs base (p = 1 × 10^{20} cm^{-3}), 2500-Å InP emitter (N = 1 × 10^{18} cm^{-3}), 500-Å InP (N = 7 × 10^{19} cm^{-3}), and 2000-Å InGaAs emitter cap (n = 7 × 10^{19} cm^{-3}). The interface between the InGaAs cap layer and the

Manuscript received February 17, 1989; revised March 28, 1989.
The authors are with AT&T Bell Laboratories, Murray Hill, NJ 07974.
IEEE Log Number 8928340.

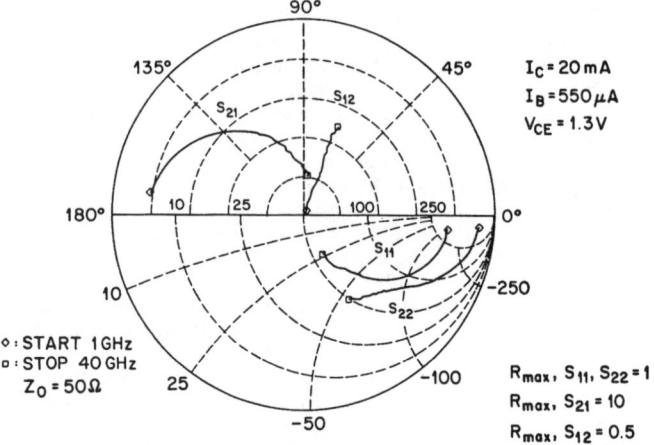

Fig. 1. Combined Smith-chart (s11 and s22) and polar (s21 and s12) graph of s-parameters from 1 to 40 GHz for a 3.6 × 3.6-μm^2 HBT at I_B = 550 μA, I_C = 20 mA, and V_{CE} = 1.3 V.

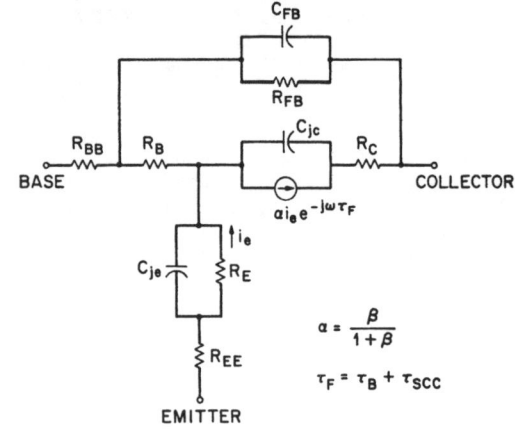

Fig. 2. Small-signal equivalent circuit model of the InP/InGaAs HBT.

InP emitter is degenerately doped to reduce emitter series resistance arising from the conduction-band discontinuity. HBT's with emitter and collector area of 3.6 × 3.6 and 7 × 11 μm^2, respectively, were fabricated. Thin-film nichrome resistors were used in the integrated circuits.

Scattering parameters were measured with an HP8510B vector network analyzer and millimeter-wave wafer probes from 1 to 40 GHz. The measured s-parameters for a typical HBT in a combined polar and Smith-chart graph are shown in Fig. 1. The small-signal equivalent circuit model of this device is shown in Fig. 2 with its element values listed in Table I(I).

Reprinted from *IEEE Electron Device Lett.*, vol. 10, no. 6, pp. 267–269, June 1989.

TABLE I
ELEMENT VALUES OF THE SMALL-SIGNAL EQUIVALENT CIRCUIT MODEL
Column (I): The HBT with 3000-Å collector depletion region (Z_C) in Fig. 1.
Column (II): Proposed HBT with $Z_C = 1500$ Å.

	(I)	(II)
Z_C(Å)	3000	1500
R_{BB} (Ω)	1.12	1.12
R_{EE} (Ω)	6.19	1.0
R_{FB} (Ω)	11500	11500
R_C (Ω)	2.26	1.0
R_B (Ω)	122.5	122.5
R_E (Ω)	1.39	0.46
C_{FB} (fF)	18.0	36.0
C_{jc} (fF)	1.0	2.0
C_{je} (fF)	200	346
β	44.5	44.5
τ_F (ps)	0.5	0.159
τ_e (ps)	0.278	0.159
τ_c (ps)	0.187	0.094
τ_{ec} (ps)	0.965	0.412
f_i (THz)	0.318	1.0
f_β (GHz)	3.9	8.7
f_T (GHz)	165	386
f_{MAX} (GHz)	100	200

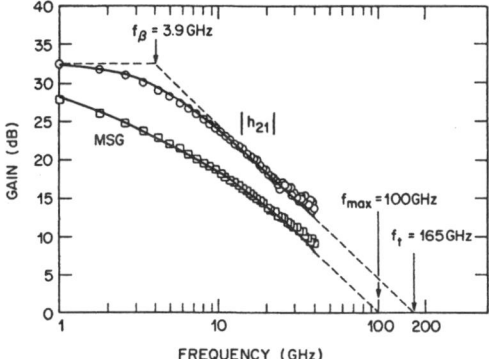

Fig. 3. Frequency dependence of the measured (points) and modeled (solid lines) current gain $|h_{21}|$ and MSG with the same bias as Fig. 1. $f_T = 165$ GHz and $f_{MAX} = 100$ GHz can be extrapolated with a -20 dB/decade slope (broken lines). The 3-dB current gain roll-off frequency f_β is 3.9 GHz.

The frequency dependence of the measured and modeled current gain $|h_{21}|$ and maximum stable power gain (MSG) are plotted in Fig. 3 at a collector current $I_C = 20$ mA and collector–emitter voltage $V_{CE} = 1.3$ V. A maximum oscillation frequency f_{MAX} of 100 GHz is obtained by extrapolating MSG to unity with -20 dB/decade slope. In our transistors, f_{MAX} is limited by extrinsic base series resistance and parasitic collector junction capacitance from the mesa structure. A unity current gain cutoff frequency f_T of 165 GHz is also extrapolated with -20 dB/decade slope from the measured h_{21}. The extrapolated f_T agrees with the product of the extrinsic low-frequency gain $|h_{21}|$ of 42.2 and the 3-dB frequency f_β of 3.9

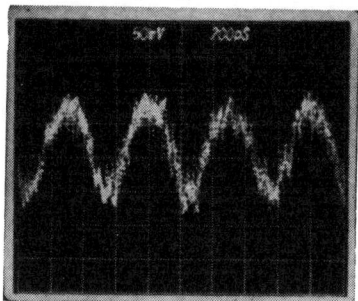

Fig. 4. Output waveform of a 17-stage NTL ring oscillator with a 50-Ω terminator. A 14.7-ps propagation delay with 5.4-mW power consumption per gate is measured.

GHz as depicted in Fig. 3 by the broken line. The emitter-collector delay time τ_{ec} of 0.97 ps is the shortest extrinsic delay time of any bipolar device reported to date.

It is well known that τ_{ec} can be expressed by

$$\tau_{ec} = R_E C_{je} + (R_E + R_{EE} + R_C)(C_{jc} + C_{FB}) + (\tau_B + \tau_{SCC}) \quad (1)$$

where $R_E = nkT/qI_E$ is the dynamic emitter junction resistance, n is the emitter junction ideality factor, R_{EE} and R_C are the emitter and collector series resistances, C_{je} and C_{jc} are the emitter and collector junction capacitances, C_{FB} is the parasitic collector junction capacitance, and τ_B and τ_{SCC} are the transit time in the base and collector depletion region, respectively. Because of the high emitter doping (10^{18} cm^{-3}) and emitter current density $J_E = 1.5 \times 10^5$ A/cm^2, a small emitter charging time $\tau_e = R_E C_{je}$ of 278 fs is realized. The collector charging time $\tau_c = (R_E + R_{EE} + R_C)(C_{jc} + C_{FB})$ is 187 fs. In these transistors, the dominant delay is the time for the injected electrons to traverse the neutral base and collector depletion region. The intrinsic transit time of 0.5 ps corresponds to an average electron velocity of 3.5×10^7 cm/s at a collector–base voltage $V_{CB} = 0.4$ V. This velocity agrees well with the Monte Carlo study of ballistic transport in InGaAs [7], and the base transit time τ_B in our structure may be very short (< 100 fs). However, a more detailed model is necessary to separate τ_B and τ_{SCC}.

Without any further lateral scaling, a reduction in the collector depletion width to 1500 Å would result in an intrinsic cutoff frequency $f_i = 1/2\pi\tau_F$ of 1 THz. This assumes an average electron velocity of 7×10^7 cm/s from [7]. By increasing emitter doping to allow a higher emitter current density $J_E = 4.5 \times 10^5$ A/cm^2 and reducing both R_{EE} and R_C to 1 Ω, an extrinsic f_T of 386 GHz may be realized with a collector depletion width of 1500 Å as tabulated in Table I(II). From Table I, both f_T and f_{MAX} can be further increased by reducing the extrinsic collector feedback capacitance C_{FB} [8].

Fig. 4 shows the output waveform of a 17-stage NTL ring oscillator. Each inverter consists of one switching transistor and a Ni–Cr load resistor. A propagation delay t_d of 14.7 ps per stage is measured with an averaged dc power consumption of 5.4 mW at a supply voltage $V_{CC} = 1.4$ V. The circuit is implemented with transistors having f_T and f_{MAX} of 115 and 71 GHz, respectively, at $I_C = 6$ mA. Without emitter pull-up resistors, the circuit is operated in the large signal mode with a logic swing of 500 mV. Because of the low base–emitter turn-on voltage $V_{BE} = 0.9$ V, these circuits operate with a power consumption as low as 667 μW per stage and $t_d = 16.6$ ps. To our knowledge, this is the first high-speed integrated circuit demonstration using InP-based HBT's and the fastest bipolar NTL circuit result reported to date. The switching speed of InP/InGaAs HBT circuits can be improved if a nonsaturating logic configuration such as emitter-coupled logic (ECL) is used. From the equivalent circuit parameters listed in Table I(I) together with the following expression for current-mode logic [9], [10]

$$t_d \approx \frac{5}{2} R_B(C_{jc} + C_{FB}) + \frac{R_B}{R_L} \tau_F + [3(C_{jc} + C_{FB}) + C_L]R_L \quad (2)$$

a switching delay of 8.8 ps can be estimated with $R_L = 50$ Ω and $C_L = 0$ pF.

In summary, we have demonstrated a subpicosecond InP/InGaAs HBT. An f_T of 165 GHz is realized with high emitter current density and extreme nonequilibrium electron transport in the InGaAs base and the collector depletion region. The high doping in the base and subcollector eliminates current crowding and an f_{MAX} of 100 GHz is achieved with a 3.6 $\times$ 3.6-μm^2 emitter size. The first digital InP/InGaAs bipolar integrated circuit is illustrated to demonstrate the high speed and low power switching capability of this new technology.

ACKNOWLDEGMENT

The authors thank A. F. J. Levi for helpful discussions.

REFERENCES

[1] P. M. Solomon, "A comparison of semiconductor devices for high-speed logic," *Proc. IEEE,* vol. 70, pp. 489–509, 1982.
[2] H. Kroemer, "Heterostructure bipolar transistors and integrated circuits," *Proc. IEEE,* vol. 70, pp. 13–25, 1982.
[3] A. F. J. Levi and Y. Yafet, "Non-equilibrium electron transport in bipolar devices," *Appl. Phys. Lett.,* vol. 51, pp. 42–44, 1987.
[4] R. N. Nottenburg, Y. K. Chen, M. B. Panish, R. Hamm, and D. A. Humphrey, "High-gain submicrometer InGaAs/InP heterostructure bipolar transistors," *IEEE Electron Device Lett.,* vol. 9, pp. 524–526, 1988.
[5] R. Hamm, M. B. Panish, R. N. Nottenburg, Y. K. Chen, and D. A. Humphrey, "Ultrahigh Be doping of Ga$_{0.47}$In$_{0.53}$As by low temperature molecular beam epitaxy," to be published in *Appl. Phys. Lett.*
[6] R. N. Nottenburg, Y. K. Chen, M. B. Panish, D. A. Humphrey, and R. Hamm, "Hot-electron InGaAs/InP heterostructure bipolar transistors with f_T of 110 GHz," *IEEE Electron Device Lett.,* vol. 10, pp. 30–32, 1989.
[7] L. W. Massengill, T. H. Glisson, J. R. Hauser, and M. A. Littlejohn, "Transient transport in central-valley-dominated ternary III-V alloys," *Solid-State Electron.,* vol. 29, pp. 725–734, 1986.
[8] P. M. Asbeck, D. L. Miller, R. J. Anderson, and F. H. Eisen, "GaAs/(Ga,Al)As heterojunction bipolar transistors with buried oxygen-implanted isolation layers," *IEEE Electron Device Lett.,* vol. EDL-5, pp. 310–312, 1984.
[9] K. G. Ashar, "The method of estimating delay in switching circuits and the figure of merit of a switching transistor," *IEEE Trans. Electron Devices,* vol. ED-11, pp. 497–506, 1964.
[10] W. P. Dumke, J. M. Woodall, and V. L. Rideout, "GaAs-GaAlAs heterojunction transistor for high frequency operation," *Solid-State Electron.,* vol. 15, pp. 1339–1343, 1972.

Effect of Hot-Electron Injection on High-Frequency Characteristics of Abrupt $In_{0.52}(Ga_{1-x}Al_x)_{0.48}As/$ InGaAs HBT's

Hideki Fukano, *Member, IEEE*, Hiroki Nakajima, Tadao Ishibashi, *Member, IEEE*, Yoshifumi Takanashi, *Member, IEEE*, and Masatomo Fujimoto, *Member, IEEE*

Abstract—The effect of hot-electron injection energy (E_i) into the base on the high-frequency characteristics of $In_{0.52}(Ga_{1-x}Al_x)_{0.48}As/InGaAs$ abrupt heterojunction bipolar transistors (HBT's) is investigated by changing the composition of the emitter. It is found that there exists an optimum E_i at which a maximum current gain cutoff frequency (f_T) is obtained. Analysis of hot-electron transport in the base and collector by Monte Carlo simulation is carried out to understand the above phenomenon. The collector transit time (τ_c) increases with E_i, because electrons with higher energy transfer from the Γ valley into the upper L and X valleys. At first, the base transit time (τ_b) decreases with E_i at the low E_i region. However, τ_b does not decrease monotonically with E_i for high E_i, because of the large nonparabolicity in the energy-band structure of InGaAs. Consequently, there exists a minimum in the sum of τ_b and τ_c, in other words a maximum f_T, at an intermediate value of E_i.

I. INTRODUCTION

RECENTLY, considerable interest has been focused on heterojunction bipolar transistors (HBT's) composed of InGaAs [1]–[10]. The advantages of InGaAs HBT's can be summarized as follows: 1) a high current gain cutoff frequency (f_T) [1]–[4], 2) a low specific contact resistance [5], 3) an excellent scaling of a high-gain submicrometer HBT [6], 4) low power dissipation [7], and 5) integration with optical devices for use in longer wavelength optical communication systems. Two material systems have been used for InGaAs HBT's: InAlAs/InGaAs and InP/InGaAs. The most significant aspect of the energy-band structure is the conduction-band discontinuity (ΔE_c), which is directly related to electron injection energy (E_i) into the base in InGaAs HBT's with abrupt emitter–base junctions. The ΔE_c for InAlAs/InGaAs is larger than for InP/InGaAs. Although HBT's with a large value of E_i are expected to exhibit superior microwave characteristics, the maximum f_T so far reported for InAlAs/InGaAs HBT's is still less than that of InP/InGaAs HBT's. In GaAs HBT's with conventional collector struc-

ture, it is recognized from experimental and theoretical work [11], [12] that the reduction of the transit time due to velocity overshoot in the collector is relatively small. This is because the electrons injected from the base into the collector depletion region transfer immediately from the Γ valley to upper valleys. This phenomenon in conventional GaAs collector HBT's may occur even in InGaAs collector HBT's, in spite of their higher Γ–L separation, when E_i is high. The effects of E_i on the electron transport phenomena in the base and collector regions in InGaAs HBT's have to be clarified in much more detail to improve the characteristics of these devices.

By adopting $In_{0.52}(Ga_{1-x}Al_x)_{0.48}As$ alloy as the emitter material in the abrupt HBT and changing its composition, ΔE_c (in other words, E_i) can be varied from 0 to 0.5 eV [13]. This enables us to investigate the effect of E_i on InGaAs HBT characteristics over a wide range.

Our concern in this paper is how the hot-electron injection energy to the base influences the base and collector transit times of electrons. The effect of E_i on HBT characteristics is investigated by fabricating abrupt $In_{0.52}(Ga_{1-x}Al_x)_{0.48}As/InGaAs$ HBT's of different emitter composition. A theoretical analysis of the transit time dependence on E_i is also carried out by the Monte Carlo simulation. These simulation results are then compared with the experimental results. The device fabrication process is briefly described in Section II-A. In Section II-B, HBT characteristics of the fabricated devices are described. First, the emitter-to-base electron injection mechanism is investigated employing the temperature dependence of the current–voltage characteristics; the value of E_i is then estimated based on these experimental results. Then, the f_T dependence on collector current density for HBT's with different E_i is described. In Section III, an analysis of hot-electron motion in the base and collector by Monte Carlo simulation is carried out, and the effect of E_i on τ_b and τ_c is discussed. Finally, the conclusions are given in Section IV.

II. EXPERIMENT

A. Device Fabrication

Epitaxial layers were grown by molecular beam epitaxy (MBE); the detailed growth conditions have been reported

Manuscript received March 1, 1991; revised July 10, 1991. The review of this paper was arranged by Associate Editor N. Moll.

H. Fukano, Y. Takanashi, and M. Fujimoto are with the NTT Optoelectronics Laboratory, 3-1, Morinosato Wakamiya, Atsugi-shi, Kanagawa 243-01, Japan.

H. Nakajima and T. Ishibashi are with NTT LSI Laboratories, 3-1, Morinosato Wakamiya, Atsugi-shi, Kanagawa 243-01, Japan.

IEEE Log Number 9105323.

Reprinted from *IEEE Trans. Electron Devices*, vol. 39, no. 3, pp. 500–506, March 1992.

TABLE I
EPITAXIAL LAYER PARAMETERS OF FABRICATED HBT's

Layer	Type	Thickness (Å)	Doping (cm^{-3})
Cap 1	n$^+$ InGaAs	1000	2×10^{19}
Cap 2	n$^+$ InAlAs	700	2×10^{19}
Emitter	n InGaAlAs	1000	5×10^{17} E_g^e = 1.45, 1.19, 1.0 eV
Spacer	n$^-$ InGaAs	50	unodped ($< 1 \times 10^{16}$)
Base	p$^+$ InGaAs	500	2×10^{19}
Collector	n$^-$ InGaAs	3000	undoped ($< 1 \times 10^{16}$)
Buffer	n$^+$ InGaAs	7000	1×10^{19}
Substrate	S. I. InP		

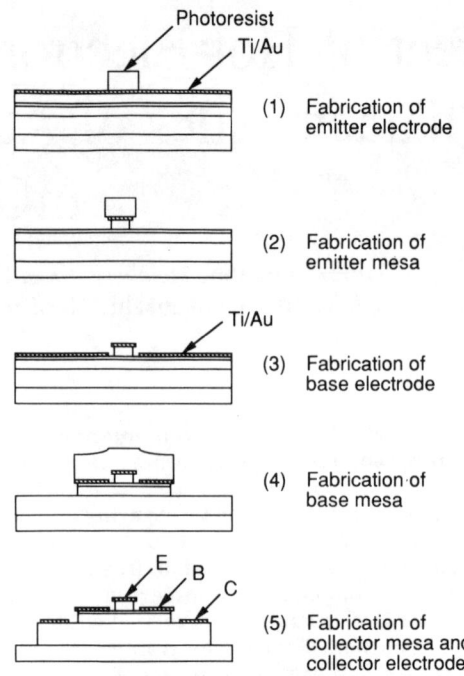

Fig. 1. Schematic HBT processing steps.

elsewhere [14]. Epitaxial layer parameters are shown in Table I. Employing quaternary In$_{0.52}$(Ga$_{1-x}$Al$_x$)$_{0.48}$As layers with different Al composition produces variable conduction-band discontinuities (ΔE_c's) at the abrupt emitter–base junction. The energy bandgap in the emitter (E_g^e) was estimated from an undoped quaternary layer, which was grown prior to the HBT wafer growth. The E_g^e of InAlAs ($x = 1$) was estimated from the wavelength providing the peak intensity of photoluminescence, and those of quaternary InGaAlAs layers from the squared absorption coefficient dependence on incident light wavelength. X values of aluminum composition for these quaternary alloys, which are calculated from E_g^e's, are 1, 0.62, and 0.35 for E_g^e's of 1.45, 1.19, and 1.0 eV, respectively. The base layer is heavily doped (2×10^{19} cm^{-3}) and thin (500 Å) InGaAs. The collector layer is undoped InGaAs with a background donor concentration of less than 1×10^{16} cm^{-3}. Heavily doped (2×10^{19} cm^{-3}) InGaAs/InAlAs cap layers were grown to obtain a low emitter contact resistance [5]. The dopants for p- and n-type were Be and Si, respectively.

An outline of the device fabrication process is shown in Fig. 1, and summarized here briefly as follows:

1) Ti/Au for the emitter electrode was deposited on the wafer by electron beam (EB) evaporation. After the photoresist pattern for the emitter was formed, metal etching was carried out. Au was etched by a KI solution and Ti by SF$_6$ reactive gas. The step was finished employing a diluted HF solution (Fig. 1—(1)).

2) An emitter mesa was formed by etching in dilute phosphoric acid until the base layer was exposed (Fig. 1—(2)).

3) Ti/Au for the base electrode was deposited on the HBT wafer by EB evaporation and defined by a lift-off technique (Fig. 1–(3)).

4) After the photoresist for the base-mesa etching was formed, Ti/Au was etched off in the same way as mentioned above, and the base mesa was formed by etching epitaxial layers under the base layer in dilute phosphoric acid until the appearance of the n$^+$ InGaAs buffer layer (Fig. 1—(4)).

5) After the n$^+$ InGaAs buffer layer was etched off for device isolation, a collector contact composed of Ti/PtAu was formed by the lift-off method (Fig. 1—(5)). Finally, Ti/Au electrode pads were formed on the silicon oxide

isolation film which was deposited after the collector contact was formed. Good specific contact resistance for each layer was obtained without any heat treatment. For example, the specific emitter contact resistance lies in the range of $1-3 \times 10^{-7}$ $\Omega \cdot$ cm^2. These results can be attributed to two factors: 1) low Schottky-barrier height for both p- and n-type InGaAs, and 2) heavy doping of each layer.

DC characteristics of the fabricated devices were measured by an HP4145 semiconductor parameter analyzer. High-frequency measurements were carried out using an HP8510 network analyzer with a Cascade Microtech probe card.

B. HBT Characteristics

Room-temperature emitter current (I_e) dependence on the base–emitter voltage (V_{be}) for HBT's of different emitter composition with an emitter–base junction area of 47×47 μm^2 are shown in Fig. 2. It is clear that the turn-on voltage of HBT's increases with increasing E_g^e. This is attributed to an increase in ΔE_c with E_g^e, and obviously implies that the electron injection energy to the base increases with E_g^e. To investigate the correlation between the electron injection mechanism from the emitter to the base and E_i, the temperature dependence of the emitter current density (J_e) versus V_{be} was measured. The result for the HBT with an E_g^e of 1.19 eV is depicted in Fig. 3, as typical. As shown in this figure, the value of J_e, where the broken line drawn by extrapolating the experimental data at the low J_e region intersects with the J_e axis, is referred to as a J_{es}. An ideality factor (n value) is derived from the slope of this line. Fig. 4 shows ln (J_{es}/T^2) versus q/nkT for each HBT. The solid lines in the figure are

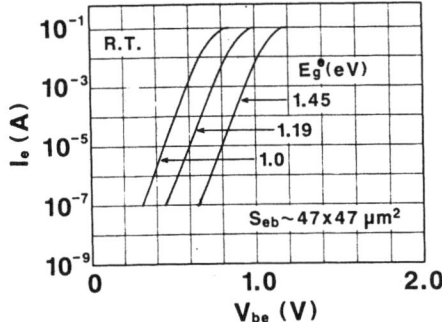

Fig. 2. Emitter current (I_e) dependence on base–emitter voltage (V_{be}) for three types of HBT's.

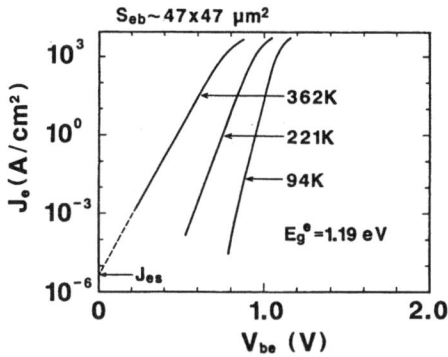

Fig. 3. Temperature dependence of emitter current density (J_e) versus base–emitter voltage (V_{be}) for an HBT with an emitter energy bandgap (E_g^e) of 1.19 eV.

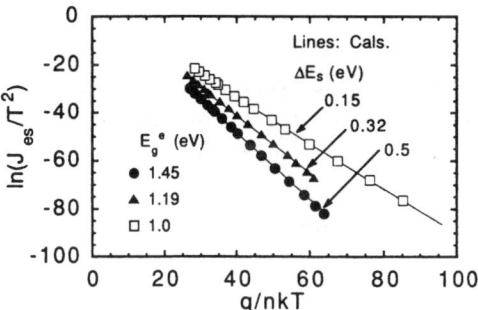

Fig. 4. Normalized emitter current density (ln (J_{es}/T^2)) versus inverse thermal energy divided by ideality factor ($1/nkT$) for three types of HBT's. The solid lines are the theoretical curves calculated from the thermionic field emission model.

theoretical curves calculated using the thermionic field emission (TFE) model [15], in which the thermally excited electrons tunnel through a hetero-energy-spike (ΔE_s) as shown schematically in Fig. 5.

J_{es} can be expressed approximately in the TFE model as [15]

$$J_{es} = (I_{ma}/I_m)A^* T^2 \exp\left[-(E_{b0} + qn\phi_e)/nkT\right]. \quad (1)$$

Here, (I_{ma}/I_m) is a normalized flat-band current density, A^* is the Richardson constant, E_{b0} is the band bending in the emitter depletion region at zero bias, and $q\phi_e$ is the potential energy of the electrons relative to the Fermi energy (E_F) in the emitter ($E_c - E_F$). E_{b0} is

$$E_{b0} = \Delta E_s - q\phi_e - q\phi_b + E_g^b. \quad (2)$$

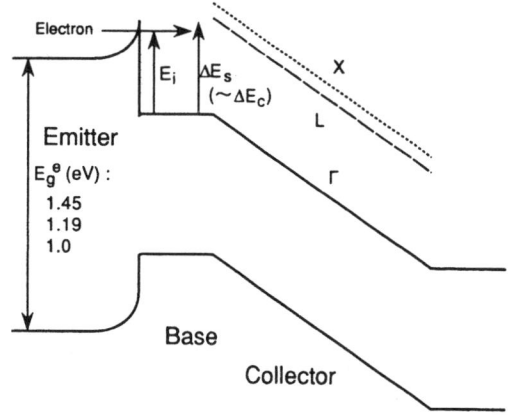

Fig. 5. Schematic energy band diagram of fabricated HBT's.

ΔE_s is the energy from the conduction-band minimum at the base neutral region to the top of the hetero-barrier, assuming that the dip in electrostatic potential on the base side is negligibly small. $q\phi_b$ is the potential energy of the holes relative to E_F in the base ($E_F - E_v$), and E_g^b is the energy bandgap of the base. ΔE_s is chosen to give the best fit between theory and experiment, while taking into account the temperature dependence of E_g^b [16]. Estimated values for ΔE_s are 0.5, 0.32, and 0.15 eV in order of decreasing E_g^e.

Fig. 6 shows the n-value dependence on temperature for each HBT. The calculated curves almost agree with the experimental data. The increase of the n value at low temperature is attributed to the relative increase in the number of electrons tunneling through the energy spike near the bottom of the conduction band due to decreasing thermally excited electrons at high energy. At room temperature, on the other hand, the electrons tunneling just below the top of the potential barrier are the dominant components of the current. The energy at which the largest number of electrons tunnel is defined as E_i (see Fig. 5) and is expressed as

$$E_i = \Delta E_s - E_b \tanh^2(E_{00}/kT) \quad (3)$$

where E_b is the band bending in the emitter depletion region and E_{00} the characteristic energy defined in [15]. The results for ΔE_s and E_i at a J_e of 2×10^4 A/cm^2 are summarized in Table II. In these calculations, image-force and bandgap shrinkage effects are not taken into account. As a matter of course, both effects should be included in the calculation to determine E_i more precisely. These factors, however, compensate each other. How they do so is explained as follows: image lowering $\Delta\phi$ is estimated to be about 40 meV at a J_e of 2×10^4 A/cm^2 from [17, eq. (3)]. On the other hand, there are no available experimental results on bandgap shrinkage ΔE_g^b in InGaAs to our knowledge. The ΔE_g^b due to the carrier–carrier interaction, however, can be derived by using theory [18]. ΔE_g^b is estimated to be 33 meV from [18, eq. (56)]. $\Delta\phi$ works so as to reduce ΔE_s (i.e., the E_i) whereas ΔE_g^b works so as to increase it.

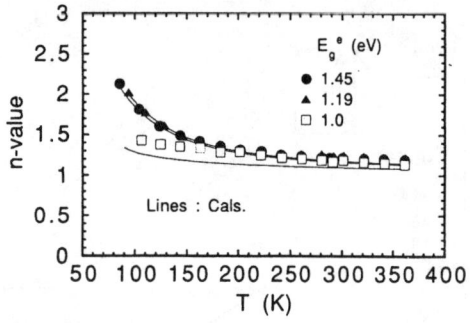

Fig. 6. Ideality factor (n-value) dependence on temperature for three types of HBT's. The solid lines are the theoretical curves calculated from the thermionic field emission model.

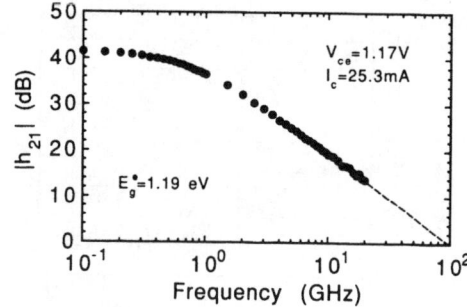

Fig. 7. Frequency dependence of the current gain $|h_{21}|$ for the device with an emitter energy bandgap (E_g^e) of 1.19 eV.

TABLE II

ESTIMATED HETERO-ENERGY-SPIKE (ΔE_s) AND ELECTRON INJECTION ENERGY (E_i) AT AN EMITTER CURRENT DENSITY OF 2×10^4 A/cm^2 FOR EACH HBT WITH DIFFERENT EMITTER ENERGY BANDGAP (E_g^e)

Energy Gap	E_g^e (eV) :	1.45	1.19	1.0
Hetero-spike	ΔE_s (eV) :	0.5	0.32	0.15
Injection energy	E_i (eV) :	0.46	0.28	0.10

To clarify the effect of E_i on base and collector transit times, the current gain cutoff frequency (f_T), which directly reflects τ_b and τ_c, was estimated for HBT's with an emitter of about 4.7×8.7 μm^2. The frequency dependence of the common emitter current gain (h_{21}) for an E_g^e of 1.19 eV, which is calculated from the measured S-parameter data, is shown in Fig. 7 as an example. f_T is estimated to be 95 GHz. The dependence of f_T on collector current density (J_c) for the three types of HBT's is shown in Fig. 8. Two noteworthy features are revealed in this figure: 1) the J_c giving the maximum f_T (J_c^{peak}) decreases as E_g^e (and E_i) increases and 2) the largest value of f_T is obtained for the intermediate bandgap ($E_g^e = 1.19$ eV), not for the highest one ($E_g^e = 1.45$ eV).

The decrease in f_T at the high current-density region can be explained in terms of the space-charge effect, which is also known as the Kirk effect, as shown in a previous paper [19]. The variation of the J_c^{peak} with E_g^e can be understood qualitatively as follows.

Space charge (n_s) in the collector is roughly expressed by

$$n_s \approx J_c / q v_{ave} \qquad (4)$$

where v_{ave} is an average electron velocity in the collector. As n_s becomes sufficiently large compared to the donor density in the collector, the collector begins to act in the same manner as the neutral base region; the base transit and collector charging times increase, thereby resulting in a decrease in f_T. By assuming the space charge at J_c^{peak} (defined here as n_s^{peak}) to be almost constant, v_{ave} can be written $J_c^{peak}/q n_s^{peak}$ from (4). Consequently, the decrease in J_c^{peak} with increasing E_g^e implies a decrease in v_{ave}.

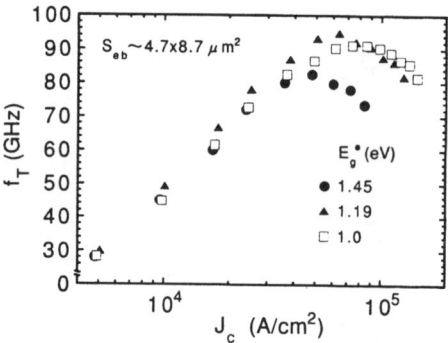

Fig. 8. Cutoff frequency (f_T) dependence on collector current density (J_c) for three types of HBT's with an emitter area of about 4.7×8.7 μm^2.

III. DISCUSSION

Analysis of hot-electron transport in the base and collector by Monte Carlo simulation was carried out to quantitatively elucidate the experimental results. Base and collector delay times, which are referred to as τ_b(cal.) and τ_c(cal.), respectively, are defined by the following equations:

$$\tau_b(\text{cal.}) = \Delta Q_b / \Delta J_c \qquad (5)$$

$$\tau_c(\text{cal.}) = \Delta Q_c / \Delta J_c \qquad (6)$$

where ΔQ_b and ΔQ_c are the variation in the total electron charge per unit area in the base and collector region, respectively, and ΔJ_c is the variation of the collector current density accompanied by a slight change in base-emitter bias voltage. The details of the simulation will be reported elsewhere. In the simulation, pure thermionic emission is used as the electron injection mechanism from emitter to base. The tunneling process for electrons across the hetero-energy-spike is taken into account by making the potential barrier for injection equal to E_i. In the following discussion, E_i is used as a unique parameter. The simulation was done at appropriate current density ($J_c \sim 2 \times 10^4$ A/cm^2) so as to avoid any influence from the space-charge effect.

Fig. 9(a) shows the electron energy distribution for an E_i of 0.18 eV. The majority of the electrons remains in the Γ valley in most of the collector region. The effective electron velocity (V_{eff}), which is defined as $W_c/2\tau_c$(cal.), is as fast as 4.2×10^7 cm/s. Here, W_c is the collector

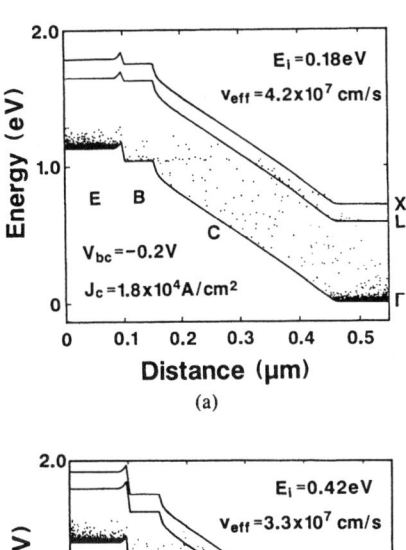

(a)

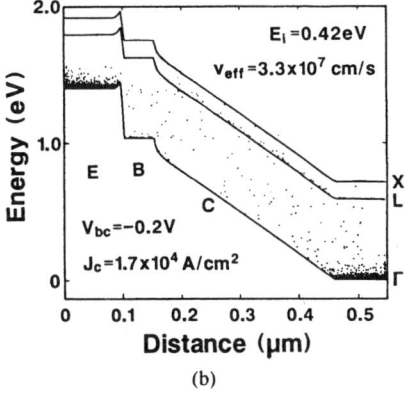

(b)

Fig. 9. Electron energy distribution. (a) In the case of electron injection energy (E_i) of 0.18 eV. (b) In the case of an E_i of 0.42 eV.

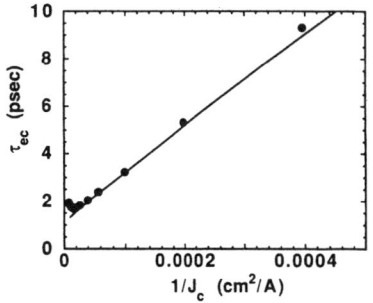

Fig. 10. Effective emitter-to-collector transit time (τ_{ec}) versus reciprocal collector current density ($1/J_c$). The solid line is a fitting curve calculated by (7).

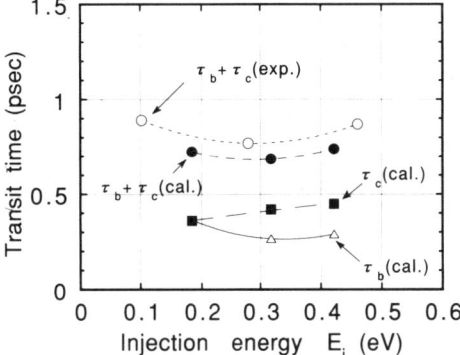

Fig. 11. Base and collector transit times (τ_b and τ_c, respectively) and their sum versus electron injection energy (E_i).

depletion layer width. Fig. 9(b) shows the result for a device with an injection energy of 0.42 eV. Some electrons transfer from the Γ valley to the upper L and X valleys even near the base–collector junction, because of their higher energy. Since electrons in the upper valleys are heavier than in the Γ valley, v_{eff} for $E_i = 0.42$ eV is lower than that for $E_i = 0.18$ eV. The value of v_{eff} in the former is about 3.3×10^7 cm/s.

The cutoff frequency is generally expressed as

$$1/f_T = 2\pi\tau_{ec} = 2\pi(\tau_e + \tau_b + \tau_c + \tau_{cc}) \qquad (7)$$

where τ_{ec} is the effective emitter-to-collector delay time, τ_e the emitter charging time, τ_b the base transit time, τ_c the collector depletion-layer transit time, and τ_{cc} the collector charging time. Each time constant is expressed by an equation given in [5]. As shown in Fig. 8, f_T varies with J_c. By fitting (7) to the experimental data for the τ_{ec} dependence on $1/J_c$ up to a slightly lower value than J_c^{peak}, $\tau_b + \tau_c$ can be estimated. The result for an E_g^e of 1.19 eV device is shown in Fig. 10, as an example. The sum of $\tau_b + \tau_c + \tau_{cc}$ for HBT shown in Fig. 10 is 1.07 ps. By subtracting the estimated τ_{cc} of 0.3 ps from the above value, $\tau_b + \tau_c$ of 0.77 ps is obtained. Note that $\tau_b + \tau_c$ obtained in this way is not affected by the space-charge effect. In our devices, τ_{cc} is not sufficiently small. This is because a large base–collector capacitance (84 fF) is arising from a base–collector junction area as large as 13×14 μm^2. τ_{cc} can be reduced by decreasing the size

of the base mesa and improving the collector resistance: The latter can be realized by adopting a self-aligned collector contact fabrication process and increasing the doping level in the buffer layer.

Fig. 11 shows τ_b, τ_c, and their sum as a function of E_i, as obtained by Monte Carlo simulation. Also shown in this figure are the experimental values for the sum. In the simulation, τ_c increases monotonically with E_i, because of an increase in the transition probability of electrons from the Γ valley to the upper valleys. This agrees with the experimental results described earlier. On the other hand, τ_b decreases with E_i till about 0.3 eV and then tends to saturate or increase slightly. This behavior of τ_b in the region over an E_i of 0.3 eV is thought to be due to the large conduction band nonparabolicity of InGaAs.

Using a nonparabolicity parameter (α), energy dispersion for the conduction band is described as

$$\hbar^2 k^2 / 2m^* = E(1 + \alpha E) \qquad (8)$$

where $\hbar$ is Planck's constant, k the magnitude of wave-vector k, m^* the band-edge effective mass, E the electron energy. A large α induces an increase in electron effective mass at a high energy state. The α is usually defined by

$$\alpha = (1 - m^*/m_0)^2 / E_g \qquad (9)$$

where E_g is the bandgap energy and m_0 the free-electron mass [20]. Narrow-bandgap InGaAs has an α larger than 1 [21] whereas the α of GaAs is 0.69 [22]. In this simulation, a value of α, obtained by linearly interpolating the binary data for InAs, GaAs, and AlAs, was used for every composition of InGaAlAs, with α of 1.06 used for InGaAs. To clarify the influence of large α, a simulation of the parabolic band structure in InGaAs (i.e., $\alpha = 0$) was carried out. Calculated τ_b at an E_i of 0.42 eV is as short as 0.2 ps. This is about 30% shorter than for the nonparabolic case. The increase in τ_b for large α is caused by an increase in the scattering probability of electrons at high energy. From a different viewpoint, the above phenomenon can be understood qualitatively to arise from an increase in the electron effective mass at high energy. As a result, τ_b does not decrease monotonically at the high-energy region. Consequently, the sum of τ_b and τ_c has a minimum as shown in Fig. 11. This agrees with the experimental results, and explains why the largest f_T is obtained at the intermediate E_i, not the highest one.

IV. CONCLUSIONS

The effect of hot-electron injection energy on the current gain cutoff frequency (f_T) of abrupt $In_{0.52}(Ga_{1-x}Al_x)_{0.48}As$/InGaAs HBT's was investigated. It was shown that there is an optimum injection energy which yields the maximum f_T. This implies that sum of base transit time (τ_b) and collector transit time (τ_c) has a minimum for a particular E_i. The existence of this minimum is based on the following two reasons. One, τ_c increases with E_i. This occurs because higher energy electrons transfer from the Γ valley into the upper L and X valleys in the collector. Two, τ_b decreases initially with E_i at the low E_i region, but tends to saturate or increase slightly at the high E_i region. This is caused by the large conduction band nonparabolicity of InGaAs. Consequently, a minimum exists in the sum of τ_b and τ_c.

ACKNOWLEDGMENT

The authors wish to thank M. Naganuma and T. Sugeta for their encouragement throughout this work. Thanks are also due to Y. Kawamura and H. Asai for fabricating the HBT wafers.

REFERENCES

[1] Y. K. Chen, A. F. J. Levi, R. N. Nottenburg, P. H. Beton, and M. B. Panish, "High-frequency study of nonequilibrium transport in heterostructure bipolar transistors," *Appl. Phys. Lett.*, vol. 55, pp. 1789–1791, 1989.

[2] C. W. Farley, M. F. Chang, P. M. Asbeck, N. H. Sheng, R. Pierson, G. J. Sullivan, K. C. Wang, and R. B. Nubling, "High-speed ($f_T = 78$ GHz) AlInAs/GaInAs single heterojunction HBT," *Electron. Lett.*, vol. 25, pp. 846–847, 1989.

[3] B. Jalali, R. N. Nottenburg, W. S. Hobson, Y. K. Chen, T. Fullowan, S. J. Pearton, and A. S. Jordan, "AlInAs/GaInAs Hetero-structure bipolar transistors grown by metalorganic chemical vapour deposition," *Electron. Lett.*, vol. 25, pp. 1496–1497, 1989.

[4] H. Fukano, Y. Kawamura, H. Asai, Y. Takanashi, and M. Fujimoto, "Improving the characteristics of InAlAs/InGaAs heterojunction bipolar transistors by employing thin base and collector layers," *Electron. Lett.*, vol. 26, pp. 1101–1102, 1990.

[5] H. Fukano, Y. Kawamura, and Y. Takanashi, "High-speed InAlAs/InGaAs Heterojunction Bipolar Transistors," *IEEE Electron. Device Lett.*, vol. 9, pp. 312–314, 1988.

[6] R. N. Nottenburg, Y. K. Chen, M. B. Panish, R. Hamm, and D. A. Humphrey, "High-current-gain submicrometer InGaAs/InP heterostructure bipolar transistors," *IEEE Electron. Device Lett.*, vol. 9, pp. 524–526, 1988.

[7] C. W. Farley, K. C. Wang, M. F. Chang, P. M. Asbeck, R. B. Nubling, N. H. Sheng, R. Pierson, and G. J. Sullivan, "A high-speed, low-power divided-by-4 frequency divider implemented with AlInAs/GaInAs HBT's," *IEEE Electron. Device Lett.*, vol. 10, pp. 377–379, 1989.

[8] Y. K. Chen, R. N. Nottenburg, M. B. Panish, R. A. Hamm, and D. A. Humphrey, "Subpicosecond InP/InGaAs heterostructure bipolar transistors," *IEEE Electron Device Lett.*, vol. 10, pp. 267–269, 1989.

[9] H. Yamada, T. Futatsugi, Y. Yamaguchi, K. Ishii, Y. Bamba, T. Fujii, and N. Yokoyama, "Emitter-coupled logic circuits implemented using InAlAs/InGaAs HBTs with improved emitter-collector breakdown voltage," presented at the 48th Device Research Conf., Santa Barbara, CA, June 1990.

[10] J. F. Jensen, W. E. Stanchina, R. A. Metzger, D. B. Rensch, R. J. Ferro, P. F. Lou, M. W. Pierce, T. V. Kargodorian, and Y. K. Allen, "Improved AlInAs/GaInAs HBTs for high-speed circuits," in *Proc. SPIE Conf. on High Speed Electronics and Device Scaling*, 1990, San Diego, CA, pp. 57–68.

[11] Y. Yamauchi and T. Ishibashi, "Electron velocity overshoot in the collector depletion layer of AlGaAs/GaAs HBT's," *IEEE Electron Device Lett.*, vol. EDL-7, pp. 655–657, 1986.

[12] C. M. Maziar, M. E. Klausmeier-Brown, and M. S. Lundstrom, "A proposed structure for collector transit time reduction in AlGaAs/GaAs bipolar transistors," *IEEE Electron Device Lett.*, vol. EDL-7, pp. 483–485, 1986.

[13] Y. Sugiyama, T. Inata, T. Fujii, Y. Nakata, S. Muto, and S. Hiyamizu, "Conduction band edge discontinuity of $In_{0.52}Ga_{0.48}As$/$In_{0.52}(Ga_{1-x}Al_x)_{0.48}As$ ($0 \leq x \leq 1$) heterostructures," *Japan. J. Appl. Phys.*, vol. 25, pp. L648–L650, 1986.

[14] Y. Kawamura, K. Wakita, and H. Asahi, "Observation of heavy-hole and light-hole excitons in InGaAs/InAlAs MQW structures at room temperature," *Electron. Lett.*, vol. 21, pp. 371–372, 1985.

[15] C. R. Crowell and V. L. Rideout, "Normalized thermionic-field (T-F) emission in metal-semiconductor (Schottky) barriers," *Solid-State Electron.*, vol. 12, pp. 89–105, 1969.

[16] P. W. Yu and E. Kuphal, "Photoluminescence of Mn- and un-doped $Ga_{0.47}In_{0.53}As$ on InP," *Solid State Commun.* vol. 49, pp. 907–910, 1984.

[17] F. A. Padovani, "Thermionic emission in Au-GaAs Schottky barriers," *Solid-State Electron.*, vol. 11, pp. 193–200, 1968.

[18] O. Gunnarsson and B. I. Lundqvist, "Exchange and correlation in atoms, molecules, and solids by the spin-density-functional formalism," *Phys. Rev. B*, vol. 13, pp. 4274–4298, 1976.

[19] H. Fukano, Y. Kawamura, H. Asai, and Y. Takanashi, "High frequency characteristics of InAlAs/InGaAs HBT's," *Japan. J. Appl. Phys.*, vol. 28, pp. L1737–L1739, 1989.

[20] T. P. Pearsall, Ed., *GaInAsP Alloy Semiconductors.* New York: Wiley, 1982.

[21] A. P. Long, P. H. Beton, and M. J. Kelly, "Hot-electron transport in $In_{0.53}Ga_{0.47}As$," *J. Appl. Phys.*, vol. 62, pp. 1842–1849, 1987.

[22] K. F. Brennan, D. H. Park, K. Hess, and M. A. Littlejohn, "Theory of the velocity-field relation in AlGaAs," *J. Appl. Phys.*, vol. 63, pp. 5004–5008, 1988.

A New Self-Alignment Technology Using Bridged Base Electrode for Small-Scaled AlGaAs/GaAs HBT's

Koichi Nagata, Osaake Nakajima, Takumi Nittono, Yoshiki Yamauchi, *Member, IEEE*, and Tadao Ishibashi, *Member, IEEE*

Abstract—The fabrication and characterization of a new self-aligned HBT utilizing Bridged Base-electrode Technology (BBT) are presented. This new technology simplifies the fabrication process and relaxes the limitations in device size scaling, thus decreasing the emitter size to 1 μm × 1 μm. In spite of a large junction periphery/area ratio, a good current gain of more than 10 is obtained in an HBT with an emitter size of 1 μm × 1 μm. A series of fabricated HBT's shows excellent high-speed performance. The highest values of f_T = 90 GHz and f_{max} = 63 GHz are obtained in an HBT with an emitter size of 1 μm × 5 μm. The realization of HBT's with small emitters and excellent high-frequency characteristics demonstrates the effectiveness of this new technology.

I. INTRODUCTION

HETEROJUNCTION bipolar transistors (HBT's) are very attractive devices for high-speed digital circuits. Ring oscillators with gate delay times of less than 10 ps/gate and frequency dividers with toggle frequencies of more than 20 GHz have already been reported [1]–[5]. Not only high-speed performance, but also low power dissipation is strongly demanded for HBT applications to integrated circuits. For the applications to MSI and LSI complexities, the power dissipation has not been lowered enough. One way to ensure both high speed and low power dissipation is to decrease device size, because this enables it to operate at a high current density while the current level is kept low.

In decreasing device size, it is necessary to develop a self-aligned structure that is easy to fabricate and maintains high-speed performance. Emitter and collector sizes have to be decreased simultaneously here. Especially in the base electrode fabrication process, fairly difficult techniques, such as angled ion milling, dual lift-off, and selective removal, have been used [1], [3], [6], [7]. The improvement of the base electrode fabrication process is quite important for relaxing the limitations in device size.

This paper describes the fabrication and characterization of a new self-aligned AlGaAs/GaAs HBT utilizing Bridged Base-electrode Technology (BBT). This new technology simplifies the fabrication process, thus decreasing the emitter size to 1 μm × 1 μm. The emitter size dependencies of dc and RF characteristics are presented and discussed.

II. DEVICE STRUCTURE AND FABRICATION

A. Device Structure

Schematics of the self-aligned HBT fabricated using BBT are shown in Fig. 1. A top view and two cross-sectional views are shown in the figure. The most important feature is that the base electrode is laid over the emitter electrode and the emitter mesa, which are surrounded by SiO$_2$ film. This structure is very simple and easy to fabricate and it does not require an extremely high alignment accuracy or any special techniques. Therefore, the device size can easily be decreased with high yield and reproducibility. The size limitation in the fabrication process is only the resolution of photolithography patterning.

Though the capacitance between the emitter electrode and the base electrode increases slightly, the influence on the characteristics is actually small because of comparatively large E-B junction capacitance. For example, supposing a SiO$_2$ film thickness of 0.2 μm, the capacitance between the emitter electrode and the base electrode is calculated to be 0.2 fF/μm^2. It is about 4% compared with the E-B junction capacitance of about 5 fF/μm^2.

The E-B junction is defined by H$^+$ implantation and emitter mesa etching. On the other hand, the B-C junction is defined only by H$^+$ implantation. The collector electrode is buried in the semi-insulating layer, so that the surface is almost planar except for the emitter.

B. Fabrication Process

Devices were fabricated using MBE-grown wafers. The epitaxial layer structure is shown in Table I. The structure is basically the same as the previously reported one [5]. The n$^+$-InGaAs emitter cap layer was used to form a non-alloyed ohmic contact with very small emitter contact resistance [8]. This allows us to use WSi/W, which can be etched by RIE, as the emitter electrode. The base layer had a linearly graded Al composition. This graded-band-

Manuscript received October 2, 1991. The review of this paper was arranged by Associate Editor M. Shur.

The authors are with NTT LSI Laboratories, 3-1, Morinosato Wakamiya, Atsugi-shi, Kanagawa 243-01, Japan.

IEEE Log Number 9201214.

Reprinted from *IEEE Trans. Electron Devices*, vol. 39, no. 8, pp. 1786–1792, Aug. 1992.

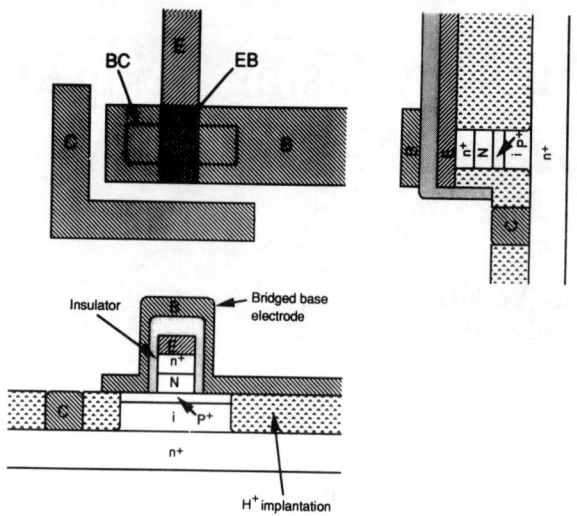

Fig. 1. Schematic view of the self-aligned HBT utilizing BBT.

TABLE I
EPITAXIAL LAYER STRUCTURE USED FOR HBT FABRICATION

Layer	Material	Doping (cm^{-3})	Thickness (Å)	InAs or AlAs Fraction
	n$^+$-InGaAs	2×10^{19}	500	0.5
Cap	n$^+$-InGaAs	2×10^{19}	500	0.5–0
	n$^+$-GaAs	2×10^{19}	500	
	N-AlGaAs	5×10^{17}	300	0–0.3
Emitter	N-AlGaAs	5×10^{17}	900	0.3
	N-AlGaAs	5×10^{17}	300	0.3–0.12
Base	P$^+$-AlGaAs	4×10^{19}	800	0.12–0
Collector	i-GaAs	undoped	3000	
Buffer	n$^+$-GaAs	1×10^{19}	6000	
S.I. GaAs substrate				

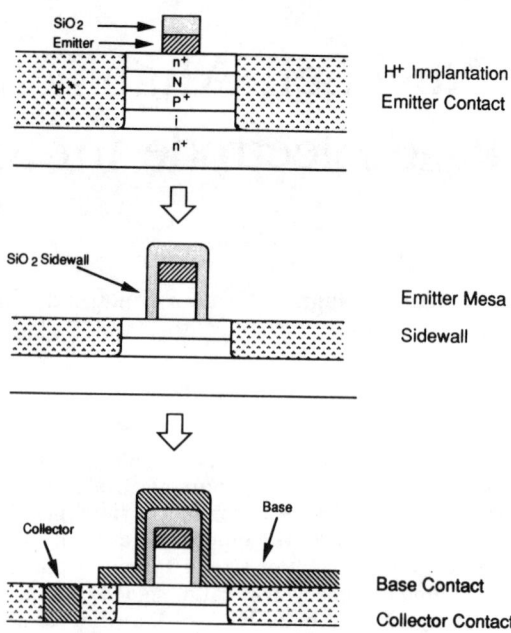

Fig. 2. Fabrication steps for the self-aligned HBT.

Fig. 3. Cross-sectional view of a fabricated HBT, observed by SEM.

gap base is effective in reducing the base transit time. The i-GaAs collector layer was used to reduce the collector transit time through extending electron velocity overshoot [5].

The main steps in the fabrication process are shown in Fig. 2. First, H$^+$ implantation for isolation of the base layer was carried out. The implant dose and energy were 2×10^{15} cm^{-2} and 100 keV, respectively. This implanted layer kept the high resistive state even after fairly high-temperature processes, such as plasma-enhanced chemical vapor deposition and ohmic sintering. The sheet resistances of the H$^+$-implanted AlGaAs and GaAs layers were measured to be over 1 MΩ/□ by a TLM method. A nonalloyed WSi/W emitter electrode with a SiO$_2$ cap was formed by RIE. A C$_2$F$_6$ gas and a SF$_6$ gas were used for RIE of the SiO$_2$ and WSi/W films, respectively. Then, the emitter mesa was etched by RIBE with BCl$_3$ gas using the emitter electrode as a mask. With these RIE and RIBE techniques, the width of the photoresist mask was accurately transferred to the emitter stripe. The SiO$_2$ sidewall was formed by plasma-enhanced chemical vapor deposition and sequential etch-back. The thickness of the sidewall was about 0.2 μm. The Ti/Pt/Au base electrode

(2000 or 3000 Å) was formed by a lift-off technique. Since the SiO$_2$ sidewall was fairly rounded in the etch-back process, the base metal could be connected over the emitter step as shown in Fig. 3. The thickness of the metal on the sidewall was estimated to be about 1/3 compared with that evaporated on the plane surface, by measuring the resistance of a step coverage test pattern which had a base metal line across 768 steps. After the recess for the collector contact was formed by RIBE, the AuGe/Ni/Ti/Pt/Au collector electrode was sequentially formed by lift-off. The thickness of the collector electrode was 4500 Å, which was almost equal to the recess depth. Sintering for ohmic contacts was carried out at 380°C for 30 s. To complete the devices, dielectric film deposition, H$^+$ implantation for device isolation, and metallization followed.

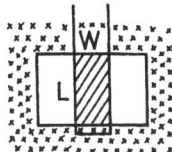

Emitter length, L	1, 2, 3, 5, 10 μm
Emitter width, W	1, 2 μm
External base width	1 μm

Fig. 4. Device dimensions.

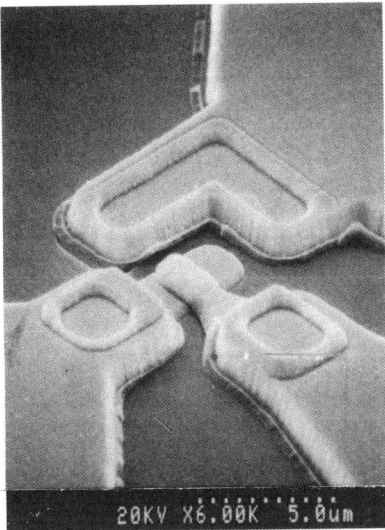

Fig. 5. SEM photograph of a fabricated HBT with an emitter size of 1 μm × 1 μm.

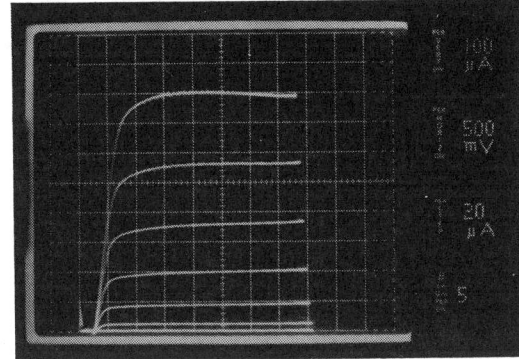

Fig. 6. I–V characteristics of a fabricated HBT with an emitter size of 1 μm × 1 μm.

Devices with various emitter sizes were fabricated for investigation of size dependencies. The dimensions of the fabricated HBT's are shown in Fig. 4. The emitters had lengths of 1, 2, 3, 5, and 10 μm, and widths of 1 and 2 μm. The width of the external base was 1 μm in all the devices.

A SEM photograph of an HBT with an emitter size of 1 μm × 1 μm is shown in Fig. 5. The collector electrode was designed to be L-shape in order to minimize the whole device area by maintaining a low collector resistance. The junction definition by H^+ implantation cannot be seen in this photograph.

III. DEVICE PERFORMANCE AND DISCUSSION

A. DC Characteristics

Current gains and transconductances were evaluated for devices with various emitter sizes. Fig. 6 shows typical I–V characteristics of an HBT with the smallest emitter, that is, 1 μm × 1 μm. A current gain, $\beta (= \Delta I_C / \Delta I_B)$, of 12 was obtained at a collector current density J_C of 7 × 10^4 A/cm^2. This was about half the value for a device with a 1 μm × 10 μm emitter. Also, the current gain was lower for smaller devices, which is known as the emitter size effect [9]. On the other hand, a transconductance per

unit area of about 7 mS/μm^2 was obtained for all the devices.

In order to find the cause of the current gain reduction, its dependency on the emitter size was studied. Our analysis was focused on the leakage current at the edge of the emitter/base junction [10]. It was assumed that this current had two components: one at the edge of the mesa and the other at the edge of the H^+-implanted region. Therefore, the base current I_B is written as

$$I_B = S_{EB}J_{Bi} + 2LI_L + 2WI_W \tag{1}$$

where J_{Bi} is the base current density in the intrinsic base region, and I_L and I_W are the leakage currents per unit length at the mesa-etched edge and H^+-implanted edge, respectively. Since the collector current I_C is expressed as $S_{EB}J_C$, one obtains

$$\frac{1}{h_{FE}} = \frac{I_B}{I_C} = \frac{S_{EB}J_{Bi} + 2LI_L + 2WI_W}{S_{EB}J_C}$$

$$= \frac{1}{h_{FEi}} + \frac{I_L}{J_C}\left(\frac{2L}{S_{EB}}\right) + \frac{I_W}{J_C}\left(\frac{2W}{S_{EB}}\right) \tag{2}$$

where h_{FEi} is the intrinsic current gain. When the collector current density is constant, the excess base current is proportional to $2L/S_{EB}$ and $2W/S_{EB}$. Fig. 7 shows the dependencies of $1/h_{FE}$ on $2W/S_{EB}$ at $J_C = 2.5 \times 10^4$ A/cm^2 with the parameters of $2L/S_{EB}$. These results are explained quite well by the equation mentioned above, that is, the relationship is linear. The leakage currents of $I_L = 10.8$ μA/μm and $I_W = 27.5$ μA/μm, and the intrinsic current gain of $h_{FEi} = 77$ were obtained from these data. In the same way, the dependencies of I_L and I_W on collector current densities were derived as shown in Fig. 8. It is found that I_W is about three times larger than I_L at all the measured collector current densities. This means that the reduction in the current gain is mainly due to the leakage current at the edge of the H^+-implanted region. Fig. 9 shows how the leakage currents vary with junction voltage. The junction voltage is given by $V_{BE} - R_{EE}I_E$, where V_{BE} is the bias voltage between the emitter pad and the base pad. Since the n value for both I_L and I_W is about 2, it is thought that these leakage currents result from gen-

299

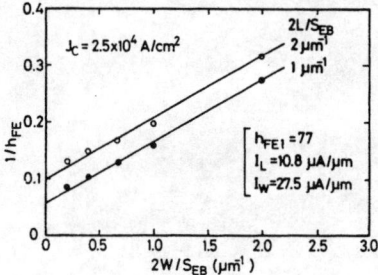

Fig. 7. $1/h_{FE}$ versus $2W/S_{EB}$ at $J_C = 2.5 \times 10^4$ A/cm².

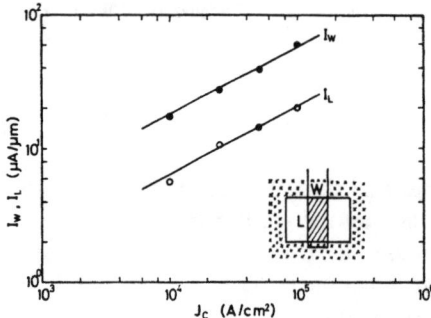

Fig. 8. Dependencies of I_L and I_W on collector current density.

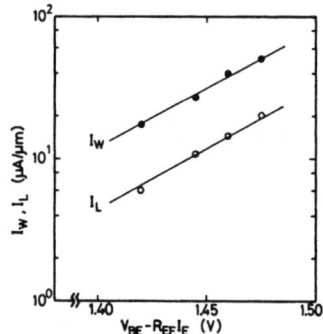

Fig. 9. Dependencies of I_L and I_W on E-B junction voltage.

eration–recombination currents at the edge of the emitter. Improving the quality of the implanted semi-insulating layer is very important for increasing the current gain. This will be achieved by the adoption of O$^+$ implantation followed by annealing, which induces less damage [11] instead of the H$^+$ implantation. In addition, I_L is also expected to be reduced by employing the thin AlGaAs passivation layer on the external base layer in emitter mesa etching [12], [13].

B. High-Frequency Characteristics

High-frequency performance of the devices was investigated by means of the *S*-parameter measurements using on-wafer RF probes. The calibration was done at the end of the RF probes. Analyses based on an equivalent circuit [14] were also carried out to explain the characteristics. The best result was obtained for an emitter size of 1 μm $\times$ 5 μm. The curves of the cutoff frequency f_T, and the maximum oscillation frequency $f_{\max}$, versus the collector

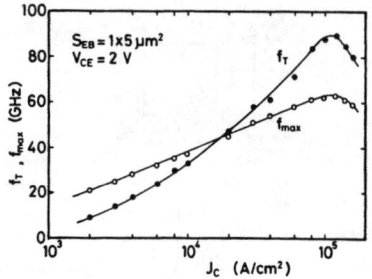

Fig. 10. RF characteristics of an HBT with an emitter size of 1 μm $\times$ 5 μm.

TABLE II
TRANSISTOR PARAMETERS DERIVED FROM DC AND RF CHARACTERISTICS

Emitter–base junction area	$S_{EB} = 1 \times 5 \ \mu$m²
Base–collector junction area	$S_{BC} = 3 \times 5 \ \mu$m²
Emitter resistance	$R_{EE} = 18 \ \Omega$
Base resistance	$R_B = 45 \ \Omega$
Collector resistance	$R_C = 8 \ \Omega$
Emitter–base capacitance	$C_{EB} = 56$ fF
Base–collector capacitance	$C_{BC} = 19$ fF
Forward transit time	$\tau_b + \tau_c = 0.95$ ps

current density are shown in Fig. 10. The peak values of $f_T = 90$ GHz and $f_{\max} = 63$ GHz were obtained at $J_C = 1.2 \times 10^5$ A/cm². These values are comparable to those obtained in other self-aligned HBT's [1], [6], [7]. The transistor parameters, which were calculated from dc and RF characteristics and verified by fitting to the equivalent circuit, are shown in Table II. It is noted that the $f_{\max}$ is mainly limited by the fairly large B-C capacitance.

To clarify the features of the new device structure, the device size dependencies of the high-frequency characteristics were studied. Figs. 11 and 12 show how f_T and $f_{\max}$ vary with emitter size. For 1 μm $\times$ 1 μm–1 μm $\times$ 5 μm emitter sizes, both f_T and $f_{\max}$ become smaller with decreasing size. Since the deviation of f_T was larger than that of $f_{\max}$, our analysis was focused on f_T. The relationship between f_T and $f_{\max}$ depends on the well-known equation

$$f_{\max} = \sqrt{\frac{f_T}{8\pi R_B C_{BC}}}.$$

The change in f_T values is basically associated with parasitic capacitance and/or resistance. At low collector current density region, the emitter charging time is dominant. In the present measurements, the device *S*-parameters included the base pad capacitance of about 30 fF, while the intrinsic E-B junction capacitance is only 5 fF/μm² for the tested devices. This comparatively large extrinsic capacitance is responsible for lower f_T values in smaller devices in the low collector current density region. In actual circuits, such effect is much less expected. Next we consider the collector charging time $\tau_{cc} = (R_{EE} + R_C)C_{BC}$. The device size dependencies of the emitter resistance (R_{EE}) and the B-C capacitance (C_{BC}) are shown in Figs. 13 and 14, respectively. R_{EE} was derived from dc

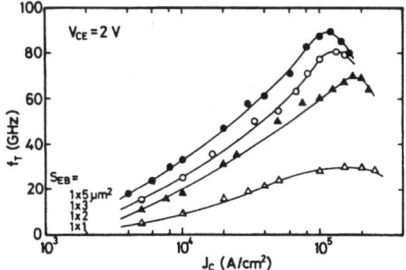

Fig. 11. Dependencies of f_T on collector current density for various emitter size HBT's.

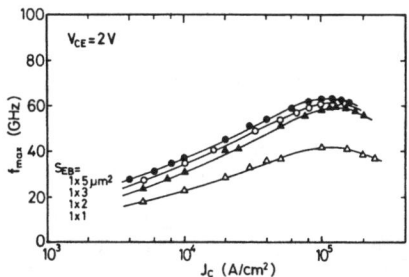

Fig. 12. Dependencies of f_{max} on collector current density for various emitter size HBT's.

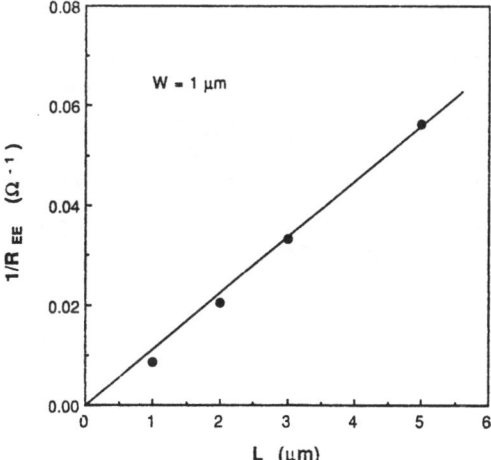

Fig. 13. Dependencies of $1/R_{EE}$ on emitter size.

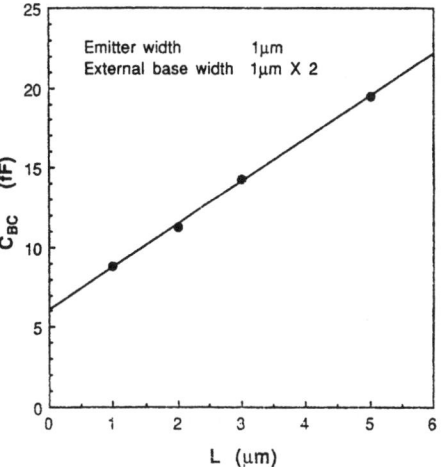

Fig. 14. Dependencies of C_{BC} on device size.

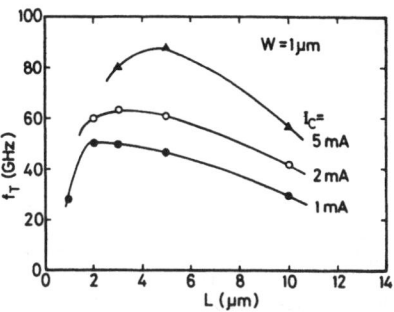

Fig. 15. Dependencies of f_T on emitter size with parameters of I_C.

characteristics including Gummel plots and C_{BC} was calculated from the measured S-parameters. The collector resistance (R_C) was small compared with R_{EE}. Since $1/R_{EE}$ is almost proportional to the emitter length, as shown in Fig. 13, the contribution of R_{EE} is about the same for all the devices. The emitter contact resistivity was about $9 \times 10^{-7} \ \Omega \cdot cm^2$. On the other hand, it has been found that a fairly large excess capacitance C_{ex} of about 6 fF remains at $L = 0 \ \mu m$ as shown in Fig. 14. This excess capacitance includes the collector pad capacitance and the B-C junction periphery capacitance. The increments in τ_{cc} resulting from C_{ex} are calculated to be 0.11, 0.18, 0.30, and 0.70 ps for 1×5, 1×3, 1×2, and $1 \times 1 \ \mu m^2$ emitter devices, respectively. This change of τ_{cc} fairly well explains the behavior of the observed maximum f_T. Al-

though the influence of the extrinsic B-C capacitance is not negligible, the value is still quite small for a variety of practical circuit applications. In the present devices, the base electrode overlapped on the H^+-implanted semi-insulating layer with a width of $0.8 \ \mu m$ to assure the process yield. Further process optimization is expected to reduce the C_{ex}.

In the application of HBT's to integrated circuits, the level of the operating current is the most important factor determining the power dissipation. The dependencies of f_T on the emitter length were investigated with the operating current as a parameter. The results are shown in Fig. 15. The emitter width was $1 \ \mu m$. It is noted that the emitter length giving the peak f_T becomes smaller as the operating current is reduced. This is mainly due to the reduction of the current density in larger HBT's. In addition, the f_T of a device with a $1 \ \mu m \times 10 \ \mu m$ emitter deteriorated because of the relatively large emitter metal resistance of $1 \ \Omega / \square$. At a current of 1 mA, a relatively high peak f_T of 50 GHz was obtained for an emitter size of $1 \ \mu m \times 2 \ \mu m$. It is promising for the realization of a 10-Gb/s master-slave flip-flop IC with a power dissipation of less than 100 mW. These results clearly show that scaling down a device is a very effective way to lower power dissipation while maintaining high-speed performance.

IV. Conclusion

A new self-alignment technology using a bridged base electrode for small-scaled AlGaAs/GaAs HBT's has been developed. This technology simplified the fabrication process and successfully decreased the emitter size to 1 μm $\times$ 1 μm. In spite of a large junction periphery/area ratio, a good current gain of more than 10 was obtained in an HBT with an emitter size of 1 μm $\times$ 1 μm. In the measurements of high-frequency characteristics, the highest values of f_T = 90 GHz and f_{max} = 63 GHz were obtained in an HBT with an emitter size of 1 μm $\times$ 5 μm. These results demonstrate the effectiveness of the new self-aligned HBT for high-speed and low-power devices.

Acknowledgment

The authors would like to thank H. Ito for MBE growth of HBT wafers. They also wish to thank K. Hirata and M. Hirayama for their valuable suggestions.

References

[1] K. Nagata, O. Nakajima, Y. Yamauchi, T. Nittono, H. Ito, and T. Ishibashi, "Self-aligned AlGaAs/GaAs HBT with low emitter resistance utilizing InGaAs cap layer," *IEEE Trans. Electron Devices*, vol. 35, p. 2, 1988.

[2] T. Ishibashi, O. Nakajima, K. Nagata, Y. Yamauchi, H. Ito, and T. Nittono, "Ultra-high speed AlGaAs/GaAs heterojunction bipolar transistors," in *IEDM Tech. Dig.*, 1988, p. 826.

[3] N. Hayama, M. Madihian, A. Okamoto, H. Toyoshima, and K. Honjo, "Fully self-aligned AlGaAs/GaAs hetero-junction bipolar transistors for high-speed integrated circuits applications," *IEEE Trans. Electron Devices*, vol. 35, p. 1771, 1988.

[4] R. B. Nubling, N. H. Sheng, K. C. Wang, M. F. Chang, W. J. Ho, G. J. Sullivan, C. W. Farley, and P. M. Asbeck, "25 GHz HBT frequency dividers," in *Tech. Dig. 1989 GaAs IC Symp.*, 1989, p. 125.

[5] Y. Yamauchi, O. Nakajima, K. Nagata, H. Ito, and T. Ishibashi, "A 34.8 GHz 1/4 static frequency divider using AlGaAs/GaAs HBTs," in *Tech. Dig. 1989 GaAs IC Symp.*, 1989, p. 121.

[6] M. F. Chang, P. M. Asbeck, K. C. Wang, G. J. Sullivan, N. H. Sheng, J. A. Higgins, and D. L. Miller, "AlGaAs/GaAs heterojunction bipolar transistors fabricated using a self-aligned dual-lift-off process," *IEEE Electron Device Lett.*, vol. EDL-8, p. 303, 1987.

[7] K. Morizuka, M. Asaka, N. Iizuka, K. Tsuda, and M. Obara, "AlGaAs/GaAs HBT's fabricated by a self-alignment technology using polyimide for electrode separation," *IEEE Electron Device Lett.*, vol. 9, p. 598, 1988.

[8] T. Nittono, H. Ito, O. Nakajima, and T. Ishibashi, "Non-alloyed ohmic contacts to n-GaAs using compositionally graded $In_xGa_{1-x}As$ layers," *Japan. J. Appl. Phys.*, vol. 27, p. 1718, 1988.

[9] O. Nakajima, K. Nagata, H. Ito, T. Ishibashi, and T. Sugeta, "Emitter-base junction size effect on current gain H_{fe} of AlGaAs/GaAs heterojunction bipolar transistors," *Japan. J. Appl. Phys.*, vol. 24, p. L596, 1985.

[10] Y. S. Hiraoka, J. Yoshida, and M. Azuma, "Two-dimensional analysis of emitter-size effect on current gain for GaAlAs/GaAs HBT's," *IEEE Trans. Electron Devices*, vol. ED-34, p. 721, 1987.

[11] K. Watanabe, K. Nagata, H. Yamazaki, S. Ishida, and T. Ichijo, "Effect of oxygen-implant isolation on the recombination leakage current of n-p$^+$ AlGaAs graded heterojunction diodes," *Appl. Phys. Lett.*, vol. 57, p. 1892, 1990.

[12] R. J. Malik, L. M. Lunardi, R. W. Ryan, S. C. Shunk, and M. D. Feuer, "Submicron scaling of AlGaAs/GaAs self-aligned thin emitter heterojunction bipolar transistors (SATE-HBT) with current gain independent of emitter area," *Electron. Lett.*, vol. 25, p. 1175, 1989.

[13] N. Hayama and K. Honjo, "Emitter size effect on current gain in fully self-aligned AlGaAs/GaAs HBT's with AlGaAs surface passivation layer," *IEEE Electron Device Lett.*, vol. 11, p. 388, 1990.

[14] Y. Yamauchi and T. Ishibashi, "Equivalent circuit and ECL ring oscillators of graded-bandgap base GaAs/AlGaAs HBTs," *Electron. Lett.*, vol. 22, p. 18, 1986.

A MULTIFUNCTIONAL HBT TECHNOLOGY

W.J. Ho, M.F. Chang, N.H. Sheng, N.L. Wang, P.M. Asbeck,
K.C. Wang, R.B. Nubling, G.J. Sullivan, and J.A.Higgins
Rockwell International Science Center
Thousand Oaks, CA 91360

ABSTRACT

A self-aligned AlGaAs/GaAs HBT technology has been developed for the fabrication of both high performance microwave circuits and high performance digital circuits. The technology provides transistors with high f_{max} (up to 218 GHz), high f_t (up to 98 GHz), and high efficiency (power added efficiency higher than 67.8% at 10 GHz) microwave amplification to at least 20 GHz. An MMIC amplifier has demonstrated 7 dB gain between 14 and 24 GHz. Excellent performance of high speed digital circuits such as ring oscillators, frequency dividers, MUXs, DEMUXs and phase detectors has also been achieved on the same wafer. A multifunction chip covering entire microwave band becomes feasible.

INTRODUCTION

The interaction between microwave and digital components in communications, electronic warfare and avionics systems is steadily increasing as the systems advance toward higher frequency and higher performance. Processing of high speed digital signals or digitally-modulated microwave signals often requires both microwave and digital circuits. Monolithic integration of both circuit types permits substantial reduction of size, weight, cost of bonding and assembly, and improved reliability.

Heterojunction bipolar transistors (HBTs) are attractive candidates for combining the high performance microwave and digital functions. AlGaAs/GaAs HBTs, by virtue of their structure, have a number of advantages for high speed and high power operation over other transistors (1,2). In comparison with Si biopolar transistors, they benefit from higher cutoff frequency, f_t, reduced base resistance, lower base-emitter capacitance, higher frequency of maximum oscillation, f_{max}, higher early voltage and semi-insulating substrate. In comparison with GaAs field effect transistors (FETs), HBTs have higher transconductance, higher current and power density, relaxed demand for fine-line lithography, better threshold voltage mataching, easier control over breakdown voltage, lower 1/f noise, and reduced trap-induced effects (2).

This paper describes the technology requirements and self-aligned AlGaAs/GaAs HBT process for multifunction chip application. The performance of HBTs, MMIC and high speed digital circuits fabricated on the same wafer are reported. The multifunction capability of self-aligned HBT technology opens up new features for direct digital signal processing at microwave frequencies.

REQUIREMENTS FOR MULTIFUNCTIONAL HBT TECHNOLOGY

Application to combined microwave/digital circuits places a set of unique demands on transistors. The requirements can be summarized in Table I.

Table I. Requirements for Multifunctional HBT Technology

Parameters	Digital	Analog	Microwave Power
ft	high	very high	very high
fmax	high	very high	very high
Rb	low	low	very low
Cc	low	low	very low
Cbe	very low	low	low
Beta*	20	10-30	10-20
BVceo	>2V	>10V	>10V
BVcbo	>3V	>5V	>20V
Vbe control	good	good	good
Yield	very high	high	high
Transistor size	small	moderate	large
Signal level	small	small	large

* for A/D and few other applications, the beta should be even higher.

For all the three applications, high ft, as well as high f_{max}, are needed. Low collector capacitance and low base resistance are essential. Low Cc reduces the capacitive loading in digital application, increases the slow rate in analog circuit and reduces the device output Q factor in microwave amplification. Low base resistance has the same positive effect in all three applications. In digital circuits, however, Rb is less critical provided it is small compared with the load resistance and the operating current is relatively low. Cbe must be low for digital application because it is part of the capacitive load to the previous stage, and cannot be resonated out with impedance matching circuits.

Although a high current gain can be achieved with HBTs, it is not required for microwave performance inasmuch as high current gain tends to decrease BV_{ceo}. In digital circuits, low power operation and high fanout requirements typically imply a need for high dc current gain at low operating current levels. A dc gain higher than 10 is adequate for most digital circuit applications with proper design care. It is beneficial to limit gain between 10-20. The breakdown voltage is easily tailored by the design of the collector thickness and doping level.

The uniformity of Vbe is crucial for digital circuit to maintain noise margin. In analog circuit, the offset

Reprinted from *Tech. Dig. of GaAs IC Symp.*, pp. 67–70, New Orleans, LA. Oct. 7-10, 1990.

voltage in a differential amplifier is determined by the V_{be} matching. In power HBTs, each cell must have the same V_{be} so that the bias current is distributed evenly and local heating is avoided. The uniformity of V_{be} for HBTs is intrinsically determined by the epitaxial growth.

In digital circuits, the transistor size must be aggressively scaled to minimize the parasitic capacitances and increase packing density. These factors are critical to low power operation. For a small signal operation, the transistor needs the high performance only over a small range of bias condition. For large signal microwave performance, not only high gain is needed along the load line, but also the model elements must be maintained fairly constant.

DEVICE AND CIRCUIT FABRICATION

The HBTs presented in this report are npn bipolar transistors, with AlGaAs/GaAs emitter, GaAs or InGaAs base, and GaAs collector. The device schematic cross section is shown in Fig. 1.

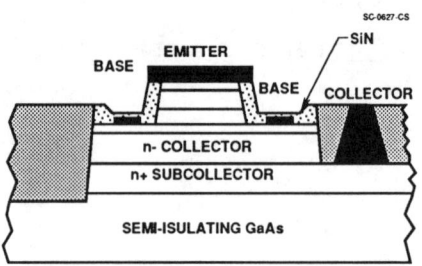

SC-0627-CS

Fig. 1   Schematic cross section of self-aligned HBTs.

To achieve high f_{max} and f_t, a self-aligned dual liftoff process hass been developed. Details have been shown in reference (3). Self-aligned process reduces the emitter-base gap to 0.2 μm and lowers the base resistance by 30% to 50% (compared with 0.8 μm gap using nonself-aligned process). The extrinsic base area is implanted with protons to reduce the base-collector capacitance. Typically, this decreases the base-collector capacitance to half of the original values. Be-doped base with doping levels ranging from 5×10^{19} to 1×10^{20} cm^{-3} and thickness of 800A have been employed. Ti/Pt/Au is used for a p-type base ohmic contact and AuGe/Ni/Au is for n-type contact. Ohmic contact alloying is carried out in a rapid thermal heating system. Representative process characteristics are shown in Table II.

Table II. HBT PROCESS CHARACTERISTICS

Base sheet resistance:	130-250 ohm square
Sub collector sheet resistance:	12 ohms/square
Emitter specific contact resistance:	$1-10 \times 10^{-7}$ ohm-cm^2
Base contact resistance:	0.1-0.3 ohm.mm
Collector contact resistance:	0.03 ohm.mm
Base-emitter capacitance:	1.6 fF/μm^2
Extrinsic base-collector capacitance:	0.2 fF/μm^2
Minimum emitter width:	1.0 μm

Wafers with HBT epitxial layers grown by MBE are qualified with a simplified process using three steps. Digital circuits were completed with nichrome resistors, two levels of interconnect metal, polyimide interlevel dielectric, and MIM capacitors. For microwave devices

and MMIC, substrate was thinned to 4 mils and additional ballasting resistors and through-substrates via holes were provided. The photograph of fabricated digital circuits, microwave MMIC and high efficiency power transistors side-by-side on the same wafer is shown in Fig. 2.

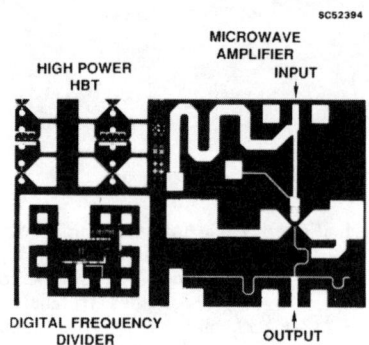

Fig. 2   Photograph of fabricated microwave devices, MMIC and digital circuits side by side on the same wafer.

MULTIFUNCTIONAL HBT DEVICE AND CIRCUIT PERFORMANCE

A number of microwave and digital circuits have been designed and tested as a vehicle for combining the high performance microwave and digital functions on the same chip.

A.   High performance GaAs/GaAlAs HBT

High frequency characteristics of the transistors are determined from S-parameter measurements over the frequency range 0.1-40 GHz, using wafer probing (with Cascade coplanar transmission line probes) and an HP8510 network analyzer. Record f_{max} of 218 GHz has been demonstrated (2) for a common-base transistor with three emitter fingers of dimension 1 μm × 1 μm. Figure 3 shows results obtained for a common-emitter transistor with two emitter fingers of dimension 2 μm by 10 μm on a separate wafer. The extrapolated values for both f_{max} and f_t (assuming 6 dB/oct falloff of U and H_{21}) are 98 GHz. The uniformity of beta, V_{be}, f_t, and f_{max} for a representative wafer on the common emitter devices is shown in Fig. 4 with yield higher than 90%.

B.   High Efficiency Microwave Power Transistor

For system application, the combination of high power-added efficiency with high power and high gain is desirable. As shown in Fig. 5, the record power-added efficiency of 67.8% has been achieved with 11.6 dB gain, 5.6 W/mm at 10 GHz. At 18 GHz the power added efficiency is 48% with 11.4 dB gain and 3.58W/mm (4). Power density of 2 mW/μm^2 and over 10 dB gain with 55% power-added efficiency were routinely obtained with CE HBTs of emitter fingers ranging from 1.2 μm to 2.0 μm at 10GHz. There is no noticeable difference in gain and power density among the devices.

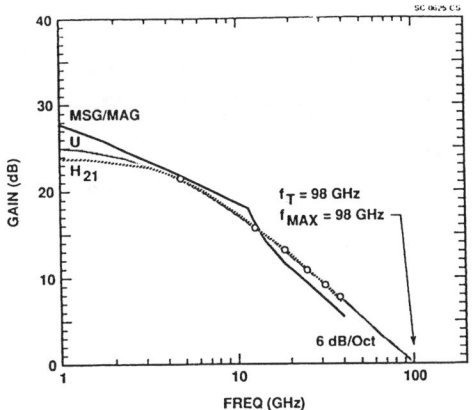

Fig. 3 Unilateral gain, U, and current gain, h_{21}, vs frequency.

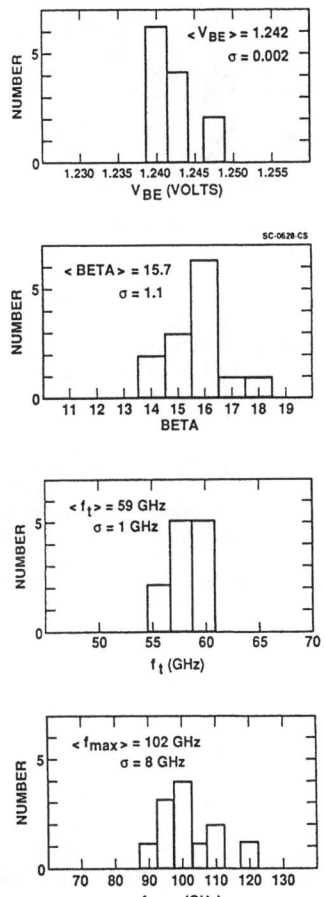

Fig. 4 Histogram of Vbe, beta, ft and fmax measured for HBTs with 2-emitter fingers of dimension 2 μm × 20 μm, distributed across a wafer.

C.   AlGaAs/GaAs HBT MMIC Wide Band Amplifier

As an example of the microwave performance achievable, a simple MMIC amplifier was designed and fabricated. A schematic circuit diagram is shown in Fig. 6. This circuit is designed as a gain block. The

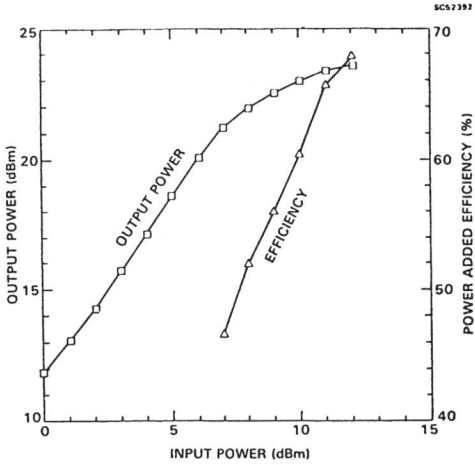

Fig. 5 Power saturation curves of common-emitter HBT at 10 GHz.

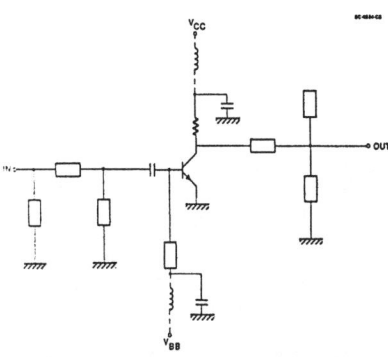

Fig. 6 Schematic circuit diagram of a HBT MMIC wide-band amplifier.

matching circuit design follows the band-pass filter principle. Gain compensation with respect to frequency is also included. This MMIC amplifier demonstrated over 7 dB gain between 14 GHz and 24 GHz, with input return loss less than 10 dB (Fig. 7).

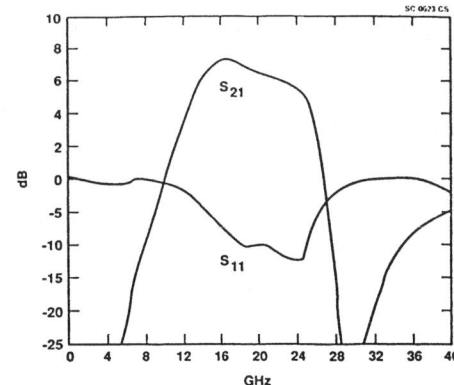

Fig. 7 Performance of a wide-band HBT MMIC amplifier.

305

D. High Speed Digital Circuits

To explore high speed circuit operation, ring oscillator, frequency divider, MUX, DEMUX and phase detector have been fabricated and tested.

Seventeen-stage NTL ring oscillators operated with propagation delays down to 10.3 ps per stage, with power consumption of about 4.3 mW per stage. Nineteen-stage CML ring oscillators operated at 14.2 ps/stage and 17 mW/stage.

Divide-by four circuit contains 35 transistors, distributed in approximately 15 equivalent NOR-gates. Recently, divider operation with a circuit of this type, implemented with microwave HBTs, was up to 26.9 GHz input frequency (5). The total power consumption of the divide-by-four circuit was 250 mW, approximately 17 mW per equivalent NOR gate. The propagation delay per complex CML gate was approximately 19 pS, with an average fan-out of 2.5.

High-speed multiplexers (MUX) and demultiplexers (DEMUX) are among the key circuits for data communication. Performance of the HBT 2:1 MUX at 8.4 Gbit/s and that of the HBT 1:2 DEMUX at 10 Gbit/s have been achieved at 6 mW per gate (6).

The phase detector is a key circuit for clock recovery in receivers of lightwave communication systems. The schematic diagram is shown in Fig. 8. The detector compares the phase difference between input data and input clock. It generates a digital PDOUT signal with its width proportional to the phase difference. The microphotograph of a fabricated phase detector is shown in Fig. 9. It contains 134 HBTs and has a chip size of 1400 μm × 950 μm. The outputs of the phase detector operated at 1 GHz clock rate are illustrated in Fig. 10. The speed performance was limited by available input signals. Output signal rise and fall times are in the range of 40-60 ps. The power consumption on chip was 750 mW, equivalent to 10mW per gate.

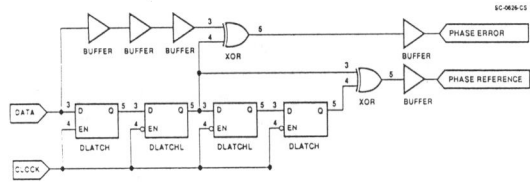

Fig. 8 Schematic diagram of a phase detector.

The excellent performance of digital and microwave circuits fabricated using self-aligned AlGaAs/GaAs HBT technology has demonstrated the potential applications for a variety of combined microwave/digital systems. An initial system with the phase-locked loop incorporating voltage controlled oscillator and digital frequency divider on the same chip is under investigation.

CONCLUSIONS

Self-aligned HBT technology has successfully demonstrated high performance microwave and high speed digital circuits fabricated on the same wafer. Such high performance and ease of production substantiate the

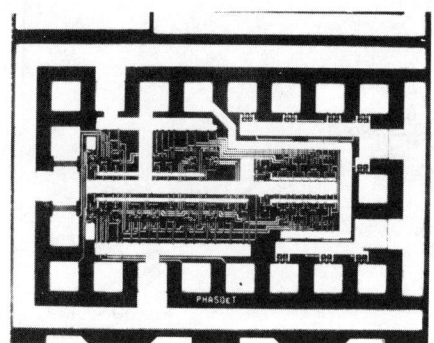

Fig. 9 Microphotograph of a fabricated Phase detector incorporated 134 transistors with chip size of 1.4 mm by .95 mm.

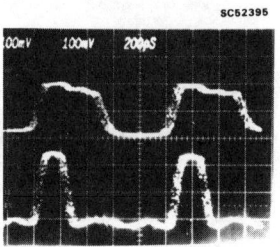

Fig. 10 Output waveform of phase detector operated at 1 GHz clock rate.

versatility of the present HBT technology. New direct signal processing opportunities become available as a result of this unique technology. A multifunction chip covering entire microwave band opens up a new area for communication, electronic wafare and avionic systems and allows a redefinition of system structure.

ACKNOWLEDGEMENTS

We thank K. Steckbauer, S. Pittman, S. Skylstad, R.L. Bernescut, E. Peterson, R. Anderson, R.L. Pierson, P. Richardson, J. Robinson, B. Rehe, and M. Szwed for their contributions in processing. We also express our thanks to Dr. Derek Cheung's technical assistance.

REFERENCES

1. H. Kroemer, "Heterostructure Bipolar Transistors and Integrated Circuits," Proc. IEEE 70, 13 (1982).
2. P. Asbeck et al, "GaAlAs/GaAs Heterojunction Bipolar Transistors: Issues and Prospects for Application" IEEE Transactions on Electronic Devices, vol. 36, no. 10, Oct. 1989.
3. F. Chang et al, "AlGaAs/GaAs Heterojunction Bipolar Transistors Fabricated Using a Self-Aligned Dual-Lift-Off Process", IEEE EDL, Vol. EDL-8, No. 7, July 1987.
4. N.L. Wang et al, "Ultrahigh Power Efficiency Operation of Common-Emitter and Common-Base HBT's at 10 GHz", to be published in IEEE Trans. MTT Oct. 1990.
5. R.B. Nubling et al, "25 GHz HBT Frequency Dividers", in 1989 IEEE GaAs IC Symp. Dig., pp 125-128.
6. K. C. Wang et al, "High speed circuitrs for lightwave communication systems implemented with (AlGaAs/GaAs) heterojunction bipolar transistors", IEEE 1987 BCTM, pp. 142-145.

Author Index

Subject Index

NOTE: Bold page numbers refer to figures, illustrations, tables

H

X

Y

Editor's Biography

Sandip Tiwari attended the Indian Institute of Technology at Kanpur, India, Rensselaer Polytechnic Institute, and Cornell University, where he received his Ph.D. degree in 1980. His contributions to the understanding and development of electronic and optical semiconductor devices is recorded in over 50 publications and 10 patents. For his contributions to the understanding and development of compound semiconductor devices, he received the Young Scientist Award of the 1991 International Symposium on Gallium Arsenide and Related Compounds. He is research staff member and manager at IBM Thomas J. Watson Research Center, Yorktown Heights, NY. During 1988–1989 he was a visiting associate professor at the University of Michigan and is currently adjunct professor at Columbia University. He has been associate and guest editor of the *IEEE Transactions on Electron Devices,* and is author of the textbook *Compound Semiconductor Device Physics,* published by Academic Press.